Student Solutions Manual

to accompany

Calculus for Business, Economics, and the Social and Life Sciences

Brief Eleventh Edition

Laurence Hoffmann
Morgan Stanley Smith Barney

Gerald Bradley
Claremont McKenna College

Dave Sobecki
Miami University – Hamilton

Michael Price
University of Oregon

Prepared by

Devilyna Nichols
Purdue University

The McGraw-Hill Companies

Student Solutions Manual to accompany
CALCULUS FOR BUSINESS, ECONOMICS, AND THE SOCIAL AND LIFE SCIENCES, BRIEF ELEVENTH EDITION
LAURENCE HOFFMANN, GERALD BRADLEY, DAVE SOBECKI AND MICHAEL PRICE

Published by McGraw-Hill Higher Education, an imprint of The McGraw-Hill Companies, Inc., 1221 Avenue of the Americas, New York, NY 10020. Copyright © 2013, 2010, 2007, and 2004 by The McGraw-Hill Companies, Inc. All rights reserved. Printed in the United States of America.

This book is printed on acid-free paper.

1 2 3 4 5 6 7 8 9 0 QDB/QDB 1 0 9 8 7 6 5 4 3 2

ISBN: 978-0-07-742738-2
MHID: 0-07-742738-6

CONTENTS

Chapter 1

Functions, Graphs, and Limits

1.1 Functions

1. $f(x) = 3x + 5$,
$f(0) = 3(0) + 5 = 5$,
$f(-1) = 3(-1) + 5 = 2$,
$f(2) = 3(2) + 5 = 11$.

3. $f(x) = 3x^2 + 5x - 2$,
$f(0) = 3(0)^2 + 5(0) - 2 = -2$,
$f(-2) = 3(-2)^2 + 5(-2) - 2 = 0$,
$f(1) = 3(1)^2 + 5(1) - 2 = 6$.

5. $g(x) = x + \dfrac{1}{x}$,
$g(-1) = -1 + \dfrac{1}{-1} = -2$,
$g(1) = 1 + \dfrac{1}{1} = 2$,
$g(2) = 2 + \dfrac{1}{2} = \dfrac{5}{2}$.

7. $h(t) = \sqrt{t^2 + 2t + 4}$,
$h(2) = \sqrt{2^2 + 2(2) + 4} = 2\sqrt{3}$,
$h(0) = \sqrt{0^2 + 2(0) + 4} = 2$,
$h(-4) = \sqrt{(-4)^2 + 2(-4) + 4} = 2\sqrt{3}$.

9. $f(t) = (2t - 1)^{-3/2} = \dfrac{1}{\left(\sqrt{2t-1}\right)^3}$,
$f(1) = \dfrac{1}{\left[\sqrt{2(1) - 1}\right]^3} = 1$,
$f(5) = \dfrac{1}{\left[\sqrt{2(5) - 1}\right]^3} = \dfrac{1}{\left[\sqrt{9}\right]^3} = \dfrac{1}{27}$,
$f(13) = \dfrac{1}{\left[\sqrt{2(13) - 1}\right]^3} = \dfrac{1}{\left[\sqrt{25}\right]^3} = \dfrac{1}{125}$.

11. $f(x) = x - |x - 2|$,
$f(1) = 1 - |1 - 2| = 1 - |-1| = 1 - 1 = 0$,
$f(2) = 2 - |2 - 2| = 2 - |0| = 2$,
$f(3) = 3 - |3 - 2| = 3 - |1| = 3 - 2 = 2$.

13. $h(x) = \begin{cases} -2x + 4 & \text{if} \quad x \le 1 \\ x^2 + 1 & \text{if} \quad x > 1 \end{cases}$

$h(3) = (3)^2 + 1 = 10$
$h(1) = -2(1) + 4 = 2$
$h(0) = -2(0) + 4 = 4$
$h(-3) = -2(-3) + 4 = 10$

15. $g(x) = \dfrac{x}{1 + x^2}$

Since $1 + x^2 \ne 0$ for any real number, the domain is the set of all real numbers.

17. $f(t) = \sqrt{1 - t}$

Since negative numbers do not have real square roots, the domain is all real numbers such that $1 - t \ge 0$, or $t \le 1$. Therefore, the domain is not the set of all real numbers.

19. $g(x) = \dfrac{x^2 + 5}{x + 2}$

Since the denominator cannot be 0, the domain consists of all real numbers such that $x \ne -2$.

21. $f(x) = \sqrt{2x + 6}$

Since negative numbers do not have real square roots, the domain is all real numbers such that $2x + 6 \ge 0$, or $x \ge -3$.

23. $f(t) = \dfrac{t+2}{\sqrt{9-t^2}}$

Since negative numbers do not have real square roots and denominators cannot be zero, the domain is the set of all real numbers such that $9 - t^2 > 0$, namely $-3 < t < 3$.

25. $f(u) = 3u^2 + 2u - 6$ and $g(x) = x + 2$, so

$\begin{aligned} f(g(x)) &= f(x+2) \\ &= 3(x+2)^2 + 2(x+2) - 6 \\ &= 3x^2 + 14x + 10. \end{aligned}$

27. $f(u) = (u-1)^3 + 2u^2$ and $g(x) = x + 1$, so

$\begin{aligned} f(g(x)) &= f(x+1) \\ &= [(x+1)-1]^3 + 2(x+1)^2 \\ &= x^3 + 2x^2 + 4x + 2. \end{aligned}$

29. $f(u) = \dfrac{1}{u^2}$ and $g(x) = x - 1$, so

$f(g(x)) = f(x-1) = \dfrac{1}{(x-1)^2}.$

31. $f(u) = \sqrt{u+1}$ and $g(x) = x^2 - 1$, so

$\begin{aligned} f(g(x)) &= f(x^2 - 1) \\ &= \sqrt{(x^2-1)+1} \\ &= \sqrt{x^2} \\ &= |x|. \end{aligned}$

33. $f(x) = 4 - 5x$

$\dfrac{f(x+h)-f(x)}{h} = \dfrac{4-5(x+h)-(4-5x)}{h}$

$\dfrac{4-5x-5h-4+5x}{h} = \dfrac{-5h}{h} = -5$

35. $f(x) = 4x - x^2$

$\dfrac{f(x+h)-f(x)}{h}$

$= \dfrac{4(x+h)-(x+h)^2-(4x-x^2)}{h}$

$= \dfrac{4x+4h-(x^2+2xh+h^2)-4x+x^2}{h}$

$= \dfrac{4x+4h-x^2-2xh-h^2-4x+x^2}{h}$

$= \dfrac{4h-2xh-h^2}{h}$

$= \dfrac{h(4-2x-h)}{h}$

$= 4 - 2x - h$

37. $f(x) = \dfrac{x}{x+1}$

$\dfrac{f(x+h)-f(x)}{h}$

$= \dfrac{\frac{x+h}{(x+h)+1} - \frac{x}{x+1}}{h}$

$= \dfrac{\frac{x+h}{x+h+1} - \frac{x}{x+1}}{h} \cdot \dfrac{(x+h+1)(x+1)}{(x+h+1)(x+1)}$

$= \dfrac{(x+h)(x+1) - x(x+h+1)}{h(x+1)(x+h+1)}$

$= \dfrac{x^2 + hx + x + h - x^2 - xh - x}{h(x+1)(x+h+1)}$

$= \dfrac{h}{h(x+1)(x+h+1)}$

$= \dfrac{1}{(x+1)(x+h+1)}$

39. $f(g(x)) = f(1-3x) = \sqrt{1-3x}$

$g(f(x)) = g\left(\sqrt{x}\right) = 1 - 3\sqrt{x}$

To solve $\sqrt{1-3x} = 1 - 3\sqrt{x}$, square both sides, so

$1 - 3x = 1 - 6\sqrt{x} + 9x$

$-3x = -6\sqrt{x} + 9x$

$6\sqrt{x} = 12x$

$\sqrt{x} = 2x$

Squaring both sides again,

$x = 4x^2$

$0 = 4x^2 - x$

$0 = x(4x - 1)$

$x = 0, \ x = \dfrac{1}{4}$

Since squaring both sides can introduce extraneous solutions, one needs to check these values.

$\sqrt{1 - 3(0)} \overset{?}{=} 1 - 3\sqrt{0}$

$1 = 1$

$\sqrt{1 - 3\left(\dfrac{1}{4}\right)} \overset{?}{=} 1 - 3\sqrt{\dfrac{1}{4}}$

$\dfrac{1}{2} \overset{?}{=} 1 - \dfrac{3}{2}$

$\dfrac{1}{2} \neq -\dfrac{1}{2}$

Also check remaining value to see if it is in the domain of f and g functions. Since $f(0)$ and $g(0)$ are both defined, $f(g(x)) = g(f(x))$ when $x = 0$.

41. $f(g(x)) = f\left(\dfrac{x+3}{x-2}\right) = \dfrac{2\left(\frac{x+3}{x-2}\right) + 3}{\frac{x+3}{x-2} - 1} = x$

$g(f(x)) = g\left(\dfrac{2x+3}{x-1}\right) = \dfrac{\frac{2x+3}{x-1} + 3}{\frac{2x+3}{x-1} - 2} = x$

Answer will be all real numbers for which f and g are defined. So, $f(g(x)) = g(f(x))$ for all real numbers except $x = 1$ and $x = 2$.

43. $f(x) = 2x^2 - 3x + 1$,

$\begin{aligned} f(x - 2) &= 2(x-2)^2 - 3(x-2) + 1 \\ &= 2x^2 - 11x + 15. \end{aligned}$

45. $f(x) = (x+1)^5 - 3x^2$,

$\begin{aligned} f(x-1) &= [(x-1)+1]^5 - 3(x-1)^2 \\ &= x^5 - 3x^2 + 6x - 3. \end{aligned}$

47. $f(x) = \sqrt{x}$,

$f(x^2 + 3x - 1) = \sqrt{x^2 + 3x - 1}.$

49. $f(x) = \dfrac{x-1}{x}$,

$f(x+1) = \dfrac{(x+1)-1}{x+1} = \dfrac{x}{x+1}.$

51. $f(x) = (x-1)^2 + 2(x-1) + 3$ can be rewritten as $g(h(x))$ with

$g(u) = u^2 + 2u + 3$ and

$h(x) = x - 1.$

53. $f(x) = \dfrac{1}{x^2 + 1}$ can be rewritten as $g(h(x))$

with $g(u) = \dfrac{1}{u}$ and $h(x) = x^2 + 1.$

55. $f(x) = \sqrt[3]{2-x} + \dfrac{4}{2-x}$ can be rewritten as

$g(h(x))$ with $g(u) = \sqrt[3]{u} + \dfrac{4}{u}$ and

$h(x) = 2 - x.$

57. $C(q) = 0.01q^2 + 0.9q + 2$

(a) Total cost of 10 units = $C(10)$

$C(10) = 0.01(10)^2 + 0.9(10) + 2 = 12$,
or \$12,000.
Average cost per unit when 10 units are

manufactured is $\dfrac{C(10)}{10}$, or \$1,200 per

unit.

(b) Cost of the 10th unit is

$C(10) - C(9)$

$= 12 - \left[0.01(9)^2 + 0.9(9) + 2\right]$

$= 12 - 10.91 = 1.09 \quad \text{or} \quad \$1,090$

59. $D(x) = -0.02x + 29$;

$C(x) = 1.43x^2 + 18.3x + 15.6$

(a) $R(x) = xD(x) = x(-0.02x + 29)$
$$= -0.02x^2 + 29x$$
$$P(x) = R(x) - C(x)$$
$$= (-0.02x^2 + 29x)$$
$$- (1.43x^2 + 18.3x + 15.6)$$
$$= -1.45x^2 + 10.7x - 15.6$$

(b) $P(x) > 0$ when

$-1.45x^2 + 10.7x - 15.6 > 0.$ Using the quadratic formula, the zeros of P are

$$x = \frac{-10.7 \pm \sqrt{(10.7)^2 - (4)(-1.45)(-15.6)}}{2(-1.45)}$$

$x = 2, 5.38$

so, $P(x) > 0$ when $2 < x < 5.38$.

61. $D(x) = -0.5x + 39;$
$C(x) = 1.5x^2 + 9.2x + 67$

(a) $R(x) = xD(x) = x(-0.5x + 39)$
$$= -0.5x^2 + 39x$$
$$P(x) = R(x) - C(x)$$
$$= (-0.5x^2 + 39x) - (1.5x^2 + 9.2x + 67)$$
$$= -2x^2 + 29.8x - 67$$

(b) $P(x) > 0$ when $-2x^2 + 29.8x - 67 > 0.$ Using the quadratic formula, the zeros of P are

$$x = \frac{-29.8 \pm \sqrt{(29.8)^2 - (4)(-2)(-67)}}{2(-2)}$$

$x \approx 2.76, 12.14$

so, $P(x) > 0$ when $2.76 < x < 12.14$.

63. $W(x) = \dfrac{600x}{300 - x}$

(a) $300 - x \neq 0$
$$x \neq 300$$
The domain is all real numbers except 300.

(b) Typically, the domain would be restricted to the first quadrant. That is, $x \geq 0$. However, since x is a

percentage, the restriction should be $0 \leq x \leq 100$.

(c) When $x = 50,$
$$W(50) = \frac{600(50)}{300 - 50}$$
$$= 120 \text{ worker-hours.}$$

(d) To distribute all of the households, $x = 100$ and
$$W(100) = \frac{600(100)}{300 - 100}$$
$$= 300 \text{ worker-hours.}$$

(e) Need to find x when $W(x) = 150.$
$$150 = \frac{600x}{300 - x}$$
$$(150)(300 - x) = (1)(600x)$$
$$300 - x = 4x$$
$$x = 60$$
After 150 worker-hours, 60% of the households have received a new telephone book.

65. $S(t) = \begin{cases} 14.7 + 0.6t & \text{if } t \leq 4 \\ 14.2t^2 - 128t + 304 & \text{if } t > 4 \end{cases}$

(a) In 1990, share price $= S(-10)$
$$S(-10) = 14.7 + 0.6(-10) = 8.7 \quad \text{or} \quad \$8.70$$
In 2006, share price $= S(6)$
$$S(6) = 14.2(6)^2 - 128(6) + 304$$
$$= 47.2 \quad \text{or} \quad \$47.20$$

(b) Share price $= \$200$ when
$$200 = 14.2t^2 - 128t + 304$$
$$0 = 14.2t^2 - 128t + 104$$
$$t = \frac{128 \pm \sqrt{(-128)^2 - 4(14.2)(104)}}{2(14.2)}$$
$$t \approx \frac{128 \pm 102.36}{28.4} \approx 0.90, 8.11$$
However, this branch of function applies when $t \geq 4$, so reject answer
$$t \approx 0.90.$$

Share price = \$200 during the year 2008. (Note: $200 = 14.7 + 0.6t$ gives $t \approx 308.83$ but $t \leq 4$ for this branch.)

(c) In the year 2012, $t = 12$

$$S(12) = 14.2(12)^2 - 128(12) + 304$$
$$= 812.8 \quad \text{or} \quad \$812.80$$

67. $Q(p) = \dfrac{4{,}374}{p^2}$ and $p(t) = 0.04t^2 + 0.2t + 12$

(a) $Q(t) = \dfrac{4{,}374}{(0.04t^2 + 0.2t + 12)^2}$

(b) $Q(10) = \dfrac{4{,}374}{(4 + 2 + 12)^2}$
$$= \dfrac{4{,}374}{324}$$
$$= 13.5 \text{ kg/week}$$

(c) $30.375 = \dfrac{4{,}374}{(0.04t^2 + 0.2t + 12)^2}$

$$(0.04t^2 + 0.2t + 12)^2 = \dfrac{4{,}374}{30.375}$$
$$= 144$$
$$= 12^2$$

So $0.04t^2 + 0.2t + 12 = \pm 12$.
The positive root leads to $t(0.04t + 0.2) = 0$ or $t = 0$. (Disregard $t < 0$.) The negative root produces imaginary numbers. $t = 0$ now.

69. $C(x) = \dfrac{150x}{200 - x}$

(a) All real numbers except $x = 200$.

(b) All real numbers for which $0 \leq x \leq 100$. If $x < 0$ or $x > 200$ then $C(x) < 0$ but cost is non-negative. $x > 100$ means more than 100%.

(c) $C(50) = \dfrac{150(50)}{200 - 50} = 50$ million dollars.

(d) $C(100) = \dfrac{150(100)}{200 - 100} = 150$
$$C(100) - C(50) = 100 \text{ million dollars.}$$

(e) $\dfrac{150x}{200 - x} = 37.5$
$$187.5x = 37.5(200)$$
$$x = \dfrac{7{,}500}{187.5} = 40\%$$

71. $P(t) = 20 - \dfrac{6}{t + 1}$

(a) $P(9) = 20 - \dfrac{6}{9 + 1}$ or 19,400 people.

(b) $P(8) = 20 - \dfrac{6}{8 + 1}$

$$P(9) - P(8) = 20 - \dfrac{3}{5} - \left(20 - \dfrac{2}{3}\right) = \dfrac{1}{15}$$

This accounts for about $\dfrac{1}{15}$ of 1,000 people, or 67 people.

(c) $P(t)$ approaches 20, or 20,000 people. Writing exercise—Answers will vary.

73. $S(r) = C(R^2 - r^2)$
$$= 1.76 \times 10^5 (1.2^2 \times 10^{-4} - r^2)$$

(a) $S(0) = (1.76 \times 10^5)(1.44 \times 10^{-4})$
$$= 25.344 \text{ cm/sec}$$

(b) $S(0.6 \times 10^{-2})$
$$= 1.76 \times 10^5 (1.44 \times 10^{-4} - 0.6^2 \times 10^{-4})$$
$$= 1.76 \times 10^5 (1.08 \times 10^{-4})$$
$$= 19.008 \text{ cm/sec}$$

75. $s(A) = 2.9\sqrt[3]{A}$

(a) $s(8) = 2.9\sqrt[3]{8} = 2.9 \times 2 = 5.8$
Since the number of species should be an integer, you would expect to find approximately 6 species.

(b) $s_1 = 2.9\sqrt[3]{A}$ and $s_2 = 2.9\sqrt[3]{2A}$

$$s_2 = 2.9\sqrt[3]{2}\sqrt[3]{A} = \sqrt[3]{2}\left(2.9\sqrt[3]{A}\right) = \sqrt[3]{2}s_1.$$

(c)
$$100 = 2.9\sqrt[3]{A}$$
$$\frac{100}{2.9} = \sqrt[3]{A}$$
$$\left(\frac{100}{2.9}\right)^3 = \left(\sqrt[3]{A}\right)^3$$
$$\left(\frac{100}{2.9}\right)^3 = A$$

Need an area of approximately 41,000 square miles.

77. $H(t) = -16t^2 + 256$

(a) Height after 2 seconds $= H(2)$
$$H(2) = -16(2)^2 + 256 = 192 \text{ feet}$$

(b) Distance traveled during 3rd second
$$= H(3) - H(2)$$
$$H(3) - H(2)$$
$$= \left[-16(3)^2 + 256\right] - 192$$
$$= -80$$
or 80 feet (negative denotes moving downward).

(c) Height of building $= H(0)$
$$H(0) = -16(0)^2 + 256 = 256 \text{ feet}$$

(d) Hits ground when height $= 0$.
$$0 = -16t^2 + 256$$
$$t = -4, 4; \text{ reject } t = -4$$
Hits ground after 4 seconds.

79. To find the domain of $f(x) = \dfrac{4x^2 - 3}{2x^2 + x - 3}$,

press $\boxed{y =}$.
Enter $(4x \wedge 2 - 3) \div (2x \wedge 2 + x - 3)$ for
$y_1 =$
Press $\boxed{\text{graph}}$.

For a better view of the vertical asymptotes, press $\boxed{\text{zoom}}$ and enter Zoom In. Use arrow buttons to move cross-hair to the left-most vertical asymptote. When it appears cross-hair is on the line, zoom in again for a more accurate reading. Move cross-hair again to be on the line. It appears that $x = -1.5$ is not in the domain of f. Zoom out once to move cross-hair to the rightmost vertical asymptote and repeat the procedure of zoom in to find that $x = 1$ is not the domain of f. The domain consists of all values except $x = -1.5$ and $x = 1$.

81. For $f(x) = 2\sqrt{x-1}$ and $g(x) = x^3 - 1.2$, to find $f(g(2.3))$, we must find $g(2.3)$ first and then input that answer into f.
Press $\boxed{y =}$. Input $2\sqrt{(x-1)}$ for $y_1 =$ and press $\boxed{\text{enter}}$. Input $x \wedge 3 - 1.2$ for $y_2 =$.
Use window dimensions $[-15, 15]1$ by $[-10, 10]1$. Use the value function under the calc menu, input 2.3, and press $\boxed{\text{enter}}$.
Use $\uparrow$ and $\downarrow$ arrows to be sure that $y_2 = x \wedge 3 - 1.2$ is displayed in the upper left corner. The lower right corner display should read $y = 10.967$.
Use the value function again and input 10.967. Verify $y_1 = 2\sqrt{(x-1)}$ is displayed in the upper left corner.
The answer of $y = 6.31$ is displayed in lower right corner.

1.2   The Graph of a Function

1. Since x-coordinate is positive and y-coordinate is positive, point is in quadrant I.

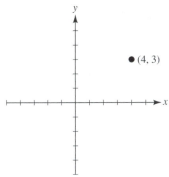

3. Since x-coordinate is positive and y-coordinate is negative, point is in quadrant IV.

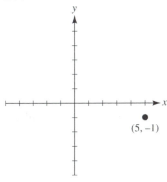

5. Since x-coordinate is zero and y-coordinate is negative, point is on y-axis, below the x-axis.

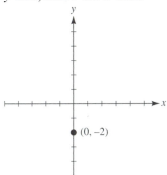

7. $(3, -1), (7, 1)$

$$
\begin{aligned}
D &= \sqrt{(x_2 - x_1)^2 + (y_2 - y_1)^2} \\
&= \sqrt{(1 - (-1))^2 + (7 - 3)^2} \\
&= \sqrt{4 + 16} \\
&= \sqrt{20} \\
&= \sqrt{4 \cdot 5} \\
&= 2\sqrt{5}
\end{aligned}
$$

9. $(7, -3), (5, 3)$

$$
\begin{aligned}
D &= \sqrt{(x_2 - x_1)^2 + (y_2 - y_1)^2} \\
&= \sqrt{(3 - (-3))^2 + (5 - 7)^2} \\
&= \sqrt{36 + 4} \\
&= \sqrt{40} \\
&= \sqrt{4 \cdot 10} \\
&= 2\sqrt{10}
\end{aligned}
$$

11. **(a)** Of the form x^n, where n is non-integer real number; is a power function.

    (b) Of the form
    $a_n x^n + a_{n-1} x^{n-1} + \cdots + a_1 x + a_0$, where n is nonnegative integer; is a polynomial function.

    (c) Can multiply out and simplify to form
    $a_n x^n + a_{n-1} x^{n-1} + \cdots + a_1 x + a_0$; is a polynomial function.

    (d) Since is quotient of two polynomial functions is a rational function.

13. $f(x) = x$

A function of the form $y = f(x) = ax + b$ is a linear function, and its graph is a line. Two points are sufficient to draw that line. The x-intercept is 0, as is the y-intercept, and $f(1) = 1$.

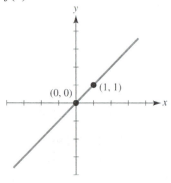

15. $f(x) = \sqrt{x}$

A function of the form $y = \sqrt{x}$ is the positive half of the function $y^2 = x$ (a parabola with vertex (0, 0), a horizontal axis and opening to the right). The x-intercept and y-intercept are the same, namely (0, 0). Choosing two more points on $y = \sqrt{x}$ (for example $P(1, 1)$ and $Q(2, 4)$), helps outline the shape of the half-parabola.

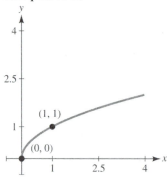

17. $f(x) = 2x - 1$

A function of the form $y = f(x) = ax + b$ is a linear function, and its graph is a line. Two points are sufficient to draw that line. The x-intercept is $\dfrac{1}{2}$ and the y-intercept is -1.

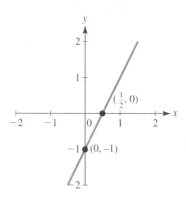

19. Since function is of the form

$y = Ax^2 + Bx + C$ (where $C = 0$), the graph is a parabola; its vertex is

$\left(-\dfrac{5}{4}, -\dfrac{25}{8} \right)$, it opens up ($A$ is positive),

and its intercepts are (0, 0) and $\left(-\dfrac{5}{2}, 0 \right)$.

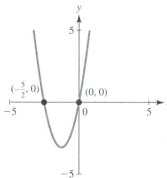

21. Since function is of the form

$y = Ax^2 + Bx + C$, the graph is a parabola which opens down (A is negative) and its vertex is $(-1, 16)$. Further,

$$f(x) = -x^2 - 2x + 15$$
$$= -(x^2 + 2x - 15)$$
$$= -(x + 5)(x - 3).$$

So the x-intercepts are $(-5, 0)$ and $(3, 0)$,

and the *y*-intercept is (0, 15).

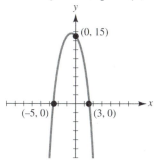

half line $y = x + 1$ has no *y*-intercept.

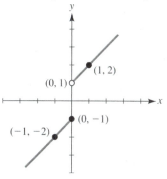

23. $f(x) = x^3$

Note that if $x > 0$ then $f(x) > 0$ and if $x < 0$, then $f(x) < 0$. This means that the curve will only appear in the first and third quadrants. Since x^3 and $(-x)^3$ have the same absolute value, only their signs are opposite, the curve will be symmetric with respect to (wrt) the origin. The *x*-intercept is 0, as is the *y*-intercept.

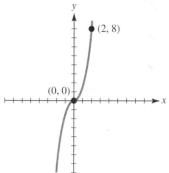

25. $f(x) = \begin{cases} x - 1 & \text{if} \quad x \le 0 \\ x + 1 & \text{if} \quad x > 0 \end{cases}$

Note that the graph consists of two half lines on either side of $x = 0$. There is no *x*-intercept for either half line. The half line $y = x - 1$ has a *y*-intercept of -1, while the

27. Graph consists of part of parabola $y = x^2 + x - 3$, namely the portion corresponding to $x < 1$, and a half line for $x \ge 1$; for the parabola portion of the graph, the vertex is $\left(-\dfrac{1}{2}, -\dfrac{13}{4} \right)$, and the parabola opens up (*A* is positive); $\left(\dfrac{-1 - \sqrt{13}}{2}, 0 \right)$ and $(0, -3)$ are its intercepts; the half line starts at $(1, -1)$ and includes the point $(2, -3)$.

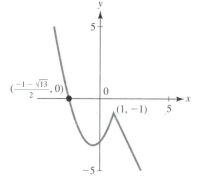

29. $y = 3x + 5$ and $y = -x + 3$

Add 3 times the second equation to the first. Then $4y = 14$ or $y = \dfrac{7}{2}$. Substitute in the first, then $x = 3 - y = -\dfrac{1}{2}$. The point of intersection is $P\left(-\dfrac{1}{2}, \dfrac{7}{2}\right)$.

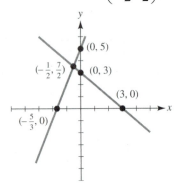

31. $y = x^2$ and $y = 3x - 2$

Setting the expressions equal to each other,

$$x^2 = 3x - 2$$
$$x^2 - 3x + 2 = 0$$
$$(x - 1)(x - 2) = 0$$
$$x = 1, 2$$

So points of intersection are $P_1(1, 1)$ and $P_2(2, 4)$.

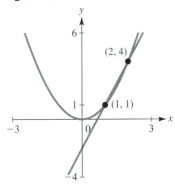

33. $3y - 2x = 5$ and $y + 3x = 9$.

Multiply the second equation by -3 and add it to the first one. Then,

$-2x - 9x = 5 - 27$,

$x = 2$, $y = 9 - 3(2) = 3$.

The point of intersection is $P(2, 3)$.

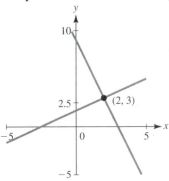

35. **(a)** Crosses y-axis at $y = -1$, y-intercept is $(0, -1)$.

(b) Crosses x-axis at $x = 1$, x-intercept is $(1, 0)$.

(c) Largest value of f is 3 and occurs at $x = 4$ (highest point on graph).

(d) Smallest value of f is -3 and occurs at $x = -2$ (lowest point on graph).

37. **(a)** Crosses y-axis at $y = 2$, y-intercept is $(0, 2)$.

(b) Crosses x-axis at $x = -1$ and 3.5; x-intercepts are $(-1, 0)$ and $(3.5, 0)$.

(c) Largest value of f is 3 and occurs at $x = 2$ (highest point on graph).

(d) Smallest value of f is -3 and occurs at $x = 4$ (lowest point on graph).

39. The monthly profit is
$P(p)$
$= (\text{number of recorders sold})(\text{price} - \text{cost})$
$= (120 - p)(p - 40)$
So, the intercepts are $(40, 0)$, $(120, 0)$, and $(0, -4800)$. The graph suggests a maximum profit when $p \approx 80$, that is, when 80 recorders are sold.
$P(80) = (120 - 80)(80 - 40) = 1600$

So estimated max profit is $1600.

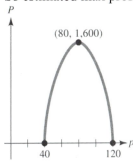

41. The weekly profit is
$P(x)$ = (number of sets sold)
$\cdot$ (price-cost per set)
$= 5(27 - x)(x - 15)$
So, the intercepts are $(27, 0)$, $(15, 0)$ and
$(0, -2025)$. The graph suggests a maximum
weekly profit when $x \approx 21$. That
is, when the price per set is $21.
$P(21) = 5(27 - 21)(21 - 15) = 180$
So, estimated max profit is $180.

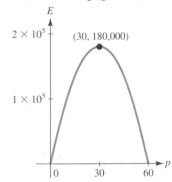

The number of sets corresponding to the
max profit is $5(27 - 21) = 30$ sets.

43. (a) $E(p) =$ (price per unit)(demand)
$= -200p(p - 60)$

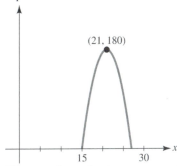

(b) The p intercepts represent prices at
which consumers do not buy
commodity.

(c) The graph suggests a maximum
expenditure when $p \approx 30$.
$E(30) = -200(30)(30 - 60) = 180,000$
So estimated max expenditure is
$180,000.

45. (a) profit = revenue − cost
= (# sold)(selling price) − cost
$P(x) = x(-0.05x + 38)$
$- (0.02x^2 + 3x + 574.77)$
$= -0.07x^2 + 35x - 574.77$
hundred dollars

P ($100)

(250, 3,800.23)

4,000

2,000

x (100)

0   250   500

(b) Average profit $= \dfrac{P(x)}{x}$

$AP(x) = -0.07x + 35 - \dfrac{574.77}{x}$
when $p = $37, $-0.05x + 38 = 37$ and
$x = 20$.
$AP(20) = -0.07(20) + 35 - \dfrac{574.77}{20}$
$= 4.86 per unit

(c) The graph suggests a maximum profit
when $x = 250$, that is, when
25,000 units are purchased. Note that
the max profit is
$P(250) = -0.07(250)^2 + 35(20) - 574.77$
≈ 3800.23 hundred,
or $380,023. For the unit price,
$p = -0.05(250) + 38 = 25.50.

47. (a)

Days of Training	Mowers per Day
2	6
3	7.23
5	8.15
10	8.69
50	8.96

(b) The number of mowers per day approaches 9.

(c) To graph $N(t) = \dfrac{45t^2}{5t^2 + t + 8}$, press

$\boxed{y =}$

Input $(45x \wedge 2) \div (5x \wedge 2 + x + 8)$ for $Y_1 =$.

Use window dimensions $[-10, 10]1$ by $[-10, 10]1$ (z standard).

Press $\boxed{\text{graph}}$.

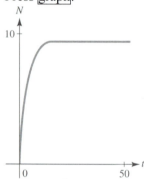

49. (a) When $m \le 200$, $C(m) = 19$ dollars. For each additional minute, that is $m - 200$, the additional cost is \$0.04 until a maximum of $m = 1,000$ is reached; so

$C(m)$

$= \begin{cases} 19 & \text{if } m \le 200 \\ 19 + 0.04(m - 200) & \text{if } 200 < m \le 1,000 \end{cases}$

(b)

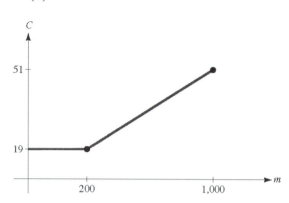

51. (a) revenue = (#apts)(rent per apt)

Since $\dfrac{p - 1200}{100}$ represents the number of \$100 increases.

$150 - 5\left(\dfrac{p - 1200}{100}\right) = 210 - 0.05p$

represents the number of apartments that will be leased. So,

$R(p) = 210p - 0.05p^2$.

(b)

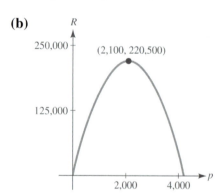

(c) The graph suggests a maximum profit when $p = 2100$; that is, when the rental price is \$2,100. The max profit is

$R(2100) = 210(2100) - 0.05(2100)^2$

$\approx \$220,500.$

53. $N(t) = -35t^2 + 299t + 3,347$

(a)

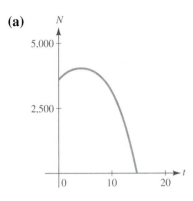

(b) Since the year 1995 is represented by $t = 5$, the amount predicted was
$$N(5) = -35(5)^2 + 299(5) + 3,347$$
$$= 3,967 \text{ thousand tons.}$$

(c) Based on the formula, the maximum lead emission would occur at the vertex, or when
$$t = -\frac{299}{2(-35)} \approx 4.27 \text{ years.}$$
This would be during March of the year 1994.

(d) No; from the graph, $N(t) < 0$ when $t \approx 15$, or during the year 2005.

55. $D(v) = 0.065v^2 + 0.148v$

For practical domain, graph is part of parabola corresponding to $v \geq 0$.

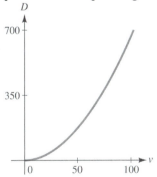

57. $H(t) = -16t^2 + 160t = -16t(t - 10)$

(a) The intercepts of the graph are (0, 0) and (10, 0). Due to symmetry, the

vertex is when $t = 5$ and
$$y = H(5) = -16(5)^2 + 160(5) = 400.$$

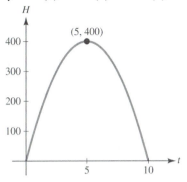

(b) Aside from when it is initially thrown, the height of the projectile is zero (ground level) when $t = 10$ seconds.

(c) The high point of the graph, which corresponds to the max height, is when y-coordinate is 400, or 400 feet.

59. The graph is a function because no vertical line intersects the graph more than once.

61. The graph is not a function because there are vertical lines intersecting the graph at more than one point; for example, the y-axis.

63. $f(x) = -9x^2 + 3600x - 358,200$

Answers will vary, but one viewing window has the following dimensions: [180, 200]10 by [−500, 1850]500.

65. (a) The graph of $y = x^2 + 3$ is graph of $y = x^2$ translated up 3 units.

(b)

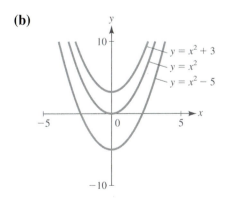

(c) When $c > 0$, the graph of g is the graph of f translated up c units. When $c < 0$, the graph is translated down $|c|$ units.

67. (a) The graph of $y = (x - 2)^2$ is the graph of $y = x^2$ translated two units to the right.

(b)

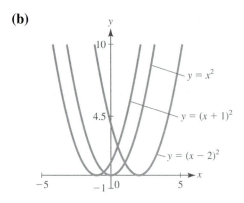

(c) When $c > 0$, the graph of g is the graph of f translated c units to the right. When $c < 0$, the graph is translated $|c|$ units to the left.

69. To graph $f(x) = \dfrac{-9x^2 - 3x - 4}{4x^2 + x - 1}$,

Press $\boxed{y=}$
Input $(-9x \wedge 2 - 3x - 4) \div (4x \wedge 2 + x - 1)$
for $y_1 =$
Press $\boxed{\text{graph}}$
Use the Zoom in function under the Zoom menu to find the vertical asymptotes to be $x_1 \approx -0.65$ and $x_2 \approx 0.39$. The function f is defined for all real x except $x_3 \approx -0.65$

and $x \approx 0.39$.

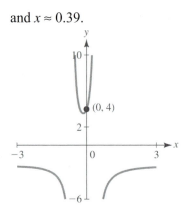

71. To graph $g(x) = -3x^3 + 7x + 4$ and find x-intercepts,
Press $\boxed{y=}$
Input $-3x \wedge 3 + 7x + 4$ for $y_1 =$
Press $\boxed{\text{graph}}$
Press $\boxed{\text{trace}}$
Use left arrow to move cursor to the left most x-intercept. When the cursor appears to be at the x-intercept, use the Zoom In feature under the Zoom menu twice. It can be seen that there are two x-intercepts in close proximity to each other. These x-intercepts appear to be $x_1 \approx -1$ and $x_2 \approx -0.76$. To estimate the third x-intercept, use the z-standard function under the Zoom menu to view the original graph. Use right arrow and zoom in to estimate the third x-intercept to be $x_3 \approx 1.8$.

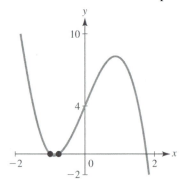

73. $(x-a)^2 + (y-b)^2 = R^2$

(a) Since the center of the circle is $(2, -3)$, $a = 2$ and $b = -3$. Since its radius is 4, $R = 4$.

$$(x-2)^2 + (y-(-3))^2 = 4^2$$
$$(x-2)^2 + (y+3)^2 = 16$$

(b) $x^2 + y^2 - 4x + 6y = 11$

First, group the x terms together and the y terms together.

$$(x^2 - 4x) + (y^2 + 6y) = 11$$

Next, complete the square for each grouping

$$(x^2 - 4x + 4) - 4 + (y^2 + 6y + 9) - 9 = 11$$
$$(x-2)^2 + (y+3)^2 = 11 + 4 + 9$$
$$(x-2)^2 + (y+3)^2 = 24$$
$$(x-2)^2 + (y-(-3))^2 = \left(\sqrt{24}\right)^2$$

center: $(2, -3)$
radius: $\sqrt{24} = \sqrt{4 \cdot 6} = 2\sqrt{6}$

(c) Proceeding as in part (b),

$$(x^2 - 2x) + (y^2 + 4y) = -10$$
$$(x^2 - 2x + 1) - 1 + (y^2 + 4y + 4) - 4 = -10$$
$$(x-1)^2 + (y+2)^2 = -10 + 1 + 4$$
$$(x-1)^2 + (y+2)^2 = -5$$

Since the left-hand side is positive for all possible points (x, y) and the right-hand side is negative, the equality can never hold. That is, there are no points (x, y) that satisfy the equation.

1.3 Lines and Linear Functions

1. For $P_1(2, -3)$ and $P_2(0, 4)$ the slope is
$$m = \frac{4-(-3)}{0-2} = -\frac{7}{2}.$$

3. For $P_1(2, 0)$ and $P_2(0, 2)$ the slope is
$$m = \frac{2-0}{0-2} = -1.$$

5. For $P_1(2, 6)$ and $P_2(2, -4)$ the slope is
$$m = \frac{6-(-4)}{2-2}, \text{ which is undefined, since the}$$
denominator is 0. The line through the given points is vertical.

7. For $P_1\left(\frac{1}{7}, 5\right)$ and $P_2\left(-\frac{1}{11}, 5\right)$ the slope
is $m = \dfrac{5-5}{-\frac{1}{11} - \frac{1}{7}} = \dfrac{0}{-\frac{18}{77}} = 0.$

9. The line has slope $= 2$ and an intercept of $(0, 0)$. So, the equation of line is $y = 2x + 0$, or $y = 2x$.

11. The slope of the line is $\dfrac{-5}{3}$. The x-intercept of the line is $(3, 0)$ and the y-intercept is $(0, 5)$. The equation of the line is $y = -\dfrac{5}{3}x + 5$.

13. The line $x = 3$ is a vertical line that includes all points of the form $(3, y)$. Therefore, the x-intercept is $(3, 0)$ and there is no y-intercept. The slope of the line is undefined, since $x_2 - x_1 = 3 - 3 = 0$.

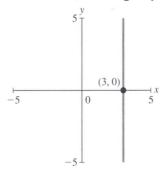

15. $y = 3x$
$m = 3$, y-intercept $b = 0$, and the

x-intercept is 0.

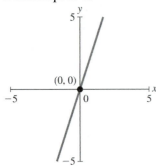

17. $3x + 2y = 6$ or $y = -\dfrac{3}{2}x + 3$

$m = -\dfrac{3}{2}$, *y*-intercept $b = 3$, and the

x-intercept is 2.

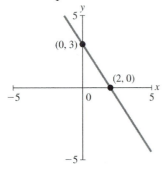

19. $\dfrac{x}{2} + \dfrac{y}{5} = 1$ or $y = -\dfrac{5}{2}x + 5$

$m = -\dfrac{5}{2}$, *y*-intercept $b = 5$, and the

x-intercept is 2.

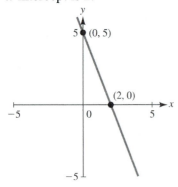

21. $m = 1$ and $P(2, 0)$, so $y - 0 = (1)(x - 2)$, or $y = x - 2$.

23. $m = -\dfrac{1}{2}$ and $P(5, -2)$, so

$$y - (-2) = -\dfrac{1}{2}(x - 5), \text{ or } y = -\dfrac{1}{2}x + \dfrac{1}{2}.$$

25. Since the line is parallel to the *x*-axis, it is horizontal and its slope is 0. For $P(2, 5)$, the line is $y - 5 = 0(x - 2)$, or $y = 5$.

27. $m = \dfrac{1 - 0}{0 - 1}$ and for $P(1, 0)$ the equation of the line is $y - 0 = -1(x - 1)$ or $y = -x + 1$. The equation would be the same if the point $(0, 1)$ had been used.

29. $m = \dfrac{1 - \left(\frac{1}{4}\right)}{-\left(\frac{1}{5}\right) - \left(\frac{2}{3}\right)} = -\dfrac{45}{52}$

For $P\left(-\dfrac{1}{5}, 1\right)$, the line is

$$y - 1 = -\dfrac{45}{52}\left(x + \dfrac{1}{5}\right), \text{ or } y = -\dfrac{45}{52}x + \dfrac{43}{52}.$$

31. The slope is 0 because the *y*-values are identical. So, $y = 5$.

33. The given line $2x + y = 3$, or $y = -2x + 3$, has a slope of -2. Since parallel lines have the same slope, $m = -2$ for the desired line. Given that the point $(4, 1)$ is on the line, $y - 1 = -2(x - 4)$, or $y = -2x + 9$.

35. The given line $x + y = 4$, or $y = -x + 4$, has a slope of -1. A perpendicular line has slope $m = -\dfrac{1}{-1} = 1$. Given that the point $(3, 5)$ is on the line, $y - 5 = 1(x - 3)$, or $y = x + 2$.

37. (a) Let *x* be the number of units manufactured. Then $60x$ is the cost of producing *x* units, to which the fixed cost must be added.

$y = 60x + 5{,}000$

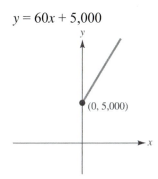

(b) Average cost $= \dfrac{y}{x}$

$$AC(x) = 60 + \frac{5{,}000}{x}$$

$$AC(20) = 60 + \frac{5{,}000}{20} = \$310 \text{ per unit}$$

39. (a) Since $t = 0$ in the year 2005, $t = 5$ is the year 2010. The given information translates to the points $(0, 7853)$ and $(5, 9127)$. The slope of a line through these points is

$$m = \frac{9127 - 7853}{5 - 0} = \frac{1274}{5} = 254.8.$$

So, the equation of the function is
$D(t) = 254.8t + 7853$.
For practical purposes, the graph is limited to quadrant I.

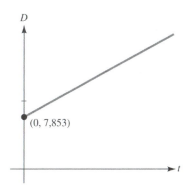

(b) In the year 2015, $t = 10$ and the predicted debt is
$D(10) = 254.8(10) + 7853 = 10{,}401$ or $\$10{,}401$.

(c) Need to find t where
$D(t) = 2(7{,}853) = 15{,}706$

$254.8t + 7853 = 15706$
$254.8t = 7853$
$t \approx 30.8$
Debt will be double the amount of 2005 during the year 2035.

41. The slope is

$$m = \frac{1{,}500 - 0}{0 - 10} = -150$$

Originally (when time $x = 0$), the value of y of the books is 1,500 (this is the y-intercept.)
$y = -150x + 1{,}500$

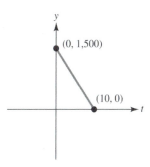

43. (a) Using the points $(0, V)$ and (N, S), the slope of the line is $\dfrac{S - V}{N}$. So, the value of an asset after t years is

$$B(t) = \frac{S - V}{N}t + V.$$

(b) For this equipment,
$B(T) = -6{,}400t + 50{,}000$. So,
$B(3) = -6{,}400(3) + 50{,}000 = 30{,}800$.
Value after three years is $\$30{,}800$.

45. Let the x-axis represent time in months and the y-axis represent price per share.

(a)

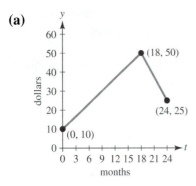

(b)

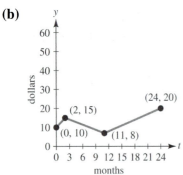

(c)

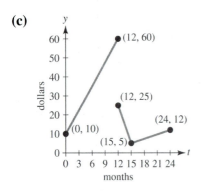

47. **(a)** Since value doubles every 10 years,
in 1910, value is $200
in 1920, value is $400
in 1930, value is $800
in 1940, value is $1,600
in 1950, value is $3,200
in 1960, value is $6,400
in 1970, value is $12,800
in 1980, value is $25,600
in 1990, value is $51,200
in 2000, value is $102,400
in 2010, value is $204,800
in 2020, value is $409,600

(b) No, it is not linear.

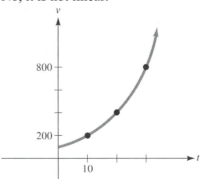

49. **(a)** Let x be the number of hours spent registering students in person. During the first 4 hours
$(4)(35) = 140$ students were registered. So,
$360 - 140 = 220$ students had pre-registered. Let y be the total number of students who register. Then,
$y = 35x + 220$.

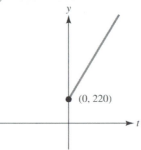

(b) $y = (3)(35) + 220 = 325$

(c) From part **(a)**, we see that 220 students had pre-registered.

51. **(a)** $H(7) = 6.5(7) + 50 = 95.5$ cm tall

(b) $150 = 6.5A + 50$, $A = 15.4$ years old

(c) $H(0) = 6.5(0) + 50 = 50$ cm tall.
This height (≈ 19.7 inches) seems reasonable.

(d) $H(20) = 6.5(20) + 50 = 180$ cm tall.
This height (≈ 5.9 feet) seems reasonable.

53. (a) Let x be the number of days. The slope is $m = \dfrac{200-164}{12-21} = -4$.

For $P(12, 200)$, $y - 200 = -4(x - 12)$, or $y = -4x + 248$.

(b) $y = 248 - (4)(8) = 216$ million gallons.

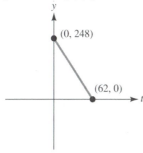

55. (a) Let C be the temperature in degrees Celsius and F the temperature in degrees Fahrenheit. The slope is

$$m = \frac{212-32}{100-0} = \frac{9}{5}.$$

So, $\dfrac{F-32}{C-0} = \dfrac{9}{5}$, or $F = \dfrac{9}{5}C + 32$.

(b) $F = \dfrac{9}{5}(15) + 32 = 59$ degrees

(c) $68 = \dfrac{9}{5}C + 32$

$36 = \dfrac{9}{5}C$

$C = 20$ degrees

(d) Solving $C = \dfrac{9}{5}C + 32$, $C = -40$. So, the temperature $-40°C$ is also $-40°F$.

57. (a) Let t represent years after 2005. Using the points $(0, 575)$ and $(5, 545)$, the slope is $m = \dfrac{545-575}{5-0} = -6$. If S represents the average SAT score, $S(t) = -6t + 575$.

(b) $S(10) = -6(10) + 575 = 515$

(c) $527 = -6t + 575$, $t = 8$, and the year would be 2013.

59. (a) Using the given points $(100, 97)$ and $(500, 110)$, slope is

$$\frac{110-97}{500-100} = 0.0325$$

$$N - 97 = 0.0325(x - 100)$$

$$N(x) = 0.0325x + 93.75$$

(b) When $x = 300$,

$$N(300) = 0.0325(300) + 93.75$$

$$\approx 104 \text{ deaths}$$

When there are 100 deaths,

$100 = 0.0325x + 93.75$

or $x \approx 192.3$ mg/m^3

(c) Writing exercise – Answers will vary.

61. The segment with a constant, positive slope for all values of time represents the tortoise moving at a constant rate. The segment with a positive slope, followed by a horizontal segment, and finished with another positive slope segment (slope of this segment matches slope of first segment) represents the hare's movement. Note that the horizontal segment represents the time during which the hare is napping.

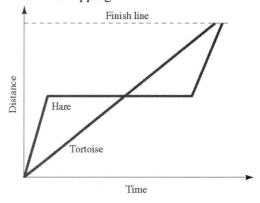

63. To graph $y = \frac{25}{7}x + \frac{13}{2}$ and

$y = \frac{144}{45}x + \frac{630}{229}$ on the same set of axes,

Press $\boxed{y =}$.

Input $\frac{(25x)}{7} + \frac{13}{2}$ for $y_1 =$ and press

$\boxed{\text{enter}}$.

Input $\frac{(144x)}{45} + \frac{630}{229}$ for $y_2 =$.

Use the window dimensions [0, 4] 0.5 by
[0, 14] 2. Press $\boxed{\text{graph}}$.
It does not appear that the lines are
parallel. To verify this, press $\boxed{\text{2ND}}$ $\boxed{\text{quit}}$.

Input $\frac{25}{7} - \frac{144}{45}$ and $\boxed{\text{enter}}$.

If the lines were parallel the difference in
their slopes would equal zero (the slopes
would be the same). The difference of
these slopes is 0.37 and therefore, the lines
are not parallel.

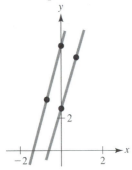

65. Lines L_1 and L_2 are given as perpendicular,

and the slope of L_1 is $m_1 = \frac{b}{a}$ while that of

L_2 is $m_2 = \frac{c}{a}$. In right triangle OAB, the

hypotenuse has length $|AB| = |b - c|$ while

the two legs have lengths $|OA| = \sqrt{a^2 + b^2}$

and $|OB| = \sqrt{a^2 + c^2}$. Thus, by the

Pythagorean theorem

$$(a^2 + b^2) + (a^2 + c^2) = (b - c)^2$$
$$a^2 + b^2 + a^2 + c^2 = b^2 - 2bc + c^2$$

$$2a^2 = -2bc$$
$$\frac{bc}{a^2} = -1$$
$$\left(\frac{b}{a}\right)\left(\frac{c}{a}\right) = -1$$
$$m_1 m_2 = -1$$

so $m_2 = \frac{-1}{m_1}$

1.4   Functional Models

1. Let x and y be the smaller and larger
numbers, respectively. Then
$$xy = 318$$
$$y = \frac{318}{x}$$
The sum is $S = x + y = x + \frac{318}{x}$.

3. Let R denote the rate of population growth
and p the population size. Since R is directly
proportional to p, $R(p) = kp$, where k is the
constant of proportionality.

5. This problem has two possible forms of the
solution. Assume the stream is along the
length, say l. Then w is the width and
$l + 2w = 1,000$ or $l = 1,000 - 2w$.
The area is $A = lw = 2w(500 - w)$ square
feet.

7. Let x be the length and y the width of the
rectangle. Then $2x + 2y = 320$ or
$y = 160 - x$.
The area is (length)(width) or
$A(x) = x(160 - x)$.
The length is estimated to be 80 meters
from the graph below, which also happens
to be the width. So the maximum area

seems to correspond to that of a square.

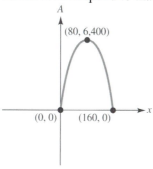

9. Let x be the length of the square base and y the height of the box. The surface area is

$$2x^2 + 4xy = 4,000. \text{ So } y = \frac{2,000 - x^2}{2x}$$

and the volume is

$$V = x^2 y = x\left(1,000 - \frac{x^2}{2}\right).$$

11. Let R denote the rate at which temperature changes, T_e the temperature of the medium, and T_0 the temperature of the object. Then $T_0 - T_e$ is the difference in the temperature between the object and the medium. Since the rate of change is directly proportional to the difference, $R = k(T_0 - T_e)$, where k is the constant of proportionality.

13. Let R denote the rate at which people are implicated, P the number of people implicated, and T the total number of people involved. Then $T - P$ is the number of people involved but not implicated. Since the rate of change is jointly proportional to those implicated and those not implicated, $R = kP(T - P)$, where k is the constant of proportionality.

15. Let R be the speed of the truck.

The cost due to wages is $\frac{k_1}{R}$, where k_1 is a constant of proportionality, and the cost due to gasoline is $k_2 R$, where k_2 is

another constant of proportionality. If C is the total cost, $C = \frac{k_1}{R} + k_2 R.$

17. **(a)** Let p be the selling price of the commodity. Then

$$\text{Profit} = \text{Revenue} - \text{Costs}$$
$$\text{Revenue} = (\text{number sold}) \cdot (\text{selling price})$$
$$R(x) = xp$$
$$\text{Costs} = (\text{cost per unit}) \cdot (\text{number units})$$
$$+ \text{fixed overhead}$$
$$C(x) = (p - 3)x + 17,000$$
$$C(x) = xp - 3x + 17,000$$
$$P(x) = xp - (xp - 3x + 17,000)$$
$$= 3x - 17,000$$

(b) $P(5,000) = 3(5,000) - 17,000$
$$= -2,000$$
or a loss of $2,000
$$P(20,000) = 3(20,000) - 17,000$$
$$= 43,000$$
or a profit of $43,000
Profit is zero (break-even point) when
$$0 = 3x - 17,000$$
$x \approx 5,666.67$ units are produced and sold so, when 5, 667 units are produced and sold, production becomes profitable.

(c) average profit $= \dfrac{P(x)}{x}$

$$AP(x) = 3 - \frac{17,000}{x}$$
when 10,000 units are produced,
$$AP(10,000) = 3 - \frac{17,000}{10,000} = 1.3$$
or $1.30 per unit.

19. Let x be the sales price per lamp. Then, $x - 50$ will be the number of $1.00 increases over the base price of $50, and $1,000(x - 50)$ is the number of unsold lamps. Therefore the number of lamps sold is $3,000 - 1,000(x - 50)$. The profit is

$$P = [3,000 - 1,000(x - 50)]x$$
$$- 29[3,000 - 1,000(x - 50)]$$
$$= [3,000 - 1,000(x - 50)](x - 29)$$
$$= (53,000 - 1,000x)(x - 29)$$

The optimal selling price is $41.

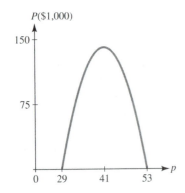

(b)

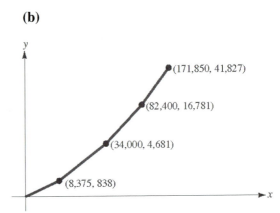

The slopes of the segments are 0.1, 0.15, 0.25 and 0.28, respectively. As taxable income increases, the slopes of the segments increase. So, as you earn more, you pay more on your earnings

21. (a) For $0 \le x \le 8,375$, the tax is
$10\%x = 0.1x$.
For $8,375 < x \le 34,000$ the tax is
$837.5 + 15\%(x - 8,375)$
$= 837.5 + 0.15(x - 8,375)$
$= 837.5 + 0.15x - 1,256.25$
$= 0.15x - 418.75$.

For $34,000 < x \le 82,400$ the tax is
$4,681.25 + 25\%(x - 34,000)$
$= 4,681.25 + 0.25(x - 34,000)$
$= 4,681.25 + 0.25x - 8,500$
$= 0.25x - 3,818.75$

For $82,400 < x \le 171,850$ the tax is
$16,781.25 + 28\%(x - 82,400)$
$= 16,781.25 + 0.28(x - 82,400)$
$= 16,781.25 + 0.28x - 23,072$
$= 0.28x - 6,290.75$
So,

$$T(x) = \begin{cases} 0.1x & \text{if } 0 \le x \le 8,375 \\ 0.15x - 418.75 & \text{if } 8,375 < x \le 34,000 \\ 0.25x - 3,818.75 & \text{if } 34,000 < x \le 82,400 \\ 0.28x - 6,290.75 & \text{if } 82,400 < x \le 171,850 \end{cases}$$

23. Let x denote the number of days after July 1 and $R(x)$ the corresponding revenue (in dollars). Then
$R(x) = $ (number of bushels sold)
        (price per bushel)
Since the crop increases at the rate of 1 bushel per day and 140 bushels were available on July 1, the number of bushels sold after x days is $140 + x$. Since the price per bushel decreases by 0.05 dollars per day and was $8 on July 1, the price per bushel after x days is
$8 - 0.05x$ dollars. Putting it all together,
$R(x) = (140 + x)(8 - 0.05x)$
$= 0.05(160 - x)(140 + x)$
The number of days to maximize revenue is approximately 10 days after July 1, or July 11. Note that
$R(10) = 0.05(160 - 10)(140 + 10) = 1,125$.
So, the estimated max revenue is $1,125.

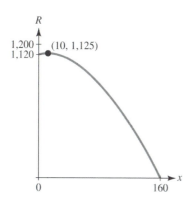

25. Let x be the number of passengers. There will be $x - 40$ additional passengers between $40 < x \le 80$ (if the total number is below 80). The price for the second category is $60 - 0.5(x - 40) = 80 - 0.5x$.
The revenue generated in this category is $80x - 0.5x^2$.

$$R(x) = \begin{cases} 2,400 & \text{if } 0 < x \le 40 \\ 80x - 0.5x^2 & \text{if } 40 < x < 80 \\ 40x & \text{if } x \ge 80 \end{cases}$$

Only the points corresponding to the integers $x = 0, 1, 2, \ldots$ are meaningful in the practical context.

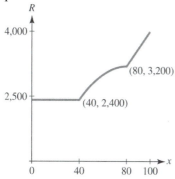

27. Royalties for publisher A are given by

$$R_A(N) = \begin{cases} 0.01(2)(N) & 0 < N \le 30,000 \\ 0.01(2)(30,000) \\ \quad + 0.035(2)(N - 30,000) & N > 30,000 \end{cases}$$

Royalties for publisher B are given by

$$R_B(N) = \begin{cases} 0 & N \le 4,000 \\ 0.02(3)(N - 4,000) & N > 4,000 \end{cases}$$

Clearly, for $N \le 4,000$, publisher A offers the better deal. When $N = 30,000$,

publisher A pays \$600, but publisher B now pays more, paying \$1,560. Therefore, the plans pay the same amount for some value of $N < 30,000$. To find the value,

$$0.01(2)(N) = 0.02(3)(N - 4,000)$$
$$0.02N = 0.06N - 240$$
$$240 = 0.04N$$
$$6,000 = N$$

So, when $N < 6,000$, publisher A offers the better deal. When $N > 6,000$, publisher B initially offers the better deal. Then, the plans again pay the same amount when

$$0.01(2)(30,000) + 0.035(2)(N - 30,000)$$
$$= 0.02(3)(N - 4,000)$$
$$0.07N - 1,500 = 0.06N - 240$$
$$0.01N = 1,260$$
$$N = 126,000$$

So, when more than 126,000 copies are sold, plan A becomes the better plan.

29. Let x be the number of machines used and t the number of hours of production. The number of kickboards produced per machine per hour is $30x$. It costs $20x$ to set up all the machines. The cost of supervision is $19.20t$. The number of kickboards produced by x machines in t hours is $30xt$ which must account for all 8,000 kickboards. Solving $30xt = 8,000$ for t leads to $t = \dfrac{800}{3x}$.

Cost of supervision: $19.20\left(\dfrac{800}{3x}\right) = \dfrac{5,120}{x}$

Total cost: $C(x) = 20x + \dfrac{5,120}{x}$

The number of machines which minimize cost is approximately 16. Note that

$$C(16) = 20(16) + \dfrac{5,120}{16} = 640. \text{ So, the}$$

estimated min cost is $640.

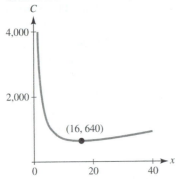

31. (a) Equilibrium occurs when $S(x) = D(x)$,

or $3x + 150 = -2x + 275$

$$5x = 125$$

$$x = 5$$

The corresponding equilibrium price is $p = S(x) = D(x)$ or

$p = 3(25) + 150 = \$225$.

(b)

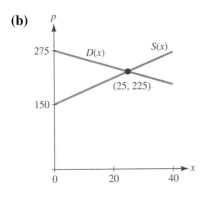

(c) There is a market shortage when demand exceeds supply. Here, a market shortage occurs when $0 < x < 25$. A market surplus occurs when supply exceeds demand. Here, a market surplus occurs when $x > 25$.

33. (a) Equilibrium occurs when $S(x) = D(x)$,

or $2x + 7.43 = -0.21x^2 - 0.84x + 50$

$$0.21x^2 + 2.84x - 42.57 = 0$$

Using the quadratic formula,

$$x = \frac{-2.84 \pm \sqrt{(2.84)^2 - 4(0.21)(-42.57)}}{2(0.21)}$$

so $x = 9$ (disregarding the negative root.) The corresponding equilibrium

price is $p = S(x) = D(x)$, or
$p = 2(9) + 7.43 = 25.43$.

(b)

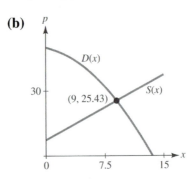

(c) There is a market shortage when demand exceeds supply. Here, a market shortage occurs when $0 < x < 9$. A market surplus occurs when supply exceeds demand. Here, a market surplus occurs when $x > 9$.

35. (a) Equilibrium occurs when $S(x) = D(x)$,

or $\qquad 2x + 15 = \dfrac{385}{x + 1}$

$$(2x + 15)(x + 1) = 385$$

$$2x^2 + 17x + 15 = 385$$

$$2x^2 + 17x - 370 = 0$$

Using the quadratic formula,

$$x = \frac{-17 \pm \sqrt{(17)^2 - 4(2)(-370)}}{2(2)}$$

so $x = 10$ (disregard the negative root). The corresponding equilibrium price is $p = S(x) = D(x)$, or
$p = 2(10) + 15 = 35$.

(b)

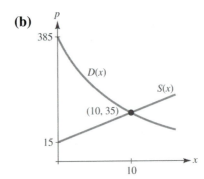

(c) The supply curve intersects the *y*-axis at $S(0) = 15$. Since this is the price at which producers are willing to supply zero units, it corresponds to their overhead at the start of production.

37. (a) Total revenue = (selling price)(# units)
Let *R* represent the total revenue and *x* represent the number of units. Then,
$$R(x) = 110x$$
Total cost = fixed overhead + (cost per unit)(#units)
Let *C* represent the total cost. Then,
$$C(x) = 7,500 + 60x$$
Total profit = total revenue – total cost
Let *P* represent the total profit. Then,
$$P(x) = 110x - \left(7,500 + 60x\right)$$
$$P(x) = 50x - 7,500$$

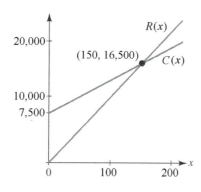

(b) To break even, need revenue = cost, or profit is zero.
$$0 = 50x - 7,500$$
$$x = 150 \text{ units}$$

(c) $P(100) = 50(100) - 7,500 = -2,500$
or a loss of \$2,500

(d) $1,250 = 50x - 7,500$
$x = 175$ units

39. (a) Volume = (length)(width)(height)
The height is given as 20 m and the perimeter is 320 m. So,

$$2(x + w) = 320$$
$$x + w = 160$$
$$w = 160 - x$$
$$V(x) = x(160 - x)(20) = 20x(160 - x)$$

(b) Since the high point of the graph occurs half-way between its intercepts, the max volume occurs when $x = 80$. The dimensions for the max volume are length = 80 m, width = $160 - 80 = 80$ m and height = 20 m.

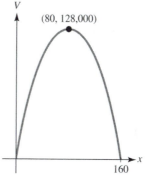

(c) Cost construction
= cost building + cost parking lot
Cost building = 75(80)(80)(20)
      = \$9,600,000
Cost parking lot
= cost top rectangle (across entire length) + cost right side rectangle (next to building)
Cost top rectangle
= 50(area rectangle)
= 50(length)(width)
= 50(length)(100 – width bldg)
= 50(120)(100 – 80)
= \$120,000
Cost right rectangle
= 50(area rectangle)
= 50(length)(width)
= 50(120 – length bldg)(width)
= 50(120 – 80)(80)
= \$160,000
Cost construction
= \$9,600,000 + \$120,000 + \$160,000
= \$9,880,000

41. $S(q) = aq + b$
$D(q) = cq + d$

(a) The graph of S is rising, while the graph of D is falling. So, $a > 0$ and $c < 0$. Further, since both y-intercepts are positive, $b > 0$ and $d > 0$.

(b) $aq + b = cq + d$
$(a - c)q = d - b$
$$q_e = \frac{d - b}{a - c}$$
$p_e = aq_e + b$
$$= a\left(\frac{d - b}{a - c}\right) + b$$
$$= \frac{ad - ab}{a - c} + b$$
$$= \frac{ad - ab + b(a - c)}{a - c}$$
$$= \frac{ad - bc}{a - c}$$

(c) As a increases, the denominator in the expression for q_e increases. This results in a decrease in q_e. As d increases, the numerator in the expression for q_e increases. This results in an increase in q_e.

43. Since I is proportional to the area, A, of the pupil, $I = kA$, where k is a constant of proportionality. Since the pupil of the eye is circular and the area of a circle is $A = \pi r^2$, $I = k\pi r^2$.

45. (a) For a newborn child, the points $(0, 46)$ and $(100, 77)$ define the linear function. Its slope is
$$m = \frac{77 - 46}{100 - 0} = 0.31 \text{ and the function is}$$
$B(t) = 0.31t + 46$.
For a 65-year-old, the points $(0, 76)$ and $(100, 83)$ define the linear function. Its slope is

$$m = \frac{83 - 76}{100 - 0} = 0.07 \text{ and the function is}$$
$E(t) = 0.07t + 76$.

(b) Need to find $B(t) = E(t)$.
$0.31t + 46 = 0.07t + 76$
$0.24t = 30$
$t = 125$ years
Note that this is where the graphs intersect.

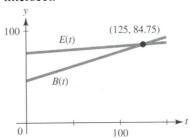

(c) Writing exercise—Answers will vary.

47. $C = \left(\dfrac{N + 1}{24}\right)(300) = \left(\dfrac{N + 1}{2}\right)(25)$

$$C = \frac{2N \cdot 300}{25} = 24N$$

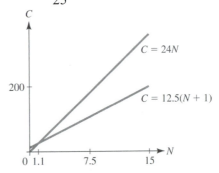

49. (a) The estimated surface area of the child is:
$$S = 0.0072(18)^{0.425}(91)^{0.725} \approx 0.6473$$
so, $C = \dfrac{(0.6473)(250)}{1.7} \approx 95.2$ mg.

(b) Using $2H$ and $2W$ for the larger child,
$$C = \frac{0.0072(2W)^{0.425}(2H)^{0.725}\,A}{1.7}.$$
Comparing to drug dosage for the smaller child,

$$\frac{\dfrac{0.0072(2W)^{0.425}(2H)^{0.725}A}{1.7}}{\dfrac{0.0072W^{0.425}H^{0.725}A}{1.7}} = (2)^{0.425}(2)^{0.725}$$

$$\approx 2.22$$

So, the drug dosage for larger child is approximately 2.22 times the dosage for the smaller child.

51. $R = \dfrac{R_m[S]}{K_m + [S]}$

$y = \dfrac{1}{R} = \dfrac{K_m + [S]}{R_m[S]}$

Dividing numerator and denominator by $[S]$ gives

$$y = \frac{K_m\left(\dfrac{1}{[S]}\right) + 1}{R_m} = \frac{K_m\left(\dfrac{1}{[S]}\right)}{R_m} + \frac{1}{R_m}$$

$$= \frac{K_m}{R_m}\left(\frac{1}{[S]}\right) + \frac{1}{R_m}$$

Letting $x = \dfrac{1}{[S]}$,

$$y = \frac{K_m}{R_m}x + \frac{1}{R_m}$$

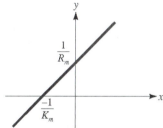

53. Let r be the radius and h the height of the cylinder. The surface area of the closed cylinder is $S = 120\pi = 2\pi r^2 + 2\pi rh$ or $h = \dfrac{60 - r^2}{r}$. So

$V(r) = \pi r^2 h = \pi r(60 - r^2)$.

55. Let r be the radius and h the height of the cylinder. Since the volume is

$V = \pi r^2 h = 4\pi$, or $h = \dfrac{4}{r^2}$.

The cost of the top or bottom is $C_t = C_b = 2(0.02)\pi r^2$, while the cost of the side is $2\pi rh(0.02) = \dfrac{0.16\pi}{r}$.

The total cost is $C(x) = 0.08\pi r^2 + \dfrac{0.16\pi}{r}$

57. Let x denote the width of the printed portion and y the length of the printed portion. Then $x + 4$ is the width of the poster and $y + 8$ is its length. The area A of the poster is $A = (x + 4)(y + 8)$ which is a function of two variables.

$A = 25$ leads to $xy = 25$ or $y = \dfrac{25}{x}$.

So $A(x) = (x + 4)\left(\dfrac{25}{x} + 8\right) = 8x + 57 + \dfrac{100}{x}$.

59. Let x be the side of the square base and y the height of the open box. The area of the base is x^2 square meters and that of each side is xy square meters. The total cost is $4x^2 + 3(4xy) = 48$.
Solving for y in terms of x,

$12xy = 48 - 4x^2$

$3xy = 12 - x^2$

$y = \dfrac{12 - x^2}{3x}$

The volume of the box is

$V = x^2 y$

$= \dfrac{x(12 - x^2)}{3}$

$= 4x - \dfrac{x^3}{3}$ cubic meters.

61. Let t be the number of hours the second plane has been flying. Since distance = (rate)(time), the equation for its distance is $d = 650t$.

The first plane has been flying for $t+\dfrac{1}{2}$ hours, so the equation for its distance is $d=550\left(t+\dfrac{1}{2}\right)$.

The planes will meet when

$650t=550\left(t+\dfrac{1}{2}\right)$

$650t=550t+275$

$100t=275$

$t=2.75$

Since three-quarters of an hour is 45 minutes, the second plane passes the first plane after it has been flying 2 hours and 45 minutes.

1.5 Limits

1. $\lim\limits_{x\to a} f(x)=b$, even though $f(a)$ is not defined.

3. $\lim\limits_{x\to a} f(x)=b$ even though $f(a)=c$.

5. $\lim\limits_{x\to a} f(x)$ does not exist since as x approaches a from the left, the function becomes unbounded.

7. $\lim\limits_{x\to 2}(3x^2-5x+2)$

$=3\lim\limits_{x\to 2} x^2-5\lim\limits_{x\to 2}x+\lim\limits_{x\to 2}2$

$=3(2)^2-5(2)+2$

$=4$

9. $\lim\limits_{x\to 0}(x^5-6x^4+7)$

$=\lim\limits_{x\to 0}x^5-6\lim\limits_{x\to 0}x^4+\lim\limits_{x\to 0}7$

$=7$

11. $\lim\limits_{x\to 3}(x-1)^2(x+1)=\lim\limits_{x\to 3}(x-1)^2\lim\limits_{x\to 3}(x+1)$

$=(3-1)^2(3+1)$

$=16$

13. $\lim\limits_{x\to 1/3}\dfrac{x+1}{x+2}=\dfrac{\lim\limits_{x\to 1/3}x+1}{\lim\limits_{x\to 1/3}x+2}=\dfrac{\frac{4}{3}}{\frac{7}{3}}=\dfrac{4}{7}$

15. $\lim\limits_{x\to 5}\dfrac{x+3}{5-x}$ does not exist since the limit of the denominator is zero while the limit of the numerator is not zero.

17. $\lim\limits_{x\to 1}\dfrac{x^2-1}{x-1}=\lim\limits_{x\to 1}\dfrac{(x+1)(x-1)}{x-1}$

$=\lim\limits_{x\to 1}(x+1)$

$=2$

19. $\lim\limits_{x\to 5}\dfrac{x^2-3x-10}{x-5}=\lim\limits_{x\to 5}\dfrac{(x-5)(x+2)}{x-5}$

$=\lim\limits_{x\to 5}(x+2)$

$=7$

21. $\lim\limits_{x\to 4}\dfrac{(x+1)(x-4)}{(x-1)(x-4)}=\dfrac{\lim\limits_{x\to 4}(x+1)}{\lim\limits_{x\to 4}(x-1)}=\dfrac{5}{3}$

23. $\lim\limits_{x\to -2}\dfrac{x^2-x-6}{x^2+3x+2}=\lim\limits_{x\to -2}\dfrac{(x-3)(x+2)}{(x+1)(x+2)}$

$=\dfrac{\lim\limits_{x\to -2}(x-3)}{\lim\limits_{x\to -2}(x+1)}$

$=\dfrac{-5}{-1}$

$=5$

25. $\lim\limits_{x\to 4}\dfrac{\sqrt{x}-2}{x-4}=\lim\limits_{x\to 4}\dfrac{\sqrt{x}-2}{x-4}\dfrac{\sqrt{x}+2}{\sqrt{x}+2}$

$=\lim\limits_{x\to 4}\dfrac{x-4}{(x-4)\left(\sqrt{x}+2\right)}$

$=\dfrac{1}{4}$

27. $f(x)=x^3-4x^2-4$

$\lim\limits_{x\to +\infty}f(x)=\lim\limits_{x\to +\infty}x^3=+\infty$

$\lim\limits_{x\to -\infty}f(x)=\lim\limits_{x\to -\infty}x^3=-\infty$

29. $f(x) = (1-2x)(x+5) = -2x^2 - 9x + 5$

$$\lim_{x\to+\infty} f(x) = \lim_{x\to+\infty} -2x^2 = -\infty$$

$$\lim_{x\to-\infty} f(x) = \lim_{x\to-\infty} -2x^2 = -\infty$$

31. $f(x) = \dfrac{x^2 - 2x + 3}{2x^2 + 5x + 1}$

$$\lim_{x\to+\infty} f(x) = \lim_{x\to+\infty} \frac{1 - \frac{2}{x} + \frac{3}{x^2}}{2 + \frac{5}{x} + \frac{1}{x^2}} = \frac{1}{2}$$

$$\lim_{x\to-\infty} f(x) = \lim_{x\to-\infty} \frac{1 - \frac{2}{x} + \frac{3}{x^2}}{2 + \frac{5}{x} + \frac{1}{x^2}} = \frac{1}{2}$$

33. $f(x) = \dfrac{2x + 1}{3x^2 + 2x - 7}$

$$\lim_{x\to+\infty} f(x) = \lim_{x\to+\infty} \frac{\frac{2}{x} + \frac{1}{x^2}}{3 + \frac{2}{x} - \frac{7}{x^2}} = 0$$

$$\lim_{x\to-\infty} f(x) = \lim_{x\to-\infty} \frac{\frac{2}{x} + \frac{1}{x^2}}{3 + \frac{2}{x} - \frac{7}{x^2}} = 0$$

35. $f(x) = \dfrac{3x^2 - 6x + 2}{2x - 9}$

$$\lim_{x\to+\infty} f(x) = \lim_{x\to+\infty} \frac{3x^2 - 6x + 2}{2x - 9}$$

$$= \lim_{x\to+\infty} \frac{3x - 6 + \frac{2}{2x}}{2 - \frac{9}{x}}$$

$$\lim_{x\to+\infty} 3x - 6 + \frac{2}{x} = +\infty \text{ and } \lim_{x\to+\infty} 2 - \frac{9}{x} = 2$$

So, $\displaystyle\lim_{x\to+\infty} \frac{3x - 6 + \frac{2}{x}}{2 - \frac{9}{x}} = +\infty$

$$\lim_{x\to-\infty} f(x) = \lim_{x\to-\infty} \frac{3x - 6 + \frac{2}{x}}{2 - \frac{9}{x}}$$

$$\lim_{x\to-\infty} 3x - 6 + \frac{2}{x} = -\infty \text{ and } \lim_{x\to-\infty} 2 - \frac{9}{x} = 2$$

So, $\displaystyle\lim_{x\to-\infty} \frac{3x - 6 + \frac{2}{x}}{2 - \frac{9}{x}} = -\infty$

37. $\displaystyle\lim_{x\to+\infty} f(x) = 1$ and $\displaystyle\lim_{x\to-\infty} f(x) = -1$

39. The corresponding table values are:

$$f(1.9) = (1.9)^2 - 1.9 = 1.71$$
$$f(1.99) = (1.99)^2 - 1.99 = 1.9701$$
$$f(1.999) = (1.999)^2 - 1.999 = 1.997001$$
$$f(2.001) = (2.001)^2 - 2.001 = 2.003001$$
$$f(2.01) = (2.01)^2 - 2.01 = 2.0301$$
$$f(2.1) = (2.1)^2 - 2.1 = 2.31$$
$$\lim_{x\to2} f(x) = 2$$

41. The corresponding table values are

$$f(0.9) = \frac{(0.9)^3 + 1}{0.9 - 1} = -17.29$$

$$f(0.99) = \frac{(0.99)^3 + 1}{0.99 - 1} = -197.0299$$

$$f(0.999) = \frac{(0.999)^3 + 1}{0.999 - 1} = -1,997.002999$$

$$f(1.001) = \frac{(1.001)^3 + 1}{1.001 - 1} = 2,003.003001$$

$$f(1.01) = \frac{(1.01)^3 + 1}{1.01 - 1} = 203.0301$$

$$f(1.1) = \frac{(1.1)^3 + 1}{1.1 - 1} = 23.31$$

$\displaystyle\lim_{x\to1} f(x)$ does not exist.

43. $\displaystyle\lim_{x\to c}[2f(x) - 3g(x)]$

$$= \lim_{x\to c} 2f(x) - \lim_{x\to c} 3g(x)$$

$$= 2\lim_{x\to c} f(x) - 3\lim_{x\to c} g(x)$$

$$= 2(5) - 3(-2)$$

$$= 16$$

45. $\lim\limits_{x\to c} \sqrt{f(x)+g(x)}$

$= \lim\limits_{x\to c}[f(x)+g(x)]^{1/2}$

$= \left[\lim\limits_{x\to c}(f(x)+g(x))\right]^{1/2}$

$= \left[\lim\limits_{x\to c} f(x)+\lim\limits_{x\to c} g(x)\right]^{1/2}$

$= [5+(-2)]^{1/2}$

$= \sqrt{3}$

47. $\lim\limits_{x\to c} \dfrac{f(x)}{g(x)} = \dfrac{\lim\limits_{x\to c} f(x)}{\lim\limits_{x\to c} g(x)} = \dfrac{5}{-2} = -\dfrac{5}{2}$

49. $\lim\limits_{x\to\infty} \dfrac{2f(x)+g(x)}{x+f(x)}$

$= \lim\limits_{x\to\infty} \dfrac{\frac{1}{x}\cdot 2f(x)+\frac{1}{x}\cdot g(x)}{1+\frac{1}{x}\cdot f(x)}$

$= \dfrac{\lim\limits_{x\to\infty}\left[\frac{1}{x}\cdot 2f(x)+\frac{1}{x}\cdot g(x)\right]}{\lim\limits_{x\to\infty}\left[1+\frac{1}{x}\cdot f(x)\right]}$

$= \dfrac{\lim\limits_{x\to\infty}\frac{1}{x}\cdot 2f(x)+\lim\limits_{x\to\infty}\frac{1}{x}\cdot g(x)}{\lim\limits_{x\to\infty}1+\lim\limits_{x\to\infty}\frac{1}{x}\cdot f(x)}$

$= \dfrac{\lim\limits_{x\to\infty}\frac{1}{x}\cdot 2\lim\limits_{x\to\infty}f(x)+\lim\limits_{x\to\infty}\frac{1}{x}\cdot\lim\limits_{x\to\infty}g(x)}{\lim\limits_{x\to\infty}1+\lim\limits_{x\to\infty}\frac{1}{x}\cdot\lim\limits_{x\to\infty}f(x)}$

Since $\lim\limits_{x\to\infty}\dfrac{1}{x}=0$ and $\lim\limits_{x\to\infty}1=1$,

$= \dfrac{0+0}{1+0}$

$= 0$

51. $p = 0.2t + 1{,}500;\ E(t) = \sqrt{9t^2 + 0.5t + 179}$

(a) Since the units of p are thousands and the units of E are millions, the units of E/p will be thousands.

$P(t) = \dfrac{\sqrt{9t^2 + 0.5t + 179}}{0.2t + 1500}$ thousand

dollars per person

(b) Dividing each term by t (note that each term under the square root will be divided by t^2 since $\sqrt{t^2}=t$),

$\lim\limits_{t\to\infty} P(t) = \lim\limits_{t\to\infty} \dfrac{\sqrt{9+\frac{0.5}{t}+\frac{179}{t^2}}}{0.2+\frac{1500}{t}}$

$= \dfrac{\sqrt{\lim\limits_{t\to\infty}\left(9+\frac{0.5}{t}+\frac{179}{t^2}\right)}}{\lim\limits_{t\to\infty}\left(0.2+\frac{1500}{t}\right)}$

$= \dfrac{\sqrt{9}}{0.2}$

$= 15$

53. $\lim\limits_{x\to+\infty} \dfrac{7.5x+120{,}000}{x}$

$= \lim\limits_{x\to+\infty} 7.5+\dfrac{120{,}000}{x}$

$= 7.5$

As the number of units produced increases indefinitely, the average cost per unit decreases, approaching a minimum of $7.50. The average cost cannot decrease further, as the expense of materials cannot be eliminated completely.

55. $C(x) = \dfrac{8x^2 - 636x - 320}{x^2 - 68x - 960}$

Need to find $\lim\limits_{x\to 80} C(x)$. Substituting $x = 80$ into C yields $\dfrac{0}{0}$, so need to try to factor and cancel.

$\lim\limits_{x\to 80} \dfrac{4(2x+1)(x-80)}{(x+12)(x-80)}$

$= \lim\limits_{x\to 80} \dfrac{4(2x+1)}{x+12} = \dfrac{4[2(80)+1]}{80+12} = 7$

So, the operating cost will be $700.

57.

x	1	0.1	0.01	0.001	0.0001
$1{,}000(1+0.05x)^{1/x}$	1,050.00	1,051.14	1,051.26	1,051.27	1,051.27

Limit appears to be $1,051.27.

59. $C(t) = \dfrac{0.4}{t^{1.2}+1} + 0.013$

(a) $C(0) = \dfrac{0.4}{0^{1.2}+1} + 0.013$
$= 0.413$ mg/ml

(b) Need to find
$C(5) - C(4)$
$= \left[\dfrac{0.4}{5^{1.2}+1} + 0.013\right] - \left[\dfrac{0.4}{4^{1.2}+1} + 0.013\right]$
$= \dfrac{0.4}{5^{1.2}+1} - \dfrac{0.4}{4^{1.2}+1}$
$\approx 0.0506 - 0.0637$
$= -0.0131$
So, the concentration decreases approximately 0.013 mg/ml during this hour.

(c) $\lim\limits_{t\to+\infty} C(t)$
$= \lim\limits_{t\to+\infty}\left[\dfrac{0.4}{t^{1.2}+1} + 0.013\right]$
$= \lim\limits_{t\to+\infty}\dfrac{0.4}{t^{1.2}+1} + \lim\limits_{t\to+\infty} 0.013$
$= 0 + 0.013$
$= 0.013$ mg/ml

61. $P(t) = \dfrac{30}{3+t}$, $Q(t) = \dfrac{64}{4-t}$

(a) $P(0) = \dfrac{30}{3} = 10$ thousand, or 10,000

$Q(0) = \dfrac{64}{4} = 16$ thousand, or 16,000

(b) Since the function P accepts all $t \ge 0$, the function values decrease as t increases. Further,

$\lim\limits_{t\to+\infty} P(t) = \lim\limits_{t\to+\infty}\dfrac{30}{3+t} = 0.$

So, in the long run, P tends to zero. The Q function, however, only accepts values of t such that $0 \le t \le 4$. The function values increase as t increases.

Further, $\lim\limits_{t\to4^-}\dfrac{64}{4-5} = +\infty$

(a t/Q table is an easy way to see this). So, Q increases without bound.

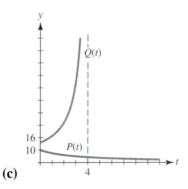

(c)

(d) Writing exercise—Answers will vary.

63. (a) $\lim\limits_{S\to\infty} \dfrac{aS}{S+c} = \lim\limits_{S\to\infty} \dfrac{a}{1+\frac{c}{S}} = a$

As bite size increases indefinitely, intake approaches a limit of a. This signifies that the animal has a limit of how much it can consume, no matter how large its bites become.

(b) Writing exercise—answers will vary.

65. $\lim\limits_{x\to+\infty} = \dfrac{a_n x^n + a_{n-1} x^{n-1} + \cdots + a_1 x + a_0}{b_m x^m + b_{m-1} x^{m-1} + \cdots + b_1 x + b_0}$

(a) When $n < m$,

$= \lim\limits_{x\to+\infty} \dfrac{a_n + \frac{a_n-1}{x} + \cdots + \frac{a_1}{x^{n-1}} + \frac{a_0}{x^n}}{b_m \frac{x^m}{x^n} + b_{m-1} \frac{x^{m-1}}{x^n} + \cdots + b_1 \frac{x}{x^n} + b_0 \frac{1}{x^n}}$

Since $\lim\limits_{x\to+\infty} \dfrac{x^m}{x^n} = +\infty$, $\lim\limits_{x\to+\infty} f(x) = 0$

(b) When $n = m$, $\dfrac{x^m}{x^n} = 1$ and

$\lim\limits_{x\to+\infty} f(x) = \dfrac{a_n}{b_m}$

(c) When $n > m$,

$= \lim\limits_{x\to+\infty} \dfrac{a_n \frac{x^n}{x^m} + a_{n-1} \frac{x^{n-1}}{x^m} + \cdots + a_1 \frac{x}{x^m} + a_0 \frac{1}{x^m}}{b_m + \frac{b_{m-1}}{x} + \cdots + \frac{b_1}{x^{m-1}} + \frac{b_0}{x^m}}$

Now,

$\lim\limits_{x\to+\infty} a_n \dfrac{x^n}{x^m} + a_{n-1} \dfrac{x^{n-1}}{x^m} + \cdots + a_1 \dfrac{x}{x^m} + a_0 \dfrac{1}{x^m}$

$= \pm\infty,$

depending on the sign of a_n. Also

$\lim\limits_{x\to+\infty} b_m + \dfrac{b_{m-1}}{x} + \cdots + \dfrac{b_1}{x^{m-1}} + \dfrac{b_0}{x^m}$

$= b_m$

So, $\lim\limits_{x\to+\infty} \dfrac{a_n \frac{x^n}{x^m} + a_{n-1} \frac{x^{n-1}}{x^m} + \cdots}{b_m + \frac{b_{m-1}}{x} + \cdots} = \pm\infty,$

depending on the signs of a_n and b_m. When a_n and b_m have the same sign, the limit is $+\infty$; when they have opposite signs, the limit is $-\infty$.

67. As the weight approaches 18 lb, displacement approaches a limit of 1.8 inches.

1.6 One-Sided Limits and Continuity

1. $\lim\limits_{x\to2^-} f(x) = -2$; $\lim\limits_{x\to2^+} f(x) = 1$

Since $-2 \neq 1$, $\lim\limits_{x\to2} f(x)$ does not exist.

3. $\lim\limits_{x\to2^-} f(x) = 2$; $\lim\limits_{x\to2^+} f(x) = 2$

Since limits are the same, $\lim\limits_{x\to2} f(x) = 2$.

5. $\lim\limits_{x\to4^+} (3x^2 - 9) = \lim\limits_{x\to4^+} 3x^2 - \lim\limits_{x\to4^+} 9$

$= 3(4)^2 - 9$

$= 39$

7. $\lim\limits_{x\to3^+} \sqrt{3x - 9} = \sqrt{3(3) - 9} = 0$

9. $\lim\limits_{x\to2^-} \dfrac{x+3}{x+2} = \dfrac{\lim\limits_{x\to2^-}(x+3)}{\lim\limits_{x\to2^-}(x+2)}$

$= \dfrac{2+3}{2+2}$

$= \dfrac{2+3}{2+2}$

$= \dfrac{5}{4}$

11. $\lim\limits_{x\to0^+} \left(x - \sqrt{x}\right) = 0 - 0 = 0$

13. $\lim\limits_{x\to3^+}\dfrac{\sqrt{x+1}-2}{x-3}$

$=\lim\limits_{x\to3^+}\dfrac{\sqrt{x+1}-2}{x-3}\cdot\dfrac{\sqrt{x+1}+2}{\sqrt{x+1}+2}$

$=\lim\limits_{x\to3^+}\dfrac{x+1-4}{(x-3)\left(\sqrt{x+1}+2\right)}$

$=\dfrac{1}{4}$

15. $\lim\limits_{x\to3^-}f(x)=\lim\limits_{x\to3^-}(2x^2-x)$

$\qquad\qquad=2(3)^2-3$

$\qquad\qquad=15$

$\lim\limits_{x\to3^+}f(x)=\lim\limits_{x\to3^+}(3-x)=3-3=0$

17. If $f(x)=5x^2-6x+1$, then $f(2)=9$ and $\lim\limits_{x\to2}f(x)=9$, so f is continuous at $x=2$.

19. If $f(x)=\dfrac{x+2}{x+1}$, then $f(1)=\dfrac{3}{2}$ and

$\lim\limits_{x\to1}f(x)=\lim\limits_{x\to1}\dfrac{x+2}{x+1}=\dfrac{\lim\limits_{x\to1}(x+2)}{\lim\limits_{x\to1}(x+1)}=\dfrac{3}{2}.$

So, f is continuous at $x=1$.

21. If $f(x)=\dfrac{x+1}{x-1}$, $f(1)$ is undefined since the denominator is zero, and so f is not continuous at $x=1$.

23. If $f(x)=\dfrac{\sqrt{x}-2}{x-4}$, $f(4)$ is undefined since the denominator is zero, and so f is not continuous at $x=4$.

25. If $f(x)=\begin{cases}x+1 & \text{if }x\le2\\2 & \text{if }x>2\end{cases}$, then $f(2)=3$

and $\lim\limits_{x\to2}f(x)$ must be determined.
As x approaches 2 from the left,
$\lim\limits_{x\to2^-}f(x)=\lim\limits_{x\to2^-}(x+1)=3$ and as x
approaches 2 from the right,

$\lim\limits_{x\to2^+}f(x)=\lim\limits_{x\to2^+}2=2.$

So the limit does not exist (since different limits are obtained from the left and the right), and f is not continuous at $x=2$.

27. If $f(x)=\begin{cases}x^2+1 & \text{if }x\le3\\2x+4 & \text{if }x>3\end{cases}$, then

$f(3)=(3)^2+1=10$ and $\lim\limits_{x\to3}f(x)$ must be determined. As x approaches 3 from the left,

$\lim\limits_{x\to3^-}f(x)=\lim\limits_{x\to3^-}(x^2+1)=(3)^2+1=10$ and

as x approaches 3 from the right,
$\lim\limits_{x\to3^+}f(x)=\lim\limits_{x\to3^+}(2x+4)=2(3)+4=10.$

So $\lim\limits_{x\to3}f(x)=10$. Since $f(x)=\lim\limits_{x\to3}f(x)$, f
is continuous at $x=3$.

29. $f(a)=3a^2-6a+9$ so f is defined for all

real numbers. $\lim\limits_{x\to a}f(a)=3(a)^2-6a+9$,

so the limit of f exists for all real numbers.
Since $f(a)=\lim\limits_{x\to a}f(a)$, there are no

values for which f is not continuous.

31. $f(x)=\dfrac{x+1}{x-2}$ is not defined at $x=2$, so f is

not continuous at $x=2$.

33. $f(x)=\dfrac{3x+3}{x+1}$ is not defined at $x=-1$, so f

is *not* continuous at $x=-1$.

35. $f(x)=\dfrac{3x-2}{(x+3)(x-6)}$ is not defined at

$x=-3$ and $x=6$, so f is *not* continuous at
$x=-3$ and $x=6$.

37. $f(x)=\dfrac{x}{x^2-x}$ is not defined at $x=0$ and

$x=1$, so f is *not* continuous at $x=0$ and
$x=1$.

39. f is defined for all real numbers. Further,

$$\lim_{x \to 1^-} f(x) = 2 + 3$$
$$= 5$$
$$= \lim_{x \to 1^+} f(x)$$
$$= 6 - 1$$
$$= f(1),$$

so there are no values for which f is not continuous.

41. f is defined for all real numbers. However,

$$\lim_{x \to 0^-} f(x) = \lim_{x \to 0^-} 3x - 2 = 3(0) - 2 = 2$$

$$\lim_{x \to 0^+} f(x) = \lim_{x \to 0^+} x^2 + x = 0 + 0 = 0$$

So $\lim_{x \to 0} f(x)$ does not exist and therefore f is *not* continuous at $x = 0$.

43. $C(x) = \dfrac{8x^2 - 636x - 320}{x^2 - 68x - 960}$

(a) $C(0) = \dfrac{-320}{-960} = \dfrac{1}{3} \approx 0.333$

$$C(100) = \dfrac{8(100)^2 - 636(100) - 320}{(100)^2 - 68(100) - 960}$$
$$\approx 7.179$$

(b) Since the denominator factors as $(x + 12)(x - 80)$, the function has a vertical asymptote when $x = 80$. This means that C is not continuous on the interval $0 \le x \le 100$, and the intermediate value theorem cannot be used.

45. $C(x) = \dfrac{12x}{100 - x}$

(a) $C(25) = \dfrac{12(25)}{100 - 25} = 4$, or \$4000

$$C(50) = \dfrac{12(50)}{100 - 50} = 12, \text{ or } \$12,000$$

(b)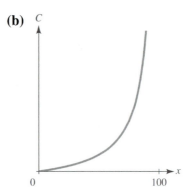

(c) From the graph, $\lim_{x \to 100^-} C(x) = \infty$.

So, it is not possible to remove all of the pollution.

47. The graph is discontinuous at $x = 10$ and $x = 25$. Sue is probably at the gas station replenishing fuel.

49. $p(t) = 20 - \dfrac{7}{t + 2}; c(p) = 0.4\sqrt{p^2 + p + 21}$

$$\lim_{t \to \infty} p(t) = \lim_{t \to \infty} 20 - \dfrac{7}{t + 2} = 20$$

$$\lim_{t \to \infty} c(p) = \lim_{t \to 20} 0.4\sqrt{p^2 + p + 21}$$
$$= 0.4\sqrt{(20)^2 + 20 + 21}$$
$$= 0.4(21) = 8.4$$

In the long run, the level of pollution will be 8.4 parts per million.

51. (a) $I(30) = 80°$ F, using the top branch of function

$$I(90) = 0.005(90)^2 - 0.65(90) + 104$$
$$= 86° \text{ F, using the bottom}$$

branch of the function

(b) First, whenever $0 \le h \le 40$, the heat index is $80°$ F.
Using the middle branch of the function,
$$83 = 80 + 0.1(h - 40)$$

$$30 = h - 40; h = 70$$
A relative humidity of 70% produces a heat index of $83°$ F.
Using the bottom branch of the

function,

$$83 = 0.005h^2 - 0.65h + 104$$

$$0 = 0.005h^2 - 0.65h + 21$$

$$h = \frac{0.65 \pm \sqrt{(-0.65) - 4(21)(0.005)}}{2(0.005)}$$

$$h = \frac{0.65 \pm \sqrt{-1.07}}{2(0.005)}$$

for which there is no solution.

(c) $\lim_{h \to 40^-} I(h) = \lim_{h \to 40^-} 80 = 80$

$\lim_{h \to 40^+} I(h) = \lim_{h \to 40^+} 80 + 0.1(h - 40)$

$\qquad = 80 + 0.1(40 - 40) = 80$

$I(40) = 80$
so, continuous at $h = 40$

$\lim_{h \to 80^-} I(h) = \lim_{h \to 80^-} 80 + 0.1(h - 40)$

$\qquad = 80 + 0.1(80 - 40) = 84$

$\lim_{h \to 80^+} I(h) = \lim_{h \to 80^+} 0.005h^2 - 0.65h + 104$

$\qquad = 0.005(80)^2 - 0.65(80) + 104$

$\qquad = 84$

$I(80) = 84$
so, continuous at $h = 80$

53. The graph will consist of horizontal line segments, with the left endpoints open and the right endpoints closed (from the inequalities). The height of each segment corresponds to the respective costs of postage.

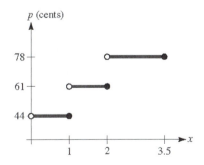

The function p is discontinuous where the price jumps or when $x = 1$ and $x = 2$.

55. On the open interval $0 < x < 1$, since $x \neq 0$,

$f(x) = x\left(1 + \dfrac{1}{x}\right) = x + 1$. So, $f(x)$, a

polynomial on $0 < x < 1$, is continuous. On the closed interval $0 \le x \le 1$, the endpoints must now be considered.

$f(x) = x\left(x + \dfrac{1}{x}\right)$ is not continuous at

$x = 0$ since $f(0)$ is not defined. However, f is continuous at $x = 1$ since

$f(1) = 1\left(1 + \dfrac{1}{1}\right) = 2$ and as x approaches 1

from the left,

$\lim_{x \to 1^-} x\left(x + \dfrac{1}{x}\right) = \lim_{x \to 1^-} x \cdot \lim_{x \to 1^-}\left(x + \dfrac{1}{x}\right)$

$\qquad = 1\left(1 + \dfrac{1}{1}\right)$

$\qquad = 2.$

57. $f(x) = \begin{cases} Ax - 3 & \text{if } x < 2 \\ 3 - x + 2x^2 & \text{if } 2 \le x \end{cases}$

f is continuous everywhere except possibly at $x = 2$, since $Ax - 3$ and $3 - x + 2x^2$ are polynomials. Since $f(2) = 3 - 2 + 2(2)^2 = 9$, in order that f be continuous at $x = 2$, A must be chosen so that $\lim_{x \to 2} f(x) = 9$.

As x approaches 2 from the right,

$\lim_{x \to 2^+} f(x) = \lim_{x \to 2^+}(3 - x + 2x^2)$

$\qquad = \lim_{x \to 2^+} 3 - \lim_{x \to 2^+} x + 2\lim_{x \to 2^+} x^2$

$\qquad = 3 - 2 + 2(2)^2$

$\qquad = 9$

and as x approaches from the left,

$\lim_{x \to 2^-} f(x) = \lim_{x \to 2^-}(Ax - 3)$

$\qquad = A\lim_{x \to 2^-} x - \lim_{x \to 2^-} 3$

$\qquad = 2A - 3.$

For $\lim_{x \to 2} f(x) = 9$, $2A - 3$ must equal 9, or

$A = 6$. f is continuous at $x = 2$ only when $A = 6$.

59. Rewrite as $\sqrt[3]{x-8} + 9x^{2/3} - 29$ and notice that at $x = 0$ this expression is negative and at $x = 8$ it is positive. Therefore, by the intermediate value property, there must be a value of x between 0 and 8 such that this expression is 0 or
$$\sqrt[3]{x-8} + 9x^{2/3} = 29.$$

61. To investigate the behavior of
$$f(x) = \frac{2x^2 - 5x + 2}{x^2 - 4}:$$
Press $\boxed{y=}$.
Input $(2x^2 - 5x + 2)/(x^2 - 4)$ for $y_1 =$
Press $\boxed{\text{graph}}$.

(a) Press $\boxed{\text{trace}}$. Using arrows to move cursor to be near $x = 2$ we see that $(1.9, 0.72)$ and $(2.1, 0.79)$ are two points on the graph. By zooming in, we find $(1.97, 0.74)$ and $(2.02, 0.76)$ to be two points on the graph. The
$$\lim_{x \to 2} f(x) = \frac{3}{4}, \text{ however, the function}$$
is not continuous at $x = 2$ since $f(2)$ is undefined. To show this, use the value function under the calc menu and enter $x = 2$. There is no y-value displayed, which indicates the function is undefined for $x = 2$.

(b) Use the z standard function under the Zoom menu to return to the original graph. We see from the graph that there is a vertical asymptote at $x = -2$.
The $\lim_{x \to 2^-} f(x) = \infty$ and
$\lim_{x \to 2^+} f(x) = -\infty$ and therefore
$\lim_{x \to -2} f(x)$ does not exist. So f is not continuous at $x = -2$.

63. Let's assume the hands of a clock move in a continuous fashion. During each hour the minute hand moves from being behind the hour to being ahead of the hour.

Therefore, at some time, the hands must be in the same place.

Checkup for Chapter 1

1. Since negative numbers do not have square roots and denominators cannot be zero, the domain of the function
$$f(x) = \frac{2x - 1}{\sqrt{4 - x^2}} \text{ is all real numbers such}$$
that $4 - x^2 > 0$ or $(2 + x)(2 - x) > 0$, namely $-2 < x < 2$.

2. $g(h(x)) = g\left(\dfrac{x+2}{2x+1}\right)$
$$= \frac{1}{2\left(\frac{x+2}{2x+1}\right) + 1}$$
$$= \frac{1}{\frac{2x+4}{2x+1} + 1}$$
$$= \frac{1}{\frac{2x+4+2x+1}{2x+1}}$$
$$= \frac{2x+1}{4x+5}, \ x \neq -\frac{1}{2}$$

3. (a) Since $m = -\dfrac{1}{2}$ and the point $(-1, 2)$ is on the line, the equation of the line is
$$y - 2 = -\frac{1}{2}(x - (-1))$$
$$y - 2 = -\frac{1}{2}(x + 1)$$
$$y - 2 = -\frac{1}{2}x - \frac{1}{2}$$
$$y = -\frac{1}{2}x - \frac{1}{2} + 2$$
$$y = -\frac{1}{2}x + \frac{3}{2}$$

(b) Since $m = 2$ and $b = -3$, the equation of the line is $y = 2x - 3$.

4. (a) The graph is a line with x-intercept $\dfrac{5}{3}$ and y-intercept -5.

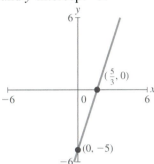

(b) The graph is a parabola which opens down (since $A < 0$). The vertex is
$$\left(-\frac{b}{2a},\, f\left(-\frac{b}{2a}\right)\right), \text{ or } \left(\frac{3}{2}, \frac{25}{4}\right).$$
The x-intercepts are
$$0 = -x^2 + 3x + 4$$
$$0 = x^2 - 3x - 4$$
$$0 = (x-4)(x+1)$$
$$x = 4,\, -1$$
The y-intercept is 4.

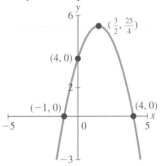

5. (a) $\displaystyle\lim_{x \to -1} \frac{x^2 + 2x - 3}{x - 1} = \frac{(-1)^2 + 2(-1) - 3}{-1 - 1}$
$$= \frac{1 - 2 - 3}{-2}$$
$$= 2$$

(b) $\displaystyle\lim_{x \to 1} \frac{x^2 + 2x - 3}{x - 1} = \lim_{x \to 1} \frac{(x+3)(x-1)}{x-1}$
$$= \lim_{x \to 1} (x + 3)$$
$$= 4$$

(c) $\displaystyle\lim_{x \to 1} \frac{x^2 - x - 1}{x - 2} = \frac{(1)^2 - 1 - 1}{1 - 2} = \frac{-1}{-1} = 1$

(d) $\displaystyle\lim_{x \to +\infty} \frac{2x^3 + 3x - 5}{-x^2 + 2x + 7} = \lim_{x \to +\infty} \frac{2x + 3 - \frac{5}{x^2}}{-1 + \frac{2}{x} + \frac{7}{x^2}}$

Since $\displaystyle\lim_{x \to +\infty} 2x + 3 - \frac{5}{x^2} = +\infty$ and
$$\lim_{x \to +\infty} -1 + \frac{2}{x} + \frac{7}{x^2} = -1,$$
$$\lim_{x \to +\infty} \frac{2x^3 + 3x - 5}{-x^2 + 2x + 7} = -\infty.$$

6. The function is defined at $x = 1$, and $f(1) = 2(1) + 1 = 3$. If $\displaystyle\lim_{x \to 1} f(x) = 3$, the function will be continuous at $x = 1$. From the left of $x = 1$,
$$\lim_{x \to 1^-} f(x) = \lim_{x \to 1^-} 2x + 1 = 2(1) + 1 = 3.$$
From the right of $x = 1$,
$$\lim_{x \to 1^+} f(x) = \lim_{x \to 1^+} \frac{x^2 + 2x - 3}{x - 1}$$
$$= \lim_{x \to 1^+} \frac{(x+3)(x-1)}{x - 1}$$
$$= \lim_{x \to 1^+} (x + 3)$$
$$= 1 + 3$$
$$= 4.$$
Since $\displaystyle\lim_{x \to 1^-} f(x) \neq \lim_{x \to 1^+} f(x)$, the limit does not exist and the function is not continuous at $x = 1$.

7. (a) Let t denote the time in months since the beginning of the year and $P(t)$ the corresponding price (in cents) of gasoline. Since the price increases at a constant rate of 2 cents per gallon per month, P is a linear function of t with slope $m = 2$. Since the price on June first (when $t = 5$) is 380 cents, the graph passes through $(5, 380)$. The equation is therefore
$$P - 380 = 2(t - 5) \text{ or}$$
$$P(t) = 2t + 370 \text{ cents},$$

$P(t) = 0.02t + 3.70$ dollars.

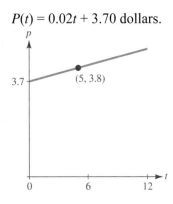

(b) When $t = 0$,
$P(0) = 0.02(0) + 3.70 = 3.70$
The price was \$3.70.

(c) On October 1st, $t = 9$ and
$P(9) = 0.02(9) + 3.70 = 3.88$
The price will be \$3.88.

8. Let t be the time, in hours, that has passed since the truck was 300 miles due east of the car. The distance the truck is from the car's original location is $300 - 30t$. The car's distance from its original location is $60t$ (due north). These two distances form the legs of a right triangle, where the distance between the car and the truck is its hypotenuse. So,

$$D(t) = \sqrt{(60t)^2 + (300 - 30t)^2}$$
$$= 30\sqrt{5t^2 - 20t + 100}$$

9. $S(x) = x^2 + A;\ D(x) = Bx + 59$

(a) Since no units are supplied until the selling price is greater than \$3 (assuming continuity), $3 = 0 + A$, or $A = 3$.
Equilibrium occurs when
$$S(7) = D(7)$$
$$(7)^2 + 3 = B(7) + 59$$
$$-1 = B$$
The equilibrium price is
$$S(7) = (7)^2 + 3 = \$52.$$

(b)

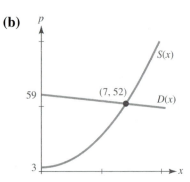

(c) When 5 units are produced the supply price is $S(5) = (5)^2 + 3 = \$28$ and the demand price is
$D(5) = -(5) + 59 = \$54.$ When 10 units are produced, the supply price is $S(10) = (10)^2 + 3 = \$103,$ and the demand price is
$D(10) = -(10) + 59 = \$49.$ The difference is $\$103 - \$49 = \$54.$ (Note that for 5 units, the demand price is higher than the supply price. However, for 10 units, the opposite is true.)

10. (a) The population is positive and increasing for $0 \le t < 5$. However, for $t \ge 5$, the population decreases. Therefore, the colony dies out when $-8t + 72 = 0$, or $t = 9$.

(b) $f(1) = 8$ and $f(7) = -56 + 72 = 16$. Since $f(5) = \lim\limits_{x \to 5} f(x) = 32$, f is continuous. Since $8 < 10 < 16$, by the intermediate value property there exists a value $1 < c < 7$ such that $f(c) = 10$.

11. Since M is a linear function of D, $M = aD + b$, for some constants a and b. Using $M = 7.7$ when $D = 3$, and $M = 12.7$ when $D = 5$, solve the system
$$a \cdot 3 + b = 7.7$$
$$a \cdot 5 + b = 12.7$$
So, $a = 2.5$ and $b = 0.2$. Thus, $M = 2.5D + 0.2$. When $D = 0$, $M = 0.2$, so 0.2% will mutate when no radiation is used.

Review Exercises

1. (a) The domain of the quadratic function
$$f(x) = x^2 - 2x + 6$$
is all real numbers x.

(b) Since denominators cannot be zero, the domain of the rational function
$$f(x) = \frac{x-3}{x^2 + x - 2} = \frac{x-3}{(x+2)(x-1)}$$
is all real numbers x except $x = -2$ and $x = 1$.

(c) Since negative numbers do not have square roots, the domain of the function
$$f(x) = \sqrt{x^2 - 9} = \sqrt{(x+3)(x-3)}$$
is all real numbers x such that $(x + 3)(x - 3) \geq 0$, that is for $x \leq -3$, or $x \geq 3$, or $|x| \geq 3$.

3. (a) If $g(u) = u^2 + 2u + 1$ and $h(x) = 1 - x$ then $g(h(x)) = g(1-x)$
$$= (1-x)^2 + 2(1-x) + 1$$
$$= x^2 - 4x + 4.$$

(b) If $g(u) = \dfrac{1}{2u+1}$ and $h(x) = x + 2$, then
$$g(h(x)) = g(x+2)$$
$$= \frac{1}{2(x+2)+1}$$
$$= \frac{1}{2x+5}.$$

5. (a) $f(3-x) = 4 - (3-x) - (3-x)^2$
$$= 4 - 3 + x - (9 - 6x + x^2)$$
$$= 1 + x - 9 + 6x - x^2$$
$$= -x^2 + 7x - 8$$

(b) $f(x^2 - 3) = (x^2 - 3) - 1 = x^2 - 4$

(c) $f(x+1) - f(x) = \dfrac{1}{(x+1)-1} - \dfrac{1}{x-1}$
$$= \frac{1}{x} - \frac{1}{x-1}$$
$$= \frac{1}{x} \cdot \frac{x-1}{x-1} - \frac{1}{x-1} \cdot \frac{x}{x}$$
$$= \frac{x-1}{x(x-1)} - \frac{x}{x(x-1)}$$
$$= \frac{x-1-x}{x(x-1)}$$
$$= -\frac{1}{x(x-1)}$$

7. (a) One of many possible solutions is
$$g(u) = u^5 \text{ and } h(x) = x^2 + 3x + 4.$$
Then, $g(h(x)) = g(x^2 + 3x + 4)$
$$= (x^2 + 3x + 4)^5$$
$$= f(x).$$

(b) One of many possible solutions is
$$g(u) = u^2 + \frac{5}{2(u+1)^3} \text{ and}$$
$$h(x) = 3x + 1.$$
Then,
$$g(h(x)) = g(3x+1)$$
$$= (3x+1)^2 + \frac{5}{2((3x+1)+1)^3}$$
$$= (3x+1)^2 + \frac{5}{2(3x+2)^3}$$
$$= f(x).$$

9. $f(x) = x^2 + 2x - 8 = (x+4)(x-2)$

The intercepts of the function are $(-4, 0)$, $(2, 0)$ and $(0, -8)$. Further, the vertex of the parabola is
$$x = -\frac{B}{2A}, \quad y = f\left(-\frac{B}{2A}\right)$$
$$x = -\frac{2}{2(1)} = -1$$
$$y = f(-1) = (-1)^2 + 2(-1) - 8$$
$$= 1 - 2 - 8$$
$$= -9.$$

So, the vertex is $(-1, -9)$.

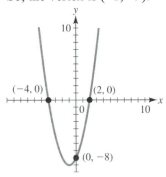

11. (a) If $y = 3x + 2$, $m = 3$ and $b = 2$.

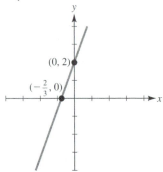

(b) If $5x - 4y = 20$ then $y = \dfrac{5}{4}x - 5$,

$m = \dfrac{5}{4}$, and $b = -5$.

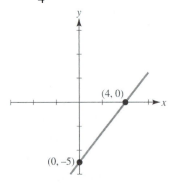

13. (a) $m = 5$ and y-intercept $b = -4$, so
$y = 5x - 4$.

(b) $m = -2$ and $P(1, 3)$, so
$y - 3 = -2(x - 1)$, or $y = -2x + 5$.

(c) $2x + y = 3 \rightarrow y = -2x + 3$, so $m = -2$ and
$P(5, 4)$.

$y - 4 = -2(x - 5)$, or $y = -2x + 14$
$\rightarrow 2x + y = 14$.

15. (a) The graphs of $y = -3x + 5$ and
$y = 2x - 10$ intersect when
$-3x + 5 = 2x - 10$, or $x = 3$.
When $x = 3$, $y = -3(3) + 5 = -4$. So
the point of intersection is $(3, -4)$.

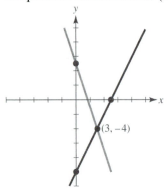

(b) The graphs of $y = x + 7$ and $y = -2 + x$
are lines having the same slope, so
they are parallel lines and there are no
points of intersection.

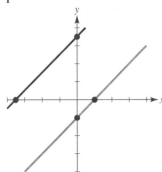

17. If the graph of $y = 3x^2 - 2x + c$ passes
through the point $(2, 4)$,
$$4 = 3(2)^2 - 2(2) + c \text{ or } c = -4.$$

19. $\displaystyle\lim_{x\to 1}\frac{x^2+x-2}{x^2-1}=\lim_{x\to 1}\frac{(x+2)(x-1)}{(x+1)(x-1)}$

$\displaystyle\qquad\qquad =\lim_{x\to 1}\frac{x+2}{x+1}$

$\displaystyle\qquad\qquad =\frac{\displaystyle\lim_{x\to 1}(x+2)}{\displaystyle\lim_{x\to 1}(x+1)}$

$\displaystyle\qquad\qquad =\frac{1+2}{1+1}$

$\displaystyle\qquad\qquad =\frac{3}{2}$

21. $\displaystyle\lim_{x\to 2}\frac{x^3-8}{2-x}=\lim_{x\to 2}\frac{(x-2)(x^2+2x+4)}{-(x-2)}$

$\displaystyle\qquad\qquad =\lim_{x\to 2}-(x^2+2x+4)$

$\displaystyle\qquad\qquad =-(2^2+2(2)+4)$

$\displaystyle\qquad\qquad =-12$

23. $\displaystyle\lim_{x\to 0}\left(2-\frac{1}{x^3}\right)=\lim_{x\to 0}2-\lim_{x\to 0}\frac{1}{x^3}$

Now, $\displaystyle\lim_{x\to 0}2=2$; but $\displaystyle\lim_{x\to 0^+}\frac{1}{x^3}=+\infty$ and

$\displaystyle\lim_{x\to 0^-}\frac{1}{x^3}=-\infty$. Since $\displaystyle\lim_{x\to 0}\frac{1}{x^3}$ does not

exist, $\displaystyle\lim_{x\to 0}\left(x^3-\frac{1}{x^3}\right)$ does not exist.

25. $\displaystyle\lim_{x\to -\infty}\frac{x}{x^2+5}=\lim_{x\to -\infty}\frac{\frac{1}{x}}{1+\frac{5}{x^2}}=0$

27. $\displaystyle\lim_{x\to -\infty}\frac{x^4+3x^2-2x+7}{x^3+x+1}$

$\displaystyle\quad =\lim_{x\to -\infty}\frac{x+\frac{3}{x}-\frac{2}{x^2}+\frac{7}{x^3}}{1+\frac{1}{x^2}+\frac{1}{x^3}}$

Since $\displaystyle\lim_{x\to -\infty}\left(x+\frac{3}{x}-\frac{2}{x^2}+\frac{7}{x^3}\right)=-\infty$ and

$\displaystyle\lim_{x\to -\infty}\left(1+\frac{1}{x^2}+\frac{1}{x^3}\right)=1$, then

$\displaystyle\lim_{x\to -\infty}\frac{x^4+3x^2-2x+7}{x^3+x+1}=-\infty.$

29. Since $\displaystyle\lim_{x\to -\infty}\left(1+\frac{1}{x}+\frac{1}{x^2}\right)=1$ and

$\displaystyle\lim_{x\to -\infty}(x^3+x+1)=-\infty,$

$\displaystyle\lim_{x\to -\infty}\frac{1+\frac{1}{x}+\frac{1}{x^2}}{x^3+x+1}=0.$

31. $\displaystyle\lim_{x\to 0^-}x\sqrt{1-\frac{1}{x}}=\left(\lim_{x\to 0^-}x\right)\left(\lim_{x\to 0^-}\sqrt{1-\frac{1}{x}}\right)$

Since $\displaystyle\lim_{x\to 0^-}x=0$, and $\displaystyle\lim_{x\to 0^-}\left(1-\frac{1}{x}\right)=\infty$

implies $\displaystyle\lim_{x\to 0^-}\sqrt{1-\frac{1}{x}}=\infty$, then

$\displaystyle\lim_{x\to 0^-}x\sqrt{1-\frac{1}{x}}=0.$

33. $f(x)=\dfrac{x^2-1}{x+3}$ is not continuous at $x=-3$

since $f(-3)=\dfrac{10}{0}$ and division by 0 is

undefined.

35. $h(x)=\begin{cases}x^3+2x-33 & \text{if }x\le 3\\[2mm]\dfrac{x^2-6x+9}{x-3} & \text{if }x>3\end{cases}$

The denominator in $\dfrac{x^2-6x+9}{x-3}$ will

never be zero, since $x=3$ is not included
in its domain. However, in checking the
break point (the only point in question),
$h(3)=(3)^3+2(3)-33=0.$
Further,

$\displaystyle\lim_{x\to 3^-}h(x)=\lim_{x\to 3^-}(x^3+2x-33)=0$ and

$$\lim_{x\to 3^+} h(x) = \lim_{x\to 3^+} \frac{x^2 - 6x + 9}{x - 3}$$
$$= \lim_{x\to 3^+} \frac{(x-3)(x-3)}{x-3}$$
$$= \lim_{x\to 3^+} (x-3)$$
$$= 3-3$$
$$= 0.$$

Since $h(3) = \lim_{x\to 3} h(x)$, h is continuous for all x.

37. $P(x) = 40 + \dfrac{30}{x+1}$

(a) $P(5) = 40 + \dfrac{30}{5+1} = 40 + 5 = \$45$

(b) Need $P(5) - P(4)$

$$P(4) = 40 + \frac{30}{4+1} = 40 + 6 = \$46$$
$$P(5) - P(4) = 45 - 46 = -1$$

Price drops one dollar during the 5th month.

(c) Find x so that $P(x) = 43$

$$40 + \frac{30}{x+1} = 43$$
$$\frac{30}{x+1} = \frac{3}{1}$$
$$3(x+1) = 30$$
$$3x + 3 = 30$$
$$3x = 27$$
$$x = 9$$

The price will be \$43 nine months from now.

(d) $\lim_{x\to +\infty} P(x) = \lim_{x\to +\infty} \left(40 + \dfrac{30}{x+1}\right)$
$$= \lim_{x\to +\infty} 40 + \lim_{x\to +\infty} \frac{30}{x+1}$$
$$= 40 + 0$$
$$= 40$$

In the long run, the price will approach \$40.

39. The number of weeks needed to reach x percent of the fundraising goal is given by $f(x) = \dfrac{10x}{150 - x}$.

(a) Since x denotes a percentage, the function has a practical interpretation for $0 \le x \le 100$.
The corresponding portion of the graph is sketched.

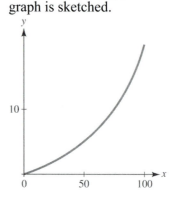

(b) The number of weeks needed to reach 50% of the goal is
$$f(50) = \frac{10(50)}{150 - 50} = 5 \text{ weeks.}$$

(c) The number of weeks needed to reach 100% of the goal is
$$f(100) = \frac{10(100)}{150 - 100} = 20 \text{ weeks.}$$

41. $S = 4\pi r^2$, or $r = \sqrt{\dfrac{S}{4\pi}} = \left(\dfrac{S}{4\pi}\right)^{1/2}$

$$V(S) = \frac{4}{3}\pi \left(\left(\frac{S}{4\pi}\right)^{1/2}\right)^3 = \frac{4}{3}\pi \left(\frac{S}{4\pi}\right)^{3/2}$$

$$V(S) = \frac{4}{3}\pi \frac{S^{3/2}}{4^{3/2}\pi^{3/2}}$$
$$= \frac{4}{3}\pi \frac{S^{3/2}}{8\pi^{3/2}}$$
$$= \frac{S^{3/2}}{6\pi^{1/2}}$$
$$= \frac{S^{3/2}}{6\sqrt{\pi}}$$

$$V(2S) = \frac{(2S)^{3/2}}{6\sqrt{\pi}} = \frac{2^{3/2}\, S^{3/2}}{6\sqrt{\pi}},\ \text{so volume}$$

increased by a factor of $2^{3/2}$, or $2\sqrt{2}$, when S is doubled.

43. Let x denote the number of machines used and $C(x)$ the corresponding cost function. Then,

$$C(x) = (\text{set up cost}) + (\text{operating cost})$$
$$= 80(\text{number of machines})$$
$$+ 5.76(\text{number of hours}).$$

Since 400,000 medals are to be produced and each of the x machines can produce 200 medals per hour,

$$\text{number of hours} = \frac{400,000}{200x} = \frac{2,000}{x}.$$

So, $C(x) = 80x + 5.76\left(\dfrac{2,000}{x}\right)$

$$= 80x + \frac{11,520}{x}.$$

The graph suggests that the cost will be smallest when x is approximately 12. Note: In chapter 3 you will learn how to use calculus to find the optimal number of machines exactly.

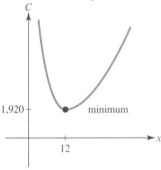

45. If p represents the selling price, the monthly profit is
$P(p)$
$= (\text{number of cameras sold})(\text{price} - \text{cost})$

Since $\dfrac{340 - p}{5}$ represents the number of

$5 decreases, $40 + 10\left(\dfrac{340 - p}{5}\right) = $

represents the number of cameras that will

sell. So, $P(p) = (720 - 2p)(p - 150)$
$$= 2(360 - p)(p - 150)$$

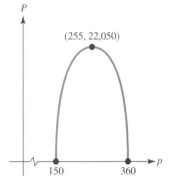

The graph suggests a maximum profit when $p = 255$, that is when the selling price is \$255.

47. Taxes under Proposition A are $100 + .08a$, where a is the assessed value of the home. Taxes under Proposition B are $1,900 + .02a$. Taxes are the same when

$$100 + .08a = 1,900 + .02a$$
$$.06a = 1,800$$
$$a = 30,000$$

or for an assessed value of \$30,000. Since both tax functions are linear, it is only necessary to test one additional assessed value to determine which proposition is best for all assessed values. For $a = 20,000$,
$100 + .08(20,000) = \$1,700$
$1,900 + .02(20,000) = \$2,300$
So, for $0 < a < 30,000$, Proposition A is preferable while for $a > 30,000$, Proposition B is preferable.

49. (a) Let x denote the number of units manufactured and sold. $C(x)$ and $R(x)$ are the corresponding cost and revenue functions, respectively.
$C(x) = 4,500 + 50x$
$R(x) = 80x$
For the manufacturer to break even, since profit = revenue − cost,
$0 = $ revenue − cost, or revenue = cost.
That is, $4,500 + 50x = 80x$ or
$x = 150$ units.

(b) Let $P(x)$ denote the profit from the manufacture and sale of x units. Then,

$$P(x) = R(x) - C(x)$$
$$= 80x - (4,500 + 50x)$$
$$= 30x - 4,500.$$

When 200 units are sold, the profit is $P(200) = 30(200) - 4,500 = \$1,500.$

(c) The profit will be \$900 when $900 = 30x - 4,500$ or $x = 180$, that is, when 180 units are manufactured and sold.

51. Let x denote the number of relevant facts recalled, n the total number of relevant facts in the person's memory, and $R(x)$ the rate of recall. Then $n - x$ is the number of relevant facts not recalled.

So, $R(x) = k(n - x)$, where k is a constant of proportionality.

53. The cost for the clear glass is (area)(cost per sq ft) $= (2xy)(3)$, and similarly, the cost for the stained glass is

$$\left(\frac{1}{2}\pi x^2\right)(10).$$

So, $C = 6xy + 5\pi x^2$. Now, the perimeter is $\frac{1}{2}(2\pi x) = 2x + 2y = 20$ so,

$$\pi x + 2x + 2y = 20, \text{ or } y = \frac{20 - \pi x - 2x}{2}.$$

Cost as a function of x is

$$C(x) = 6x\left(\frac{20 - \pi x - 2x}{2}\right) + 5\pi x^2$$
$$= 3x(20 - \pi x - 2x) + 5\pi x^2$$
$$= 60x - 3\pi x^2 - 6x^2 + 5\pi x^2$$
$$= 60x - 6x^2 + 2\pi x^2$$
$$= 60x + (2\pi - 6)x$$

55. (a) The fixed cost is \$1,500 and the cost per unit is \$2, so the cost is

$C(x) = 1,500 + 2x$, for $0 \le x \le 5,000$.

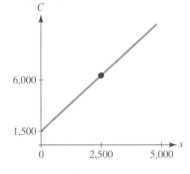

(b) The average cost is

$$AC(x) = \frac{C(x)}{x} = \frac{1,500}{x} + 2, \text{ so}$$

$$AC(3,000) = \frac{1,500}{3,000} + 2 = 2.5. \text{ The average}$$

cost of producing 3,000 units per day is \$2.50 per unit.

(c) As to the question of continuity, the answer is both yes and no. Yes, if (as we normally do) x is any real number. No, if x is discrete ($x = 0, 1, 2, ..., 5,000$). The discontinuities of $C(x)$ occur wherever x is discontinuous.

57. $w(x) = \begin{cases} Ax & \text{if } x \le 4,000 \\ \dfrac{B}{x^2} & \text{if } x > 4,000 \end{cases}$

For continuity $4,000A = \dfrac{B}{(4,000)^2}$ or

$B = A(4,000)^3.$

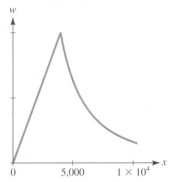

59. This limit does exist. The curve is bounded by the lines $y = mx$ and $y = -mx$. Since $-m|x| \le g(x) \le m|x|$, as x approaches 0, the bounding values on the right and the left of the inequality also approach 0. The function in the middle $g(x)$, is squeezed or sandwiched between 0 and 0. Its limit has to be 0.

Note: $\lim\limits_{x \to 0^-} |x| \sin\left(\dfrac{1}{x}\right) = \lim\limits_{x \to 0^+} |x| \sin\left(\dfrac{1}{x}\right) = 0$

since $-1 \le \sin x \le 1$.

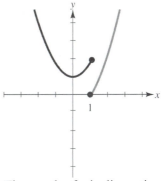

The graph of y is discontinuous at $x = 1$.

61. To graph $y = \dfrac{21}{9}x - \dfrac{84}{35}$ and

$y = \dfrac{654}{279}x - \dfrac{54}{10}$, press $\boxed{y =}$.

Input $(21x)/9 - 84/35$ for $y_1 =$ and press $\boxed{\text{enter}}$.

Input $(654x)/279 - 54/10$ for $y_2 =$.

Use the z-standard function under the zoom menu to use the window dimensions given. Press $\boxed{\text{graph}}$.

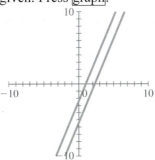

It appears from the graph that the two lines are parallel. However, the difference in the slopes is $\dfrac{21}{9} - \dfrac{654}{279} = -.01$ which shows that, in fact, the lines are not parallel since they have different slopes.

63. Press $\boxed{y =}$.

Input $(x \wedge 2 + 1)/(x \le 1)$ for $y_1 =$ and press $\boxed{\text{enter}}$. (You can obtain the $\le$ from $\boxed{\text{2ND}}$ $\boxed{\text{test}}$ and enter 6: $\le$).

Input $(x \wedge 2 - 1)/(x > 1)$ for $y_2 =$ and press $\boxed{\text{enter}}$. (You can obtain the $>$ from $\boxed{\text{2ND}}$ $\boxed{\text{test}}$ and enter 3: $>$). Press graph.

Chapter 2

Differentiation: Basic Concepts

2.1 The Derivative

1. If $f(x) = 4$, then $f(x + h) = 4$.
The difference quotient (DQ) is
$$\frac{f(x+h) - f(x)}{h} = \frac{4-4}{h} = 0.$$
$$f'(x) = \lim_{h \to 0} \frac{f(x+h) - f(x)}{h} = 0$$
The slope is $m = f'(0) = 0$.

3. If $f(x) = 5x - 3$, then
$f(x + h) = 5(x + h) - 3$.
The difference quotient (DQ) is
$$\frac{f(x+h) - f(x)}{h} = \frac{[5(x+h)-3] - [5x-3]}{h}$$
$$= \frac{5h}{h}$$
$$= 5$$
$$f'(x) = \lim_{h \to 0} \frac{f(x+h) - f(x)}{h} = 5$$
The slope is $m = f'(2) = 5$.

5. If $f(x) = 2x^2 - 3x + 5$, then
$$f(x + h) = 2(x + h)^2 - 3(x + h) + 5.$$
The difference quotient (DQ) is
$$\frac{f(x+h) - f(x)}{h}.$$
$$\frac{[2(x+h)^2 - 3(x+h)+5]}{h} - \frac{[2x^2 - 3x+5]}{h}$$
$$= \frac{4xh + 2(h)^2 - 3h}{h}$$
$$= 4x + 2h - 3$$
$$f'(x) = \lim_{h \to 0} \frac{f(x+h) - f(x)}{h} = 4x - 3$$
The slope is $m = f'(0) = -3$.

7. If $f(x) = x^3 - 1$, then
$$f(x + h) = (x + h)^3 - 1$$
$$= (x^2 + 2xh + h^2)(x + h) - 1$$
$$= x^3 + 3x^2h + 3xh^2 + h^3 - 1$$
$$\frac{f(x+h) - f(x)}{h}$$
$$= \frac{x^3 + 3x^2h + 3xh^2 + h^3 - 1 - (x^3 - 1)}{h}$$
$$= \frac{3x^2h + 3xh^2 + h^3}{h}$$
$$= \frac{h(3x^2 + 3xh + h^2)}{h}$$
$$= 3x^2 + 3xh + h^2$$
$$f'(x) = \lim_{h \to 0} \frac{f(x+h) - f(x)}{h}$$
$$= \lim_{h \to 0} 3x^2 + 3xh + h^2$$
$$= 3x^2$$
The slope is $m = f'(2) = 3(2)^2 = 12$.

9. If $g(t) = \dfrac{2}{t}$, then $g(t+h) = \dfrac{2}{t+h}$.
The difference quotient (DQ) is
$$\frac{g(t+h) - g(t)}{h} = \frac{\frac{2}{t+h} - \frac{2}{t}}{h}$$
$$= \frac{\frac{2}{t+h} - \frac{2}{t}}{h} \cdot \frac{t(t+h)}{t(t+h)}$$
$$= \frac{2t - 2(t+h)}{h(t)(t+h)}$$
$$= \frac{-2}{t(t+h)}$$
$$g'(t) = \lim_{h \to 0} \frac{g(t+h) - g(t)}{h} = -\frac{2}{t^2}$$
The slope is $m = g'\left(\dfrac{1}{2}\right) = -8$.

11. If $H(u)=\dfrac{1}{\sqrt{u}}$, then $H(u+h)=\dfrac{1}{\sqrt{u+h}}$.

The difference quotient is

$$\frac{f(x+h)-f(x)}{h}$$

$$=\frac{\frac{1}{\sqrt{u+h}}-\frac{1}{\sqrt{u}}}{h}\cdot\frac{\sqrt{u}\sqrt{u+h}}{\sqrt{u}\sqrt{u+h}}$$

$$=\frac{\sqrt{u}-\sqrt{u+h}}{h\sqrt{u}\sqrt{u+h}}\cdot\frac{\left(\sqrt{u}+\sqrt{u+h}\right)}{\left(\sqrt{u}+\sqrt{u+h}\right)}$$

$$=\frac{u-(u+h)}{h\sqrt{u}\sqrt{u+h}\left(\sqrt{u}+\sqrt{u+h}\right)}$$

$$=\frac{-h}{h\sqrt{u}\sqrt{u+h}\left(\sqrt{u}+\sqrt{u+h}\right)}$$

$$=\frac{-1}{\sqrt{u}\sqrt{u+h}\left(\sqrt{u}+\sqrt{u+h}\right)}$$

$$H'(u)=\lim_{h\to0}\frac{f(x+h)-f(x)}{h}$$

$$=\lim_{h\to0}\frac{-1}{\sqrt{u}\sqrt{u+h}\left(\sqrt{u}+\sqrt{u+h}\right)}$$

$$=\frac{-1}{\sqrt{u}\cdot\sqrt{u}\left(\sqrt{u}+\sqrt{u}\right)}$$

$$=\frac{-1}{u\left(2\sqrt{u}\right)}$$

$$=-\frac{1}{2u\sqrt{u}}$$

The slope is $m=H'(4)=-\dfrac{1}{2(4)\sqrt{4}}=-\dfrac{1}{16}$.

13. If $f(x)=2$, then $f(x+h)=2$.
The difference quotient (DQ) is

$$\frac{f(x+h)-f(x)}{h}=\frac{2-2}{h}=0.$$

$$f'(x)=\lim_{h\to0}\frac{f(x+h)-f(x)}{h}=\lim_{h\to0}0=0$$

The slope of the tangent is zero for all values of x. Since $f(13)=2$.
$y-2=0(x-13)$, or $y=2$.

15. If $f(x)=7-2x$, then
$f(x+h)=7-2(x+h)$.
The difference quotient (DQ) is

$$\frac{f(x+h)-f(x)}{h}=\frac{[7-2(x+h)]-[7-2x]}{h}$$

$$=-2$$

$$f'(x)=\lim_{h\to0}\frac{f(x+h)-f(x)}{h}=-2$$

The slope of the line is $m=f'(5)=-2$.
Since $f(5)=-3$, $(5,-3)$ is a point on the curve and the equation of the tangent line is $y-(-3)=-2(x-5)$ or $y=-2x+7$.

17. If $f(x)=x^2$, then $f(x+h)=(x+h)^2$.
The difference quotient (DQ) is

$$\frac{f(x+h)-f(x)}{h}=\frac{(x+h)^2-x^2}{h}$$

$$=\frac{2xh+h^2}{h}$$

$$=2x+h$$

$$f'(x)=\lim_{h\to0}\frac{f(x+h)-f(x)}{h}=2x$$

The slope of the line is $m=f'(1)=2$.
Since $f(1)=1$, $(1,1)$ is a point on the curve and the equation of the tangent line is $y-1=2(x-1)$ or $y=2x-1$.

19. If $f(x)=-\dfrac{2}{x}$, then $f(x+h)=\dfrac{-2}{x+h}$.
The difference quotient (DQ) is

$$\frac{f(x+h)-f(x)}{h}=\frac{\frac{-2}{x+h}-\frac{-2}{x}}{h}$$

$$=\frac{\frac{-2}{x+h}+\frac{2}{x}}{h}\cdot\frac{x(x+h)}{x(x+h)}$$

$$=\frac{-2x+2(x+h)}{h(x)(x+h)}$$

$$=\frac{2}{x(x+h)}$$

$$f'(x)=\lim_{h\to0}\frac{f(x+h)-f(x)}{h}=\frac{2}{x^2}$$

The slope of the line is $m=f'(-1)=2$.
Since $f(-1)=2$, $(-1,2)$ is a point on the curve and the equation of the tangent line

is $y - 2 = 2(x - (-1))$
$$y = 2x + 4$$

21. First we obtain the derivative of
$g(x) = \sqrt{x}$.
The difference quotient is
$$\frac{g(x+h) - g(x)}{h}$$
$$= \frac{\sqrt{x+h} - \sqrt{x}}{h}$$
$$= \frac{\sqrt{x+h} - \sqrt{x}}{h} \cdot \frac{\sqrt{x+h} + \sqrt{x}}{\sqrt{x+h} + \sqrt{x}}$$
$$= \frac{x+h-x}{h\left(\sqrt{x+h} + \sqrt{x}\right)}$$
$$= \frac{h}{h\left(\sqrt{x+h} + \sqrt{x}\right)}$$
$$= \frac{1}{\sqrt{x+h} + \sqrt{x}}$$

Then $g'(x) = \lim\limits_{h \to 0} \dfrac{1}{\sqrt{x+h} + \sqrt{x}} = \dfrac{1}{2\sqrt{x}}$.

Now since $\dfrac{d}{dx} k \cdot f(x) = k \cdot \dfrac{d}{dx} f(x)$,

$$f'(x) = 2\left(\frac{1}{2\sqrt{x}}\right) = \frac{1}{\sqrt{x}}.$$

The slope is $m = f'(4) = \dfrac{1}{2}$, $f(4) = 4$, the
equation of the tangent line is
$$y - 4 = \frac{1}{2}(x - 4), \text{ or } y = \frac{1}{2}x + 2.$$

23. If $f(x) = \dfrac{1}{x^3}$, then $f(x+h) = \dfrac{1}{(x+h)^3}$.
The difference quotient (DQ) is

$$\frac{f(x+h) - f(x)}{h}$$
$$= \frac{\frac{1}{(x+h)^3} - \frac{1}{x^3}}{h} \cdot \frac{x^3(x+h)^3}{x^3(x+h)^3}$$
$$= \frac{x^3 - (x+h)^3}{hx^3(x+h)^3}$$
$$= \frac{x^3 - (x^3 + 3x^2h + 3xh^2 + h^3)}{hx^3(x+h)^3}$$
$$= \frac{-3x^2h - 3xh^2 - h^3}{hx^3(x+h)^3}$$
$$= \frac{h(-3x^2 - 3xh - h^2)}{hx^3(x+h)^3}$$
$$= \frac{-3x^2 - 3xh - h^2}{x^3(x+h)^3}$$

$$f'(x) = \lim_{h \to 0} \frac{f(x+h) - f(x)}{h}$$
$$= \lim_{h \to 0} \frac{-3x^2 - 3xh - h^2}{x^3(x+h)^3}$$
$$= \frac{-3x^2}{x^3(x)^3}$$
$$= -\frac{3}{x^4}$$

The slope is $m = f'(1) = -\dfrac{3}{(1)^4} = -3$.
Further, $f(1) = 1$ so the equation of the line
is $y - 1 = -3(x - 1)$, or $y = -3x + 4$.

25. If $y = f(x) = 3$, then $f(x + h) = 3$.
The difference quotient (DQ) is
$$\frac{f(x+h) - f(x)}{h} = \frac{3-3}{h} = 0.$$
$$\frac{dy}{dx} = \lim_{h \to 0} \frac{f(x+h) - f(x)}{h} = 0$$
$$\frac{dy}{dx} = 0 \text{ when } x = 2.$$

27. If $y = f(x) = 3x + 5$, then
$f(x + h) = 3(x + h) + 5 = 3x + 3h + 5$.
The difference quotient (DQ) is

$$\frac{f(x+h)-f(x)}{h}=\frac{3x+3h+5-(3x+5)}{h}$$
$$=\frac{3h}{h}$$
$$=3$$
$$\frac{dy}{dx}=\lim_{h\to 0}\frac{f(x+h)-f(x)}{h}=\lim_{h\to 0}3=3$$
$$\frac{dy}{dx}=3 \text{ when } x=-1.$$

29. If $y=f(x)=x(1-x)$, or $f(x)=x-x^2$,

then $f(x+h)=(x+h)-(x+h)^2$.
The difference quotient (DQ) is
$$\frac{f(x+h)-f(x)}{h}$$
$$=\frac{[(x+h)-(x+h)^2]-[x-x^2]}{h}$$
$$=\frac{h-2xh-h^2}{h}$$
$$=1-2x-h$$
$$\frac{dy}{dx}=\lim_{h\to 0}\frac{f(x+h)-f(x)}{h}=1-2x$$
$$\frac{dy}{dx}=3 \text{ when } x=-1.$$

31. If $y=f(x)=x-\dfrac{1}{x}$, then

$$f(x+h)=x+h-\frac{1}{x+h}.$$
The difference quotient (DQ) is
$$\frac{f(x+h)-f(x)}{h}=\frac{x+h-\frac{1}{x+h}-\left(x-\frac{1}{x}\right)}{h}$$
$$=\frac{h-\frac{1}{x+h}+\frac{1}{x}}{h}\cdot\frac{x(x+h)}{x(x+h)}$$
$$=\frac{hx(x+h)-x+x+h}{h}$$
$$=\frac{hx^2+h^2x+h}{h}$$
$$=\frac{h(x^2+hx+1)}{h}$$
$$=x^2+hx+1$$

$$\frac{dy}{dx}=\lim_{h\to 0}\frac{f(x+h)-f(x)}{h}$$
$$=\lim_{h\to 0}x^2+hx+1$$
$$=x^2+1$$
When $x=1$, $\dfrac{dy}{dx}=(1)^2+1=2.$

33. **(a)** If $f(x)=x^2$, then $f(-2)=(-2)^2=4$

and $f(-1.9)=(-1.9)^2=3.61$. The slope of the secant line joining the points $(-2, 4)$ and $(-1.9, 3.61)$ on the graph of f is
$$m_{\text{sec}}=\frac{y_2-y_1}{x_2-x_1}=\frac{3.61-4}{-1.9-(-2)}=-3.9.$$

(b) If $f(x)=x^2$, then

$$f(x+h)=(x+h)^2=x^2+2xh+h^2.$$
The difference quotient (DQ) is
$$\frac{f(x+h)-f(x)}{h}=\frac{x^2+2xh+h^2-x^2}{h}$$
$$=\frac{2xh+h^2}{h}$$
$$=\frac{h(2x+h)}{h}$$
$$=2x+h$$
$$f'(x)=\lim_{h\to 0}\frac{f(x+h)-f(x)}{h}$$
$$=\lim_{h\to 0}2x+h$$
$$=2x$$
The slope of the tangent line at the point $(-2, 4)$ on the graph of f is
$$m_{\text{tan}}=f'(-2)=2(-2)=-4.$$

35. **(a)** If $f(x)=x^3$, then $f(1)=1$,

$$f(1.1)=(1.1)^3=1.331.$$
The slope of the secant line joining the points $(1, 1)$ and $(1.1, 1.331)$ on the graph of f is
$$m_{\text{sec}}=\frac{y_2-y_1}{x_2-x_1}=\frac{1.331-1}{1.1-1}=3.31.$$

(b) If $f(x) = x^3$, then $f(x+h) = (x+h)^3$.
The difference quotient (DQ) is

$$\frac{f(x+h) - f(x)}{h} = \frac{(x+h)^3 - x^3}{h}$$

$$= \frac{3x^2 h + 3xh^2 + h^3}{h}$$

$$= 3x^2 + 3xh + h^2$$

$$f'(x) = \lim_{h \to 0} \frac{f(x+h) - f(x)}{h} = 3x^2$$

The slope is $m_{\tan} = f'(1) = 3$.
Notice that this slope was approximated by the slope of the secant in part (a).

37. (a) If $f(x) = 3x^2 - x$, the average rate of change of f is $\dfrac{f(x_2) - f(x_1)}{x_2 - x_1}$.

Since $f(0) = 0$ and

$$f\left(\frac{1}{16}\right) = 3\left(\frac{1}{16}\right)^2 - \frac{1}{16} = -\frac{13}{256},$$

$$\frac{f(x_2) - f(x_1)}{x_2 - x_1} = \frac{-\frac{13}{256} - 0}{\frac{1}{16} - 0}$$

$$= -\frac{13}{16}$$

$$= -0.8125.$$

(b) If $f(x) = 3x^2 - x$, then

$$f(x+h) = 3(x+h)^2 - (x+h).$$

The difference quotient (DQ) is

$$\frac{f(x+h) - f(x)}{h}$$

$$= \frac{3(x+h)^2 - (x+h) - (3x^2 - x)}{h}$$

$$= \frac{3x^2 + 6xh + 3h^2 - x - h - 3x^2 + x}{h}$$

$$= \frac{6xh + 3h^2 - h}{h}$$

$$= 6x + 3h - 1.$$

$$f'(x) = \lim_{h \to 0} (6x + 3h - 1) = 6x - 1$$

The instantaneous rate of change at

$x = 0$ is $f'(0) = -1$. Notice that this rate is estimated by the average rate in part **(a)**.

39. (a) If $s(t) = \dfrac{t-1}{t+1}$, the average rate of change of s is $\dfrac{s(t_2) - s(t_1)}{t_2 - t_1}$.

Since $s\left(-\dfrac{1}{2}\right) = \dfrac{\frac{1}{2} - 1}{-\frac{1}{2} + 1} = -3$ and

$$s(0) = \frac{0-1}{0+1} = -1, \quad \frac{-3+1}{-\frac{1}{2} - 0} = 4.$$

(b) If $s(t) = \dfrac{t-1}{t+1}$, then

$$s(t+h) = \frac{(t+h) - 1}{(t+h) + 1}.$$

The difference quotient (DQ) is

$$\frac{s(t+h) - s(t)}{h} = \frac{\frac{t+h-1}{t+h+1} - \frac{t-1}{t+1}}{h}.$$

Multiplying numerator and denominator by $(t + h + 1)(t + 1)$.

$$= \frac{(t+h-1)(t+1) - (t-1)(t+h+1)}{h(t+h+1)(t+1)}$$

$$= \frac{t^2 + th - t + t + h - 1 - t^2 - th - t + t + h + 1}{h(t+h+1)(t+1)}$$

$$= \frac{2h}{h(t+h+1)(t+1)}$$

$$= \frac{2}{(t+h+1)(t+1)}$$

$$s'(t) = \lim_{h \to 0} \frac{2}{(t+h+1)(t+1)} = \frac{2}{(t+1)^2}$$

The instantaneous rate of change when $t = -\dfrac{1}{2}$ is

$$s'\left(-\frac{1}{2}\right) = \frac{2}{\left(-\frac{1}{2} + 1\right)^2} = 8.$$

Notice that the estimate given by the average rate in part **(a)** differs significantly.

41. (a) The average rate of temperature change between t_0 and $t_0 + h$ hours after midnight. The instantaneous rate of temperature change t_0 hours after midnight.

(b) The average rate of change in blood alcohol level between t_0 and $t_0 + h$ hours after consumption. The instantaneous rate of change in blood alcohol level t_0 hours after consumption.

(c) The average rate of change of the 30-year fixed mortgage rate between t_0 and $t_0 + h$ years after 2005. The instantaneous rate of change of 30-year fixed mortgage rate t_0 years after 2005.

43. $P(x) = 4,000(15 - x)(x - 2)$

(a) The difference quotient (DQ) is

$$\frac{P(x+h) - P(x)}{h}$$

$$= \frac{[4,000(15 - (x+h))((x+h) - 2)]}{h}$$

$$\quad - \frac{[4,000(15 - x)(x - 2)]}{h}$$

$$= \frac{4,000[(15 - x - h)(x + h - 2) - (15 - x)(x - 2)]}{h}$$

$$= \frac{4,000(17h - 2xh - h^2)}{h}$$

$$= 4,000(17 - 2x - h)$$

$$P'(x) = \lim_{h \to 0} \frac{P(x+h) - P(x)}{h}$$
$$= 4,000(17 - 2x)$$

(b) $P'(x) = 0$ when $4,000(17 - 2x) = 0$.

$$x = \frac{17}{2} = 8.5, \text{ or } 850 \text{ units.}$$

When $P'(x) = 0$, the line tangent to the graph of P is horizontal. Since the graph of P is a parabola which opens

down, this horizontal tangent indicates a maximum profit.

45. $C(x) = 0.04x^2 + 2.1x + 60$

(a) As x increases from 10 to 11, the average rate of change is

$$\frac{C(11) - C(10)}{11 - 10}$$

$$C(11) = 0.04(11)^2 + 2.1(11) + 60 = 87.94$$
$$C(10) = 0.04(10)^2 + 2.1(10) + 60 = 85$$

$$\frac{87.94 - 85}{11 - 10} = 2.94$$
or $2,940 per unit.

(b)
$$C(x+h)$$
$$= 0.04(x+h)^2 + 2.1(x+h) + 60$$
So, the difference quotient (DQ) is
$$\frac{C(x+h) - C(x)}{h}$$

$$= \frac{\left[0.04(x+h)^2 + 2.1(x+h) + 60 \right.}{h}$$
$$\frac{\left. - \left(0.04x^2 + 2.1x + 60\right) \right]}{h}$$

$$= \frac{\left[0.04x^2 + 0.08xh + 0.04h^2 + 2.1x \right.}{h}$$
$$\frac{\left. +2.1h + 60 - 0.04x^2 - 2.1x - 60 \right]}{h}$$

$$= \frac{0.08xh + 0.04h^2 + 2.1h}{h}$$
$$= 0.08x + 0.04h + 2.1$$
$$C'(x) = \lim_{h \to 0} \left(0.08x + 0.04h + 2.1 \right)$$
$$= 0.08x + 2.1$$
$$C'(10) = 0.08(10) + 2.1 = 2.90$$

or \$2,900 per unit

The average rate of change is close to this value and is an estimate of this instantaneous rate of change.

Since $C'(10)$ is positive, the cost will increase.

47. Writing Exercise—Answers will vary.

49. When $t = 30$, $\dfrac{dV}{dt} \approx \dfrac{65-50}{50-30} = \dfrac{3}{4}$.

In the "long run," the rate at which V is changing with respect to time is getting smaller and smaller, decreasing to zero.

51. When $h = 1,000$ meters,

$$\dfrac{dT}{dh} \approx \dfrac{-6-0}{2,000-1,000}$$

$$= \dfrac{-6}{1,000}$$

$$= -0.006°C/\text{meter}$$

When $h = 2,000$ meters, $\dfrac{dT}{dh} = 0°C/\text{meter}$.

Since the line tangent to the graph at $h = 2,000$ is horizontal, its slope is zero.

53. $H(t) = 4.4t - 4.9t^2$

(a) $H(t+h)$

$= 4.4(t+h) - 4.9(t+h)^2$

$= 4.4t + 4.4h - 4.9(t^2 + 2th + h^2)$

$= 4.4t + 4.4h - 4.9t^2 - 9.8th - 4.9h^2$

The difference quotient (DQ) is

$\dfrac{H(t+h) - H(t)}{h}$

$= \dfrac{4.4t + 4.4h - 4.9t^2 - 9.8th}{h}$

$\quad \dfrac{-4.9h^2 - (4.4t - 4.9t^2)}{h}$

$= \dfrac{4.4h - 9.8th - 4.9h^2}{h}$

$= \dfrac{h(4.4 - 9.8t - 4.9h)}{h}$

$= 4.4 - 9.8t - 4.9h$

$H'(t)$

$= \lim_{h \to 0} \dfrac{H(t+h) - H(t)}{h}$

$= \lim_{h \to 0} 4.4 - 9.8t - 4.9h$

$= 4.4 - 9.8t$

After 1 second, H is changing at a rate of $H'(1) = 4.4 - 9.8(1) = -5.4$ m/sec, where the negative represents that H is decreasing.

(b) $H'(t) = 0$ when $4.4 - 9.8t = 0$, or

$t \approx 0.449$ seconds.

This represents the time when the height is not changing (neither increasing nor decreasing). That is, this represents the highest point in the jump.

(c) When the flea lands, the height $H(t)$ will be zero (as it was when $t = 0$).

$4.4t - 4.9t^2 = 0$

$(4.4 - 4.9t)t = 0$

$4.4 - 4.9t = 0$

$t = \dfrac{44}{49} \approx 0.898$ seconds

At this time, the rate of change is

$H'\left(\dfrac{44}{49}\right) = 4.4 - 9.8\left(\dfrac{44}{49}\right)$

$= -4.4$ m/sec.

Again, the negative represents that H is decreasing.

55. $D(p) = -0.0009p^2 + 0.13p + 17.81$

(a) The average rate of change is

$\dfrac{D(p_2) - D(p_1)}{p_2 - p_1}$.

Since

$D(60)$

$= -0.0009(60)^2 + 0.13(60) + 17.81$

$= 22.37$

and

$D(61)$

$= -0.0009(61)^2 + 0.13(61) + 17.81$

$= 22.3911,$

$= \dfrac{22.3911 - 22.37}{61 - 60}$

$= 0.0211$ mm per mm of mercury

(b) $D(p+h) = -0.0009(p+h)^2 + 0.13$

$(p+h) + 17.81$

So, the difference quotient (DQ) is

$\dfrac{D(p+h) - D(p)}{h}$

$= \dfrac{\begin{array}{l}[-0.0009(p+h)^2 + 0.13(p+h) \\ \quad + 17.81 \\ \quad - (-0.0009p^2 + 0.13p + 17.81)]\end{array}}{h}$

$= \dfrac{\begin{array}{l}[-0.0009p^2 - 0.0018ph - 0.0009h^2 \\ \quad + 0.13p + 0.13h + 17.81 \\ \quad + 0.0009p^2 - 0.13p - 17.81]\end{array}}{h}$

$= \dfrac{-0.0018ph - 0.0009h^2 + 0.13h}{h}$

$= -0.0018p - 0.0009h + 0.13$

$D'(x)$

$= \lim_{h \to 0}(-0.0018p - 0.0009h + 0.13)$

$= -0.0018p + 0.13$

The instantaneous rate of change when $p = 60$ is

$D'(60) = -0.0018(60) + 0.13$

$= 0.022$ mm

per mm of mercury. Since $D'(60)$ is positive, the pressure is increasing when $p = 60$.

(c) $-0.0018p + 0.13 = 0$

$p \approx 72.22$ mm of

mercury

At this pressure, the diameter is neither increasing nor decreasing.

57. $s(t) = 4\sqrt{t+1} - 4$

$= 4(t+1)^{1/2} - 4$

(a) $s(t+h) = 4\left[(t+h)+1\right]^{1/2} - 4$

So, the difference quotient (DQ) is

$\dfrac{4(t+h+1)^{1/2} - 4 - \left[4(t+1)^{1/2} - 4\right]}{h}$

$= \dfrac{4(t+h+1)^{1/2} - 4 - 4(t+1)^{1/2} + 4}{h}$

$= \dfrac{4(t+h+1)^{1/2} - 4(t+1)^{1/2}}{h}$

Multiplying the numerator and

denominator by $4(t+h+1)^{1/2} + 4(t+1)^{1/2}$

gives

$\dfrac{16(t+h+1) - 16(t+1)}{h\left[4(t+h+1)^{1/2} + 4(t+1)^{1/2}\right]}$

$= \dfrac{16t + 16h + 16 - 16t - 16}{4h\left[(t+h+1)^{1/2} + (t+1)^{1/2}\right]}$

$= \dfrac{16h}{4h\left[(t+h+1)^{1/2} + (t+1)^{1/2}\right]}$

$= \dfrac{4}{(t+h+1)^{1/2} + (t+1)^{1/2}}$

$s'(t) = \lim_{h \to 0} \dfrac{4}{(t+h+1)^{1/2} + (t+1)^{1/2}}$

$= \dfrac{4}{(t+1)^{1/2} + (t+1)^{1/2}}$

$= \dfrac{4}{2(t+1)^{1/2}}$

$v_{\text{ins}}(t) = \dfrac{2}{(t+1)^{1/2}} = \dfrac{2}{\sqrt{t+1}}$

(b) $v_{ins}(0) = \dfrac{2}{(0+1)^{1/2}} = \dfrac{2}{\sqrt{1}} = 2 \text{ m/sec}\sqrt{2}$

(c) $s(3) = 4\sqrt{3+1} - 4 = 8 - 4 = 4 \text{ m}$

$v_{ins}(3) = \dfrac{2}{\sqrt{3+1}} = \dfrac{2}{2} = 1 \text{ m/sec}$

59. **(a)** For $y = f(x) = x^2$,

$f(x+h) = (x+h)^2$.

The difference quotient (DQ) is

$\dfrac{f(x+h) - f(x)}{h} = \dfrac{(x+h)^2 - x^2}{h}$

$= \dfrac{2xh + h^2}{h}$

$= 2x + h$

$\dfrac{dy}{dx} = f'(x)$

$= \lim\limits_{h \to 0} \dfrac{f(x+h) - f(x)}{h}$

$= 2x$

For $y = f(x) = x^2 - 3$,

$f(x+h) = (x+h)^2 - 3$.

The difference quotient (DQ) is

$\dfrac{[(x+h)^2 - 3] - (x^2 - 3)}{h} = \dfrac{2xh + h^2}{h}$

$= 2x + h$

$\dfrac{dy}{dx} = f'(x)$

$= \lim\limits_{h \to 0} \dfrac{f(x+h) - f(x)}{h}$

$= 2x$

The graph of $y = x^2 - 3$ is the graph of $y = x^2$ shifted down 3 units. So the graphs are parallel and their tangent lines have the same slopes for any value of x. This accounts geometrically for the fact that their derivatives are identical.

(b) Since $y = x^2 + 5$ is the parabola $y = x^2$ shifted up 5 units and the constant appears to have no effect on the derivative, the derivative of the function $y = x^2 + 5$ is also $2x$.

61. **(a)** For $y = f(x) = x^2$,

$f(x+h) = (x+h)^2$.

The difference quotient (DQ) is

$\dfrac{f(x+h) - f(x)}{h} = \dfrac{(x+h)^2 - x^2}{h}$

$= \dfrac{2xh + h^2}{h}$

$= 2x + h$

$\dfrac{dy}{dx} = f'(x)$

$= \lim\limits_{h \to 0} \dfrac{f(x+h) - f(x)}{h}$

$= 2x$

For $y = f(x) = x^3$, $f(x+h) = (x+h)^3$.

The difference quotient (DQ) is

$\dfrac{(x+h)^3 - x^3}{h} = \dfrac{3x^2 h + 3xh^2 + h^3}{h}$

$= 3x^2 + 3xh + h^2$

$\dfrac{dy}{dx} = f'(x)$

$= \lim\limits_{h \to 0} \dfrac{f(x+h) - f(x)}{h}$

$= 3x^2$

(b) The pattern seems to be that the derivative of x raised to a power (x^n) is that power times x raised to the power decreased by one (nx^{n-1}). So, the derivative of the function $y = x^4$ is $4x^3$ and the derivative of the function $y = x^{27}$ is $27x^{26}$.

63. When $x < 0$, the difference quotient (DQ)
is $\dfrac{f(x+h)-f(x)}{h} = \dfrac{-(x+h)-(-x)}{h}$

$$= \dfrac{-h}{h}$$

$$= -1$$

So, $f'(x) = \lim_{h \to 0} -1 = -1$.

When $x > 0$, the difference quotient (DQ)
is $\dfrac{f(x+h)-f(x)}{h} = \dfrac{(x+h)-x}{h} = 1$.

So, $f'(x) = \lim_{h \to 0} 1 = 1$.

Since there is a sharp corner at $x = 0$ (graph changes from $y = -x$ to $y = x$), the graph makes an abrupt change in direction at $x = 0$. So, f is not differentiable at $x = 0$.

65. To show that $f(x) = \dfrac{\left|x^2 - 1\right|}{x-1}$ is not

differentiable at $x = 1$,

press $\boxed{y =}$ and input $(\text{abs}(x^2 - 1))/(x-1)$

for $y_1 =$

The abs is under the NUM menu in the math application.

Use window dimensions $[-4, 4]1$ by $[-4, 4]1$

Press $\boxed{\text{Graph}}$

We see that f is not defined at $x = 1$. There can be no point of tangency.

$$\lim_{x \to 1^+} \dfrac{\left|x^2 - 1\right|}{x-1} = \lim_{x \to 1^+} \dfrac{\left|(x-1)(x+1)\right|}{x-1} = 2$$

$$\lim_{x \to 1^-} \dfrac{\left|x^2 - 1\right|}{x-1} = \lim_{x \to 1^-} \dfrac{\left|(x-1)(x+1)\right|}{x-1} = -2$$

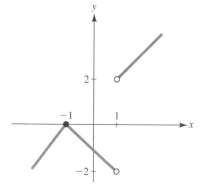

67. To find the slope of line tangent to the graph of $f(x) = \sqrt{x^2 + 2x} - \sqrt{3x}$ at $x = 3.85$, fill in the table below. The $x + h$ row can be filled in manually. For $f(x)$, press $\boxed{y =}$ and input

$$\sqrt{\ }\left(x \wedge 2 + 2x - \sqrt{\ }(3x)\right) \text{ for } y_1 =$$

Use window dimensions $[-1, 10]1$ by $[-1, 10]1$.

Use the value function under the calc menu and enter $x = 3.85$ to find $f(x) = 4.37310$.

For $f(x + h)$, use the value function under the calc menu and enter $x = 3.83$ to find $f(x + h) = 4.35192$. Repeat this process for $x = 3.84, 3.849, 3.85, 3.851, 3.86,$ and 3.87.

The $\dfrac{f(x+h)-f(x)}{h}$ can be filled in

manually given that the rest of the table is now complete. So, slope $= f'(3.85) \approx 1.059$.

h		-0.02	-0.01	-0.001
$x + h$		3.83	3.84	3.849
$f(x)$		4.37310	4.37310	4.37310
$f(x + h)$		4.35192	4.36251	4.37204
$\frac{f(x+h)-f(x)}{h}$		1.059	1.059	1.059

0		0.001	0.01	0.02
3.85		3.851	3.86	3.87
4.37310		4.37310	4.37310	4.37310
4.37310		4.37415	4.38368	4.39426
undefined		1.058	1.058	1.058

2.2   Techniques of Differentiation

1. Since the derivative of any constant is zero,
$$y = -2$$
$$\frac{dy}{dx} = 0$$
(Note: $y = -2$ is a horizontal line and all horizontal lines have a slope of zero, so $\frac{dy}{dx}$ must be zero.)

3. $y = 5x - 3$
$$\frac{dy}{dx} = \frac{d}{dx}(5x) - \frac{d}{dx}(3)$$
$$\frac{dy}{dx} = 5 - 0 = 5$$

5. $y = x^{-4}$
$$\frac{dy}{dx} = -4x^{-4-1} = -4x^{-5} = -\frac{4}{x^5}$$

7. $y = x^{3.7}$
$$\frac{dy}{dx} = 3.7x^{3.7-1} = 3.7x^{2.7}$$

9. $y = \pi r^2$
$$\frac{dy}{dx} = \pi(2r^{2-1}) = 2\pi r$$

11. $y = \sqrt{2x} = \sqrt{2} \cdot x^{1/2}$
$$\frac{dy}{dx} = \sqrt{2}\left(\frac{1}{2}x^{1/2-1}\right)$$
$$= \sqrt{2}\left(\frac{1}{2}x^{-1/2}\right)$$
$$= \sqrt{2} \cdot \frac{1}{2x^{1/2}} \text{ or } \frac{\sqrt{2}}{2\sqrt{x}}$$

13. $y = \dfrac{9}{\sqrt{t}} = 9t^{-1/2}$
$$\frac{dy}{dx} = 9\left(-\frac{1}{2}t^{-1/2-1}\right)$$
$$= 9\left(-\frac{1}{2}t^{-3/2}\right)$$
$$= -\frac{9}{2t^{3/2}} \text{ or } -\frac{9}{2\sqrt{t^3}}$$

15. $y = x^2 + 2x + 3$
$$\frac{dy}{dx} = \frac{d}{dx}(x^2) + \frac{d}{dx}(2x) + \frac{d}{dx}(3)$$
$$\frac{dy}{dx} = 2x + 2$$

17. $y = x^9 - 5x^8 + x + 12$
$$\frac{dy}{dx} = \frac{d}{dx}(x^9) - \frac{d}{dx}(5x^8) + \frac{d}{dx}(x) + \frac{d}{dx}(12)$$
$$\frac{dy}{dx} = 9x^8 - 40x^7 + 1$$

19. $f(x) = -0.02x^3 + 0.3x$
$$f'(x) = \frac{d}{dx}(-0.02x^3) + \frac{d}{dx}(0.3x)$$
$$f'(x) = -0.02(3x^2) + 0.3 = -0.06x^2 + 0.3$$

21. $y = \dfrac{1}{t} + \dfrac{1}{t^2} - \dfrac{1}{\sqrt{t}} = t^{-1} + t^{-2} - t^{-1/2}$
$$\frac{dy}{dt} = \frac{d}{dt}(t^{-1}) + \frac{d}{dt}(t^{-2}) - \frac{d}{dt}(t^{-1/2})$$
$$= -1t^{-1-1} + -2t^{-2-1} - \left(-\frac{1}{2}t^{-1/2-1}\right)$$
$$= -1t^{-2} - 2t^{-3} + \frac{1}{2}t^{-3/2}$$
$$= -\frac{1}{t^2} - \frac{2}{t^3} + \frac{1}{2t^{3/2}},$$
$$\text{or } -\frac{1}{t^2} - \frac{2}{t^3} + \frac{1}{2\sqrt{t^3}}$$

23. $f(x) = \sqrt{x^3} + \dfrac{1}{\sqrt{x^3}} = x^{3/2} + x^{-3/2}$,

$$f'(x) = \frac{d}{dx}(x^{3/2}) + \frac{d}{dx}(x^{-3/2})$$

$$= \frac{3}{2}x^{3/2-1} + \frac{-3}{2}x^{-3/2-1}$$

$$= \frac{3}{2}x^{1/2} - \frac{3}{2}x^{-5/2}$$

$$= \frac{3}{2}x^{1/2} - \frac{3}{2x^{5/2}}, \text{ or } \frac{3}{2}\sqrt{x} - \frac{3}{2\sqrt{x^5}}$$

25. $y = -\dfrac{x^2}{16} + \dfrac{2}{x} - x^{3/2} + \dfrac{1}{3x^2} + \dfrac{x}{3}$

$$= -\frac{1}{16}x^2 + 2x^{-1} - x^{3/2} + \frac{1}{3}x^{-2} + \frac{1}{3}x,$$

$$\frac{dy}{dx} = \frac{d}{dx}\left(-\frac{1}{16}x^2\right) + \frac{d}{dx}(2x^{-1}) - \frac{d}{dx}(x^{3/2})$$

$$+ \frac{d}{dx}\left(\frac{1}{3}x^{-2}\right) + \frac{d}{dx}\left(\frac{1}{3}x\right)$$

$$= -\frac{1}{16}(2x) + 2(-1x^{-1-1}) - \frac{3}{2}x^{3/2-1}$$

$$+ \frac{1}{3}(-2x^{-2-1}) + \frac{1}{3}$$

$$= -\frac{1}{8}x - 2x^{-2} - \frac{3}{2}x^{1/2} - \frac{2}{3}x^{-3} + \frac{1}{3}$$

$$= -\frac{1}{8}x - \frac{2}{x^2} - \frac{3}{2}x^{1/2} - \frac{2}{3x^3} + \frac{1}{3},$$

$$\text{or } -\frac{1}{8}x - \frac{2}{x^2} - \frac{3}{2}\sqrt{x} - \frac{2}{3x^3} + \frac{1}{3}$$

27. $y = \dfrac{x^5 - 4x^2}{x^3}$

$$= \frac{x^5}{x^3} - \frac{4x^2}{x^3}$$

$$= x^2 - \frac{4}{x}$$

$$= x^2 - 4x^{-1}$$

$$\frac{dy}{dx} = \frac{d}{dx}(x^2) - \frac{d}{dx}(4x^{-1})$$

$$= 2x - 4(-1x^{-1-1})$$

$$= 2x + 4x^{-2}$$

$$= 2x + \frac{4}{x^2}$$

29. $y = -x^3 - 5x^2 + 3x - 1$

$$\frac{dy}{dx} = -3x^2 - 10x + 3$$

At $x = -1$, $\dfrac{dy}{dx} = 10$. The equation of the tangent line at $(-1, -8)$ is
$y + 8 = 10(x + 1)$, or $y = 10x + 2$.

31. $y = 1 - \dfrac{1}{x} + \dfrac{2}{\sqrt{x}} = 1 - x^{-1} + 2x^{-1/2}$

$$\frac{dy}{dx} = x^{-2} - x^{-3/2} = \frac{1}{x^2} - \frac{1}{x^{3/2}}$$

At $\left(4, \dfrac{7}{4}\right)$, $\dfrac{dy}{dx} = -\dfrac{1}{16}$. The equation of the tangent line is $y - \dfrac{7}{4} = -\dfrac{1}{16}(x - 4)$, or

$$y = -\frac{1}{16}x + 2.$$

33. $y = (x^2 - x)(3 + 2x) = 2x^3 + x^2 - 3x$

$$\frac{dy}{dx} = 6x^2 + 2x - 3$$

At $x = -1$, $\dfrac{dy}{dx} = 1$. The equation of the tangent line at $(-1, 2)$ is $y - 2 = 1(x + 1)$, or $y = x + 3$.

35. $f(x) = -2x^3 + \dfrac{1}{x^2} = -2x^3 + x^{-2}$

$f'(x) = -6x^2 - \dfrac{2}{x^3}$

At $x = -1$, $f'(-1) = -4$. Further,
$y = f(-1) = 3$. The equation of the tangent
line at $(-1, 3)$ is $y - 3 = -4(x + 1)$, or
$y = -4x - 1$.

37. $f(x) = x - \dfrac{1}{x^2} = x - x^{-2}$

$f'(x) = 1 + \dfrac{2}{x^3}$

At $x = 1$, $f'(1) = 3$. Further, $y = f(1) = 0$.
The equation of the tangent line at $(1, 0)$ is
$y - 0 = 3(x - 1)$, or $y = 3x - 3$.

39. $f(x) = -\dfrac{1}{3}x^3 + \sqrt{8x} = -\dfrac{1}{3}x^3 + \sqrt{8} \cdot x^{1/2}$

$f'(x) = -x^2 + \dfrac{\sqrt{8}}{2x^{1/2}}$

At $x = 2$, $f'(2) = -4 + \dfrac{\sqrt{8}}{2\sqrt{2}}$

$= -4 + \dfrac{1}{2}\sqrt{\dfrac{8}{2}}$

$= -4 + \dfrac{1}{2} \cdot 2$

$= -3.$

Further, $y = f(2) = -\dfrac{8}{3} + 4 = \dfrac{4}{3}$. The

equation of the tangent line at $\left(2, \dfrac{4}{3}\right)$ is

$y - \dfrac{4}{3} = -3(x - 2)$, or $y = -3x + \dfrac{22}{3}$.

41. $f(x) = 2x^4 + 3x + 1$
$f'(x) = 8x^3 + 3$
The rate of change of f at $x = -1$ is
$f'(-1) = -5.$

43. $f(x) = x - \sqrt{x} + \dfrac{1}{x^2} = x - x^{1/2} + x^{-2}$

$f'(x) = 1 - \dfrac{1}{2x^{1/2}} - \dfrac{2}{x^3}$

The rate of change of f at $x = 1$ is
$f'(1) = -\dfrac{3}{2}.$

45. $f(x) = \dfrac{x + \sqrt{x}}{\sqrt{x}}$

$= \dfrac{x}{\sqrt{x}} + \dfrac{\sqrt{x}}{\sqrt{x}}$

$= \sqrt{x} + 1$

$= x^{1/2} + 1$

$f'(x) = \dfrac{1}{2x^{1/2}}$

The rate of change of f at $x = 1$ is
$f'(1) = \dfrac{1}{2}.$

47. $f(x) = 2x^3 - 5x^2 + 4$
$f'(x) = 6x^2 - 10x$
The relative rate of change is
$\dfrac{f'(x)}{f(x)} = \dfrac{6x^2 - 10x}{2x^3 - 5x^2 + 4}.$

When $x = 1$, $\dfrac{f'(1)}{f(1)} = \dfrac{6 - 10}{2 - 5 + 4} = -4.$

49. $f(x) = x\sqrt{x} + x^2$

$= x \cdot x^{1/2} + x^2$

$= x^{3/2} + x^2$

$f'(x) = \dfrac{3}{2}x^{1/2} + 2x = \dfrac{3}{2}\sqrt{x} + 2x$

The relative rate of change is

$\dfrac{f'(x)}{f(x)} = \dfrac{\frac{3}{2}\sqrt{x} + 2x}{x\sqrt{x} + x^2} \cdot \dfrac{2}{2} = \dfrac{3\sqrt{x} + 4x}{2\left(x\sqrt{x} + x^2\right)}.$

When $x = 4$, $\dfrac{f'(x)}{f(4)} = \dfrac{3\sqrt{4} + 4(4)}{2\left(4\sqrt{4} + 4^2\right)} = \dfrac{11}{24}.$

51. (a) $A(t) = 0.1t^2 + 10t + 20$

$A'(t) = 0.2t + 10$

In the year 2008, the rate of change is
$A'(4) = 0.8 + 10$ or $10,800 per year.

(b) $A(4) = (0.1)(16) + 40 + 20 = 61.6$, so
the percentage rate of change is
$$\frac{(100)(10.8)}{61.6} = 17.53\%.$$

53. (a) $T(x) = 20x^2 + 40x + 600$ dollars

The rate of change of property tax is
$T'(x) = 40x + 40$ dollars/year.
In the year 2008, $x = 0$,
$T'(0) = 40$ dollars/year.

(b) In the year 2012, $x = 4$ and
$T(4) = \$1,080$. In the year 2008, $x = 0$
and $T(0) = \$600$.
The change in property tax is
$T(4) - T(0) = \$480$.

55. (a) Cost = cost driver + cost gasoline

cost driver = 20(#hrs)
$$= 20\left(\frac{250 \text{ mi}}{x}\right)$$
$$= \frac{5,000}{x}$$

cost gasoline
$= 4.0(\#\text{gals})$
$$= 4.0(250)\left[\frac{1}{250}\left(\frac{1,200}{x} + x\right)\right]$$
$$= \frac{4,800}{x} + 4.0x \text{ dollars}$$

So, the cost function is
$$C(x) = \frac{9,800}{x} + 4x.$$

(b) The rate of change of the cost is
$C'(x)$.

$C(x) = 9,800x^{-1} + 4x$
$$C'(x) = -\frac{9,800}{x^2} + 4 \text{ dollars/mi per hr}$$
When $x = 40$,
$C'(40) = -2.125$ dollars/mi per hr.

Since $C'(40)$ is negative, the cost is
decreasing.

57. (a) Since Gary's starting salary is $45,000
and he gets a raise of $2,000 per year,
his salary t years from now will be
$S(t) = 45,000 + 2,000t$ dollars.

The percentage rate of change of this
salary t years from now is
$$100\left[\frac{S'(t)}{S(t)}\right] = 100\left(\frac{2,000}{45,000 + 2,000t}\right)$$
$$= \frac{200}{45 + 2t} \text{ percent per year}$$

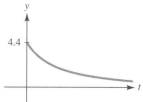

(b) The percentage rate of change after 1
year is
$$\frac{200}{47} \approx 4.26\%$$

(c) In the long run, $\dfrac{200}{45 + 2t}$ approaches 0.
That is, the percentage rate of Gary's
salary will approach 0 (even though
Gary's salary will continue to increase
at a constant rate.)

59. (a) $f(x) = -6x + 582$
The rate of change of SAT scores is
$f'(x) = -6$.

(b) The rate of change is constant, so the
drop will not vary from year to year.
The rate of change is negative, so the
scores are declining.

61. (a) $P(x) = 2x + 4x^{3/2} + 5,000$ is the
population x months from now. The
rate of population growth is

$$P'(x) = 2 + 4\left(\frac{3x^{1/2}}{2}\right)$$

$$= 2 + 6x^{1/2} \text{ people per month.}$$

Nine months from now, the population will be changing at the rate of $P'(9) = 2 + 6(9^{1/2}) = 20$ people per month.

(b) The percentage rate at which the population will be changing 9 months from now is

$$100\frac{P'(9)}{P(9)} = \frac{100(20)}{2(9) + 4(9^{3/2}) + 5,000}$$

$$= \frac{2,000}{5,126}$$

$$= 0.39\%.$$

63. $N(t) = 10t^3 + 5t + \sqrt{t} = 10t^3 + 5t + t^{1/2}$

The rate of change of the infected population is

$$N'(t) = 30t^2 + 5 + \frac{1}{2t^{1/2}} \text{ people/day.}$$

On the 9th day, $N'(9) = 2,435$ people/day.

65. (a) $T(t) = -68.07t^3 + 30.98t^2 + 12.52t + 37.1$

$$T'(t) = -204.21t^2 + 61.96t + 12.52$$

$T'(t)$ represents the rate at which the bird's temperature is changing after t days, measured in °C per day.

(b) $T'(0) = 12.52$°C/day since $T'(0)$ is positive, the bird's temperature is increasing.

$$T'(0.713) \approx -47.12°\text{C/day}$$

Since $T'(0.713)$ is negative, the bird's temperature is decreasing.

(c) Find t so that $T'(t) = 0$.

$$0 = -204.21t^2 + 61.96t + 12.52$$

$$t = \frac{-61.96 \pm \sqrt{(61.96)^2 - 4(-204.21)(12.52)}}{2(-204.21)}$$

$t \approx 0.442$ days

The bird's temperature when $t = 0.442$ is $T(0.442) \approx 42.8$°C.
The bird's temperature starts at $T(0) = 37.1$°C, increases to $T(0.442) = 42.8$°C, and then begins to decrease.

67. (a) $Q(t) = 0.05t^2 + 0.1t + 3.4$ PPM
$Q'(t) = 0.1t + 0.1$ PPM/year
The rate of change of Q at $t = 1$ is
$Q'(1) = 0.2$ PPM/year.

(b) $Q(1) = 3.55$ PPM, $Q(0) = 3.40$, and $Q(1) - Q(0) = 0.15$ PPM.

(c) $Q(2) = 0.2 + 0.2 + 3.4 = 3.8$,
$Q(0) = 3.4$, and
$Q(2) - Q(0) = 0.4$ PPM.

69. Since g represents the acceleration due to gravity for the planet our spy is on, the formula for the rock's height is

$$H(t) = -\frac{1}{2}gt^2 + V_0 t + H_0$$

Since he throws the rock from ground level, $H_0 = 0$. Also, since it returns to the ground after 5 seconds,

$$0 = -\frac{1}{2}g(5)^2 + V_0(5)$$

$$0 = -12.5g + 5V_0$$

$$V_0 = \frac{12.5g}{5} = 2.5g$$

The rock reaches its maximum height halfway through its trip, or when $t = 2.5$. So,

$$37.5 = -\frac{1}{2}g(2.5)^2 + V_0(2.5)$$

$$37.5 = -3.125g + 2.5V_0$$

Substituting $V_0 = 2.5g$

$$37.5 = -3.125g + 2.5(2.5g)$$
$$37.5 = -3.125g + 6.25g$$
$$37.5 = 3.125g$$
$$g = 12 \text{ ft/sec}^2$$

So, our spy is on Mars.

71. (a) $s(t) = 3t^2 + 2t - 5$ for $0 \le t \le 1$
$v(t) = 6t + 2$ and $a(t) = 6$

(b) $6t + 2 = 0$ at $t = -3$. The particle is not stationary between $t = 0$ and $t = 1$.

73. (a) $s(t) = t^4 - 4t^3 + 8t$ for $0 \le t \le 4$
$v(t) = 4t^3 - 12t^2 + 8$ and
$a(t) = 12t^2 - 24t$

(b) To find all time in given interval when stationary,
$$4t^3 - 12t^2 + 8 = 0$$
$$4(t^3 - 3t^2 + 2) = 0$$
$$t^3 - 3t^2 + 2 = 0$$
$$(t-1)(t^2 - 2t - 2) = 0$$
$$t = 1 \text{ or } t = \frac{2 \pm \sqrt{4 - 4 \cdot 1 \cdot (-2)}}{2}$$
Since $0 \le t \le 4$, $t = 1$ or $t = 1 + \sqrt{3}$.

75. (a) If after 2 seconds the ball passes you on the way down, then $H(2) = H_0$,
where $H(t) = -16t^2 + V_0 t + H_0$.
So, $-16(2^2) + (V_0)(2) + H_0 = H_0$,
$$-64 + 2V_0 = 0, \text{ or } V_0 = 32 \frac{\text{ft}}{\text{sec}}.$$

(b) The height of the building is H_0 feet. From part **(a)** you know that
$H(t) = -16t^2 + 32t + H_0$. Moreover, $H(4) = 0$ since the ball hits the ground after 4 seconds. So,
$$-16(4)^2 + 32(4) + H_0 = 0, \text{ or }$$
$$H_0 = 128 \text{ feet.}$$

(c) From parts **(a)** and **(b)** you know that
$H(t) = -16t^2 + 32t + 128$ and so the speed of the ball is
$$H'(t) = -32t + 32 \frac{\text{ft}}{\text{sec}}.$$
After 2 seconds, the speed will be
$H'(2) = -32$ feet per second, where the minus sign indicates that the direction of motion is down.

(d) The speed at which the ball hits the ground is $H'(4) = -96 \dfrac{\text{ft}}{\text{sec}}$.

77. $f(x) = ax^2 + bx + c$

Since $f(0) = 0$, $c = 0$ and $f(x) = ax^2 + bx$.
Since $f(5) = 0$, $0 = 25a + 5b$.
Further, since the slope of the tangent is 1 when $x = 2$, $f'(2) = 1$.
$$f'(x) = 2ax + b$$
$$1 = 2a(2) + b = 4a + b$$
Now, solve the system: $0 = 25a + 5b$ and $1 = 4a + b$. Since $1 - 4a = b$, using substitution
$$0 = 25a + 5(1 - 4a)$$
$$0 = 25a + 5 - 20a$$
$$0 = 5a + 5$$
or $a = -1$ and $b = 1 - 4(-1) = 5$.
So, $f(x) = -x^2 + 5x$.

79. $(f+g)'(x)$

$$= \lim_{h \to 0} \frac{(f+g)(x+h)-(f+g)(x)}{h}$$

$$= \lim_{h \to 0} \frac{f(x+h)+g(x+h)-[f(x)+g(x)]}{h}$$

$$= \lim_{h \to 0} \frac{f(x+h)-f(x)+g(x+h)-g(x)}{h}$$

$$= \lim_{h \to 0} \frac{f(x+h)-f(x)}{h} + \lim_{h \to 0} \frac{g(x+h)-g(x)}{h}$$

$$= f'(x)+g'(x)$$

2.3 Product and Quotient Rules; Higher-Order Derivatives

1. $f(x) = (2x+1)(3x-2)$,

$$f'(x) = (2x+1)\frac{d}{dx}(3x-2)$$

$$+ (3x-2)\frac{d}{dx}(2x+1)$$

$$= (3x+1)(3)+(3x-2)(2)$$

$$= 12x-1$$

3. $y = 10(3u+1)(1-5u)$,

$$\frac{dy}{du}$$

$$= 10\frac{d}{du}(3u+1)(1-5u)$$

$$= 10\left[(3u+1)\frac{d}{du}(1-5u)+(1-5u)\frac{d}{du}(3u+1)\right]$$

$$= 10[(3u+1)(-5)+(1-5u)(3)]$$

$$= -300u-20$$

5. $f'(x) = \frac{1}{3}\left[(x^5-2x^3+1)\frac{d}{dx}\left(x-\frac{1}{x}\right)\right.$

$$\left. + \left(x-\frac{1}{x}\right)\frac{d}{dx}(x^5-2x^3+1)\right]$$

$$= \frac{1}{3}\left[(x^5-2x^3+1)\left(1+\frac{1}{x^2}\right)\right.$$

$$\left. + \left(x-\frac{1}{x}\right)(5x^4-6x^2)\right]$$

$$= 2x^5-4x^3+\frac{4}{3}x+\frac{1}{3x^2}+\frac{1}{3}$$

7. $y = \dfrac{x+1}{x-2}$,

$$\frac{dy}{dx} = \frac{(x-2)\frac{d}{dx}(x+1)-(x+1)\frac{d}{dx}(x-2)}{(x-2)^2}$$

$$= \frac{(x-2)(1)-(x+1)(1)}{(x-2)^2}$$

$$= -\frac{3}{(x-2)^2}$$

9. $f(t) = \dfrac{t}{t^2-2}$,

$$f'(t) = \frac{(t^2-2)\frac{d}{dt}(t)-t\frac{d}{dt}(t^2-2)}{(t^2-2)^2}$$

$$= \frac{(t^2-2)(1)-(t)(2t)}{(t^2-2)^2}$$

$$= \frac{-t^2-2}{(t^2-2)^2}$$

11. $y = \dfrac{3}{x+5}$,

$$\frac{dy}{dx} = \frac{(x+5)\frac{d}{dx}(3)-3\frac{d}{dx}(x+5)}{(x+5)^2}$$

$$= \frac{(x+5)(0)-3(1)}{(x+5)^2}$$

$$= -\frac{3}{(x+5)^2}$$

13. $f(x) = \dfrac{x^2 - 3x + 2}{2x^2 + 5x - 1}$,

$$f'(x) = \frac{(2x^2 + 5x - 1)\frac{d}{dx}(x^2 - 3x + 2)}{(2x^2 + 5x - 1)^2}$$

$$-\frac{(x^2 - 3x + 2)\frac{d}{dx}(2x^2 + 5x - 1)}{(2x^2 + 5x - 1)^2}$$

$$= \frac{(2x^2 + 5x - 1)(2x - 3)}{(2x^2 + 5x - 1)^2}$$

$$-\frac{(x^2 - 3x + 2)(4x + 5)}{(2x^2 + 5x - 1)^2}$$

$$= \frac{11x^2 - 10x - 7}{(2x^2 + 5x - 1)^2}$$

15. $f(x) = (2 + 5x)^2 = (2 + 5x)(2 + 5x)$

$$f'(x) = (2 + 5x)\frac{d}{dx}(2 + 5x)$$

$$+ (2 + 5x)\frac{d}{dx}(2 + 5x)$$

$$= 2(2 + 5x)\frac{d}{dx}(2 + 5x)$$

$$= 2(2 + 5x)(5)$$

$$= 20 + 50x$$

$$= 10(2 + 5x)$$

17. $g(t) = \dfrac{t^2 + \sqrt{t}}{2t + 5} = \dfrac{t^2 + t^{1/2}}{2t + 5}$

$$g'(t) = \frac{(2t + 5)\frac{d}{dt}(t^2 + t^{1/2})}{(2t + 5)^2}$$

$$\frac{-(t^2 + t^{1/2})\frac{d}{dt}(2t + 5)}{(2t + 5)^2}$$

$$= \frac{(2t + 5)\left(2t + \frac{1}{2t^{1/2}}\right) - (t^2 + t^{1/2})(2)}{(2t + 5)^2}$$

$$= \frac{2t^2 + 10t - t^{1/2} + \frac{5}{2t^{1/2}}}{(2t + 5)^2} \cdot \frac{2t^{1/2}}{2t^{1/2}}$$

$$= \frac{4t^{5/2} + 20t^{3/2} - 2t + 5}{2t^{1/2}(2t + 5)^2}$$

$$= \frac{4\sqrt{t^5} + 20\sqrt{t^3} - 2t + 5}{2\sqrt{t}(2t + 5)^2}$$

19. $y = (5x - 1)(4 + 3x)$

$$\frac{dy}{dx} = 30x + 17$$

When $x = 0$, $y = -4$ and $\dfrac{dy}{dx} = 17$. The equation of the tangent line at $(0, -4)$ is $y + 4 = 17(x - 0)$, or $y = 17x - 4$.

21. $y = \dfrac{x}{2x + 3}$

$$\frac{dy}{dx} = \frac{3}{(2x + 3)^2}$$

When $x = -1$, $y = -1$ and $\dfrac{dy}{dx} = 3$. The equation of the tangent line at $(-1, -1)$ is $y + 1 = 3(x + 1)$, or $y = 3x + 2$.

23. $y = \left(3\sqrt{x} + x\right)(2 - x^2)$

$$= (3x^{1/2} + x)(2 - x^2)$$

$$\frac{dy}{dx} = -3x^2 - \frac{15}{2}x^{3/2} + \frac{3}{x^{1/2}} + 2$$

When $x = 1$, $y = 4$ and $\dfrac{dy}{dx} = -\dfrac{11}{2}$.
The equation of the tangent line at $(1, 4)$ is
$$y - 4 = -\frac{11}{2}(x - 1), \text{ or } y = -\frac{11}{2}x + \frac{19}{2}.$$

25. $f(x) = (x + 1)(x^2 - x - 2)$

$$f'(x) = (x + 1)(2x - 1) + (x^2 - x - 2)(1)$$

$$= 3x^2 - 3$$

Since $f'(x)$ represents the slope of the tangent line and the slope of a horizontal line is zero, need to solve

$$0 = 3x^2 - 3 = 3(x + 1)(x - 1) \text{ or } x = -1, 1.$$

When $x = -1$, $f(-1) = 0$ and when $x = 1$, $f(1) = -4$. So, the tangent line is horizontal at the points $(-1, 0)$ and $(1, -4)$.

27. $f(x) = \dfrac{x+1}{x^2+x+1}$

$f'(x) = \dfrac{-x^2-2x}{(x^2+x+1)^2}$

Since $f'(x)$ represents the slope of the tangent line and the slope of a horizontal line is zero, need to solve

$0 = \dfrac{-x^2-2x}{(x^2+x+1)^2}$

$0 = -x^2-2x = -x(x+2)$ or $x=0,\ -2$.

When $x=0, f(0)=1$ and when $x=-2$,

$f(-2) = -\dfrac{1}{3}$. So, the tangent line is

horizontal at the points $(0, 1)$ and

$\left(-2, -\dfrac{1}{3}\right)$.

29. $y = (x^2+3)(5-2x^3)$

$\dfrac{dy}{dx} = (x^2+3)(-6x^2)+(5-2x^3)(2x)$

When $x=1$,

$\dfrac{dy}{dx} = (1+3)(-6)+(5-2)(2) = -18$.

31. $y = x + \dfrac{3}{2-4x}$

$\dfrac{dy}{dx} = 1 + \dfrac{(2-4x)(0)-3(-4)}{(2-4x)^2}$

When $x=0$, $\dfrac{dy}{dx} = 1 + \dfrac{12}{(2)^2} = 4$.

33. $y = \dfrac{2}{x} - \sqrt{x} = 2x^{-1} - x^{1/2}$

$\dfrac{dy}{dx} = \dfrac{-2}{x^2} - \dfrac{1}{2x^{1/2}}$

When $x=1$, $\dfrac{dy}{dx} = -2 - \dfrac{1}{2} = -\dfrac{5}{2}$.

The slope of a line perpendicular to the

tangent line at $x=1$ is $\dfrac{2}{5}$.

The equation of the normal line at $(1, 1)$ is

$y-1 = \dfrac{2}{5}(x-1)$, or $y = \dfrac{2}{5}x + \dfrac{3}{5}$.

35. $y = \dfrac{5x+7}{2-3x}$

$\dfrac{dy}{dx} = \dfrac{(2-3x)(5)-(5x+7)(-3)}{(2-3x)^2}$

When $x=1$,

$\dfrac{dy}{dx} = \dfrac{(2-3)(5)-(5+7)(-3)}{(2-3)^2} = 31$.

The slope of a line perpendicular to the

tangent line at $x=1$ is $-\dfrac{1}{31}$.

The equation of the normal line at

$(1, -12)$ is $y+12 = -\dfrac{1}{31}(x-1)$, or

$y = -\dfrac{1}{31}x - \dfrac{371}{31}$.

37. $h(x) = \left[3x^2 - 2g(x)\right]\left[g(x)+5x\right]$

Using the product rule,

$h'(x) = \left[3x^2 - 2g(x)\right]\left[g'(x)+5\right]$
$\qquad + \left[g(x)+5x\right]\left[6x - 2g'(x)\right]$

Substituting $x=-3$,

$h'(-3) = \left[3(-3)^2 - 2g(-3)\right]\left[g'(-3)+5\right]$
$\qquad + \left[g(-3)+5(-3)\right]\left[6(-3) - 2g'(-3)\right]$
$\quad = \left[27 - 2g(-3)\right]\left[g'(-3)+5\right]$
$\qquad + \left[g(-3)-15\right]\left[-18 - 2g'(-3)\right]$

Since $g(-3)=1$ and $g'(-3)=2$,

$h'(-3) = \left[27 - 2(1)\right]\left[2+5\right]$
$\qquad + \left[1-15\right]\left[-18 - 2(2)\right]$
$\quad = (25)(7)+(-14)(-22) = 483$

39. $h(x) = \dfrac{x^3 + xg(x)}{3x - 5}$

Using the quotient rule, and noting that the derivative of the numerator requires the product rule for the term $x \cdot g(x)$,

$$h'(x) = \frac{\Big([3x-5]\big[3x^2 + xg'(x) + g(x) \cdot 1\big]\Big)}{(3x-5)^2}$$
$$- \frac{\Big(\big[x^3 + xg(x)\big][3]\Big)}{(3x-5)^2}$$

Substituting $x = -1$,

$$h'(-1) = \frac{\big[3(-1)-5\big]\big[3(-1)^2 + (-1)g'(-1) + g(-1)\big]}{\big[3(-1)-5\big]^2}$$
$$- \frac{\big[(-1)^3 + (-1)g(-1)\big][3]}{\big[3(-1)-5\big]^2}$$

$$h'(-1) = \frac{\big([-8]\big[3 - g'(-1) + g(-1)\big]\big)}{(-8)^2}$$
$$- \frac{\big[-1 - g(-1)\big][3]\big)}{(-8)^2}$$

Since $g(-1) = 0$ and $g'(-1) = 1$,

$$h'(-1) = \frac{(-8)(3 - 1 + 0) - (-1 - 0)(3)}{64}$$
$$= \frac{-16 + 3}{64} = -\frac{13}{64}$$

41. (a) $y = \dfrac{2x - 3}{x^3}$

$$\frac{dy}{dx} = \frac{(x^3)(2) - (2x-3)(3x^2)}{x^6}$$
$$= \frac{-4x^3 + 9x^2}{x^6}$$
$$= \frac{-4x + 9}{x^4}$$

(b) $y = (2x - 3)(x^{-3})$

$$\frac{dy}{dx} = (2x-3)(-3x^{-4}) + (x^{-3})(2)$$
$$= \frac{-3(2x-3) + 2x}{x^4}$$
$$= \frac{-4x + 9}{x^4}$$

(c) $y = 2x^{-2} - 3x^{-3}$

$$\frac{dy}{dx} = -4x^{-3} + 9x^{-4}$$
$$= \frac{-4}{x^3} + \frac{9}{x^4}$$
$$= \frac{-4x + 9}{x^4}$$

43. $f(x) = \dfrac{2}{5}x^5 - 4x^3 + 9x^2 - 6x - 2$

$$f'(x) = 2x^4 - 12x^2 + 18x - 6$$
$$f''(x) = 8x^3 - 24x + 18$$

45. $y = \dfrac{2}{3}x^{-1} - \sqrt{2}x^{1/2} + \sqrt{2}x - \dfrac{1}{6}x^{-1/2}$

$$\frac{dy}{dx} = y'$$
$$= -\frac{2}{3}x^{-2} - \frac{\sqrt{2}}{2}x^{-1/2} + \sqrt{2} + \frac{1}{12}x^{-3/2}$$

$$\frac{d^2y}{dx^2} = y'' = \frac{4}{3}x^{-3} + \frac{\sqrt{2}}{4}x^{-3/2} - \frac{1}{8}x^{-5/2}$$

$$= \frac{4}{3x^3} + \frac{\sqrt{2}}{4x^{3/2}} - \frac{1}{8x^{5/2}}$$

47. $y = (x^3 + 2x - 1)(3x + 5)$

$$\frac{dy}{dx} = y'$$

$$= (x^3 + 2x - 1)(3) + (3x + 5)(3x^2 + 2)$$

$$= 12x^3 + 15x^2 + 12x + 7$$

$$\frac{d^2y}{dx^2} = y'' = 36x^2 + 30x + 12$$

49. $S(t) = \dfrac{2000t}{4 + 0.3t}$

(a) $S'(t) = \dfrac{(4 + 0.3t)(2000) - (2000t)(0.3)}{(4 + 0.3t)^2}$

The rate of change in the year 2010 is

$$S'(2) = \frac{(4 + 0.6)(2,000) - (4,000)(0.3)}{(4 + 0.6)^2}$$

$$\approx \$378,072 \text{ per year.}$$

(b) Rewrite the function as

$$S(t) = \frac{2,000}{\frac{4}{t} + 0.3}.$$

Since $\dfrac{4}{t} \to 0$ as $t \to +\infty$, sales

approach $\dfrac{2,000}{0.3} \approx 6,666.67$ thousand,

or approximately \$6,666,667 in the long run.

51. $P(t) = 100 \left[\dfrac{t^2 + 5t + 5}{t^2 + 10t + 30} \right]$

(a) $P'(t) = 100 \dfrac{(t^2 + 10t + 30)(2t + 5) - (t^2 + 5t + 5)(2t + 10)}{(t^2 + 10t + 30)^2}$

The rate of change after 5 weeks is

$$P'(5) = 100 \frac{(25 + 50 + 30)(10 + 5) - (25 + 25 + 5)(10 + 10)}{(25 + 50 + 30)^2}$$

$P'(5) = 4.31\%$ per week.

Since $P'(5)$ is positive, the percentage is increasing.

(b) Rewrite the function as

$$p(t) = 100 \frac{1 + \frac{5}{t} + \frac{5}{t^2}}{1 + \frac{10}{t} + \frac{30}{t^2}}.$$

Since $\dfrac{5}{t}, \dfrac{5}{t^2}, \dfrac{10}{t}$ and $\dfrac{30}{t^2}$ all go to

zero as $t \to +\infty$, the percentage approaches 100% in the long run, so the rate of change approaches 0.

53. (a) Revenue = (# units)(selling price)

$x = \#$ units $= 1000 - 4t$, where t is measured in weeks

$p =$ selling price $= 5 + 0.05t$

So,

$$R(t) = x(t)\,p(t)$$

To find $R(x)$, solve the first equation for t

$$t = \frac{1000 - x}{4} = 250 - 0.25x$$

Substitute this into the second equation to get

$$p = 5 + 0.05(250 - 0.25x)$$

$$p = 5 + 12.5 - 0.0125x$$

$$= 17.5 - 0.0125x$$

Then,

$$R(x) = x(17.5 - 0.0125x)$$
$$= 17.5x - 0.0125x^2$$
$$R'(x) = 17.5 - 0.025x$$

Currently, when $x = 0$,
$$R'(0) = 17.5$$
or \$17.50 per unit
Since $R'(0)$ is positive, the revenue is currently increasing.

(b) $AR(x) = \dfrac{R(x)}{x}$

$$AR(x) = 17.5 - 0.0125x$$
$$AR'(x) = -0.0125$$
$$AR'(0) = -0.0125$$

Since $AR'(0)$ is negative, the average revenue is currently decreasing.

55. $P(x) = \dfrac{100\sqrt{x}}{0.03x^2 + 9} = 100\dfrac{x^{1/2}}{0.03x^2 + 9}$

(a) $P'(x) = 100\dfrac{(0.03x^2 + 9)\left(\frac{1}{2}x^{-1/2}\right) - (x^{1/2})(0.06x)}{(0.03x^2 + 9)^2}$

The rate of change of percentage pollution when 16 million dollars are spent is
$P'(16)$

$$= 100\left(\frac{[0.03(16)^2 + 9]\left[\frac{1}{2}(16)^{-1/2}\right]}{[0.03(16)^2 + 9]^2}\right.$$
$$\left. - \frac{(16)^{1/2}[0.06(16)]}{[0.03(16)^2 + 9]^2}\right)$$

$= -0.63$ percent
Since $P'(16)$ is negative, the percentage is decreasing.

(b) $P'(x) = 0$ when

$$0 = (0.03x^2 + 9)\left(\frac{1}{2}x^{-1/2}\right) \quad \text{or}$$
$$- (x^{1/2})(0.06x)$$

$x = 10$ million dollars.
Testing one value less than 10 and one value greater than 10 shows $P'(x)$ is increasing when $0 < x < 10$, and decreasing when $x > 10$.

57. $F = \dfrac{1}{3}(KM^2 - M^3)$

(a) $S = \dfrac{dF}{dM}$
$$= \frac{1}{3}(2KM - 3M^2)$$
$$= \frac{2}{3}KM - M^2$$

(b) $\dfrac{dS}{dM} = \dfrac{1}{3}(2K - 6M) = \dfrac{2}{3}K - 2M$ is
the rate the sensitivity is changing.

59. $P(t) = 20 - \dfrac{6}{t+1}$

(a) $P'(t) = 0 - \dfrac{(t+1)(0) - (6)(1)}{(t+1)^2}$
$$= \frac{6}{(t+1)^2}$$

(b) $P'(1) = \dfrac{6}{(1+1)^2} = \dfrac{6}{4} = 1.5$

or increasing at a rate of 1,500 people per year

(c) Actual change $= P(2) - P(1)$

$$P(2) = 20 - \frac{6}{2+1} = 20 - 2 = 18$$

$$P(1) = 20 - \frac{6}{1+1} = 20 - 3 = 17$$

So, the population will actually increase by 1 thousand people.

(d) $P'(9) = \dfrac{6}{(9+1)^2} = \dfrac{60}{100} = 0.06$

or increasing at a rate of 60 people per year

(e) In the long run,

$$\lim_{t\to\infty} P'(t) = \lim_{t\to\infty} \dfrac{6}{(t+1)^2} = 0$$

So, the population growth approaches zero.

61. (a) $s(t) = 3t^5 - 5t^3 - 7$

$v(t) = 15t^4 - 15t^2 = 15(t^4 - t^2)$

$a(t) = 15(4t^3 - 2t) = 30t(2t^2 - 1)$

(b) $a(t) = 0$ when $30t(2t^2 - 1) = 0$, or $t = 0$

and $t = \dfrac{\sqrt{2}}{2}$.

63. $s(t) = -t^3 + 7t^2 + t + 2$

(a) $v(t) = -3t^2 + 14t + 1$
$a(t) = -6t + 14$

(b) $a(t) = 0$ when $-6t + 14 = 0$, or $t = \dfrac{7}{3}$.

65. $D(t) = 10t + \dfrac{5}{t+1} - 5$

(a) Speed = rate of change of distance with respect to time.

$\dfrac{dD}{dt} = 10 + \dfrac{(t+1)(0) - (5)(1)}{(t+1)^2}$

$= 10 - \dfrac{5}{(t+1)^2}$

When $t = 4$,

$\dfrac{dD}{dt} = 10 - \dfrac{5}{25} = \dfrac{49}{5} = 9.8$ meters/minute.

(b) $D(5) = 10(5) + \dfrac{5}{5+1} - 5 = 45 + \dfrac{5}{6}$

$D(4) = 10(4) + \dfrac{5}{4+1} - 5 = 36$

$D(5) - D(4) = 9 + \dfrac{5}{6} \approx 9.83$ meters.

67. $H(t) = -16t^2 + V_0 t + H_0$

(a) $H'(t) = -32t + V_0$ and the acceleration is $H''(t) = -32$.

(b) Since the acceleration is a constant, it does not vary with time.

(c) The only acceleration acting on the object is due to gravity. The negative sign signifies that this acceleration is directed downward.

69. $y = x^{1/2} - \dfrac{1}{2}x^{-1} + \dfrac{1}{\sqrt{2}}x$

$\dfrac{dy}{dx} = \dfrac{1}{2}x^{-1/2} + \dfrac{1}{2}x^{-2} + \dfrac{1}{\sqrt{2}}$

$\dfrac{d^2 y}{dx^2} = -\dfrac{1}{4}x^{-3/2} - x^{-3}$

$\dfrac{d^3 y}{dx^3} = \dfrac{3}{8}x^{-5/2} + 3x^{-4} = \dfrac{3}{8x^{5/2}} + \dfrac{3}{x^4}$

71. (a)

$$\frac{d}{dx}\left(\frac{fg}{h}\right) = \frac{h\frac{d}{dx}(fg) - (fg)\frac{d}{dx}h}{h^2}$$

$$= \frac{h\left(f\frac{d}{dx}g + g\frac{d}{dx}f\right) - fg\frac{d}{dx}h}{h^2}$$

(b) $y = \dfrac{(2x+7)(x^2+3)}{3x+5}$

$$\frac{dy}{dx} = \frac{\begin{array}{c}(3x+5)[(2x+7)(2x)\\ +(x^2+3)(2)]\end{array}}{(3x+5)^2}$$

$$-\frac{(2x+7)(x^2+3)(3)}{(3x+5)^2}$$

$$= \frac{(3x+5)(6x^2+14x+6)}{(3x+5)^2}$$

$$-\frac{3(2x^3+7x^2+6x+21)}{(3x+5)^2}$$

$$= \frac{12x^3+51x^2+70x-33}{(3x+5)^2}$$

73. For f/g the difference quotient (DQ) is

$$= \frac{\left(\frac{f}{g}\right)(x+h) - \left(\frac{f}{g}\right)(x)}{h}$$

$$= \frac{1}{h}\left[\frac{f(x+h)}{g(x+h)} - \frac{f(x)}{g(x)}\right]$$

$$= \frac{1}{h}\left[\frac{f(x+h)g(x) - f(x)g(x+h)}{g(x+h)g(x)}\right]$$

$$= \frac{1}{h}\left[\frac{f(x+h)g(x) - f(x)g(x) + f(x)g(x)}{g(x+h)g(x)} - \frac{f(x)g(x+h)}{g(x+h)g(x)}\right]$$

$$= \frac{1}{h}\left[\frac{g(x)[f(x+h) - f(x)]}{g(x+h)g(x)} - \frac{f(x)[g(x+h) - g(x)]}{g(x+h)g(x)}\right]$$

$$= \frac{1}{g(x+h)g(x)}\cdot\left[\frac{g(x)[f(x+h) - f(x)]}{h} - \frac{f(x)[g(x+h) - g(x)]}{h}\right]$$

$$\frac{d}{dx}\left(\frac{f}{g}\right) = \lim_{h\to 0}\frac{1}{g(x+h)g(x)}\cdot\left[\frac{g(x)[f(x+h) - f(x)]}{h} - \frac{f(x)[g(x+h) - g(x)]}{h}\right]$$

$$= \frac{1}{g(x)g(x)}[g(x)f'(x) - f(x)g'(x)]$$

$$= \frac{g(x)f'(x) - f(x)g'(x)}{[g(x)]^2}$$

75. To use a graphing utility to sketch

$f(x) = x^2(x-1)$ and find where

$f'(x) = 0$,

Press $\boxed{y =}$

Input $x^2(x-1)$ for $y_1 =$

Use window dimensions [−2, 3].5 by [−2, 2].5

Press graph

Press $\boxed{2nd}$ $\boxed{Draw}$ and enter the tangent function

Enter $x = 1$

The calculator draws the line tangent to the graph of f at $x = 1$ and gives $y = 1.000001x − 1.000001$ as the equation of that line. $f'(x) = 0$ when the slope of the line tangent to the graph of f is zero. This happens where the graph of f has a local high or low point. Use the trace button to move cross-hairs to the local low point on the graph of f. Use the zoom-in function under the zoom menu to find $f'(x) = 0$ when $x \approx 0.673$. Repeat this process to find where the local high point occurs. We see $f'(x) = 0$ also for $x = 0$.

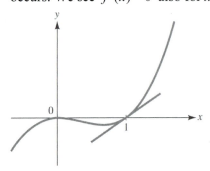

77. To use a graphing utility to graph

$f(x) = x^4 + 2x^3 − x + 1$ and to find minima and maxima,

Press $\boxed{y =}$ and input $x^4 + 2x^3 − x + 1$ for $y_1 =$

Use window dimensions [−5, 5]1 by [0, 2].5

Press $\boxed{Graph}$

We see from the graph that there are two minimums and one maximum.

To find the first minimum, use trace and zoom-in for a more accurate reading. Alternatively, use the minimum function under the calc menu. Using trace, enter a value to the left of (but close to) the minimum for the left bound. Enter a value to the right of (but close to) the minimum for the right bound. Finally, enter a guess in between the bounds and the minimum is displayed.

One minimum occurs at (−1.37, 0.75). Repeat this process for the other minimum and find it to be at (0.366, 0.75).

Repeat again for the maximum (using the maximum function) to find it at (−0.5, 1.31).

$f'(x) = 4x^3 + 6x^2 − 1$

Press $\boxed{y =}$ and input $4x^\wedge 3 + 6x^2 − 1$ for $y_2 =$

Change window dimensions to [−5, 5]1 by [−2, 2].5.

Use trace and zoom-in to find the x-intercepts of $f'(x)$ or use the zero function under the calc menu. The three x-intercepts of $f'(x)$ are $x \approx −1.37$, −0.5, and 0.366.

The x-values extrema occur at the x-intercepts of f' because the tangent line at the corresponding points on the curve are horizontal and so, the slopes are zero.

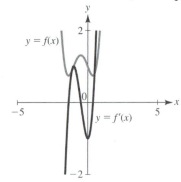

2.4 The Chain Rule

1. $y = u^2 + 1,\ u = 3x - 2,$

$$\frac{dy}{du} = 2u, \frac{du}{dx} = 3,$$

$$\frac{dy}{dx} = \frac{dy}{du} \cdot \frac{du}{dx} = (2u)(3) = 6(3x - 2).$$

3. $y = \sqrt{u} = u^{1/2},\ u = x^2 + 2x - 3,$

$$\frac{dy}{du} = \frac{1}{2} u^{-1/2} = \frac{1}{2u^{1/2}},$$

$$\frac{du}{dx} = 2x + 2 = 2(x + 1),$$

$$\frac{dy}{dx} = \frac{dy}{du} \cdot \frac{du}{dx} = \frac{x + 1}{(x^2 + 2x - 3)^{1/2}}.$$

5. $y = \dfrac{1}{u^2} = u^{-2},\ u = x^2 + 1,$

$$\frac{dy}{du} = -2u^{-3} = -\frac{2}{u^3},\ \frac{du}{dx} = 2x,$$

$$\frac{dy}{dx} = \frac{dy}{du} \cdot \frac{du}{dx} = -\frac{4x}{(x^2 + 1)^3}$$

7. $y = \dfrac{1}{u - 1} = (u - 1)^{-1},\ u = x^2$

$$\frac{dy}{du} = -(u - 1)^{-2} = -\frac{1}{(u - 1)^2},$$

$$\frac{du}{dx} = 2x,\ \frac{dy}{dx} = \frac{dy}{du} \cdot \frac{du}{dx} = -\frac{2x}{(x^2 - 1)^2}.$$

9. $y = u^2 + 2u - 3,\ u = \sqrt{x} = x^{1/2}$

$$\frac{dy}{du} = 2u + 2, \frac{du}{dx} = \frac{1}{2} x^{-1/2} = \frac{1}{2\sqrt{x}}$$

$$\frac{dy}{dx} = \frac{dy}{du} \cdot \frac{du}{dx} = (2u + 2) \cdot \frac{1}{2\sqrt{x}}$$

$$= \left(2\sqrt{x} + 2\right) \cdot \frac{1}{2\sqrt{x}}$$

$$= 1 + \frac{1}{\sqrt{x}}$$

11. $y = u^2 + u - 2,\ u = \dfrac{1}{x} = x^{-1}$

$$\frac{dy}{du} = 2u + 1, \frac{du}{dx} = -x^{-2} = \frac{-1}{x^2}$$

$$\frac{dy}{dx} = \frac{dy}{du} \cdot \frac{du}{dx} = (2u + 1) \cdot \frac{-1}{x^2}$$

$$= \left(\frac{2}{x} + 1\right) \cdot \frac{-1}{x^2}$$

$$= -\frac{2}{x^3} - \frac{1}{x^2}$$

$$= \frac{-2}{x^3} + \frac{-1}{x^2} \cdot \frac{x}{x}$$

$$= \frac{-2}{x^3} + \frac{-x}{x^3}$$

$$= -\frac{2 + x}{x^3}$$

13. $y = u^2 - u,\ u = 4x + 3$

$$\frac{dy}{du} = 2u - 1, \frac{du}{dx} = 4$$

$$\frac{dy}{dx} = \frac{dy}{du} \cdot \frac{du}{dx} = (2u - 1) \cdot 4$$

When $x = 0,\ u = 4(0) + 3 = 3$, so

$$\frac{dy}{dx} = (2(3) - 1) \cdot 4 = 20.$$

15. $y = 3u^4 - 4u + 5,\ u = x^3 - 2x - 5$

$$\frac{dy}{du} = 12u^3 - 4, \frac{du}{dx} = 3x^2 - 2,$$

$$\frac{dy}{dx} = \frac{dy}{du} \cdot \frac{du}{dx} = (12u^3 - 4)(3x^2 - 2).$$

When $x = 2,\ u = 2^3 - 2(2) - 5 = -1,$ so

$$\frac{dy}{dx} = [12(-1)^3 - 4][3(2^2)) - 2] = -160$$

17. $y = \sqrt{u} = u^{1/2},\ u = x^2 - 2x + 6,$

$$\frac{dy}{du} = \frac{1}{2} u^{-1/2} = \frac{1}{2u^{1/2}},$$

$$\frac{du}{dx} = 2x - 2,\ \frac{dy}{dx} = \frac{dy}{du} \cdot \frac{du}{dx} = \frac{x - 1}{u^{1/2}}.$$

When $x = 3$, $u = 3^2 - 2(3) + 6 = 9$, so

$$\frac{dy}{dx} = \frac{3-1}{9^{1/2}} = \frac{2}{3}.$$

19. $y = \dfrac{1}{u} = u^{-1}$, $u = 3 - \dfrac{1}{x^2} = 3 - x^{-2}$,

$$\frac{dy}{du} = -u^{-2} = -\frac{1}{u^2}, \quad \frac{du}{dx} = 2x^{-3} = \frac{2}{x^3}$$

$$\frac{dy}{dx} = \frac{dy}{du} \cdot \frac{du}{dx} = -\frac{1}{u^2} \cdot \frac{2}{x^3}$$

When $x = \dfrac{1}{2}$, $u = 3 - \dfrac{1}{\left(\frac{1}{2}\right)^2} = 3 - 4 = -1$,

$$\frac{dy}{dx} = \frac{-1}{(-1)^2} \cdot \frac{2}{\left(\frac{1}{2}\right)^3} = -16$$

21. $f(x) = (2x + 3)^{1.4}$

$$f'(x) = 1.4(2x + 3)^{0.4} \frac{d}{dx}(2x + 3)$$

$$= 1.4(2x + 3)^{0.4} \cdot 2$$

$$= 2.8(2x + 3)^{0.4}$$

23. $f(x) = (2x + 1)^4$,

$$f'(x) = 4(2x + 1)^3 \frac{d}{dx}(2 + 1) = 8(2x + 1)^3$$

25. $f(x) = (x^5 - 4x^3 - 7)^8$

$$f'(x) = 8(x^5 - 4x^3 - 7)^7 \frac{d}{dx}(x^5 - 4x^3 - 7)$$

$$= 8(x^5 - 4x^3 - 7)^7 (5x^4 - 12x^2)$$

$$= 8x^2(x^5 - 4x^3 - 7)^7 (5x^2 - 12)$$

27. $f(t) = \dfrac{1}{5t^2 - 6t + 2} = (5t^2 - 6t + 2)^{-1}$,

$$f'(t) = -(5t^2 - 6t + 2)^{-2} \frac{d}{dt}(5t^2 - 6t + 2)$$

$$= -\frac{10t - 6}{(5t^2 - 6t + 2)^2}$$

$$= -\frac{2(5t - 3)}{(5t^2 - 6t + 2)^2}$$

29. $g(x) = \dfrac{1}{\sqrt{4x^2 + 1}} = (4x^2 + 1)^{-1/2}$

$$g'(x) = -\frac{1}{2}(4x^2 + 1)^{-3/2} \frac{d}{dx}(4x^2 + 1)$$

$$= \frac{-8x}{2(4x^2 + 1)^{3/2}}$$

$$= \frac{-4x}{(4x^2 + 1)^{3/2}}$$

31. $f(x) = \dfrac{3}{(1 - x^2)^4} = 3(1 - x^2)^{-4}$,

$$f'(x) = -12(1 - x^2)^{-5} \frac{d}{dx}(1 - x^2)$$

$$= \frac{24x}{(1 - x^2)^5}$$

33. $h(s) = \left(1 + \sqrt{3s}\right)^5$

$$h'(s) = 5\left(1 + \sqrt{3s}\right)^4 \frac{d}{ds}\left(1 + \sqrt{3s}\right)$$

$$= 5\left(1 + \sqrt{3s}\right)^4 \frac{d}{ds}\left(1 + \sqrt{3}s^{1/2}\right)$$

$$= 5\left(1 + \sqrt{3s}\right)^4 \cdot \frac{\sqrt{3}}{2s^{1/2}}$$

$$= \frac{5\sqrt{3}\left(1 + \sqrt{3s}\right)^4}{2\sqrt{s}}\left(\frac{\sqrt{3}}{\sqrt{3}}\right)$$

$$= \frac{15(1 + \sqrt{3s})^4}{2\sqrt{3s}}$$

35. $f(x) = (x + 2)^3 (2x - 1)^5$

$$f'(x) = (x + 2)^3 \frac{d}{dx}(2x - 1)^5$$

$$+ (2x - 1)^5 \frac{d}{dx}(x + 2)^3$$

Now, $\dfrac{d}{dx}(2x - 1)^5 = 5(2x - 1)^4 \dfrac{d}{dx}(2x - 1)$

$$= 10(2x - 1)^4$$

and $\dfrac{d}{dx}(x+2)^3 = 3(x+2)^2\dfrac{d}{dx}(x+2)$

$\qquad\qquad = 3(x+2)^2.$

So, $f'(x) = 10(x+2)^3(2x-1)^4$

$\qquad\qquad + 3(2x-1)^5(x+2)^2$

$\qquad = (x+2)^2(2x-1)^4[10(x+2)$

$\qquad\qquad + 3(2x-1)]$

$\qquad = (x+2)^2(2x-1)^4(16x+17)$

37. $f(x) = \dfrac{(x+1)^5}{(1-x)^4}$

$f'(x) = \dfrac{(1-x)^4\frac{d}{dx}(x+1)^5 - (x+1)^5\frac{d}{dx}(1-x)^4}{[(1-x)^4]^2}$

Now, $\dfrac{d}{dx}(x+1)^5 = 5(x+1)^4\dfrac{d}{dx}(x+1)$

$\qquad\qquad\qquad = 5(x+1)^4$

and $\dfrac{d}{dx}(1-x)^4 = 4(1-x)^3\dfrac{d}{dx}(1-x)$

$\qquad\qquad\qquad = -4(1-x)^3.$

So, $f'(x)$

$= \dfrac{5(1-x)^4(x+1)^4 + 4(x+1)^5(1-x)^3}{(1-x)^8}$

$= \dfrac{(1-x)^3(x+1)^4[5(1-x)+4(x+1)]}{(1-x)^8}$

$= \dfrac{(x+1)^4(9-x)}{(1-x)^5}.$

39. $f(x) = \sqrt{3x+4} = (3x+4)^{1/2}$

$f'(x) = \dfrac{1}{2}(3x+4)^{-1/2}\cdot 3 = \dfrac{3}{2\sqrt{3x+4}}$

$m = f'(0) = \dfrac{3}{2\sqrt{3(0)+4}} = \dfrac{3}{4}$ and $f(0) = 2$

So, the equation of the tangent line at

$(0, 2)$ is $y = \dfrac{3}{4}x+2.$

41. $f(x) = (3x^2+1)^2$

$f'(x) = 2(3x^2+1)(6x)$

$m = f'(-1) = -48$ and $f(-1) = 16$, so the

equation of the tangent line at $(-1, 16)$ is

$y - 16 = -48(x+1)$, or $y = -48x - 32$.

43. $f(x) = \dfrac{1}{(2x-1)^6} = (2x-1)^{-6}$

$f'(x) = -6(2x-1)^{-5}(2) = -\dfrac{12}{(2x-1)^5}$

$m = f'(1) = -12$ and $f(1) = 1$, so the

equation of the tangent line at $(1, 1)$ is

$y - 1 = -12(x-1)$, or $y = -12x + 13$.

45. $f(x) = \sqrt[3]{\dfrac{x}{x+2}} = \left(\dfrac{x}{x+2}\right)^{1/3}$

$f'(x) = \dfrac{1}{3}\left(\dfrac{x}{x+2}\right)^{-2/3}\cdot\dfrac{(x+2)(1)(x)(1)}{(x+2)^2}$

$= \dfrac{(x+2)^{2/3}}{3x^{2/3}}\cdot\dfrac{2}{(x+2)^2}$

$= \dfrac{2}{3x^{2/3}(x+2)^{4/3}}$

$m = f'(-1) = \dfrac{2}{3}$ and $f(-1) = -1$, so the

equation of the tangent line at $(-1, -1)$ is

$y + 1 = \dfrac{2}{3}(x+1)$, or $y = \dfrac{2}{3}x - \dfrac{1}{3}.$

47. $f(x) = (x^2+x)^2$

$f'(x) = 2(x^2+x)(2x+1)$

$\qquad = 2x(x+1)(2x+1)$

$\qquad = 0$

when $x = -1$, $x = 0$, and $x = -\dfrac{1}{2}.$

49. $f(x) = \dfrac{x}{(3x-2)^2}$

$$f'(x) = \dfrac{(3x-2)^2(1) - (x)[2(3x-2)(3)]^2}{[(3x-2)^2]^2}$$

$$= \dfrac{(3x-2)[(3x-2) - 6x]}{(3x-2)^4}$$

$$= \dfrac{-3x-2}{(3x-2)^3}$$

$0 = \dfrac{-3x-2}{(3x-2)^3}$ when $-3x - 2 = 0$, or

$x = -\dfrac{2}{3}$.

51. $f(x) = \sqrt{x^2 - 4x + 5} = (x^2 - 4x + 5)^{1/2}$

$$f'(x) = \dfrac{1}{2}(x^2 - 4x + 5)^{-1/2}(2x - 4)$$

$$= \dfrac{2x-4}{2(x^2 - 4x + 5)^{1/2}}$$

$$= \dfrac{x-2}{(x^2 - 4x + 5)^{1/2}}$$

$0 = \dfrac{x-2}{(x^2 - 4x + 5)^{1/2}}$ when $x - 2 = 0$, or

$x = 2$.

53. $f(x) = (3x + 5)^2$

(a) $f'(x) = 2(3x + 5)(3) = 6(3x + 5)$

(b) $f(x) = (3x + 5)(3x + 5)$
$\quad f'(x) = (3x + 5)(3) + (3x + 5)(3)$
$\qquad = 6(3x + 5)$

55. $f(x) = (3x + 1)^5$

$f'(x) = 5(3x + 1)^4(3) = 15(3x + 1)^4$,

$f''(x) = 60(3x + 1)(3)^3 = 180(3x + 1)^3$

57. $h = (t^2 + 5)^8$

$\dfrac{dh}{dt} = 8(t^2 + 5)^7(2t) = 16t(t^2 + 5)^7$,

$\dfrac{d^2 h}{dt^2}$

$= 16t\left[7(t^2 + 5)^6(2t)\right] + (t^2 + 5)^7(16)$

$= 16(t^2 + 5)^6\left[14t^2 + (t^2 + 5)\right]$

$= 16(t^2 + 5)^6(15t^2 + 5)$

$= 80(t^2 + 5)^6(3t^2 + 1)$

59. $f(x) = \sqrt{1 + x^2} = (1 + x^2)^{1/2}$

$f'(x) = \dfrac{1}{2}(1 + x^2)^{-1/2}(2x)$

$\quad = \dfrac{x}{(1 + x^2)^{1/2}}$

$f''(x) = \dfrac{(1 + x^2)^{1/2}(1) - (x)}{1 + x}$

$\qquad \dfrac{\left[\dfrac{1}{2}(1 + x^2)^{-1/2}(2x)\right]}{1 + x}$

$= \dfrac{(1 + x^2)^{1/2} - \dfrac{x^2}{(1 + x^2)^{1/2}}}{1 + x^2} \cdot \dfrac{(1 + x^2)^{1/2}}{(1 + x^2)^{1/2}}$

$= \dfrac{1 + x^2 - x^2}{(1 + x^2)^{3/2}} = \dfrac{1}{(1 + x^2)^{3/2}}$

61. $h(x) = \sqrt{5x^2 + g(x)} = \left[5x^2 + g(x)\right]^{1/2}$

Using the general power rule,

$$h'(x) = \frac{1}{2}\left[5x^2 + g(x)\right]^{-1/2}\left[10x + g'(x)\right]$$

$$= \frac{10x + g'(x)}{2\sqrt{5x^2 + g(x)}}$$

Substituting $x = 0$, $g(0) = 4$ and

$g'(0) = 2$,

$$h'(0) = \frac{0+2}{2\sqrt{0+4}} = \frac{2}{4} = \frac{1}{2}$$

63. $h(x) = \left[3x + \dfrac{1}{g(x)}\right]^{3/2}$

Using the general power rule and noting

that the derivative of the term $\dfrac{1}{g(x)}$

requires the quotient rule (or a second application of the general power rule),

$$h'(x) = \frac{3}{2}\left[3x + \frac{1}{g(x)}\right]^{1/2}\left[3 + \frac{g(x)(0) - (1)g'(x)}{\left[g(x)\right]^2}\right]$$

$$= \frac{3}{2}\sqrt{3x + \frac{1}{g(x)}}\left(3 - \frac{g'(x)}{\left[g(x)\right]^2}\right)$$

Substituting $x = 1$, $g(1) = g'(1) = 1$,

$$h'(x) = \frac{3}{2}\sqrt{3(1) + \frac{1}{1}}\left(3 - \frac{1}{1^2}\right)$$

$$= (3)(2) = 6$$

65. (a) $f(t) = \sqrt{10t^2 + t + 229}$

$$= (10t^2 + t + 229)^{1/2}$$

The rate at which the earnings are growing is

$$f'(t) = \frac{1}{2}(10t^2 + t + 229)^{-1/2}(20t + 1)$$

$$= \frac{20t + 1}{2(10t^2 + t + 229)^{1/2}}\text{ thousand}$$

dollars per year.
The rate of growth in 2015 ($t = 5$) is

$$f'(5) = \frac{20(5) + 1}{2(10(5)^2 + 5 + 229)^{1/2}} \approx 2.295$$

or \$2,295 per year.

(b) The percentage rate of the earnings increases in 2015 was

$$100\frac{f'(5)}{f(5)} = \frac{100(2.295)}{\sqrt{10(5^2) + 5 + 229}}$$

$$\approx 10.4\% \text{ per year.}$$

67. $D(p) = \dfrac{4,374}{p^2} = 4,374p^{-2}$

(a) $\dfrac{dD}{dp} = -8,748p^{-3} = \dfrac{-8,784}{p^3}$

When the price is \$9,

$$\frac{dD}{dp} = \frac{-8,748}{(9)^3} = -12 \text{ pounds per}$$

dollar.

(b) $\dfrac{dD}{dt} = \dfrac{dD}{dp} \cdot \dfrac{dp}{dt}$

Now, $p(t) = 0.02t^2 + 0.1t + 6$

$$\frac{dp}{dt} = 0.04t + 0.1 \text{ dollars per week}$$

$$\frac{dD}{dt} = \frac{-8,748}{p^3}(0.04t + 0.1) \text{ pounds per}$$

week
When $t = 10$,

$$p(10) = 0.02(10)^2 = 0.1(10) + 6 = 9.$$

So, $\dfrac{dD}{dt} = \dfrac{-8748}{9^3}[0.04(10) + 0.1]$

$$= -6 \text{ pounds per week}$$

Since the rate is negative, demand will be decreasing.

69. $N(t) = \sqrt{t^2 + 3t + 6} = \left(t^2 + 3t + 6\right)^{1/2}$

Using the general power rule,

$N'(t) = \frac{1}{2}\left(t^2 + 3t + 6\right)^{-1/2}(2t + 3)$

$= \frac{2t + 3}{2\sqrt{t^2 + 3t + 6}}$

When $t = 2$,

$N'(2) = \frac{2(2) + 3}{2\sqrt{(2)^2 + 3(2) + 6}} = \frac{7}{2\sqrt{16}} = 0.875$

or 875 units per month.
Since $N'(2)$ is positive, production is increasing at this time.

71. $Q(K) = 500K^{2/3}$

$K(t) = \frac{2t^4 + 3t + 149}{t + 2}$

(a) $K(3) = \frac{2(3)^4 + 3(3) + 149}{3 + 2} = 64$ or $\$64,000$.

$Q(64) = 500(64)^{2/3} = 8,000$ units

(b) $\frac{dQ}{dt} = \frac{dQ}{dK} \cdot \frac{dK}{dt}$

$\frac{dQ}{dK} = 500\left(\frac{2}{3}K^{-1/3}\right) = \frac{1000}{3K^{1/3}}$

$\frac{dK}{dt}$

$= \frac{(t + 2)(8t^3 + 3) - (2t^4 + 3t + 149)(1)}{(t + 2)^2}$

$= \frac{6t^4 + 16t^3 - 143}{(t + 2)^2}$

When $t = 5$,

$K(5) = \frac{2(5)^4 + 3(5) + 149}{5 + 2} = 202.$

So,

$\frac{dQ}{dt} = \frac{1000}{3(202)^{1/3}} \cdot \frac{6(5)^4 + 16(5)^3 - 143}{(5 + 2)^2}$

$\approx 6,501$ units per month

Since $\frac{dQ}{dt}$ is positive when $t = 5$,
production will be increasing.

73. $A = 10,000\left(1 + \frac{0.01r}{12}\right)^{120}$

$= 10,000\left(1 + \frac{1}{1200}r\right)^{120}$

(a) $A' = 1,200,000\left(1 + \frac{1}{1200}r\right)^{119}\left(\frac{1}{1200}\right)$

$A' = 1,000\left(1 + \frac{1}{1200}(5)\right)^{119}$

$A'(5) = 1,000\left[1 + \frac{1}{1200}(5)\right]^{119}$

$\approx \$1,640.18$ per percent

(b)   When r goes from 5 to 6, the actual change in the amount is $A(6) - A(5)$.

$A(6) = 10,000\left[1 + \frac{0.01(6)}{12}\right]^{120} \approx 18,193.9673$

$A(5) = 10,000\left[1 + \frac{0.01(5)}{12}\right]^{120} \approx 16,470.0950$

The actual change is approximately $\$1,723.87$.

75. $p(t) = 20 - \dfrac{6}{t+1} = 20 - 6(t+1)^{-1}$

$$c(p) = 0.5\sqrt{p^2 + p + 58}$$
$$= 0.5(p^2 + p + 58)^{1/2}$$

(a) $\dfrac{dc}{dp} = \dfrac{1}{4}(p^2 + p + 58)^{-1/2}(2p+1)$

$$= \dfrac{2p+1}{4\sqrt{p^2 + p + 58}}$$

When $p = 18$,

$$\dfrac{dc}{dp} = \dfrac{2(18)+1}{4\sqrt{18^2 + 18 + 58}}$$
$$= \dfrac{37}{80}$$
$$= 0.4625 \text{ ppm/thous people}$$

(b) $\dfrac{dc}{dt} = \dfrac{dc}{dp} \cdot \dfrac{dp}{dt}$

$$\dfrac{dp}{dt} = 0 + 6(t+1)^{-2} \cdot 1 = \dfrac{6}{(t+1)^2}$$

$$\dfrac{dc}{dt} = \dfrac{2p+1}{4\sqrt{p^2 + p + 58}} \cdot \dfrac{6}{(t+1)^2}$$

When $t = 2$, $p(2) = 20 - \dfrac{6}{2+1} = 18$ and

$$\dfrac{dc}{dt} = (0.4625) \cdot \dfrac{6}{(2+1)^2}$$
$$\approx 0.308 \text{ ppm/year}$$

Since $\dfrac{dc}{dt}$ is positive, the level is increasing.

77. $L = 0.25 w^{2.6}$; $w = 3 + 0.21A$

(a) $\dfrac{dL}{dw} = 0.65 w^{1.6}$ mm per kg

When $w = 60$,

$$\dfrac{dL}{dw} = 0.65(60)^{1.6} \approx 455 \text{ mm per kg.}$$

(b) When $A = 100$,

$w = 3 + 0.21(100) = 24$ and

$L(24) = 0.25(24)^{2.6} \approx 969$ mm long.

$$\dfrac{dL}{dA} = \dfrac{dL}{dw} \cdot \dfrac{dw}{dA}$$

Since $\dfrac{dw}{dA} = 0.21$,

$$\dfrac{dL}{dA} = (0.65 w^{1.6})(0.21).$$

When $A = 100$, since $w = 24$,

$$\dfrac{dL}{dA} = 0.65(24)^{1.6}(0.21) \approx 22.1.$$

The tiger's length is increasing at the rate of about 22.1 mm per day.

79. $P(t) = 1 - \dfrac{12}{t+12} + \dfrac{144}{(t+12)^2}$

(a) $P(t) = 1 - 12(t+12)^{-1} + 144(t+12)^{-2}$

$P' = 0 + 12(t+12)^{-2} \cdot 1 - 288(t+12)^{-3} \cdot 1$

$$= \dfrac{12}{(t+12)^2} - \dfrac{288}{(t+12)^3}$$

When $t = 10$,

$$P'(10) = \dfrac{12}{(10+12)^2} - \dfrac{288}{(10+12)^3}$$
$$\approx -0.002254$$
$$= -0.2254\% \text{ per day}$$

where the negative sign indicates that the proportion is decreasing.

(b) $P'(15) = \dfrac{12}{(15+12)^2} - \dfrac{288}{(15+12)^3}$

$$\approx 0.001829$$

Since this value is positive, the proportion is increasing.

(c) $\displaystyle\lim_{t \to +\infty} P(t) = \lim_{t \to +\infty} 1 - \dfrac{12}{t+12} + \dfrac{144}{(t+12)^2}$

$$= 1 - 0 + 0$$
$$= 1$$

Since $P(0) = 1$, this is the normal level in the lake.

81. $V(T) = 0.41(-0.01T^2 + 0.4T + 3.52)$

$m(V) = \dfrac{0.39V}{1 + 0.09V}$

(a) $\dfrac{dV}{dt} = 0.41(-0.02T + 0.4)$ cm^3 per °C

(b) $\dfrac{dm}{dV} = \dfrac{(1 + 0.09V)(0.39) - (0.39V)(0.09)}{(1 + 0.09V)^2}$

$= \dfrac{0.39}{(1 + 0.09V)^2}$ gm per cm^3

(c) When $T = 10$,

$V(10)$

$= 0.41[-0.01(10)^2 + 0.4(10) + 3.52]$

$= 2.6732$ cm^3

$\dfrac{dm}{dT} = \dfrac{dm}{dV} \cdot \dfrac{dV}{dt}$

$= \dfrac{0.39}{(1 + 0.09V)^2}$

$\cdot 0.41(-0.02T + 0.4)$

When $T = 10$,

$\dfrac{dm}{dT} = \dfrac{0.39}{[1 + 0.09(2.6732)]^2}$

$\cdot 0.41[-0.02(10) + 0.4]$

$= 0.02078$ gm per °C

83. $s(t) = (3 + t - t^2)^{3/2}$, $0 \le t \le 2$

(a) $v(t) = s'(t) = \dfrac{3}{2}(3 + t - t^2)^{1/2}(1 - 2t)$

$a(t) = v'(t)$

$= \dfrac{3}{2}\left[(3 + t - t^2)^{1/2}(-2) + (1 - 2t)\dfrac{1}{2}(3 + t - t^2)^{-1/2}(1 - 2t) \right]$

$= \dfrac{3}{2}\left[-2(3 + t - t^2)^{1/2}\dfrac{2(3 + t - t^2)^{1/2}}{2(3 + t - t^2)^{1/2}} + \dfrac{(1 - 2t)^2}{2(3 + t - t^2)^{1/2}} \right]$

$= \dfrac{3}{2}\left[\dfrac{-4(3 + t - t^2) + (1 - 2t)^2}{2(3 + t - t^2)^{1/2}} \right]$

$= \dfrac{3}{2}\left[\dfrac{-12 - 4t + 4t^2 + 1 - 4t + 4t^2}{2(3 + t - t^2)^{1/2}} \right]$

$= \dfrac{24t^2 - 24t - 33}{4\sqrt{3 + t - t^2}}$

(b) To find when object is stationary for

$0 \le t \le 2,\ \dfrac{3}{2}\sqrt{3+t-t^2}\,(1-2t) = 0.$

Press $\boxed{y =}$ and input

$1.5\sqrt{\ \ }\,(3+x-x^2)*(1-2x)$ for $y_1 =$

Use window dimensions [−5, 5]1 by [−5, 5]1
Use the zero function under calc menu to find the only x-intercept occurs at

$x = \dfrac{1}{2}.$

(Note: algebraically, $\sqrt{3+t-t^2} = 0$

when $t = \dfrac{1+\sqrt{13}}{2}$, but this value is not

in the domain.)

Object is stationary when $t = \dfrac{1}{2}.$

$$s\!\left(\frac{1}{2}\right) = \left[3 + \frac{1}{2} - \left(\frac{1}{2}\right)^2\right]^{3/2}$$

$$= \frac{\sqrt{2197}}{8}$$

$$\approx 5.859$$

$$a\!\left(\frac{1}{2}\right) = \frac{24\left(\frac{1}{2}\right)^2 - 24\left(\frac{1}{2}\right) - 33}{4\sqrt{3 + \frac{1}{2} - \left(\frac{1}{2}\right)^2}}$$

$$= \frac{-39}{2\sqrt{13}}$$

$$= \frac{-3\sqrt{13}}{2}$$

$$\approx -5.4083$$

For $a\!\left(\dfrac{1}{2}\right)$ you can use the $\dfrac{dy}{dx}$

function under the calc menu and enter $x = .5$ to find

$$v'\!\left(\frac{1}{2}\right) = a\!\left(\frac{1}{2}\right) \approx -5.4083.$$

(c) To find when the acceleration is zero

for $0 \le t \le 2,\ \dfrac{24t^2 - 24t - 33}{4\sqrt{3+t-t^2}} = 0.$

Press $\boxed{y =}$ and input

$(24x^2 - 24x - 33)/(4\sqrt{\ \ }\,(3+x-x^2))$

for $y_2 =$
Press $\boxed{\text{Graph}}$
You may wish to deactivate y_1 so only the graph of y_2 is shown.
Use zero function under the calc menu to find the x-intercepts are $x \approx -0.775$ and $x \approx 1.77$. (Disregard $x = -0.775$.)
The acceleration is zero for $t = 1.77$,

$s(1.77) = (3+1.77-(1.77)^2)^{3/2} \approx 2.09$

Reactivate y_1 and use the value function under the calc menu. Make sure that y_1 is displayed in the upper left corner and enter $x = 1.77$ to find $v(1.77)$ $\approx -4.87.$

(d) We already have $v(t)$ inputted for $y_1 =$ and $a(t)$ inputted for $y_2 =$
Pres $\boxed{y =}$ and input
$(3+x-x^2)\wedge(3/2)$ for $y_3 =$
Use window dimensions [0, 2]1 by [−5, 5]1
Press $\boxed{\text{Graph}}$

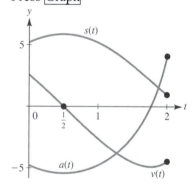

(e) To determine when $v(t)$ and $a(t)$ have opposite signs, press $\boxed{y =}$ and deactivate $y_3 =$ so only $v(t)$ and $a(t)$ are shown. Press graph. We see from the graph, $v(t)$ and $a(t)$ have opposite signs in two intervals. We know the t-intercept of $v(t)$ is $t = \dfrac{1}{2}$ and the t-intercept of $a(t)$ is $t = 1.77$. The object is slowing down for $0 \le t < 0.5$ and $1.77 < t \le 2$.

85. To prove that $\dfrac{d}{dx}[h(x)]^2 = 2h(x)h'(x)$, use the product rule to get

$$\frac{d}{dx}[h(x)]^2 = \frac{d}{dx}[h(x)h(x)]$$
$$= h(x)h'(x) + h'(x)h(x)$$
$$= 2h(x)h'(x).$$

87. To use numeric differentiation to calculate $f'(1)$ and $f'(-3)$, press y= and input $(3.1x^2 + 19.4)^{\wedge}(1/3)$ for $y_1 =$
Use the window dimensions $[-5, 5]1$ by $[-3, 8]1$
Press Graph
Use the dy/dx function under the calc menu and enter $x = 1$ to find $f'(1) \approx 0.2593$.
Repeat this for $x = -3$ to find $f'(-3) \approx -0.474$
Since there is only one minimum, we can conclude the graph has only one horizontal tangent.

89. $f'(x) = \dfrac{1}{1+x^2}$

$g(x) = f(2x+1)$

$g'(x) = f'(2x+1) \cdot (2x+1)'$

$g'(x) = \dfrac{1}{1+(2x+1)^2} \cdot 2 = \dfrac{2}{1+(2x+1)^2}$

2.5   Marginal Analysis and Approximations Using Increments

1. $C(x) = \dfrac{1}{5}x^2 + 4x + 57;$

$p(x) = \dfrac{1}{4}(48 - x) = 12 - \dfrac{1}{4}x$

(a) Marginal cost $= C'(x) = \dfrac{2}{5}x + 4$

Revenue = (# sold)(selling price)

$$R(x) = x\left(12 - \frac{1}{4}x\right) = 12x - \frac{1}{4}x^2$$

Marginal revenue $= R'(x) = 12 - \dfrac{1}{2}x$

(b) Estimated cost of 21st unit
$$= C'(20) = \frac{2}{5}(20) + 4 = \$12$$
Actual cost of 21st unit is
$C(21) - C(20)$
$$C(21) = \frac{1}{5}(21)^2 + 4(21) + 57$$
$$= 229.20$$
$$C(20) = \frac{1}{5}(20)^2 + 4(20) + 57$$
$$= 217.00$$
So, actual cost is $12.20

(c) Estimated revenue from sale of 21st unit
$$= R'(20) = 12 - \frac{1}{2}(20) = \$2$$

Actual revenue from sale of 21st unit
$$= R(21) - R(20)$$
$$R(21) = 12(21) - \frac{1}{4}(21)^2 = 141.75$$
$$R(20) = 12(20) - \frac{1}{4}(20)^2 = 140.00$$
So, actual revenue is $1.75

3. $C(x) = \frac{1}{3}x^2 + 2x + 39$;
$$p(x) = -x^2 - 10x + 4,000$$

(a) Marginal cost $= C'(x) = \frac{2}{3}x + 2$

Revenue = (# sold)(selling price)
$$R(x) = x\left(-x^2 - 10x + 4,000\right)$$
$$= -x^3 - 10x^2 + 4,000x$$
Marginal revenue
$$= R'(x) = -3x^2 - 20x + 4,000$$

(b) Estimated cost of 21st unit
$$= C'(20) = \frac{2}{3}(20) + 2 \approx \$15.33$$
Actual cost of 21st unit
$$= C(21) - C(20)$$
$$C(21) = \frac{1}{3}(21)^2 + 2(21) + 39 = 228.00$$
$$C(20) = \frac{1}{3}(20)^2 + 2(20) + 39 \approx 212.33$$
So, actual cost is $15.67

(c) Estimated revenue from sale of 21st unit
$$= R'(20) = -3(20)^2 - 20(20) + 4,000$$
$$= \$2,400$$
Actual revenue from sale of 21st unit

$$= R(21) - R(20)$$
$$R(21) = -(21)^3 - 10(21)^2 + 4,000(21)$$
$$= 70,329$$
$$R(20) = -(20)^3 - 10(20)^2 + 4,000(20)$$
$$= 68,000$$
So, actual revenue is $2,329.

5. $C(x) = \frac{1}{4}x^2 + 43$; $\quad p(x) = \frac{3 + 2x}{1 + x}$

(a) Marginal cost $= C'(x) = \frac{1}{2}x$

Revenue = (# sold)(selling price)
$$R(x) = x\left(\frac{3 + 2x}{1 + x}\right) = \frac{3x + 2x^2}{1 + x}$$
Marginal revenue
$$R'(x) = \frac{(1 + x)(3 + 4x) - (3x + 2x^2)(1)}{(1 + x)^2}$$
$$= \frac{3 + 7x + 4x^2 - 3x - 2x^2}{(1 + x)^2}$$
$$= \frac{2x^2 + 4x + 3}{(1 + x)^2}$$

(b) Estimated cost of 21st unit
$$= C'(20) = \frac{1}{2}(20) = \$10.00$$
Actual cost of 21st unit
$$= C(21) - C(20)$$
$$C(21) = \frac{1}{4}(21)^2 + 43 = 153.25$$
$$C(20) = \frac{1}{4}(20)^2 + 43 = 143.00$$
So, the actual cost is $10.25.

(c) Estimated revenue from sale of 21st unit
$$= R'(20) = \frac{2(20)^2 + 4(20) + 3}{(1 + 20)^2} \approx \$2.00$$

Actual revenue from sale of 21st unit

$= R(21) - R(20)$

$R(21) = \dfrac{3(21) + 2(21)^2}{1 + 21} \approx 42.95$

$R(20) = \dfrac{3(20) + 2(20)^2}{1 + 20} \approx 40.95$

So, the actual revenue is \$2.00.

7. $f(x) = x^2 - 3x + 5$; x increases from 5 to 5.3

$\Delta f \approx f'(x) \Delta x$

$f'(x) = 2x - 3$

$\Delta x = 5.3 - 5 = 0.3$

$\Delta f \approx [2(5) - 3](0.3) = 2.1$

9. $f(x) = x^2 + 2x - 9$; x increases from 4 to 4.3. Estimated percentage change is

$100 \dfrac{\Delta f}{f}$ where $\Delta f \approx f'(x) \Delta x$

$f'(x) = 2x + 2$, $\Delta x = 4.3 - 4 = 0.3$

$\Delta f \approx [2(4) + 2](0.3) = 3$

$f(4) = (4)^2 + 2(4) - 9 = 15$

$100 \dfrac{\Delta f}{f} = 100 \dfrac{3}{15} = 20\%$

11. $R(q) = 240q - 0.05q^2$

(a) Estimated revenue from 81st unit

$= R'(80)$

Now, $R'(q) = 240 - 0.1q$

So, $R'(80) = 240 - 0.1(80) = \$232$

Yes, since $R'(80)$ is positive, revenue is increasing and she should recommend increasing production.

(b) Actual revenue from 81st unit

$= R(81) - R(80)$

$R(81) = 240(81) - 0.05(81)^2 = 19,111.95$

$R(80) = 240(80) - 0.05(80)^2 = 18,880.00$

So, the actual revenue is \$231.95 and the estimate is quite accurate.

13. $C(q) = 3q^2 + q + 500$

(a) $C'(q) = 6q + 1$

$C'(40) = 6(40) + 1 = \$241$

(b) $C(41) - C(40)$

$= [3(41)^2 + 41 + 500]$

$\qquad - [3(40)^2 + 40 + 500]$

$= \$244$

15. $R(q) = 240q - 0.05q^2$

$\Delta R \approx R'(q) \Delta q$

$R'(q) = 240 - 0.1q$

Since will decrease by 0.65 unit,

$\Delta q = -0.65$

$\Delta R \approx R'(80)(-0.65)$

$\quad = [240 - 0.1(80)](-0.65)$

$\quad = -150.8,$

or a decrease of approximately \$150.80.

17. $Q(K) = 600K^{1/2}$

$\Delta Q \approx Q'(K) \Delta K$

$Q'(K) = 300K^{-1/2} = \dfrac{300}{\sqrt{K}}$

Since K is measured in thousands of dollars, the current value of K is 900 and

$\Delta K = \dfrac{800}{1000} = 0.8$

$\Delta Q \approx Q'(900)(0.8)$

$\quad = \left(\dfrac{300}{\sqrt{900}} \right)(0.8)$

$\quad = 8,$

or an increase of approximately 8 units.

19. $Q = 3,000K^{1/2}L^{1/3}$

If the labor force is kept constant,

$Q = 3,000K^{1/2}(1,331)^{1/3}$

$\quad = 33,000K^{1/2}$

Increasing the capital expenditure causes a change in weekly output that can be estimated using

$\Delta Q \approx Q'(K_0)\Delta K$

Now, substituting $K_0 = 400$, $\Delta K = 10$ and using

$Q'(K) = 16,500K^{-1/2} = \dfrac{16,500}{\sqrt{K}}$

$\Delta Q \approx \dfrac{16,500}{\sqrt{400}}(10) = 8,250$ units

By comparison, if the capital expenditure is kept constant,

$Q = 3,000(400)^{1/2}L^{1/3}$

$\quad = 60,000L^{1/3}$

Increasing the labor force causes a change in weekly output that can be estimated using

$\Delta Q \approx Q'(L_0)\Delta L$

Now, substituting $L_0 = 1,331$,
$\Delta L = 10$, and using

$Q' = 20,000L^{-2/3} = \dfrac{20,000}{L^{2/3}}$

$\Delta Q = \dfrac{20,000}{(1,331)^{2/3}}(10) \approx 1,653$ units

So, capital expenditure should be increased by $10,000.

21. $C(q) = \dfrac{1}{6}q^3 + 642q + 400$

$\Delta C \approx C'(q)\Delta q$

We want to approximate Δq, so

$\Delta q \approx \dfrac{\Delta C}{C'(q)}$

$C'(q) = \dfrac{1}{2}q^2 + 642,$

$C'(4) = \dfrac{1}{2}(4)^2 + 642 = 650,$ and

$\Delta C = -130.$ So, $\Delta q \approx \dfrac{-130}{650} = -0.2,$ or increase production by 0.2 units.

23. $C(t) = 100t^2 + 400t + 5,000$

$\Delta C \approx C'(t)\Delta t$

$C'(t) = 200t + 400$

Since t is measured in years, the next six

months $= \dfrac{1}{2}$ year $= \Delta t$

$\Delta C \approx C'(0)\left(\dfrac{1}{2}\right)$

$\quad = [200(0) + 400]\left(\dfrac{1}{2}\right)$

$\quad = 200,$

or an increase of approximately 200 newspapers.

25. $P(t) = -t^3 + 9t^2 + 48t + 200$

(a) $R(t) = P'(t) = -3t^2 + 18t + 48$

(b) $R'(t) = -6t + 18$

(c) $\Delta R \approx R'(t_0)\Delta t$

Substituting $t_0 = 3$ and $\Delta t = \dfrac{1}{12}$,

$\Delta R \approx \left[-6(3) + 18\right]\left[\dfrac{1}{12}\right] = 0$

or no change in the growth rate is expected during the 1st month of the 4th year. The actual change in the growth rate during the 1st month

(b) For $\dfrac{dy}{dx}$ to be undefined, $x - 32y = 0$,
or $x = 32y$. Substituting into the original equation,

$$(32y)y = 16y^2 + 32y$$
$$16y^2 - 32y = 0$$
$$16y(y - 2) = 0$$
$$y = 0 \text{ or } y = 2$$

When $y = 0$, $x = 32(0) = 0$ and when $y = 2$, $x = 32(2) = 64$. So, there are vertical tangents at $(0, 0)$ and $(64, 2)$.

35. $x^2 + xy + y^2 = 3$

$$2x + x\frac{dy}{dx} + y \cdot 1 + 2y\frac{dy}{dx} = 0$$
$$\frac{dy}{dx} = \frac{-2x - y}{x + 2y}$$

(a) $\dfrac{-2x - y}{x + 2y} = 0$ when $-2x - y = 0$, or

$y = -2x$.
Substituting in the original equation,

$$x^2 - 2x^2 + 4x^2 = 3$$
$$3x^2 = 3$$
$$x = \pm 1$$

When $x = -1$, $y = -2(-1) = 2$, and when $x = 1$, $y = -2(1) = -2$. So, there are horizontal tangents at $(-1, 2)$ and $(1, -2)$.

(b) $x + 2y = 0$ when $x = -2y$.
Substituting in the original equation,

$$4y^2 - 2y^2 + y^2 = 3$$
$$3y^2 = 3$$
$$y = \pm 1$$

When $y = -1$, $x = -2(-1) = 2$, and when $y = 1$, $x = -2(1) = -2$. So, there are vertical tangents at $(-2, 1)$ and $(2, -1)$.

37. $x^2 + 3y^2 = 5$

$$2x + 6y\frac{dy}{dx} = 0$$
$$\frac{dy}{dx} = -\frac{x}{3y}$$

$$\frac{d^2y}{dx^2} = \frac{(3y)(-1) - (-x)\left(3\frac{dy}{dx}\right)}{(3y)^2}$$
$$= \frac{-3y + 3x\frac{dy}{dx}}{9y^2}$$
$$= \frac{-3y + 3x\left(\frac{-x}{3y}\right)}{9y^2}$$
$$= \frac{-3y - \frac{x^2}{y}}{9y^2} \cdot \frac{y}{y}$$
$$= \frac{-3y^2 - x^2}{9y^3}$$
$$= \frac{-(x^2 + 3y^2)}{9y^3}$$
$$= -\frac{5}{9y^3}$$

39. Need to find $\Delta y \approx \dfrac{dy}{dx}$.

Since Q is to remain constant, let c be the constant value of Q. Then

$$c = 0.08x^2 + 0.12xy + 0.03y^2$$
$$0 = 0.16x + 0.12x\frac{dy}{dx} + y \cdot 0.12 + 0.06t\frac{dy}{dx}$$
$$\frac{dy}{dx} = \frac{-0.16x - 0.12y}{0.12x + 0.06y}$$

Since $x = 80$ and $y = 200$, $\dfrac{dy}{dx} \approx -1.704$, or a decrease of 1.704 hours of unskilled labor.

41. $3p^2 - x^2 = 12$

$$6p\frac{dp}{dt} - 2x\frac{dx}{dt} = 0$$

When $p = 4$, $48 - x^2 = 12$, $x^2 = 36$, or

$x = 6$.

Substituting, $6(4)(0.87) - 2(6)\dfrac{dx}{dt} = 0$,

$20.88 - 12\dfrac{dx}{dt} = 0$

$\dfrac{dx}{dt} = \dfrac{20.88}{12} = 1.74$ or increasing at a rate of 174 units/month.

43. $D(p) = \dfrac{32{,}670}{2p+1} = 32{,}670(2p+1)^{-1}$

$p(t) = 0.04t^{3/2} + 44$

Need $\dfrac{dD}{dt}$ when $t = 25$.

$\dfrac{dD}{dt} = \dfrac{dD}{dp} \cdot \dfrac{dp}{dt}$

Now,

$\dfrac{dD}{dp} = -32{,}670(2p+1)^{-2}(2) = -\dfrac{65{,}340}{(2p+1)^2}$

$\dfrac{dp}{dt} = 0.06t^{1/2}$

When $t = 25$, $p = 0.04(25)^{3/2} + 44 = 49$,

so $\dfrac{dD}{dt} = -\dfrac{65{,}340}{[2(49)+1]^2} \cdot 0.06(25)^{1/2} = -2$,

or the demand will be decreasing at a rate of 2 toasters per month.

45. Since Q is to remain constant,

$C = 60K^{1/3}L^{2/3}$.

$0 = 60K^{1/3} \cdot \dfrac{2}{3}L^{-1/3}\dfrac{dL}{dt} + L^{2/3} \cdot 20K^{-2/3}\dfrac{dK}{dt}$

Substituting,

$0 = \dfrac{40(8)^{1/3}}{(1{,}000)^{1/3}}(25) + \dfrac{20(1{,}000)^{2/3}}{(8)^{2/3}}\dfrac{dK}{dt}$

$0 = 200 + 500\dfrac{dK}{dt}$

$\dfrac{dK}{dt} = -0.4$,

or decreasing at a rate of \$400 per week.

47. $V = \dfrac{4}{3}\pi r^3$

$\dfrac{dV}{dt} = 4\pi r^2\dfrac{dr}{dt}$

Substituting, $0.002\pi = 4\pi(0.005)^2\dfrac{dr}{dt}$

$\dfrac{dr}{dt} = \dfrac{0.002}{4(0.005)^2} = 20$,

or increasing at a rate of 20 mm per min.

49. $V = \dfrac{4}{3}\pi R^3$

$\dfrac{dV}{dt} = 4\pi R^2\dfrac{dR}{dt}$

Substituting,

$\dfrac{dV}{dt} = 4\pi(0.54)^2(0.13) \approx 0.476$ or increasing

at a rate of 0.476 cm^3 per month.

51. $M = 70w^{3/4}$

(a) $\dfrac{dM}{dt} = 52.5w^{-1/4}\dfrac{dw}{dt}$

Substituting,

$\dfrac{dM}{dt} = \dfrac{52.5}{(80)^{1/4}}(0.8) \approx 14.04$, or

increasing at a rate of 14.04 kcal per day^2.

(b) $\dfrac{dM}{dt} = \dfrac{52.5}{(50)^{1/4}}(-0.5) \approx -9.87$, or

decreasing at a rate of 9.87 kcal per day^2.

53. $v = \dfrac{K}{L}(R^2 - r^2)$

At the center of the vessel, $r = 0$ so

$v = \dfrac{K}{L}R^2 = KL^{-1}R^2$.

Using implicit differentiation with t as the variable,

$$\frac{dv}{dt} = K\left[L^{-1}\left(2R\frac{dR}{dt}\right) + R^2\left(-L^{-2}\frac{dL}{dt}\right)\right].$$

Since the speed is unaffected, $\frac{dv}{dt} = 0$ and

$$0 = K\left[\frac{2R}{L}\cdot\frac{dR}{dt} - \frac{R^2}{L^2}\cdot\frac{dL}{dt}\right]$$

$$0 = \frac{2R}{L}\cdot\frac{dR}{dt} - \frac{R^2}{L^2}\cdot\frac{dL}{dt}$$

Solving for the relative rate of change of L,

$$\frac{R^2}{L^2}\cdot\frac{dL}{dt} = \frac{2R}{L}\cdot\frac{dR}{dL}$$

$$\frac{\frac{dL}{dt}}{L} = 2\frac{\frac{dR}{dL}}{R}$$

or double the relative rate of change of R.

55. $F = kD^2\sqrt{A-C} = kD^2\left(A-C\right)^{1/2}$

(a) Treating A and D as constants,

$$\frac{dF}{dC} = \frac{1}{2}kD^2\left(A-C\right)^{-1/2}(-1)$$

$$= \frac{-kD^2}{2\sqrt{A-C}}$$

As C increases, the denominator increases, so F decreases.

(b) We need $100\dfrac{dF/dA}{F}$

Treating C and D as constants,

$$\frac{dF}{dA} = \frac{1}{2}kD^2\left(A-C\right)^{-1/2}(1)$$

$$= \frac{kD^2}{2\sqrt{A-C}}$$

$$100\frac{\frac{dF}{dA}}{F} = 100\frac{\frac{kD^2}{2\sqrt{A-C}}}{kD^2\sqrt{A-C}}$$

$$= \frac{50}{\left(A-C\right)}\%$$

57. Let x be the distance between the man and the base of the street light and L the length of the shadow. Because of similar triangles, $\dfrac{L}{6} = \dfrac{x+L}{12}$ or $L = x$.

So, $\dfrac{dL}{dt} = \dfrac{dx}{dt} = 4$, or increasing at a rate of 4 feet per second.

59. $PV = C$

Differentiating with respect to time,

$$P\cdot\frac{dV}{dt} + V\cdot\frac{dP}{dt} = 0$$

Solving for $\dfrac{dP}{dt}$ gives $\dfrac{dP}{dt} = \dfrac{-P\cdot\dfrac{dV}{dt}}{V}$

Substituting $P = 70$, $V = 40$ and $\dfrac{dV}{dt} = 12$,

$$\frac{dP}{dt} = \frac{-70(12)}{40} = -21 \text{ lb/in}^2/\text{sec}$$

Since $\dfrac{dP}{dt}$ s negative, the pressure is decreasing.

61. $\dfrac{x^2}{a^2} + \dfrac{y^2}{b^2} = 1$

$$\frac{2x}{a^2} + \frac{2y}{b^2}\frac{dy}{dx} = 0$$

$$\frac{dy}{dx} = \frac{\frac{-2x}{a^2}}{\frac{2y}{b^2}} = \frac{-b^2 x}{a^2 y}$$

Substituting, $\dfrac{dy}{dx} = \dfrac{-b^2 x_0}{a^2 y_0}$ and the equation of the tangent line is

$$y - y_0 = \frac{-b^2 x_0}{a^2 y_0}(x - x_0)$$

$$\frac{y_0 y}{b^2} - \frac{y_0^2}{b^2} = \frac{-x_0 x}{a^2} + \frac{x_0^2}{a^2}$$

$$\frac{x_0 x}{a^2} + \frac{y_0 y}{b^2} = \frac{x_0^2}{a^2} + \frac{y_0^2}{b^2}$$

So, $\dfrac{x_0 x}{a^2} + \dfrac{y_0 y}{b^2} = 1$.

63. $y = x^{r/s}$ or $y^s = x^r$

$$sy^{s-1}\frac{dy}{dx} = rx^{r-1}$$

$$\frac{dy}{dx} = \frac{rx^{r-1}}{sy^{s-1}}$$

$$= \frac{rx^{r-1}}{s(x^{r/s})^{s-1}}$$

$$= \frac{rx^{r-1}}{sx^{r-r/s}}$$

$$= \frac{r}{s}x^{(r-1)-(r-r/s)}$$

$$= \frac{r}{s}x^{r/s-1}$$

65. To use the graphing utility to graph $11x^2 + 4xy + 14y^2 = 21$, we must express y in terms of x.

$$11x^2 + 4xy + 14y^2 = 21$$

$$14y^2 + 4xy + 11x^2 - 21 = 0$$

Using the quadratic formula,

$$y = \frac{-4x \pm \sqrt{16x^2 - 4(14)(11x^2 - 21)}}{28}$$

$$= \frac{-2x \pm \sqrt{294 - 150x^2}}{14}$$

Press $\boxed{y=}$ and input

$(-2x + \sqrt{\ }(294 - 150x^2))/14$ for $y_1 =$

Input $(-2x - \sqrt{\ }(294 - 150x^2))/14$ for

$y_2 =$

Use window dimensions [−1.5, 1.5].5 by [−1.5, 1.5].5

Press $\boxed{\text{Graph}}$

Press $\boxed{\text{2nd}}$ $\boxed{\text{Draw}}$ and select tangent function. Enter $x = -1$. We see the equation of the line tangent at $(-1, 1)$ is approximately $y = .75x + 1.75$. To find the horizontal tangents.

$$22x + 4x\left(\frac{dy}{dx}\right) + 4y + 28y\left(\frac{dy}{dx}\right) = 0$$

$$\frac{dy}{dx} = \frac{-22x - 4y}{4x + 28y}$$

$$\frac{dy}{dx} = 0 \text{ when } y = -\frac{11}{2}x$$

$$11x^2 + 4x11x^2 + 4x\left(-\frac{11}{2}x\right) + 14\left(-\frac{11}{2}x\right)^2$$

$$= 21$$

Solving yields $x = \pm 0.226$ and $y = \mp 1.241$. The two horizontal tangents are at $y = -1.241$ and $y = 1.241$.

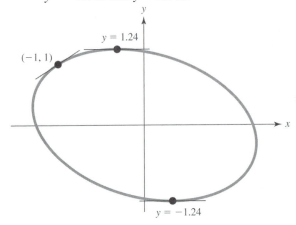

67. to use the graphing utility to graph curve

$$x^2 + y^2 = \sqrt{x^2 + y^2} + x,$$

it is best to graph $x^2 + y^2 = \sqrt{x^2 + y^2} + x$ using polar coordinates.

Given that $r^2 = x^2 + y^2$ and $r\cos\theta = x$, we change the equation to

$$r^2 = r + r\cos\theta$$

$$r(r - 1 - \cos\theta) = 0$$

$r = 0$ gives the origin and thus, we graph $r = 1 + \cos\theta$ using the graphing utility.

Press $\boxed{\text{2nd}}$ $\boxed{\text{format}}$ and select Polar Gc

Press $\boxed{\text{mode}}$ and select Pol

Pres $\boxed{y=}$ and input $1 + \cos\theta$

In the viewing window, use $\theta\min = 0$, $\theta\max = 2\pi$, $\theta\text{step} = \pi/24$ and dimensions [−1, 2].5 by [−1.5, 1.5].5

Using trace and zoom, it appears that a horizontal tangent is approximately

$y = \pm 1.23$.

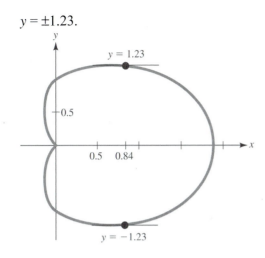

Checkup for Chapter 2

1. (a) $y = 3x^4 - 4\sqrt{x} + \dfrac{5}{x^2} - 7$

$y = 3x^4 - 4x^{1/2} + 5x^{-2} - 7$

$\dfrac{dy}{dx} = 12x^3 - 2x^{-1/2} - 10x^{-3} - 0$

$\dfrac{dy}{dx} = 12x^3 - \dfrac{2}{\sqrt{x}} - \dfrac{10}{x^3}$

(b) $y = (3x^3 - x + 1)(4 - x^2)$

$\dfrac{dy}{dx} = (3x^3 - x + 1)(-2x) + (4 - x^2)(9x^2 - 1)$

$\dfrac{dy}{dx}$

$= -6x^4 + 2x^2 - 2x + 36x^2 - 9x^4 - 4 + x^2$

$\dfrac{dy}{dx} = -15x^4 + 39x^2 - 2x - 4$

(c) $y = \dfrac{5x^2 - 3x + 2}{1 - 2x}$

$\dfrac{dy}{dx} = \dfrac{(1 - 2x)(10x - 3) - (5x^2 - 3x + 2)(-2)}{(1 - 2x)^2}$

$\dfrac{dy}{dx} = \dfrac{10x - 20x^2 - 3 + 6x + 10x^2 - 6x + 4}{(1 - 2x)^2}$

$\dfrac{dy}{dx} = \dfrac{-10x^2 + 10x + 1}{(1 - 2x)^2}$

(d) $y = (3 - 4x + 3x^2)^{3/2}$

$\dfrac{dy}{dx} = \dfrac{3}{2}(3 - 4x + 3x^2)^{1/2}(-4 + 6x)$

$\dfrac{dy}{dx} = (9x - 6)(3 - 4x + 3x^2)^{1/2}$

2. $f(t) = t(2t + 1)^2$

$f'(t) = t \cdot 2(2t + 1)(2) + (2t + 1)^2(1)$

$f'(t) = (2t + 1)(4t + 2t + 1)$

$f'(t) = (2t + 1)(6t + 1) = 12t^2 + 8t + 1$

$f''(t) = 24t + 8$

3. $y = x^2 - 2x + 1$

$\text{Slope} = \dfrac{dy}{dx} = 2x - 2$

When $x = -1$, $y = (-1)^2 - 2(-1) + 1 = 4$

and $\dfrac{dy}{dx} = 2(-1) - 2 = -4$. The equation of

the tangent line is $y - 4 = -4(x + 1)$, or

$y = -4x$.

4. $f(x) = \dfrac{x + 1}{1 - 5x}$

$f'(x) = \dfrac{(1 - 5x)(1) - (x + 1)(-5)}{(1 - 5x)^2}$

$f'(x) = \dfrac{1 - 5x + 5x + 5}{(1 - 5x)^2} = \dfrac{6}{(1 - 5x)^2}$

$f'(1) = \dfrac{6}{(1 - 5)^2} = \dfrac{3}{8}$

5. $T(x) = 3x^2 + 40x + 1800$

(a) $T'(x) = 6x + 40$
In 2013, $x = 3$ and
$T'(3) = 6(3) + 40 = \$58$ per year.

(b) Need $100 \dfrac{T'(3)}{T(3)}$

$T(3) = 3(3)^2 + 40(3) + 1800 = 1947$

$100\dfrac{T'(3)}{T(3)}=100\dfrac{58}{1947}\approx 2.98\%$ per year

6. $s(t)=2t^3-3t^2+2,\ t\ge 0$

(a) $v(t)=s'(t)=6t^2-6t$

$a(t)=s''(t)=12t-6$

(b) When stationary, $v(t)=0$

$6t^2-6t=0$

$6t(t-1)=0$, or $t=0,\ 1$

When retreating, $v(t)<0$

$6t^2-6t<0$

$6t(t-1)<0$, or $0<t<1$

When retreating, $v(t)>0$

$6t^2-6t>0$

$6t(t-1)>0$, or $t>1$

(c) $\left|s(1)-s(0)\right|+\left|s(2)-s(1)\right|=1+5=6$

7. $C(x)=0.04x^2+5x+73$

$C'(x)=0.08x+5$

$C'(4.09)=0.08(4.09)+5=5.3272$

From the marginal cost,

$\Delta C\approx C'\left(x_0\right)\Delta x$

$C'(4.09)(0.01)=0.053272$ thousand dollars

The actual cost is

$C(4.10)-C(4.09)$

$=94.1724-94.119124$

$=0.053276$ thousand dollars

8. $Q=500L^{3/4}$

$\Delta Q\approx Q'(L)\Delta L$

$Q'(L)=375L^{-1/4}=\dfrac{375}{L^{1/4}}$

$Q'(2401)=\dfrac{375}{(2401)^{1/4}}=\dfrac{375}{7}$

Since $\Delta L=200$,

$\Delta Q\approx\dfrac{375}{7}(200)=\dfrac{75,000}{7}$, or an increase of approximately 10,714.29 units.

9. $S=0.2029w^{0.425}$

$\dfrac{dS}{dt}=(0.2029)(0.425)w^{-0.575}\dfrac{dw}{dt}$

$=\dfrac{(0.2029)(0.425)}{(30)^{0.575}}(0.13)$

$\approx 0.001586,$

or an increasing at a rate of 0.001586 m^2 per week.

10. a. $V(r)=\dfrac{4}{3}\pi r^3$

$V'(r)=4\pi r^2$

$V'(0.75)=4\pi(0.75)^2$

$=2.25\pi$

≈ 7.069 cm^3 per cm

(b) $V=\dfrac{4}{3}\pi r^3$

Want $100\dfrac{\Delta V}{V}\le 8$, where

$\Delta V\approx V'(r)\Delta r,\ V'(r)=4\pi r^2$ and

$\Delta r=a\cdot r$, where a represents the % error in the measure of r (as a decimal).

$100\dfrac{\Delta V}{V}\le 8$

$100\dfrac{4\pi r^2\cdot ar}{\frac{4}{3}\pi r^3}\le 8$

$100a\le\dfrac{8}{3}$

or $\dfrac{8}{3}\%$ error in the measurement of r.

Review Exercises

1. $f(x) = x^2 - 3x + 1$

$$\frac{f(x+h) - f(x)}{h}$$

$$= \frac{[(x+h)^2 - 3(x+h) + 1] - (x^2 - 3x + 1)}{h}$$

$$= \frac{x^2 + 2xh + h^2 - 3x - 3h + 1 - x^2 + 3x + 1}{h}$$

$$= \frac{2xh + h^2 - 3h}{h}$$

$$= 2x + h - 3$$

$$f'(x) = \lim_{h \to 0} 2x + h - 3 = 2x - 3$$

3. $f(x) = 6x^4 - 7x^3 + 2x + \sqrt{2}$

$$f'(x) = 24x^3 - 21x^2 + 2$$

5. $y = \dfrac{2 - x^2}{3x^2 + 1}$

$$\frac{dy}{dx} = \frac{(3x^2 + 1)(-2x) - (2 - x^2)(6x)}{(3x^2 + 1)^2}$$

$$= \frac{-14x}{(3x^2 + 1)^2}$$

7. $f(x) = (5x^4 - 3x^2 + 2x + 1)^{10}$

$$f'(x)$$
$$= 10(5x^4 - 3x^2 + 2x + 1)^9 (20x^3 - 6x + 2)$$

9. $y = \left(x + \dfrac{1}{x}\right)^2 - \dfrac{5}{\sqrt{3x}}$

$$= (x + x^{-1})^2 - \frac{5}{\sqrt{3}} x^{-1/2}$$

$$\frac{dy}{dx} = 2(x + x^{-1})(1 - x^{-2}) + \frac{5}{2\sqrt{3}} x^{-3/2}$$

$$= 2\left(x + \frac{1}{x}\right)\left(1 - \frac{1}{x^2}\right) + \frac{5}{2\sqrt{3}x^{3/2}}$$

11. $f(x) = (3x+1)\sqrt{6x+5}$
$$= (3x+1)(6x+5)^{1/2}$$

$$f'(x) = (3x+1)\left(\frac{1}{2}\right)(6x+5)^{1/2}(6)$$
$$+ (6x+5)^{1/2}(3)$$

$$= \frac{3(3x+1)}{(6x+5)^{1/2}} + 3(6x+5)^{1/2}$$

$$= \frac{3(3x+1)}{\sqrt{6x+5}} + 3\sqrt{6x+5}$$

13. $y = \sqrt{\dfrac{1-2x}{3x+2}} = \left(\dfrac{1-2x}{3x+2}\right)^{1/2}$

$$\frac{dy}{dx} = \frac{1}{2}\left(\frac{1-2x}{3x+2}\right)^{-1/2}$$

$$\cdot \frac{(3x+2)(-2) - (1-2x)(3)}{(3x+2)^2}$$

$$= \frac{1}{2}\frac{(3x+2)^{1/2}}{(1-2x)^{1/2}} \cdot \frac{-7}{(3x+2)^2}$$

$$= \frac{-7}{2\sqrt{1-2x}\,(3x+2)^{3/2}}$$

15. $f(x) = \dfrac{4}{x-3}$

$$f'(x) = \frac{-4}{(x-3)^2}$$

$f(1) = -2$

The slope of the tangent line at $(1, -2)$ is $f'(1) = -1$. The equation of the tangent line is $y + 2 = -(x - 1)$, or $y = -x - 1$.

17. $f(x) = \sqrt{x^2 + 5} = (x^2 + 5)^{1/2}$

$$f'(x) = \frac{1}{2}(x^2 + 5)^{-1/2}(2x) = \frac{x}{\sqrt{x^2 + 5}}$$

$f(-2) = 3$

The slope of the tangent line at $(-2, 3)$ is $f'(-2) = -\dfrac{2}{3}$. The equation of the tangent line is $y - 3 = -\dfrac{2}{3}(x + 2)$, or $y = -\dfrac{2}{3}x + \dfrac{5}{3}$.

19.(a) $f(t) = t^3(t^2 - 1)$, $t = 0$

The rate of change of f is

$f'(t) = (t^3)(2t) + (t^2 - 1)(3t^2)$.

When $t = 0$, the rate is

$f'(0) = (0^3)(2 \cdot 0) + (0^2 - 1)(3 \cdot 0^2) = 0$.

(b) $f(t) = (t^2 - 3t + 6)^{1/2}$, $t = 1$

The rate of change of f is

$f'(t) = \dfrac{1}{2}(t^2 - 3t + 6)^{-1/2}(2t - 3)$

$= \dfrac{2t - 3}{2(t^2 - 3t + 6)^{1/2}}$.

When $t = 1$, the rate is

$f'(1) = \dfrac{2(1) - 3}{2\sqrt{1^2 - 3(1) + 6}} = -\dfrac{1}{4}$.

21. (a) $f(t) = t^2(3 - 2t)^3$

$f'(t) = t^2 \cdot 3(3 - 2t)^2(-2) + (3 - 2t)^3(2t)$

$f'(1) = 1 \cdot 3(3 - 2)^2(-2) + (3 - 2)^3(2)$

$= -4$

$f(1) = 1(3 - 2)^3 = 1$

$100\dfrac{f'(1)}{f(1)} = 100\dfrac{-4}{1} = -400\%$

(b) $f(t) = \dfrac{1}{t + 1} = (t + 1)^{-1}$

$f'(t) = -(t + 1)^{-2} = \dfrac{-1}{(t + 1)^2}$

$f'(0) = \dfrac{-1}{(0 + 1)^2} = -1$

$f(0) = \dfrac{1}{0 + 1} = 1$

$100\dfrac{f'(0)}{f(0)} = 100\dfrac{-1}{1} = -100\%$

23. (a) $y = (u + 1)^2$, $u = 1 - x$

$\dfrac{dy}{du} = 2(u + 1)(1)$, $\dfrac{du}{dx} = -1$

$\dfrac{dy}{dx} = \dfrac{dy}{du} \cdot \dfrac{dy}{dx} = 2(u + 1) \cdot -1 = -2(u + 1)$

Since $u = 1 - x$,

$\dfrac{dy}{dx} = -2[(1 - x) + 1] = -2(2 - x)$.

(b) $y = \dfrac{1}{\sqrt{u}} = u^{-1/2}$, $u = 2x + 1$

$\dfrac{dy}{du} = -\dfrac{1}{2}u^{-3/2}$, $\dfrac{du}{dx} = 2$

$\dfrac{dy}{dx} = \dfrac{dy}{du} \cdot \dfrac{du}{dx} = -\dfrac{1}{2u^{3/2}} \cdot 2 = -\dfrac{1}{u^{3/2}}$

$\dfrac{dy}{dx} = -\dfrac{1}{(2x + 1)^{3/2}}$

25. (a) $y = u^3 - 4u^2 + 5u + 2$, $u = x^2 + 1$

$\dfrac{dy}{du} = 3u^2 - 8u + 5$, $\dfrac{du}{dx} = 2x$,

$\dfrac{dy}{dx} = \dfrac{dy}{du} \cdot \dfrac{du}{dx}$

When $x = 1$, $u = 2$, and so

$\dfrac{dy}{dx} = [3(2^2) - 8(2) + 5][2(1)] = 2$.

(b) $y = \sqrt{u} = u^{1/2}$, $u = x^2 + 2x - 4$,

$\dfrac{dy}{du} = \dfrac{1}{2u^{1/2}}$, $\dfrac{du}{dx} = 2x + 2$,

$\dfrac{dy}{dx} = \dfrac{dy}{du} \cdot \dfrac{du}{dx}$

When $x = 2$, $u = 4$, and so

$\dfrac{dy}{dx} = \dfrac{1}{2(4)^{1/2}} \cdot [2(2) + 2] = \dfrac{3}{2}$.

27. (a) $f(x) = 4x^3 - 3x$

$f'(x) = 12x^2 - 3$

$f''(x) = 24x$

(b) $f(x) = 2x(x+4)^3$

$$f'(x) = (2x) \cdot 3(x+4)^2(1) + (x+4)^3(2)$$
$$= 2(x+4)^2[3x + (x+4)]$$
$$= 2(x+4)^2(4x+4)$$
$$= 8(x+4)^2(x+1)$$

$f''(x)$
$$= 8[(x+4)^2(1) + (x+1) \cdot 2(x+4)(1)]$$
$$= 8(x+4)[(x+4) + 2(x+1)]$$
$$= 8(x+4)(3x+6)$$
$$= 24(x+4)(x+2)$$

(c) $f(x) = \dfrac{x-1}{(x+1)^2}$

$$f'(x) = \frac{(x+1)^2(1) - (x-1) \cdot 2(x+1)(1)}{[(x+1)^2]^2}$$
$$= \frac{(x+1)[(x+1) - 2(x-1)]}{(x+1)^4}$$
$$= \frac{3-x}{(x+1)^3}$$

$f''(x)$
$$= \frac{(x+1)^3(-1) - (3-x) \cdot 3(x+1)^2(1)}{[(x+1)^3]^2}$$
$$= \frac{(x+1)^2[-(x+1) - 3(3-x)]}{(x+1)^6}$$
$$= \frac{2x-10}{(x+1)^4}$$
$$= \frac{2(x-5)}{(x+1)^4}$$

29. (a) $x^2 y = 1$, $\quad x^2 \dfrac{dy}{dx} + y(2x) = 0$

$$\frac{dy}{dx} = -\frac{2xy}{x^2} = -\frac{2y}{x}$$

(b) $(1 - 2xy^3)^5 = x + 4y$

$$5(1-2xy^3)^4\left(-2x \cdot 3y^2 \frac{dy}{dx} + y^3 \cdot -2\right)$$
$$= 1 + 4\frac{dy}{dx}$$

$$-30xy^2(1-2xy^3)^4 \frac{dy}{dx}$$
$$\qquad -10y^3(1-2xy^3)^4$$
$$= 1 + 4\frac{dy}{dx}$$

$$\frac{dy}{dx} = \frac{1 + 10y^3(1-2xy^3)^4}{-30xy^2(1-2xy^3)^4 - 4}$$

31. (a) $x^2 + 2y^3 = \dfrac{3}{xy}$, $(1, 1)$

$$x^2 + 2y^3 = 3(xy)^{-1}$$
$$2x + 6y^2 \cdot \frac{dy}{dx} = -3(xy)^{-2}\left(x \cdot \frac{dy}{dx} + y \cdot 1\right)$$
$$2x + 6y^2 \cdot \frac{dy}{dx} = \frac{-3\left(x \cdot \frac{dy}{dx} + y\right)}{(xy)^2}$$

When $x = 1$ and $y = 1$

$$2(1) + 6(1)^2 \cdot \frac{dy}{dx} = \frac{-3\left(1 \cdot \frac{dy}{dx} + 1\right)}{(1 \cdot 1)^2}$$
$$2 + 6 \cdot \frac{dy}{dx} = -3 \cdot \frac{dy}{dx} - 3$$
$$9 \cdot \frac{dy}{dx} = -5$$
$$\frac{dy}{dx} = -\frac{5}{9}$$

The slope of the tangent to the curve at $(1, 1)$ is $-\dfrac{5}{9}$.

(b) $y = \dfrac{x+y}{x-y}$, $(6, 2)$

$$\frac{dy}{dx} = \frac{(x-y)\left(1+\frac{dy}{dx}\right)-(x+y)\left(1-\frac{dy}{dx}\right)}{(x-y)^2}$$

$$\frac{dy}{dx} = \frac{x+x\frac{dy}{dx}-y-y\frac{dy}{dx}-\left(x-x\frac{dy}{dx}+y-y\frac{dy}{dx}\right)}{(x-y)^2}$$

$$\frac{dy}{dx} = \frac{2x\frac{dy}{dx}-2y}{(x-y)^2}$$

when $x = 6$ and $y = 2$.

$$\frac{dy}{dx} = \frac{2(6)\frac{dy}{dx}-2(2)}{(6-2)^2} = \frac{12\frac{dy}{dx}-4}{16}$$

$$16\frac{dy}{dx} = 12\frac{dy}{dx} - 4$$

$$4\frac{dy}{dx} = -4$$

$$\frac{dy}{dx} = -1$$

The slope of the tangent to the curve at (6, 2) is -1.

33. $3x^2 - 2y^2 = 6$, $6x - 4y\frac{dy}{dx} = 0$, or

$$\frac{dy}{dx} = \frac{3x}{2y}.$$

$$\frac{d^2y}{dx^2} = \frac{2y(3)-3x\left(2\frac{dy}{dx}\right)}{(2y)^2} = \frac{3y-3x\frac{dy}{dx}}{2y^2}$$

Since $\frac{dy}{dx} = \frac{3x}{2y}$,

$$\frac{d^2y}{dx^2} = \frac{3y-3x\left(\frac{3x}{2y}\right)}{2y^2} = \frac{6y^2-9x^2}{4y^3}.$$

From the original equation

$$6y^2 - 9x^2 = 3(2y^2 - 3x^2)$$
$$= -3(3x^2 - 2y^2)$$
$$= -3(6)$$
$$= -18$$

and so $\frac{d^2y}{dx^2} = -\frac{18}{4y^3} = -\frac{9}{2y^3}.$

35. $P(t) = -2t^3 + 9t^2 + 8t + 200$

(a) $P'(t) = -6t^2 + 18t + 8$

$P'(3) = -6(3)^2 + 18(3) + 8 = 8$, or increasing at a rate of 8,000 people per year.

(b) Letting R be rate of population growth, where

$$R(t) = P'(t) = -6t^2 + 18t + 8$$
$$R'(t) = -12t + 18$$
$$R'(3) = -12(3) + 18 = -18$$

or the rate of population growth is decreasing at a rate of 18,000 people per year.

37. $s(t) = \dfrac{2t+1}{t^2+12}$ for $0 \le t \le 4$

(a) $v(t) = \dfrac{(t^2+12)(2)-(2t+1)(2t)}{(t^2+12)^2}$

$$= \frac{-2t^2-2t+24}{(t^2+12)^2}$$

$$= \frac{-2(t+4)(t-3)}{(t^2+12)^2}$$

$$a(t) = \frac{(t^2+12)^2(-4t-2)}{(t^2+12)^4}$$

$$-\frac{(-2t^2-2t+24)2(t^2+12)(2t)}{(t^2+12)^4}$$

$$= -2(t^2+12)\left[\frac{(t^2+12)(2t+1)}{(t^2+12)^4}\right.$$

$$\left.+\frac{(-2t^2-2t+24)(2t)}{(t^2+12)^4}\right]$$

$$= \frac{2(2t^3+3t^2-72t-12)}{(t^2+12)^3}$$

Now, for $0 \le t \le 4$, $v(t) = 0$ when $t = 3$ and $a(t) \ne 0$.

When $0 \le t < 3$, $v(t) > 0$ and $a(t) < 0$, so the object is advancing and decelerating.

When $3 < t \le 4$, $v(t) < 0$ and $a(t) < 0$, so the object is retreating and decelerating.

(b) The distance for $0 < t < 3$ is
$$|s(3) - s(0)| = \left|\frac{1}{3} - \frac{1}{12}\right| = \frac{1}{4}.$$
The distance for $3 < t < 4$ is
$$|s(4) - s(3)| = \left|\frac{9}{28} - \frac{1}{3}\right| = \frac{1}{84}.$$
So, the total distance travelled is
$$\frac{1}{4} + \frac{1}{84} = \frac{22}{84} = \frac{11}{42}.$$

39. (a) $Q(x) = 50x^2 + 9{,}000x$
$$\Delta Q \approx Q'(x) = 100x + 9{,}000$$
$Q'(30) = 12{,}000$, or an increase of 12,000 units.

(b) The actual increase in output is
$Q(31) - Q(30) = 12{,}050$ units.

41. $Q(L) = 20{,}000L^{1/2}$
$$\Delta Q \approx Q'(L)\Delta L$$
$$Q'(L) = 10{,}000L^{-1/2} = \frac{10{,}000}{\sqrt{L}}$$
$$Q'(900) = \frac{10{,}000}{\sqrt{900}} = \frac{1{,}000}{3}$$
Since L will decrease to 885,
$$\Delta L = 885 - 900 = -15$$
$$\Delta Q \approx \left(\frac{1{,}000}{3}\right)(-15) = -5{,}000, \text{ or a}$$
decrease in output of 5,000 units.

43. Let A be the level of air pollution and p be the population.

$A = kp^2$, where k is a constant of proportionality
$$\Delta A \approx A'(p)\Delta p$$
$A'(p) = 2kp$ and $\Delta p = .05p$, so
$\Delta A \approx (2kp)(0.05p) = 0.1kp^2 = 0.1A$, or a 10% increase in air pollution.

45. $D = 36m^{-1.14}$

(a) $D = 36(70)^{-1.14} \approx 0.2837$ individuals per square kilometer.

(b) $(0.2837 \text{ individuals/km}^2)$
$$(9.2 \times 10^6)\text{km}^2$$
≈ 2.61 million people

(c) The ideal population density would be $36(30)^{-1.14} \approx 0.7454$ animals/km^2. Since the area of the island is 3,000 km^2, the number of animals on the island for the ideal population density would be
$(0.7454 \text{ animals/km}^2)(3{,}000 \text{ km}^2)$
$\approx 2{,}235$ animals
Since the animal population is given by $P(t) = 0.43t^2 + 13.37t + 200$, this population is reached when
$$2236 = 0.43t^2 + 13.37t + 200$$
$$0 = 0.43t^2 + 13.37t - 2036$$
or, using the quadratic formula, when $t \approx 55$ years. The rate the population is changing at this time is $P'(55)$, where
$P'(t) = 0.86t + 13.37$, or
$0.86(55) + 13.37 = 60.67$ animals per year.

47. Need $100\dfrac{\Delta L}{L}$, given that $100\dfrac{\Delta Q}{Q} = 1\%$, where $\Delta Q \approx Q'(L)\Delta L$. Since,
$100\dfrac{Q'(L)\Delta L}{Q(L)} = 1$, solving for ΔL yields

$\Delta L = \dfrac{Q(L)}{100Q'(L)}$ and

$100\dfrac{\Delta L}{L} = 100\dfrac{\frac{Q(L)}{100Q'(L)}}{L} = \dfrac{Q(L)}{Q'(L)\cdot L}.$

Since $Q(L) = 600L^{2/3}$,

$Q'(L) = 400L^{-1/3} = \dfrac{400}{L^{1/3}}$

$100\dfrac{\Delta L}{L} = \dfrac{600L^{2/3}}{\left(\frac{400}{L^{1/3}}\right)(L)} = \dfrac{3}{2}$, or 1.5%

Increase labor by approximately 1.5%.

49. Need $\Delta A \approx A'(r)\Delta r$

Since $A = \pi r^2$, $A'(r) = 2\pi r$.

When $r = 12$, $A'(12) = 2\pi(12) = 24\pi$.

Since $\Delta r = \pm 0.03r$,

$\Delta r = \pm 0.03(12) = \pm 0.36$ and

$\Delta A \approx (24\pi)(\pm 0.36) \approx \pm 27.14$ cm^2.

When $r = 12$,

$A = \pi(12)^2 = 144\pi \approx 452.39$ square centimeters. The calculation of area is off by ± 27.14 at most, or $\dfrac{27.14}{452.39} \approx 0.06$ or 6% and $425.25 \le A \le 479.53$

51. $Q = 600K^{1/2}L^{1/3}$

Need $100\dfrac{\Delta Q}{Q}$, where $\Delta Q \approx Q'(L)\Delta L$.

Treating K as a constant

$Q'(L) = 200K^{1/2}L^{-2/3} = \dfrac{200K^{1/2}}{L^{2/3}}$ with

$\Delta L = 0.02L.$

$100\dfrac{\Delta Q}{Q} = 100\dfrac{\left(\frac{200K^{1/2}}{L^{2/3}}\right)(0.02L)}{600K^{1/2}L^{1/3}} \approx 0.67\%$

53. The error in the calculation of the tumor's surface area, due to the error in measuring its radius is

$\Delta S \approx S'(r)\Delta r$

$\quad = 8\pi r(\Delta r)$

Since $3\%r = 0.03r = 0.03(1.2) = 0.036$

$\quad = 8\pi(1.2)(\pm 0.036)$

$\quad = \pm 0.3456\pi$

The calculated surface area is

$S = 4\pi(1.2)^2 = 5.76\pi$

The true surface area is between

$S + \Delta S = 5.76\pi \pm 0.3456\pi$, or

$17.01 \le S \le 19.18$

The measurement is accurate within

$\dfrac{0.3456\pi}{5.76\pi} = 0.06$, or 6%

55. $75x^2 + 17p^2 = 5{,}300$

$150x\dfrac{dx}{dt} + 34p\dfrac{dp}{dt} = 0$

$\dfrac{dx}{dt} = \dfrac{-34p\frac{dp}{dt}}{150x}$

When $p = 7$, $75x^2 + 17(7)^2 = 5{,}300$ or $x \approx 7.717513$.

So, $\dfrac{dx}{dt} = \dfrac{-34(7)(-0.75)}{150(7.717513)}$

≈ 0.15419 hundred, or

≈ 15.1419 units/month.

57. $P(t) = 20 - \dfrac{6}{t+1} = 20 - 6(t+1)^{-1}$

Need $100\dfrac{\Delta P}{P}$, where $\Delta P \approx P'(t)\Delta t$

$P'(t) = 6(t+1)^{-2}(1) = \dfrac{6}{(t+1)^2}$

The next quarter year is from $t = 0$ to $t = \dfrac{1}{4}$, so $P(0) = 14$, $P'(0) = 6$ and

$\Delta t = \dfrac{1}{4}.$

$100\dfrac{\Delta P}{P} = 100\dfrac{(6)\left(\frac{1}{4}\right)}{14} \approx 10.7\%$

59. $s(t) = 88t - 8t^2$

$v(t) = s'(t) = 88 - 16t$

The car is stopped when $v(t) = 0$, so
$0 = 88 - 16t$, or $t = 5.5$ seconds.
The distance travelled until it stops is
$s(5.5) = 88(5.5) - 8(5.5)^2 = 242$ feet.

61. $P(t) = -t^3 + 7t^2 + 200t + 300$

(a) $P'(t) = -3t^2 + 14t + 200$

$P'(5) = -3(5)^2 + 14(5) + 200 = 195$,

or increasing at a rate of $195 per unit
per month.

(b) $P''(t) = -6t + 14$

$P''(5) = -6(5) + 14 = -16$, or

decreasing at a rate of $16 per unit per
month per month.

(c) Need $\Delta P' \approx P''(t)\Delta t$

Now, $P''(5) = -16$ and the first six
months of the sixth year corresponds
to $\Delta t = \dfrac{1}{2}$.

$\Delta P' \approx (-16)\left(\dfrac{1}{2}\right) = -8$, or a decrease of

$8 per unit per month.

(d) Need $P'(5.5) - P'(5)$

$P'(5.5) = -3(5.5)^2 + 14(5.5) + 200$
$\qquad = 186.25$

The actual change in the rate of price
increase is $186.25 - 195 = -8.75$, or
decreasing at a rate of $8.75 per unit
per month.

63. $C(x) = 0.06x + 3x^{1/2} + 20$ hundred

$\dfrac{dx}{dt} = -11$ when $x = 2,500$

$\dfrac{dC}{dt} = \dfrac{dC}{dx} \cdot \dfrac{dx}{dt}$

$\dfrac{dC}{dx} = 0.06 + 1.5x^{-1/2} = 0.06 + \dfrac{1.5}{\sqrt{x}}$

$\dfrac{dC}{dt} = \left(0.06 + \dfrac{1.5}{\sqrt{2,500}}\right)(-11)$

$\qquad = -0.99$ hundred,

or decreasing at a rate of $99 per month.

65. Consider the volume of the shell as a

change in volume, where $r = \dfrac{8.5}{2}$ and

$\Delta r = \dfrac{1}{8} = 0.125$.

$\Delta V \approx V'(r)\Delta r$

$V(r) = \dfrac{4}{3}\pi r^3$

$V'(r) = 4\pi r^2$

$V'(4.25) = 4\pi(4.25)^2 = 72.25\pi$

$\Delta V = (72.25\pi)(0.125) \approx 28.37$ in^3

67. Let the length of string be the hypotenuse
of the right triangle formed by the
horizontal and vertical distance of the kite
from the child's hand. Then,

$s = x^2 + (80)^2$

$2s\dfrac{ds}{dt} = 2x\dfrac{dx}{dt}$

$\dfrac{ds}{dt} = \dfrac{2x\frac{dx}{dt}}{2s} = \dfrac{x\frac{dx}{dt}}{s}$

When $s = 100$, $(100)^2 = x^2 + (80)^2$, or
$x = 60$

$\dfrac{ds}{dt} = \dfrac{(60)(5)}{100} = 3$, or increasing at a rate of

3 feet per second.

69. Need $\dfrac{dx}{dt}$.

$$x^2 + y^2 = (10)^2$$

$$2x\frac{dx}{dt} + 2y\frac{dy}{dt} = 0$$

$$\frac{dx}{dt} = \frac{-2y\frac{dy}{dt}}{2x} = \frac{-y\frac{dy}{dt}}{x}$$

When $y = 6$, $x^2 + 36 = 100$, or $x = 8$.

Since $\dfrac{dy}{dt} = -3$, $\dfrac{dx}{dt} = \dfrac{(-6)(-3)}{8} = 2.25$, or increasing at a rate of 2.25 feet per second.

71. Let x be the distance from the player to third base. Then,

$$s^2 = x^2 + (90)^2$$

$$2s\frac{ds}{dt} = 2x\frac{dx}{dt}$$

$$\frac{ds}{dt} = \frac{2x\frac{dx}{dt}}{2s} = \frac{x\frac{dx}{dt}}{s}$$

When $x = 15$, $s^2 = (15)^2 + (90)^2$, or $s = \sqrt{8325}$.

$$\frac{ds}{dt} = \frac{(15)(-20)}{\sqrt{8325}} \approx -3.29, \text{ or decreasing at a}$$

rate of 3.29 feet per second.

73. Let x be the distance from point P to the object.

$V = ktx$

When $t = 5$ and $x = 20$, $V = 4$, so

$4 = k(5)(20)$, or $k = \dfrac{1}{25}$.

Since $a = V'$,

$$a = k\left(t\frac{dx}{dt} + x \cdot 1\right)$$

$$a = \frac{1}{25}(5 \cdot 4 + 20) = \frac{8}{5} \text{ ft/sec}^2$$

75. Need $100\dfrac{y'}{y}$ as $x \to \infty$.

$$y = mx + b$$

$$y' = m$$

$$100\frac{y'}{y} = 100\frac{m}{mx + b}$$

As x approaches ∞, this value approaches zero.

77. To use a graphing utility to graph f and f',

Press $\boxed{y =}$ and input

$(3x + 5)(2x \wedge 3 - 5x + 4)$ for $y_1 = f'(x)$

$= (3x + 5)(6x^2 - 5) + (3)(2x^3 - 5x + 4)$

Input $f'(x)$ for $y_2 =$

Use window dimensions $[-3, 2]1$ by $[-20, 30]10$

Use trace and zoom-in to find the x-intercepts of $f'(x)$ or use the zero function under the calc menu. In either case, make sure that y_2 is displayed in the upper left corner. The three zeros are $x \approx -1.78$, $x \approx -0.35$, and $x \approx 0.88$.

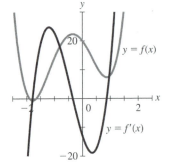

79. (a) To graph $y^2(2-x)=x^3$,

$$y^2 = \frac{x^3}{2-x}$$

$$y = \pm\sqrt{\frac{x^3}{2-x}}$$

Press $\boxed{y=}$ and input $\sqrt{}\,((x) \div (2-x))$ for $y_1 =$ and input $-y_1$ for $y_2 =$ (you can find y_1 by pressing $\boxed{\text{vars}}$ and selecting function under the y-vars menu). Use window dimensions $[-2, 5]1$ by $[-10, 10]5$ and press $\boxed{\text{graph}}$.

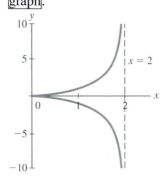

(b) With the graph shown, press $\boxed{\text{2nd}}$ $\boxed{\text{Draw}}$ and select the tangent function. Enter $x = 1$ to obtain the equation of the tangent lines to be $y = 2x - 1$ and $y = -2x + 1$.

(c) It can be seen as x approaches 2 from the left the portion of the graph above the x-axis approaches ∞ and the portion below the x-axis approaches $-\infty$.

(d) From the graph, the portion above the graph has a horizontal tangent of $y = 0$, as does the portion below the graph.

Chapter 3

Additional Applications of the Derivative

3.1 Increasing and Decreasing Functions; Relative Extrema

1. $f'(x) > 0$ when f is increasing, or $-2 < x < 2$

$f'(x) < 0$ when f is decreasing, or $x < -2$ and $x > 2$

3. $f'(x) > 0$ when f is increasing, or $x < -4$ and $0 < x < 2$

$f'(x) < 0$ when f is decreasing, or $-4 < x < -2$, $-2 < x < 0$, and $x > 2$

5. Function is decreasing, so $f'(x) < 0$ and graph of f' is below the x-axis. Function then levels, so $f'(x) = 0$ and graph of f crosses the x-axis. Function next increases for a period of time, so $f'(x) > 0$ and graph of f' is above the x-axis. Function then levels again, so $f'(x) = 0$ and graph of f' crosses the x-axis. Lastly, function decreases, so $f'(x) < 0$ and graph of f' is below the x-axis. Therefore, graph of f' is B.

7. Function is decreasing, so $f'(x) < 0$ and graph of f' is below the x-axis. Function then levels, so $f'(x) = 0$ and graph of f' crosses the x-axis. Function next increases, so $f'(x) > 0$ and graph of f' is above the x-axis. Therefore, graph of f' is D.

9. $f(x) = x^2 - 4x + 5$

$f'(x) = 2x - 4$

f is increasing when $f'(x) > 0$

$2x - 4 > 0$, or $x > 2$

f is decreasing when $f'(x) < 0$

$2x - 4 < 0$, or $x < 2$

11. $f(x) = x^3 - 3x - 4$

$f'(x) = 3x^2 - 3 = 3(x+1)(x-1)$

$f'(x) = 0$ when $x = -1, 1$

When $x < -1$, $f'(x) > 0$

$-1 < x < 1$, $f'(x) < 0$

$x > 1$, $f'(x) > 0$.

So, f is increasing when $x < -1$ and $x > 1$; f is decreasing when $-1 < x < 1$.

13. $g(t) = t^5 - 5t^4 + 100$

$g'(t) = 5t^4 - 20t^3 = 5t^3(t-4)$

$g'(t) = 0$ when $t = 0, 4$

When $t > 0$, $g'(t) > 0$

$0 < t < 4$, $g'(t) < 0$

$t > 4$, $g'(t) > 0$

So, g is increasing when $t < 0$ and $t > 4$; g is decreasing when $0 < t < 4$.

15. $f(t) = \dfrac{1}{4 - t^2} = (4 - t^2)^1$, defined for $t \neq -2, 2$

$f'(t) = -(4 - t^2)^{-2}(-2t)$

$= \dfrac{2t}{(4 - t^2)^2}$

$= \dfrac{2t}{[(2+t)(2-t)]^2}$

$f'(t) = 0$ when $t = 0$

When $t < -2$, $f'(t) < 0$

$-2 < t < 0$, $f'(t) < 0$

$0 < t < 2$, $f'(t) > 0$

$t > 2$, $f'(t) > 0$

So, f is increasing when $0 < t < 2$ and $t > 2$; f is decreasing when $t < -2$ and $-2 < t < 0$.

17. $h(u) = \sqrt{9 - u^2}$

$\quad = (9 - u^2)^{1/2}$

$\quad = [(3 + u)(3 - u)]^{1/2}$,

defined for $-3 \le u \le 3$

$h'(u) = \dfrac{1}{2}(9 - u^2)^{-1/2}(-2u) = \dfrac{-u}{\sqrt{9 - u^2}}$

$h'(u) = 0$ when $u = 0$

When $-3 < u < 0,\ h'(u) > 0$

$\quad\quad\quad 0,\ u < 3,\ h'(u) < 0$

So, h is increasing when $-3 < u < 0$; h is decreasing when $0 < u < 3$.

19. $F(x) = x + \dfrac{9}{x} = x + 9x^{-1} = \dfrac{x^2 + 9}{x}$, defined

when $x \ne 0$

$F'(x) = 1 - 9x^2$

$\quad = 1 - \dfrac{9}{x^2}$

$\quad = \dfrac{x^2 - 9}{x^2}$

$\quad = \dfrac{(x + 3)(x - 3)}{x^2}$

$F'(x) = 0$ when $x = -3,\ 3$

When $x < -3,\ F'(x) > 0$

$\quad\quad -3 < x < 0,\ F'(x) > 0$

$\quad\quad 0 < x < 3,\ F'(x) < 0$

$\quad\quad x > 3,\ F'(x) > 0$

So, F is increasing when $x < -3$ and $x > 3$; F is decreasing when $-3 < x < 0$ and $0 < x < 3$.

21. $f(x) = \sqrt{x} + \dfrac{1}{\sqrt{x}} = x^{1/2} + x^{-1/2} = \dfrac{x + 1}{\sqrt{x}}$,

defined for $x > 0$

$f'(x) = \dfrac{1}{2}x^{-1/2} - \dfrac{1}{2}x^{-3/2}$

$\quad = \dfrac{1}{2x^{1/2}} - \dfrac{1}{2x^{3/2}}$

$\quad = \dfrac{x - 1}{2x^{3/2}}$

$f'(x) = 0$ when $x = 1$

When $0 < x < 1,\ f'(x) < 0$

$\quad\quad x > 1,\ f'(x) > 0$

So, f is increasing when $x > 1$; f is decreasing when $0 < x < 1$.

23. $f(x) = 3x^4 - 8x^3 + 6x^2 + 2$

$f'(x) = 12x^3 - 24x^2 + 12x = 12x(x - 1)^2$

$f'(x) = 0$ when $x = 0,\ 1$

When $x < 0,\ f'(x) < 0$ so f decreasing

$\quad\quad 0 < x < 1,\ f'(x) > 0$ so f increasing

$\quad\quad x > 1,\ f'(x) > 0$ so f increasing

When $x = 0,\ f(0) = 2$ and the point $(0, 2)$ is a relative minimum. When $x = 1,\ f(1) = 3$, but there is no relative extremum at $(1, 3)$.

25. $f(t) = 2t^3 + 6t^2 + 6t + 5$

$f'(t) = 6t^2 + 12t + 6 = 6(t + 1)^2$

$f'(t) = 0$ when $t = -1$

When $t < -1,\ f'(t) > 0$ so f increasing

$\quad\quad t > -1,\ f'(t) > 0$ so f increasing

When $t = -1,\ f(-1) = 3$, but there is no relative extremum at $(-1, 3)$.

27. $g(x) = (x - 1)^5$

$g'(x) = 5(x - 1)^4(1)$

$g'(x) = 0$ when $x = 1$

When $x < 1,\ g'(x) > 0$ so g increasing

$\quad\quad x > 1,\ g'(x) > 0$ so g increasing

When $x = 1,\ g(1) = 0$, but there is no relative extremum at $(1, 0)$.

29. $f(t) = \dfrac{t}{t^2+3}$

$f'(t) = \dfrac{(t^2+3)(1)-(t)(2t)}{(t^2+3)^2} = \dfrac{3-t^2}{(t^2+3)^2}$

$f'(t) = 0$ when $t = \pm\sqrt{3}$

When

$t < -\sqrt{3},\ f'(t) < 0$ so f decreasing

$-\sqrt{3} < t < \sqrt{3},\ f'(t) > 0$ so f increasing

$t > \sqrt{3},\ f'(t) < 0$ so f decreasing

When $t = -\sqrt{3},\ f\left(-\sqrt{3}\right) = -\dfrac{\sqrt{3}}{6}$ and the

point $\left(-\sqrt{3}, -\dfrac{\sqrt{3}}{6}\right)$ is a relative

minimum.

When $t = \sqrt{3},\ f\left(\sqrt{3}\right) = \dfrac{\sqrt{3}}{6}$ and the point

$\left(\sqrt{3}, \dfrac{\sqrt{3}}{6}\right)$ is a relative maximum.

31. $h(t) = \dfrac{t^2}{t^2+t-2} = \dfrac{t^2}{(t+2)(t-1)}$ defined for

$t \neq -2, 1$

$h'(t) = \dfrac{(t^2+t-2)(2t)-(t^2)(2t+1)}{(t^2+t-2)^2}$

$= \dfrac{t(t-4)}{(t^2+t-2)^2}$

$h'(t) = 0$ when $t = 0, 4$

When $-2 < t < 0,\ h'(t) > 0$ so h increasing

$0 < t < 1,\ h'(t) < 0$ so h decreasing

$1 < t < 4,\ h'(t) < 0$ so h decreasing

$t > 4,\ h'(t) > 0$ so h increasing

When $t = 0,\ h(0) = 0$ and the point $(0, 0)$ is
a relative maximum.

When $t = 4,\ h(4) = \dfrac{8}{9}$ and the point

$\left(4, \dfrac{8}{9}\right)$ is a relative minimum.

33. $s(t) = (t^2-1)^4$

$s'(t) = 4(t^2-1)^3(2t) = 8t[(t+1)(t-1)]^3$

$s'(t) = 0$ when $t = -1, 0, 1$

When $t < -1,\ s'(t) < 0$ so s decreasing

$-1 < t < 0,\ s'(t) > 0$ so s increasing

$0 < t < 1,\ s'(t) < 0$ so s decreasing

$t > 1,\ s'(t) > 0$ so s increasing

When $t = -1,\ s(-1) = 0$ and the point
$(-1, 0)$ is a relative minimum. When $t = 0$,
$s(0) = 1$ and the point $(0, 1)$ is a relative
maximum. When $t = 1,\ s(1) = 0$ and the
point $(1, 0)$ is a relative minimum.

35. $f(x) = x^3 - 3x^2 = x^2(x-3)$,
intercepts: $(0, 0),\ (3, 0)$

$f'(x) = 3x^2 - 6x = 3x(x-2)$

$f'(x) = 0$ when $x = 0, 2$

When $x < 0,\ f'(x) > 0$ so f increasing

$x = 0,\ f'(x) = 0$ so f levels

$0 < x < 2,\ f'(x) < 0$ so f decreasing

$x = 2,\ f'(x) = 0$ so f levels

$x > 0,\ f'(x) > 0$ so f increasing

The point $(0, 0)$ is a relative maximum and
the point $(2, -4)$ is a relative minimum.

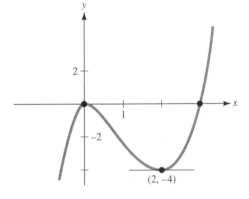

37. $f(x) = 3x^4 - 8x^3 + 6x^2 + 2$

When $x = 0,\ f(0) = 2$ so $(0, 2)$ is an

intercept.

$f(x) = 0$ is too difficult to solve.

$$f'(x) = 12x^3 - 24x^2 + 12x$$
$$= 12x(x-1)(x-1)$$

$f'(x) = 0$ when $x = 0, 1$

When $x < 0$, $f'(x) < 0$ so f decreasing

$\quad\quad x = 0$, $f'(x) = 0$ so f levels

$\quad\quad 0 < x < 1$, $f'(x) > 0$ so f increasing

$\quad\quad x = 1$, $f'(x) = 0$ so f levels

$\quad\quad x > 1$, $f'(x) > 0$ so f increasing

The point $(0, 2)$ is a relative minimum, but the point $(1, 3)$ is not a relative extremum.

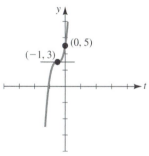

39. $f(t) = 2t^3 + 6t^2 + 6t + 5$

$$f'(t) = 6t^2 + 12t + 6 = 6(t+1)^2$$

$f'(t) = 0$ when $t = -1$

When $t < -1$, $f'(t) > 0$ so f increasing

$\quad\quad t = -1$, $f'(t) = 0$ so f levels

$\quad\quad t > -1$, $f'(t) > 0$ so f increasing

The point $(-1, 3)$ is not a relative extremum.

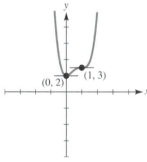

41. $g(t) = \dfrac{t}{t^2 + 3}$

$$g'(t) = \frac{(t^2+3)(1) - (t)(2t)}{t^2+3} = \frac{3 - t^2}{(t^2+3)^2}$$

$g'(t) = 0$ when $t = -\sqrt{3}, \sqrt{3}$

When

$t < -\sqrt{3}$, $f'(t) < 0$ so f decreasing

$t = -\sqrt{3}$, $f'(t) = 0$ so f levels

$-\sqrt{3} < t < \sqrt{3}$, $f'(t) > 0$ so f increasing

$t = \sqrt{3}$, $f'(t) = 0$ so f levels

$t > \sqrt{3}$, $f'(t) < 0$ so f decreasing

The point $\left(-\sqrt{3}, \dfrac{-\sqrt{3}}{6}\right)$ is a relative

minimum and the point $\left(\sqrt{3}, \dfrac{\sqrt{3}}{6}\right)$ is a

relative maximum.

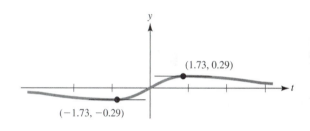

43. $f(x) = 3x^5 - 5x^3 + 4$

$$f'(x) = 15x^4 - 15x^2 = 15x^2(x+1)(x-1)$$

$f'(x) = 0$ when $x = -1, 0, 1$

When

$x < -1$, $f'(x) > 0$ so f increasing
$x = -1$, $f'(x) = 0$ so f levels
$-1 < x < 0$, $f'(x) < 0$ so f decreasing
$x = 0$, $f'(x) = 0$ so f levels
$0 < x < 1$, $f'(x) < 0$ so f decreasing
$x = 1$, $f'(x) = 0$ so f levels
$x > 1$, $f'(x) > 0$ so f increasing

The point $(-1, 6)$ is a relative maximum, the point $(0, 4)$ is not a relative extremum, and the point $(1, 2)$ is a relative minimum.

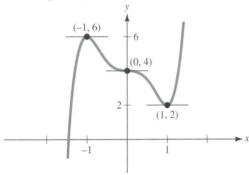

45. $f'(x) = x^2(4 - x^2) = x^2(2 + x)(2 - x)$

$f'(x) = 0$ when $x = -2, 0, 2$

When $x < -2$, $f'(x) < 0$ so f decreasing
  $-2 < x < 0$, $f'(x) > 0$ so f increasing
  $0 < x < 2$, $f'(x) > 0$ so f increasing
  $x > 2$, $f'(x) < 0$ so f decreasing

When $x = -2$, f has a relative minimum, and when $x = 0$, f does not have a relative extremum, and when $x = 2$, f has a relative maximum.

47. $f'(x) = \dfrac{(x+1)^2(4 - 3x)^3}{(x^2 + 1)^2}$

$f'(x) = 0$ when $x = -1, \dfrac{4}{3}$

When

$x < -1$, $f'(x) > 0$ so f increasing
$-1 < x < \dfrac{4}{3}$, $f'(x) > 0$ so f increasing
$x > \dfrac{4}{3}$, $f'(x) < 0$ so f decreasing

When $x = -1$, f does not have a relative extremum, and when $x = \dfrac{4}{3}$, f has a relative maximum.

49. When
$x < 1$, f is decreasing and the graph of f' is below x-axis
$x = 1$, f levels and graph of f' crosses the x-axis
$1 < x < 3$, f is increasing and graph of f' is above x-axis
$x = 3$, f levels and graph of f' crosses the x-axis
$x > 3$, f is decreasing and graph of f' is below x-axis

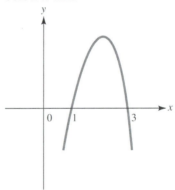

51. When
$x < 2$, f is decreasing and graph of f' is below x-axis
$x = 2$, f levels and graph of f' crosses the x-axis
$2 < x < 5$, f is increasing and graph of f'

is above *x*-axis

$x = 5$, f levels and graph of f' touches *x*-axis

$x > 5$, f is increasing and graph of f' is above *x*-axis

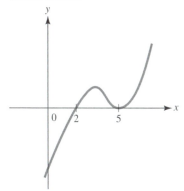

When

$0 \le x < 11$, $A'(x) < 0$ so A decreasing

$x > 11$, $A'(x) > 0$ so A increasing

(c) When $x = 11$, A has a relative minimum which is actually an absolute minimum. So the average cost is minimized when 11 units are produced. The corresponding minimum average cost is

$$A(11) = (11)^2 - 20(11) + 179 + \frac{242}{11}$$

$$= 102$$

or $102,000 per unit.

53. $C(x) = x^3 - 20x^2 + 179x + 242$

(a) $A(x) = \dfrac{C(x)}{x}$

$$= \frac{x^3 - 20x^2 + 179x + 242}{x}$$

$$= x^2 - 20x + 179 + \frac{242}{x}$$

$$= x^2 - 20x + 179 + 242x^{-1}$$

$$A'(x) = 2x - 20 - \frac{242}{x^2}$$

(b) $A'(x) = 0$ when

$$0 = 2x - 20 - \frac{242}{x^2}$$

$$0 = 2x^3 - 20x^2 - 242$$

$$x^2 - 10x^2 - 121 = 0$$

Press $\boxed{y =}$ and enter $x^3 - 10x^2 - 121$ for $y_1 =$.

Use window dimensions $[-10, 100]10$ by $[-500, 500]100$.

Press $\boxed{\text{graph}}$.

To find the zero (*x*-intercept), enter the zero function under the calc menu. Enter a left bound close to the *x*-intercept, a right bound close to the *x*-intercept and a guess. The *x*-intercept or zero is $x = 11$.

55. $R(x) = xp(x) = x(10 - 3x^2)$, $0 \le x \le 3$

$$R'(x) = x \cdot 2(10 - 3x)(-3) + (10 - 3x)^2(1)$$

$$= (10 - 3x)(-6x + 10 - 3x)$$

$$= (10 - 3x)(10 - 9x)$$

$R'(x) = 0$ when $x = \dfrac{10}{9}, \dfrac{10}{3}$

When

$0 \le x < \dfrac{10}{9}$, $R'(x) > 0$ so R increasing

$x = \dfrac{10}{9}$, $R'(x) = 0$ so R levels

$\dfrac{10}{9} < x \le 3$, $R'(x) < 0$ so R decreasing

The point $(1.11, 49.38)$ is a relative maximum, so revenue is maximized when approximately 1.11 hundred, or 111 units are produced.

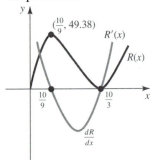

57. $S(x) = -2x^3 + 27x^2 + 132x + 207$,
$0 \le x \le 17$

(a) $S'(x) = -6x^2 + 54x + 132$
$= -6(x - 11)(x + 2)$

$S'(x) = 0$ when $x = -2, 11$
When
$0 \le x < 11$, $S'(x) > 0$ and S is increasing
$x = 11$, $S'(x) = 0$ and S is level
$11 < x \le 17$, $S'(x) < 0$ and S is decreasing
The point $(11, 2264)$ is a relative maximum.

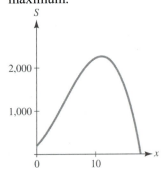

(b) $S(0) = 207$, or 207 units will sell.

(c) Since $(11, 2264)$ is a relative maximum, sales are maximized when

11 thousand, or \$11,000 are spent on advertising. The maximum number of units sold is 2,264.

59. $M(r) = \dfrac{1 + 0.05r}{1 + 0.004r^2}$

(a) $M'(r)$

$= \dfrac{(1 + 0.004r^2)(0.005) - (1 + 0.05r)(0.008r)}{(1 + 0.004r^2)^2}$

$= \dfrac{0.05 - 0.008r - 0.0002r^2}{(1 + 0.004r^2)^2}$

$= \dfrac{500 - 80r - 2r^2}{10,000(1 + 0.004r^2)^2}$

Using the quadratic formula,
$M'(r) = 0$ when

$$r = \frac{80 \pm \sqrt{(-80)^2 - 4(4)(-2)(500)}}{2(-2)}$$

$r \approx 5.495$ (rejecting the negative answer)

When

$0 \le r < 5.495$, $M'(r) > 0$ so M is increasing

$r > 5.495$, $M'(r) < 0$ so M is decreasing

(b) When $r \approx 5.495$, M has a relative maximum which is actually an absolute maximum. So, the number of mortgages is maximized when the rate is 5.495%. The corresponding maximum number of mortgages is

$$M(5.495) = \frac{1 + 0.05(5.495)}{1 + 0.004(5.495)^2}$$
$$\approx 1.137$$

or 1,137 refinanced mortgages.

61. (a) Approximately 1971, 1976, 1980, 1983, 1988, 1996

(b) Approximately 1973, 1979, 1981, 1985, 1989

(c) Approximately $\frac{1}{2}$% per year

(d) Approximately $\frac{1}{2}$% per year

63. $C(t) = \dfrac{0.15t}{t^2 + 0.81}$

Note that, since degree of numerator < degree of denominator, $y = 0$ is a horizontal asymptote.

$$C'(t) = \frac{(t^2 + 0.81)(0.15) - (0.15t)(2t)}{(t^2 + 0.81)^2}$$
$$= \frac{-0.15t^2 + 0.1215}{(t^2 + 0.81)^2}$$

$C'(t) = 0$ when $t = 0.9$.

When

$0 < t < 0.9$, $C'(t) > 0$ so C increasing

$t = 0.9$, $C'(t) = 0$ and C levels

$t > 0.9$, $C'(t) < 0$ and C decreasing

The point (0.9, 0.083) is a relative maximum, so the maximum concentration occurs when $t = 0.9$ hours.

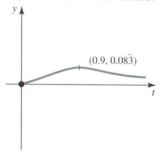

65. (a) $\text{Yield} = \begin{pmatrix} \text{orig} \\ \# \\ \text{fish} \end{pmatrix} \begin{pmatrix} \text{proportion} \\ \text{still} \\ \text{living} \end{pmatrix} \begin{pmatrix} \text{weight} \\ \text{per} \\ \text{fish} \end{pmatrix}$

$$Y(t) = 300 \left(\frac{31}{31 + t} \right)(3 + t - 0.05t^2)$$
$$= 9,300(31 + t)^{-1}(3 + t - 0.05t^2)$$

$$Y'(t) = 9,300[(31 + t)^{-1}(1 - 0.1t) + (3 + t - 0.05t^2)(31 + t)^{-2}(1)]$$
$$= 9,300\left(\frac{1 - 0.1t}{31 + t} - \frac{3 + t - 0.05t^2}{(31 + t)^2} \right)$$
$$= 9,300 \frac{28 - 3.1t - 0.05t^2}{(31 + t)^2}$$

$Y'(t) = 0$ when $t = 8$

When

$0 \le t < 8$, $Y'(t) > 0$ and Y is increasing

$t = 8$, $Y'(t) = 0$ and Y levels

$8 < t \le 10$, $Y'(t) < 0$ and Y is decreasing

The point $(8, 1860)$ is a relative maximum.

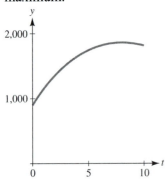

(b) Since $(8, 1860)$ is a relative maximum, the yield is maximized after 8 weeks and the maximum yield is 1,860 pounds.

67. $H(t) = -0.53T^2 + 25T - 209$, $15 \le T \le 30$

$H'(t) = -1.06T + 25$

When $H'(0) = 0$ when $t \approx 23.58$

When $15 \le T < 23.58$, $H'(T) > 0$ so H is increasing

$T = 23.58$, $H'(T) = 0$ so H levels

$23.58 < T \le 30$, $H'(t) < 0$ so H is decreasing.

The point $(23.58, 85.81)$ is a relative maximum.

So, the maximum percentage is 85.81% and it occurs at 23.58°C.

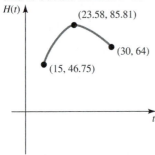

69. (a) Graph levels when $x = 0, 1, 2$.

(b) Graph is decreasing when $0 < x < 1$.

(c) Graph is increasing when $x < 0$, $1 < x < 2$, and $x > 2$.

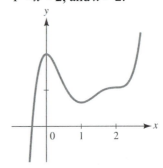

71. (a) Graph is decreasing when $x < -1$.

(b) Graph is increasing when $-1 < x < 3$ and $x > 3$.

(c) Graph levels when $x = -1, 3$.

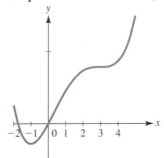

73. $f(x) = ax^3 + bx^2 + cx + d$

$f'(x) = 3ax^2 + 2bx + c$

$f'(x) = 0$ when $x = -2$, so

$0 = 3a(-2)^2 + 2b(-2) + c$

$0 = 12a - 4b + c$

$f'(x) = 0$ when $x = 1$, so

$0 = 3a(1)^2 + 2b(1) + c$

$0 = 3a + 2b + c$

So, $12a - 4b + c = 3a + 2b + c$, or $b = \dfrac{3}{2}a$.

Now, $f(-2) = 8$ so

$8 = a(-2)^3 + \dfrac{3}{2}a(-2)^2 + c(-2) + d$

$8 = -8a + 6a - 2c + d$

$8 = -2a - 2c + d$ or, $d = 8 + 2a + 2c$

Now, $f(1) = -19$ so

$$-19 = a(1)^3 + \frac{3}{2}a(1)^2 + c(1) + d$$

$$-19 = \frac{5}{2}a + c + (8 + 2a + 2c)$$

$$-27 = \frac{9}{2}a + 3c, \text{ or}$$

$$c = \frac{1}{3}\left(-27 - \frac{9}{2}a\right) = -9 - \frac{3}{2}a$$

Using

$$0 = 3a + 2b + c$$

$$0 = 3a + 2\left(\frac{3}{2}a\right) + \left(-9 - \frac{3}{2}a\right)$$

$$0 = \frac{9}{2}a - 9, \text{ or}$$

$$a = 2, \ b = \frac{3}{2}(2) = 3, \ c = -9 - \frac{3}{2}(2) = -12,$$

$$d = 8 + 2(2) + 2(-12) = -12$$

75. $f(x) = 1 - x^{3/5}$

When $x = 0$, $f(0) = 1$ so $(0, 1)$ is an intercept.

$f(x) = 0$, $x = 1$ so $(1, 0)$ is an intercept.

$$f'(x) = -\frac{3}{5}x^{-2/5} = \frac{-3}{5x^{2/5}}$$

When $x < 0$, $f'(x) < 0$ so f is decreasing

$\qquad x > 0$, $f'(x) < 0$ so f is decreasing

f' is undefined when $x = 0$, but f is defined, so this corresponds to a vertical tangent at $x = 0$.

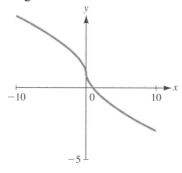

77. $y = (x - p)(x - q)$

$$y' = (x - p)(1) + (x - q)(1) = 2x - p - q$$

So, $y' = 0$ when $0 = 2x - p - q$ or

$x = \dfrac{p + q}{2}$, which is the midpoint of the

segment PQ. So any relative extremum occurs midway between its intercepts.

79. $f(x) = (x^2 + x - 1)^3(x + 3)^2$

Press $\boxed{y =}$ and input f for $y_1 =$

Use the window dimensions $[-4, 2]1$ by $[-20, 25]5$

Press $\boxed{\text{Graph}}$

$$f'(x) = 3(x^2 + x - 1)^2(2x + 1)(x + 3)^2$$
$$\qquad + (x^2 + x - 1)^3(2)(x + 3)$$
$$\qquad = (x^2 + x - 1)^2(x + 3)$$
$$\qquad\qquad \cdot [3(2x + 1)(x + 3)$$
$$\qquad\qquad + (x^2 + x - 1)(2)]$$
$$\qquad = (x^2 + x - 1)^2(x + 3)$$
$$\qquad\qquad \cdot (8x^2 + 23x + 7)$$

Press $\boxed{y =}$ and input $f'(x)$ for $y_2 =$

Press $\boxed{\text{Graph}}$

To find the values of x for which $f'(x) = 0$, it may be easiest to deactivate y_1 so only the graph of $y_2 = f'$ is shown.

Use Trace and verify

$y_2 = (x^2 + x - 1)^2 (x + 3)(8x^2 + 23x + 7)$ is shown in the upper left corner. Trace along the graph to move near an x-intercept. Use Zoom in function for more accurate readings. The values of x for which $f'(x) = 0$ are $x \approx -3$,

$x_2 \approx -2.5$, $x_3 \approx -1.6$, $x_4 \approx -0.35$,

$x_5 \approx 0.62$.

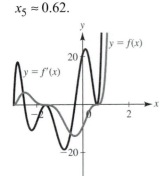

81. $f(x) = (1 - x^{1/2})^{1/2}$

Press y = and input f for $y_1 =$

Use the window dimensions [0, 1]0.5 by [−2, 1]1

(the domain of f is $0 < x < 1$)

Press Graph

$$f'(x) = \frac{1}{2}(1 - x^{1/2})^{-1/2}\left(-\frac{1}{2}x^{-1/2}\right)$$

$$f'(x) = \frac{-1}{4x^{1/2}(1 - x^{1/2})^{1/2}}$$

Press y = and input f' for $y_2 =$

Press Graph

We see from the graph that there are no values of x for which $f'(x) = 0$.

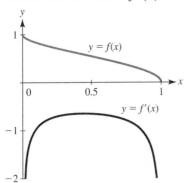

83. Let $f(x) = 4 + \sqrt{9 - 2x - x^2}$. Before graphing, f appears to be the upper half of a circle.

$$y = 4 + \sqrt{9 - 2x - x^2}$$

$$y - 4 = \sqrt{9 - 2x - x^2}$$

By squaring both sides and completing the square we obtain the equation of the whole circle with center $(-1, 4)$ and radius $\sqrt{10}$.

$$(y - 4)^2 = 9 - 2x - x^2$$

$$x^2 + 2x + 1 + (y - 4)^2 = 9 + 1$$

$$(x + 1)^2 + (y - 4)^2 = 10$$

Therefore, $f(x) = 4 + \sqrt{9 - 2x - x^2}$ should be the upper half of this circle.

Press y = and input f for $y_1 =$

Use window dimensions [−5, 5] by [−10, 10]

Press Graph

Initially, the graph appears to be the upper half of an ellipse but by using the Zsquare function, we see the graph is, in fact, the upper half of the circle.

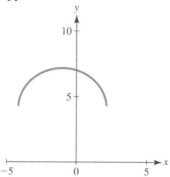

3.2 Concavity and Points of Inflection

1. The graph is:
concave downward $(f''(x) < 0)$ for $x < 2$,
and concave upward $(f''(x) > 0)$ for $x > 2$.

3. The graph is:
concave downward $(f''(x) < 0)$ for
$-1 < x < 1$, and concave upward
$(f''(x) > 0)$ for $x < -1$ and $x > 1$.

5. $f(x) = x^3 + 3x^2 + x + 1$
$f'(x) = 3x^2 + 6x = 1$
$f''(x) = 6x + 6 = 6(x + 1)$
$f''(x) = 0$ when $x = -1$
When
$x < -1$, $f''(x) < 0$ so f is concave down
$x > -1$, $f''(x) > 0$ so f is concave up
Since the concavity changes at the critical value $x = -1$, the point $(-1, 2)$ is an inflection point.

7. $f(x) = x(2x + 1)^2$
$\quad = x(4x^2 + 4x + 1)$
$\quad = 4x^3 + 4x^2 + x$
$f'(x) = 12x^2 + 8x + 1$

$f''(x) = 24x + 8 = 8(3x + 1)$
$f''(x) = 0$ when $x = -\dfrac{1}{3}$
When
$x < -\dfrac{1}{3}$, $f''(x) < 0$ so f is concave down
$x > -\dfrac{1}{3}$, $f''(x) > 0$ so f is concave up
Since the concavity changes at the critical value $x = -\dfrac{1}{3}$. the point $\left(-\dfrac{1}{3}, -\dfrac{1}{27}\right)$ is an inflection point.

9. $g(t) = t^2 - \dfrac{1}{t} = t^2 - t^{-1}$
$g'(t) = 2t + t^{-2}$
$g''(t) = 2 - 2t^{-3}$
$\quad = 2 - \dfrac{2}{t^3}$
$\quad = \dfrac{2t^3 - 2}{t^3}$
$\quad = \dfrac{2(t^3 - 1)}{t^3}$
$g''(t) = 0$ when $t = 1$
(Note that $g''(t)$ and $g(t)$ are undefined for $t = 0$.)
When
$t < 0$, $g''(t) > 0$ so g is concave up
$0 < t < 1$, $g''(t) < 0$ so g is concave down
$t > 1$, $g''(t) > 0$ so g is concave up
Since the concavity changes at the critical value $t = 1$, the point $(1, 0)$ is an inflection point.

11. $f(x) = x^4 - 6x^3 + 7x - 5$

$f'(x) = 4x^3 - 18x^2 + 7$

$f''(x) = 12x^2 - 36x = 12x(x-3)$

$f''(x) = 0$ when $x = 0, 3$

When

$x < 0$, $f''(x) < 0$ so f is concave up

$0 < x < 3$, $f''(x) < 0$ so f is concave down

$x > 3$, $f''(x) > 0$ so f is concave up

Since the concavity changes at both critical values $x = 0$ and $x = 3$, the points $(0, -5)$ and $(3, -65)$ are inflection points.

13. $f(x) = \dfrac{1}{3}x^3 - 9x + 2$

$f'(x) = x^2 - 9 = (x+3)(x-3)$

$f'(x) = 0$ when $x = -3, 3$

$f''(x) = 2x$

$f''(x) = 0$ when $x = 0$

When $x < -3$,

$f'(x) > 0$ so f is increasing

$f''(x) < 0$ so f is concave down

When $-3 < x < 0$,

$f'(x) < 0$ so f is decreasing

$f''(x) < 0$ so f is concave down

When $0 < x < 3$,

$f'(x) < 0$ so f is decreasing

$f''(x) > 0$ so f is concave up

When $x > 3$,

$f'(x) > 0$ so f is increasing

$f''(x) > 0$ so f is concave up

Overall, f is increasing for $x < -3$ and $x > 3$; decreasing for $-3 < x < 3$; concave up for $x > 0$; and concave down for $x < 0$. The critical value $x = -3$ corresponds to the point $(-3, 20)$, which is a relative maximum. The critical value $x = 3$ corresponds to the point $(3, -16)$, which is a relative minimum. Since the concavity changes at $x = 0$, the point $(0, 2)$ is an inflection

point.

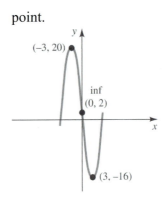

15. $f(x) = x^4 - 4x^3 + 10$

$f'(x) = 4x^3 - 12x^2 = 4x^2(x-3)$

$f'(x) = 0$ when $x = 0, 3$

$f''(x) = 12x^2 - 24x = 12x(x-2)$

$f''(x) = 0$ when $x = 0, 2$

When $x < 0$,

$f'(x) < 0$ so f is decreasing

$f''(x) > 0$ so f is concave up

When $0 < x < 2$,

$f'(x) < 0$ so f is decreasing

$f''(x) < 0$ so f is concave down

When $2 < x < 3$,

$f'(x) < 0$ so f is decreasing

$f''(x) > 0$ so f is concave up

When $x > 3$, $f'(x) > 0$ so f is increasing

$f''(x) > 0$ so f is concave up

Overall, f is increasing for $x > 3$; decreasing $x < 3$; concave up for $x < 0$ and $x > 2$; and concave down for $0 < x < 2$. The critical value $x = 0$ corresponds to the point $(0, 10)$, which is not a relative extremum. However, the concavity changes at $x = 0$, so $(0, 10)$ is an inflection point. The concavity changes again at $x = 2$, so the point $(2, -6)$ is also an inflection point. The critical value $x = 3$ corresponds to the point $(3, -17)$, which is a relative minimum.

point.

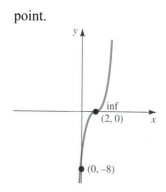

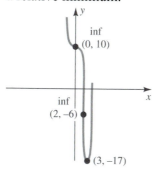

17. $f(x) = (x-2)^3$

$f'(x) = 3(x-2)^2(1)$

$f'(x) = 0$ when $x = 2$

$f''(x) = 6(x-2)$

$f''(x) = 0$ when $x = 2$

When $x < 2$,

$f'(x) > 0$ so f is increasing

$f''(x) < 0$ so f is concave down

When $x > 2$, $f'(x) > 0$ so f is increasing

$\quad\quad\quad\quad f''(x) > 0$ so f is concave up

Overall, f is increasing for all values of x; concave up for $x > 2$; and concave down for $x < 2$.

The critical value $x = 2$ corresponds to the point $(2, 0)$ which is not a relative extremum. However, the concavity changes at $x = 2$, so $(2, 0)$ is an inflection

19. $f(x) = (x^2 - 5)^3$

$f'(x) = 3(x^2 - 5)^2(2x) = 6x(x^2 - 5)^2$

$f'(x) = 0$ when $x = -\sqrt{5}, 0, \sqrt{5}$

$f''(x) = (6x)[2(x^2 - 5)(2x)] + (x^2 - 5)^2(6)$

$\quad\quad = 6(x^2 - 5)[4x^2 + x^2 - 5]$

$\quad\quad = 6(x^2 - 5)(5x^2 - 5)$

$\quad\quad = 30(x^2 - 5)(x+1)(x-1)$

$f''(x) = 0$ when $x = -\sqrt{5}, -1, 1, \sqrt{5}$

When $x < -\sqrt{5}$,

$f'(x) < 0$ so f is decreasing

$f''(x) > 0$ so f is concave up

When $-\sqrt{5} < x < -1$,

$f'(x) < 0$ so f is decreasing

$f''(x) < 0$ so f is concave down

When $-1 < x < 0$,

$f'(x) < 0$ so f is decreasing

$f''(x) > 0$ so f is concave up

When $0 < x < 1$,

$f'(x) > 0$ so f is increasing

$f''(x) > 0$ so f is concave up

When $1 < x < \sqrt{5}$,

$f'(x) > 0$ so f is increasing

$f''(x) < 0$ so f is concave down

When $x > \sqrt{5}$, $f'(x) > 0$ so f is increasing

$f''(x) > 0$ so f is concave up

Overall, f is increasing for $x > 0$; decreasing for $x < 0$; concave up for $x < -\sqrt{5}$, $-1 < x < 1$, and $x > \sqrt{5}$; and concave down for $-\sqrt{5} < x < -1$ and $1 < x < \sqrt{5}$.

The critical value $x = -\sqrt{5}$ corresponds to the point $\left(-\sqrt{5}, 0\right)$, which is not a relative extremum. However, the concavity changes at $x = -\sqrt{5}$, so $\left(-\sqrt{5}, 0\right)$ is an inflection point. The concavity changes again at $x = -1$, so the point $(-1, -64)$ is also an inflection point. The critical value $x = 0$ corresponds to the point $(0, -125)$, which is a relative minimum. The concavity next changes at $x = 1$, so the point $(1, -64)$ is an inflection point. The critical value $x = \sqrt{5}$ corresponds to the point $\left(\sqrt{5}, 0\right)$, which is not a relative extremum. However, the concavity changes at $x = \sqrt{5}$, so $\left(\sqrt{5}, 0\right)$ is an inflection point.

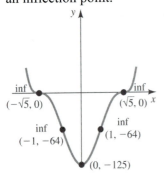

21. $f(s) = 2s(s+4)^3$

$f'(s) = (2s)[3(s+4)^2(1)] + (s+4)^2(2)$

$\qquad = 2(s+4)^2[3s + s + 4]$

$\qquad = 8(s+4)^2(s+1)$

$f'(s) = 0$ when $s = -4, -1$

$f''(s) = 8[(s+4)^2(1) + (s+1)(2(s+4)(1))]$

$\qquad = 8(s+4)[s+4 + 2(s+1)]$

$\qquad = 24(s+4)(s+2)$

$f''(s) = 0$ when $s = -4, -2$

When $s < -4$, $f'(x) < 0$ so f is decreasing

$f''(x) > 0$ so f is concave up

When $-4 < s < -2$,

$f'(x) < 0$ so f is decreasing

$f''(x) < 0$ so f is concave down

When $-2 < s < -1$,

$f'(x) < 0$ so f is decreasing

$f''(x) > 0$ so f is concave up

When $s > -1$,

$f'(x) > 0$ so f is increasing

$f''(x) > 0$ so f is concave up

Overall, f is increasing for $s > -1$; decreasing for $s < -1$; concave up for $s < -4$ and $s > -2$; and concave down for $-4 < s < -2$.

The critical value $s = -4$ corresponds to the point $(-4, 0)$, which is not a relative extremum. However, the concavity changes at $s = -4$, so $(-4, 0)$ is an inflection point. The concavity changes again at $s = -2$, so the point $(-2, -32)$ is also an inflection point. The critical value $s = -1$ corresponds to the point $(-1, -54)$, which is a relative minimum.

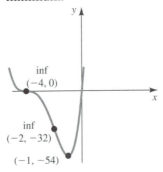

23. $g(x) = \sqrt{x^2 + 1} = (x^2 + 1)^{1/2}$

$g'(x) = \dfrac{1}{2}(x^2 + 1)^{-1/2}(2x) = \dfrac{x}{\sqrt{x^2 + 1}}$

$g'(x) = 0$ when $x = 0$

$g''(x)$

$= \dfrac{(x^2 + 1)^{1/2}(1) - (x)\left[\frac{1}{2}(x^2 + 1)^{-1/2}(2x)\right]}{\left(\sqrt{x^2 + 1}\right)^2}$

$g''(x) = \dfrac{(x^2 + 1)^{1/2} - \frac{x^2}{(x^2+1)^{1/2}}}{x^2 + 1} \cdot \dfrac{(x^2 + 1)^{1/2}}{(x^2 + 1)^{1/2}}$

$= \dfrac{x^2 + 1 - x^2}{(x^2 + 1)^{3/2}}$

$= \dfrac{1}{(x^2 + 1)^{3/2}}$

When $x < 0$, $g'(x) < 0$ so g is decreasing
$\qquad\qquad g''(x) > 0$ so g is concave up

When $x > 0$, $g'(x) > 0$ so g is increasing
$\qquad\qquad g''(x) > 0$ so g is concave up

Overall, g is increasing for $x > 0$; decreasing for $x < 0$; and concave up for all values of x. The critical value $x = 0$ corresponds to the point $(0, 1)$, which is a relative minimum.

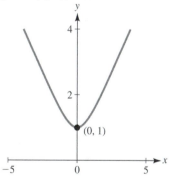

25. $f(x) = \dfrac{1}{x^2 + x + 1} = (x^2 + x + 1)^{-1}$

$f'(x) = -(x^2 + x + 1)^{-2}(2x + 1)$

$= \dfrac{-(2x + 1)}{(x^2 + x + 1)^2}$

$f'(x) = 0$ when $x = -\dfrac{1}{2}$

$f''(x) = \dfrac{1}{[(x^2 + x + 1)^2]^2}[(x^2 + x + 1)^2(-2)$

$\qquad\qquad + (2x + 1)(2)(x^2 + x + 1)(2x + 1)]$

$= \dfrac{2(x^2 + x + 1)}{[-(x^2 + x + 1) + (2x + 1)^2]}{(x^2 + x + 1)^4}$

$= \dfrac{6x(x + 1)}{(x^2 + x + 1)^3}$

$f''(x) = 0$ when $x = -1, 0$

When $x < -1$,
$f'(x) > 0$ so f is increasing
$f''(x) > 0$ so f is concave up

When $-1 < x < -\dfrac{1}{2}$,

$f'(x) > 0$ so f is increasing
$f''(x) < 0$ so f is concave down

When $-\dfrac{1}{2} < x < 0$,

$f'(x) < 0$ so f is decreasing
$f''(x) < 0$ so f is concave down

When $x > 0$, $f'(x) < 0$ so f is decreasing
$\qquad\qquad f''(x) > 0$ so f is concave up

Overall, f is increasing for $x < -\dfrac{1}{2}$;

decreasing for $x > -\dfrac{1}{2}$; concave up for

$x < -1$ and $x > 0$; and concave down for
$-1 < x < 0$.
At $x = -1$, the concavity changes, so the
point $(-1, 1)$ is an inflection point. The

critical value $x = -\dfrac{1}{2}$ corresponds to the

point $\left(-\dfrac{1}{2}, \dfrac{4}{3}\right)$, which is relative maximum.

The concavity changes again at
$x = 0$, so the point $(0, 1)$ is an inflection
point.

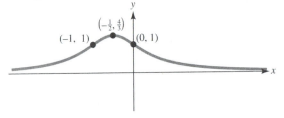

27. $f(x) = x^3 + 3x^2 + 1$

$f'(x) = 3x^2 + 6x = 3x(x + 2)$
$f'(x) = 0$ when $x = -2, 0$
$f''(x) = 6(x + 1)$
$f''(0) = 6 > 0$ and $f''(-2) = -6 < 0$, and
$f(-2) = 5$ and $f(0) = 1$.
So $(0, 1)$ is a relative minimum, and
$(-2, 5)$ is a relative maximum.

29. $f(x) = (x^2 - 9)^2$

$f'(x) = 2(x^2 - 9)(2x) = 4x(x - 3)(x + 3)$
$f'(x) = 0$ when $x = -3, 0, 3$
$f''(x) = 12(x^2 - 3)$
$f''(-3) = 72 > 0$, $f''(0) = -36 < 0$, and
$f''(3) = 72 > 0$; $f(\pm 3) = 0$ and $f(0) = 81$. So
$(0, 81)$ is a relative maximum, and $(-3, 0)$,
$(3, 0)$ are relative minima.

31. $f(x) = 2x + 1 + \dfrac{18}{x}$

$f'(x) = 2 - \dfrac{18}{x^2} = \dfrac{2(x - 3)(x + 3)}{x^2}$
$f'(x) = 0$ when $x = -3, 3$
$f''(x) = \dfrac{36}{x^3}$
$f''(-3) = -\dfrac{4}{3} < 0$ and $f''(3) = \dfrac{4}{3} > 0$;
$f(-3) = -11, f(3) = 13$.
So $(-3, -11)$ is a relative maximum,
$(3, 13)$ is a relative minimum.

33. $f(x) = x^2(x - 5)^2 = x^4 - 10x^3 + 25x^2$

$f'(x) = 4x^3 - 30x^2 + 50x$
$\qquad = 2x(x - 5)(2x - 5)$
$f'(x) = 0$ when $x = 0, 2.5, 5$
$f''(x) = 12x^2 - 60x + 50$
$f''(0) = 50 > 0$, $f''(2.5) = -25 < 0$, and
$f''(5) = 50 > 0$; $f(0) = 0, f(2.5) = 39.0625$
and $f(5) = 0$. So $(0, 0)$ and $(5, 0)$ are
relative minima and $(2.5, 39.065)$ is a
relative maximum.

35. $h(t) = \dfrac{2}{1 + t^2} = 2(1 + t^2)^{-1}$

$h'(t) = -2(1 + t^2)^{-2}(2t) = \dfrac{-4t}{(1 + t^2)^2}$

$h'(t) = 0$ when $t = 0$

$$h''(t) = \frac{-4(1+t^2)^2 - (-4t)(2)(1+t^2)(2t)}{(1+t^2)^4}$$

$$= \frac{4(1+t)[-(1+t^2)+4t^2]}{(1+t^2)^4}$$

$$= \frac{4(3t^2-1)}{(1+t^2)^3}$$

$h''(0) = -4 < 0$ and $h(0) = 2$.

So, $(0, 2)$ is a relative maximum.

37. $f(x) = \dfrac{(x-2)^3}{x^2}$

$$f'(x) = \frac{x^2[3(x-2)^2(1)] - (x-2)^3(2x)}{x^4}$$

$$= \frac{x(x-2)^2[3x - 2(x-2)]}{x^4}$$

$$= \frac{(x-2)^2(x+4)}{x^3}$$

$f'(x) = 0$ when $x = -4, 2$

$$f''(x) = \frac{1}{x^6}(x^3[(x-2)^2(1) + (x+4)(2)(x-2)] - [(x-2)^2(x+4)(3x^2)])$$

$$= \frac{x^2(x-2)(x[(x-2)+2(x+4)] + 3(x-2)(x+4))}{x^6}$$

$$= \frac{24(x-2)}{x^4}$$

$f''(-4) = -\dfrac{9}{16} < 0$ and $f(-4) = -13.5$. So, $(-4, -13.5)$ is a relative maximum. $f''(2) = 0$, so the test fails.

39. $f''(x) = x^2(x-3)(x-1)$

$f''(x) = 0$ when $x = 0, 1, 3$

When
$x < 0$, $f''(x) > 0$ so f is concave up
$0 < x < 1$, $f''(x) > 0$ so f is concave up
$1 < x < 3$, $f''(x) < 0$ so f is concave down
$x > 3$, $f''(x) > 0$ so f is concave up

Overall, f is concave up for $x < 0$, $0 < x < 1$, and $x > 3$; concave down for $1 < x < 3$. There are inflection points at $x = 1$ and $x = 3$, as the concavity changes at those values.

41. $f''(x) = (x-1)^{1/3}$

$f''(x) = 0$ when $x = 1$

When $\begin{array}{l} x < 1, f''(x) < 0 \text{ so } f \text{ is concave down} \\ x > 1, f''(x) > 0 \text{ so } f \text{ is concave up} \end{array}$

There is an inflection point at $x = 1$, as the concavity changes at that value.

43. $f'(x) = x^2 - 4x$

(a) $f'(x) = x(x - 4)$

$f'(x) = 0$ when $x = 0, 4$

When

$x < 0$, $f'(x) > 0$ so f is increasing

$0 < x < 4$, $f'(x) < 0$ so f is decreasing

$x > 4$, $f'(x) > 0$ so f is increasing

(b) $f''(x) = 2x - 4 = 2(x - 2)$

$f''(x) = 0$ when $x = 2$

When

$x < 2$, $f''(x) < 0$ so f is concave down

$x > 2$, $f''(x) > 0$ so f is concave up

(c) at $x = 0$, there is a relative maximum; at $x = 4$, there is a relative minimum; at $x = 2$, there is no inflection point

(d)

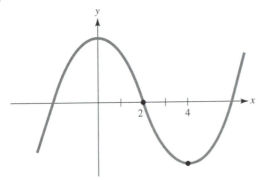

45. $f'(x) = 5 - x^2$

$f'(x) = 0$ when $-\sqrt{5}, \sqrt{5}$

(a) When

$x < -\sqrt{5}$, $f'(x) < 0$ so f is decreasing

$-\sqrt{5} < x < \sqrt{5}$, $f'(x) > 0$ so f is increasing

$x > \sqrt{5}$, $f'(x) < 0$ so f is decreasing

(b) $f''(x) = -2x$

$f''(x) = 0$ when $x = 0$

When

$x < 0$, $f''(x) > 0$ so f is concave up

$x > 0$, $f''(x) < 0$ so f is concave down

(c) at $x = -\sqrt{5}$, there is a relative minimum; at $x = \sqrt{5}$, there is a relative maximum; at $x = 0$, there is an inflection point

(d)

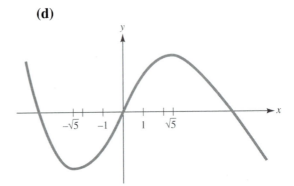

47. (a) The curve rises for $x < -1$ and $x > 3$.

(b) It falls when $-1 < x < 3$.

(c) The curve is concave down for $x < 2$.

(d) The curve is concave up for $x > 2$. Here is a possible graph.

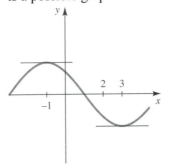

49. When $x < 2$, $f'(x) < 0$ so f is decreasing

$x = 2$, $f'(x) = 0$ and there is a relative minimum

$x > 2$, $f'(x) > 0$ so f is increasing.

Since f' is increasing for all values of x, its rate of change $f''(x) > 0$ for all x, and f

is concave up for all x.

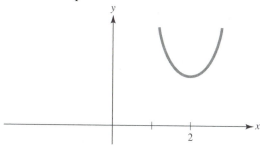

51. When $x < -3$, $f'(x) < 0$ so f is decreasing

$x = -3$, $f'(x) = 0$ so f levels

but there is not a relative extremum

$-3 < x < 2$, $f'(x) < 0$ so f is decreasing

$x = 2$, $f'(x) = 0$ and there is a relative minimum

$x > 2$, $f'(x) > 0$ so f is increasing

Since $f'(x)$ is increasing for $x < -3$ and for $x > -1$, $f''(x) > 0$ on these intervals and f is concave up.

Since $f'(x)$ is decreasing for $-3 < x < -1$, $f''(x) < 0$ on that interval and f is concave down. Since the concavity changes at $x = -3$ and $x = -1$, there are inflection points at these values.

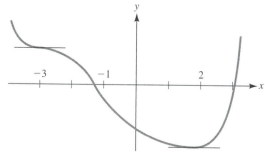

53. **(a)** $C(x) = 0.3x^3 - 5x^2 + 28x + 200$

$M(x) = C'(x) = 0.9x^2 - 10x + 28$

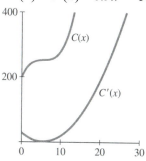

(b) $M'(x) = C''(x) = 1.8x - 10$

$C''(x) = 0$ when $x \approx 5.56$

Critical values of C'' are x values of possible extrema of C', which is the marginal cost function. $x = 5.56$ corresponds to a minimum on the graph of C'.

55. $S(x) = -x^3 + 33x^2 + 60x + 1{,}000$

(a) When $x = 0$,

$S(0) = -(0)^3 + 33(0)^2 + 60(0) + 1{,}000$

$= 1{,}000$

One thousand units will be sold.

(b) $S'(x) = -3x^2 + 66x + 60$

$= -3(x^2 - 22x - 20)$

$S'(x) = 0$ when

$x = \dfrac{22 \pm \sqrt{(-22)^2 - 4(1)(-20)}}{2(1)}$

≈ 22.9 (deleting negative root)

$S''(x) = -6x + 66 = -6(x - 11)$

$S''(x) = 0$ when $x = 11$

When $0 \le x < 11$,

$S'(x) > 0$ so S is increasing

$S''(x) > 0$ so S is concave up

When $11 < x < 22.9$,

$S'(x) > 0$ so S is increasing

$S''(x) < 0$ so S is concave down

When $x > 22.9$,

$S'(x) < 0$ so S is decreasing
$S''(x) < 0$ so S is concave down
Overall, S is increasing for
$0 \le x < 22.9$; decreasing for $x > 22.9$;
concave up for $0 \le x < 11$; concave
down for $x > 22.9$.
The critical value $x = 22.9$ corresponds
to the point $(22.9, 7671)$, which is a
relative maximum. When
$x = 11$, the corresponding point is
$(11, 4322)$, which is an inflection point.

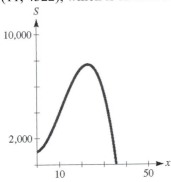

The inflection point corresponds to the
amount spent on marketing ($11,000)
related to when sales are increasing most
rapidly (since it is the critical value for
S').

$Q'(t) = -3t^2 + 9t + 15$

$Q''(t) = -6t + 9 = 3(-2t + 3)$

$Q''(t) = 0$ when $t = \dfrac{3}{2}$

So, after 1.5 hours, the point of
diminishing returns is reached. This
corresponds to 8:30 A.M.

(b) Efficiency relates to the fastest rate of
successful work. So need to minimize
the rate of work. Let R represent the
rate of work. Then,

$R(t) = Q'(t) = -3t^2 + 9t + 15$

To find the minimum rate, need to
find where $R'(t) = 0$, which is when

$Q''(t) = 0$, or $t = 1.5$. So minimum rate
occurs either when $t = 1.5$, or at one of
the interval endpoints, $t = 0$, or $t = 5$.

$R(0) = 15$

$R(1.5) = -3(1.5)^2 + 9(1.5) + 15 = 21.75$

$R(5) = -3(5)^2 + 9(5) + 15 = -15$

So, the minimum rate occurs when
$t = 5$, which corresponds to noon.

57. $Q(t) = -t^3 + \dfrac{9}{2}t^2 + 15t$

(a) The point of diminishing returns
corresponds to an inflection point,
where $Q''(t) = 0$.

59. $M(r) = \dfrac{1 + 0.02r}{1 + 0.09r^2}$

(a) $M'r = \dfrac{(1+0.009r^2)(0.02)-(1+0.02r)(0.018r)}{(1+0.009r^2)^2}$

$= \dfrac{0.02-0.018r-0.00018r^2}{(1+0.009r^2)^2}$

$M''(r) = \dfrac{[(1+0.009r^2)^2(-0.018-0.00036r)-(0.02-0.018r-0.00018r^2)\cdot 2(1+0.09r^2)(0.018r)]}{(1+0.009r^2)^4}$

$= \dfrac{[0.018(1+0.009r^2)[(1+0.009r^2)(-1-0.02r)-2r(0.02-0.018r-0.00018r^2)]]}{(1+0.009r^2)^4}$

$= \dfrac{0.018[-1-0.06r+0.0027r^2+0.00018r^3]}{(1+0.009r^2)^3}$

(b) Press $\boxed{y=}$ and input

$(1+0.02x)\div(1+0.009x^2)$ for $y_1=$.
Use window dimensions [0, 20]0.05
by [0, 2]0.25.
Press $\boxed{\text{graph}}$.

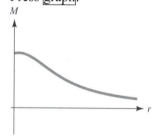

(c) To find the rate of interest at which the
rate of construction of new houses is
minimized, we must find r for which
$M''(r)=0$. $M'(r)$ gives us the rate of
construction and thus to minimize this,
we take $M''(r)$ and set it equal to zero.
Press $\boxed{y=}$ and input M'' for $y_1=$.
Use window dimensions [−10, 10]1 by
[−0.005, 0.005]0.001.
Press $\boxed{\text{graph}}$. Since we are only
concerned with the positive zero, use the
zero function under the calc menu with a
close value to the positive

x-intercept for the left bound, right
bound and guess. We find the interest
rate to be $r = 7.10\%$.

61. $S(t) = \dfrac{3}{t+2} - \dfrac{12}{(t+2)^2} + 5$

$= 3(t+2)^{-1} - 12(t+2)^{-2} + 5$

(a) $S'(t) = -3(t+2)^{-2}(1) + 24(t+2)^{-3}(1)$

$= \dfrac{-3}{(t+2)^2} + \dfrac{24}{(t+2)^3}$

$= \dfrac{-3(t+2)+24}{(t+2)^3}$

$= \dfrac{-3t+18}{(t+2)^3}$

$$S''(t) = \frac{(t+2)^3(-3) - (-3t+18) \cdot 3(t+2)^2(1)}{\left[(t+2)^3\right]^2}$$

$$= \frac{-3(t+2)^3 - 3(-3t+18)(t+2)^2}{(t+2)^6}$$

$$= \frac{-3(t+2)^2\left[(t+2) + (-3t+18)\right]}{(t+2)^6}$$

$$= \frac{-3(-2t+20)}{(t+2)^4} = \frac{6(t-10)}{(t+2)^4}$$

(b) $S'(t) = 0$ when $-3t + 18 = 0$, or $t = 6$.

$$S''(6) = \frac{6(6-10)}{(6+2)^4} < 0, \text{ so } t = 6$$

corresponds to a maximum. The corresponding sales are

$$S(6) = \frac{3}{6+2} - \frac{12}{(6+2)^2} + 5$$

$$= 5.1875$$

or approximately 519 pairs.

(c) To minimize the sales rate, S', need to find when $S''(t) = 0$.

$$0 = \frac{6(t-10)}{(t+2)^4}$$

So, $S''(t) = 0$ when $t = 10$.
When $0 \le t < 10$, $S'' < 0$ and S' is decreasing.
When $t > 10$, $S'' > 0$ and S' is increasing.
So, $t = 10$ corresponds to a minimum sales rate.
Sales level at this time are

$$S(10) = \frac{3}{10+2} - \frac{12}{(10+2)^2} + 5$$

$$\approx 5.1667$$

or approximately 517 pairs.

The sales rate at this time is

$$S'(10) = \frac{-3(10-6)}{(10+2)^3}$$

$$= \frac{-12}{(12)^3} = -\frac{1}{144} \approx -0.0069$$

or rate is decreasing by approximately 0.7 pair per month.

63. $N(t) = \dfrac{5t}{12 + t^2}$

(a) $N'(t) = \dfrac{(12+t^2)(5) - (5t)(2t)}{(12+t^2)^2}$

$$= \frac{60 - 5t^2}{(12+t^2)^2}$$

$$N''(t) = \frac{(12+t^2)^2(-10t) - (60-5t^2)}{(12+t^2)^4}$$
$$\cdot 2(12+t^2)(2t)$$

$$= \frac{2t(12+t^2)[-5(12+t^2)}{(12+t^2)^4}$$
$$\frac{-2(60-5t^2)]}{}$$

$$= \frac{2t(5t^2 - 180)}{(12+t^2)^4}$$

$$= \frac{10t(t^2 - 36)}{(12+t^2)^4}$$

(b) $N'(t) = 0$ when $t = \sqrt{12} \approx 3.46$. Since $N''(3.46) < 0$, the maximum number of reported cases occurs after 3.46 weeks. The corresponding maximum number of new cases is

$$N(3.46) = \frac{5\sqrt{12}}{12 + 12} \approx 0.7217 \text{ or}$$

722 new cases.

(c) $N''(t) = 0$ when $t = 0.6$
When
$0 < t < 6$, $N''(t) < 0$ so N' is decreasing
$t > 0$, $N''(t) > 0$ so N' is increasing

So, the rate of reported cases N' is minimized after 6 weeks. The corresponding minimum number of new cases is $N(6) = \dfrac{5(6)}{12+36} = 0.625$ or approximately 63 new cases.

65. Need to optimize the rate of population growth on the interval $0 \le t \le 5$. Since the population is $P(t) = -t^3 + 9t^2 + 48t + 50$ the rate of growth is

$R(t) = P'(t) = -3t^2 + 18t + 48$

$R'(t) = P''(t) = -6t + 18$

$R'(t) = 0$ when $t = 3$

Using the interval endpoints and this

critical value, $R(0) = 48$, $R(3) = 75$, and $R(5) = 63$.

(a) The rate of growth is greatest when $t = 3$, or 3 years from now.

(b) It is smallest when $t = 0$, or now.

(c) The rate the population growth changes most rapidly is when $R'(t)$ is a maximum. Since $R'(t) = -6t + 18$, is most rapid when $t = 0$, or now.

67. $\dfrac{dA}{dt} = k\sqrt{A(t)}[M - A(t)], \ k > 0$

(a) $R(t) = \dfrac{dA}{dt} = k[A(t)]^{1/2}[M - A(t)]$

$R'(t) = k\left[[A(t)]^{1/2}\left(-\dfrac{dA}{dt}\right) + (M - A(t))\left(\dfrac{1}{2}[A(t)]^{-1/2}\dfrac{DA}{dt}\right)\right]$

$= k\dfrac{dA}{dt}\left[-[A(t)]^{1/2} \cdot \dfrac{2[A(t)]^{1/2}}{2[A(t)]^{1/2}} + \dfrac{M - A(t)}{2[A(t)]^{1/2}}\right]$

$= k\dfrac{dA}{dt}\left[\dfrac{-2A(t) + M - A(t)}{2[A(t)]^{1/2}}\right]$

$= k\dfrac{dA}{dt}\left[\dfrac{M - 3A(t)}{2[A(t)]^{1/2}}\right]$

$R'(t) = 0$ when $M - 3A(t) = 0$, or $A(t) = \dfrac{M}{3}$.

(b) When $A(t) < \dfrac{M}{3}$, $M - 3A(t) > 0$ and $R'(t) > 0$, so R is increasing.

When $A(t) > \dfrac{M}{3}$, $M - 3A(t) < 0$ and $R'(t) < 0$, so R is decreasing.

So, when $A(t) = \dfrac{M}{3}$, the rate is the greatest.

(c) $R'(t) = A''(t)$, so graph of A has an inflection point when $A(t) = \dfrac{M}{3}$.

69. $f(x) = x^4 + x$

$f'(x) = 4x^3 + 1$

$f'(x) = 0$ when $x = \sqrt[3]{-\dfrac{1}{4}} \approx -0.63$

$f''(x) = 12x^2$

$f''(x) = 0$ when $x = 0$

When $x < -0.63$,

$f'(x) < 0$ so f is decreasing

$f''(x) > 0$ so f is concave up

When $-0.63 < x < 0$,

$f'(x) > 0$ so f is increasing

$f''(x) > 0$ so f is concave up

When $x > 0$, $f'(x) > 0$ so f is increasing

$\qquad\qquad f''(x) > 0$ so f is concave up

When $x = -0.63$, f has a relative minimum. When $x = 0$, f does not have a relative extremum, nor does f have an inflection point, as the concavity does not switch.

$f(-0.63) \approx -0.47; f(0) = 0$

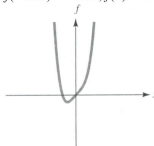

71. As shown by the following counterexample, the sum h needn't also have an inflection point at $x = c$.

$f(x) = \dfrac{1}{6}x^3 - x^2$

$g(x) = -\dfrac{1}{6}x^3 + x^2$

Then, $f'(x) = \dfrac{1}{2}x^2 - 2x$

$f''(x) = x - 2$, so f has an inflection point when $x = 2$

$g'(x) = -\dfrac{1}{2}x^2 + 2x$

$g''(x) = -x + 2$,

so g also has an inflection point when

$x = 2$. However,

$h(x) = f(x) + g(x) = 0$, so h does *not* have an inflection point when $x = 2$.

73. $f(x) = 2x^3 + 3x^2 - 12x - 7$

(a) To graph,

Press $\boxed{y =}$ and input f for $y_1 =$

Use zstandard function of zoom for viewing window.

Press $\boxed{\text{Graph}}$. The maxima and minima do not display properly. Change window dimensions to

$[-10, 10]1$ by $[-20, 20]2$

Press $\boxed{\text{Graph}}$

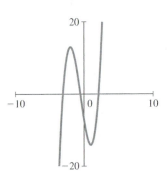

(b) $f'(x) = 6x^2 + 6x - 12$

$f''(x) = 12x + 6$

To use the TI-84 to find these values, input f for $y_1 =$, f' for $y_2 =$, and f'' for $y_3 =$.

De-select $y_2 =$ and $y_3 =$ so only $y_1 =$ is activated.

Use the value function in the calc menu. For $f(-4)$, input $x = -4$ and press $\boxed{\text{enter}}$. The display shows

$y = -39$. Repeat process for $x = -2, -1, 0, 1,$ and 2. For the $f'(x)$ values, de-select $y_1 =$ and activate $y_2 =$ and repeat process.

For the $f''(x)$ values, de-select $y_2 =$ and activate $y_3 =$ and repeat process.

x	-4	-2	-1	0	1	2
$f(x)$	-39	13	6	-7	-14	-3
$f'(x)$	60	0	-12	-12	0	24
$f''(x)$	-42	-18	-6	6	18	30

(c) To approximate the x-intercepts and y-intercept, use $z$standard function, activate $y_1 =$ and press graph.
You may use trace and zoom-in to estimate x-intercepts to be $x_1 \approx -3.08$, $x_2 \approx -0.54$, and $x_3 \approx 2.11$.
An alternative is to use the zero function under the calc menu. Press 2nd calc and enter zero function. The graph is displayed with left bound? For the left-most x-intercept, trace the graph to a value close to the intercept, but to the left of it and press enter. For the right bound? Enter a value close to the x-intercept but to the right of it.
To guess a value, enter an x-value in between the bounds and press enter. The display shows the zero value of $x_1 \approx -3.08$. Repeat this process for the other two x-intercepts.
For the y-intercept, use $z$standard and read the y-intercept as $y = -7$ (also given from the table in part (b)).

(d) To find the relative maximum and relative minimum points,
Use $z$standard function with $y_1 =$ activated. Press graph. Trace graph left until off the screen (near the relative maximum) and press enter. This will move the viewing window to the relative maximum. Use trace and zoom functions to estimate the maximum point to be $(-2, 13)$.
As an alternative, use the maximum function under the calc menu. Enter a left bound, right bound, and guess. For the relative minimum, use $z$standard to view the original graph and move cross-hair so relative

minimum is in window. Use the minimum function under the calc menu to find the relative minimum to be $(1, -14)$.

(e) Using the graph and the information from part (d), f is increasing on $x < -2$ and $x > 1$.

(f) Using the graph and the information from part (d), f is decreasing on $-2 < x < 1$.

(g) There is an inflection value on $-2 < x < 1$, since the concavity changes from downward to upward.
On the TI-84, de-select $y_1 =$ and activate y_3. Press graph. Use the zero function under the calc menu to find the zero of f'' to be $x = -0.5$. Activate $y_1 =$ and use the value function under the calc menu to find $f(-0.5) = -0.5$. The inflection point is $(-0.5, -0.5)$.

(h) Using the graph of f and the information from the previous parts, f is concave upward on $x > -\dfrac{1}{2}$.

(i) Using the graph of f and the information from the previous parts, f is concave downward on $x < -\dfrac{1}{2}$.

(j) To verify f changes from concave downward to concave upward, use the value function under the calc menu to show
$f''(-0.6) = -1.2$
$f''(-0.4) = 1.2$
(Make sure that you have $y_3 = 12x + 6$ activated.)

(k) Relative minimum point: $(1, -14)$
Relative maximum point: $(-2, 13)$
Both of the x-values are within the specified interval. Check the

endpoints of the interval. From part **(a)**,
$f(-4) = -39$ and $f(2) = -3$.
Absolute maximum value $= 13$
Absolute minimum value $= -39$

3.3 Curve Sketching

1. $\lim_{x \to \infty} f(x) = +\infty$, so $x = 0$ is a vertical
asymptote. $\lim_{x \to \pm\infty} f(x) = 0$, so $y = 0$ is a
horizontal asymptote.

3. There are no vertical asymptotes.
$\lim_{x \to -\infty} f(x) = 0$, so $y = 0$ is a horizontal
asymptote.

5. $\lim_{x \to -2} f(x) = +\infty$, so $x = -2$ is a vertical
asymptote. $\lim_{x \to 2^-} f(x) = -\infty$ and
$\lim_{x \to 2^+} f(x) = +\infty$, so $x = 2$ is a vertical
asymptote. $\lim_{x \to -\infty} f(x) = 0$, so $y = 0$ is a horizontal
asymptote. $\lim_{x \to +\infty} f(x) = 2$, so $y = 2$ is a
horizontal asymptote.

7. $\lim_{x \to 2^+} f(x) = +\infty$, so $x = 2$ is a vertical
asymptote. $\lim_{x \to \pm\infty} f(x) = 0$, so $y = 0$ is a
horizontal asymptote.

9. Since the denominator is zero when
$x = -2$, $x = -2$ is a vertical asymptote.
$\lim_{x \to \pm\infty} \dfrac{3x - 1}{x + 2} = \lim_{x \to \pm\infty} \dfrac{3 - \frac{1}{x}}{1 + \frac{2}{x}} = 3$, so $y = 3$ is a
horizontal asymptote.

11. Since the denominator cannot be zero for
any value of x, there are no vertical
asymptotes.
$\lim_{x \to \pm\infty} \dfrac{x^2 + 2}{x^2 + 1} = \lim_{x \to \pm\infty} \dfrac{1 + \frac{2}{x^2}}{1 + \frac{1}{x^2}} = 1$, so $y = 1$ is a
horizontal asymptote.

13. $f(t) = \dfrac{t^2 + 3t - 5}{t^2 - 5t + 6} = \dfrac{t^2 + 3t - 5}{(t - 2)(t - 3)}$
Since the denominator is zero when $t = 2$,
3, the vertical asymptotes are $t = 2$ and
$t = 3$.
$\lim_{x \to \pm\infty} \dfrac{t^2 + 3t - 5}{t^2 - 5t + 6} = \lim_{x \to \pm\infty} \dfrac{1 + \frac{3}{t} - \frac{5}{t^2}}{1 - \frac{5}{t} + \frac{6}{t^2}} = 1$, so
$y = 1$ is a horizontal asymptote.

15. $h(x) = \dfrac{1}{x} - \dfrac{1}{x-1} = \dfrac{-1}{x(x-1)} = \dfrac{-1}{x^2 - x}$
Since the denominator is zero when
$x = 0, 1$, the vertical asymptotes are $x = 0$
and $x = 1$.
$\lim_{x \to \pm\infty} \dfrac{-1}{x^2 - x} = 0$, so $y = 0$ is a horizontal
asymptote.

17. $f(x) = x^3 + 3x^2 - 2$
domain: all real numbers
intercepts:
when $x = 0, f(0) = -2$; point $(0, -2)$
$f(x) = 0$ is too difficult to solve
asymptotes: no vertical or horizontal
asymptotes
$f'(x) = 3x^2 + 6x = 3x(x + 2)$
$f'(x) = 0$ when $x = -2, 0$
$f''(x) = 6x + 6 = 6(x + 1)$
$f''(x) = 0$ when $x = -1$
When $x < -2$,
$f'(x) > 0$ so f is increasing
$f''(x) < 0$ so f is concave down
When $-2 < x < -1$,
$f'(x) < 0$ so f is decreasing
$f''(x) < 0$ so f is concave down
When $-1 < x < 0$,
$f'(x) < 0$ so f is decreasing
$f''(x) > 0$ so f is concave up
When $x > 0$, $f'(x) > 0$ so f is increasing
$f''(x) > 0$ so f is concave up
$(-2, 2)$ is a relative maximum, $(-1, 0)$ is an
inflection point, and $(0, -2)$ is a relative

minimum.

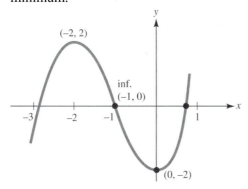

point, and (0, 0) is a relative minimum.

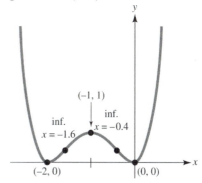

19. $f(x) = x^4 + 4x^3 + 4x^2 = x^2(x+2)^2$

domain: all real numbers
intercepts:
when $x = 0$, $f(0) = 0$; point $(0, 0)$
$f(x) = 0$, $x = 0, -2$; point $(-2, 0)$
asymptotes: no vertical or horizontal
asymptotes

$f'(x) = 4x^3 + 12x^2 + 8x$
$\quad\quad = 4x(x+1)(x+2)$

$f'(x) = 0$ when $x = -2, -1, 0$

$f''(x) = 12x^2 + 24x + 8 = 4(3x^2 + 6x + 2)$

$f''(x) = 0$ when $x = -1.6, -0.4$

When $x < -2$, $f'(x) < 0$ so f is decreasing
$\quad\quad\quad\quad\quad\quad f''(x) > 0$ so f is concave up

When $-2 < x < -1.6$,
$f'(x) > 0$ so f is increasing
$f''(x) > 0$ so f is concave up

When $-1.6 < x < -1$,
$f'(x) > 0$ so f is increasing
$f''(x) < 0$ so f is concave down

When $-1 < x < -0.4$,
$f'(x) < 0$ so f is decreasing
$f''(x) < 0$ so f is concave down

When $-0.4 < x < 0$,
$f'(x) < 0$ so f is decreasing
$f''(x) > 0$ so f is concave up

When $x > 0$, $f'(x) > 0$ so f is increasing
$\quad\quad\quad\quad\quad f''(x) > 0$ so f is concave up

$(-2, 0)$ is a relative minimum, $(-1.6, 0.4)$
is an inflection point, $(-1, 1)$ is a relative
maximum, $(-0.4, 0.4)$ is an inflection

21. $f(x) = (2x-1)^2(x^2 - 9)$
$\quad\quad = (2x-1)^2(x+3)(x-3)$

domain: all real numbers
intercepts:
when $x = 0$, $f(0) = -9$; point $(0, -9)$

$f(x) = 0$, $x = \dfrac{1}{2}, -3, 3$; points $\left(\dfrac{1}{2}, 0\right)$,

$(-3, 0)$, $(3, 0)$
asymptotes: no vertical or horizontal
asymptotes.

$f'(x)$
$\quad = (2x-1)^2(2x) + (x^2-9)[2(2x-1)(2)]$
$\quad = 2(2x-1)(4x^2 - x - 18)$
$\quad = 2(2x-1)(4x-9)(x+2)$

$f'(x) = 0$ when $x = -2, \dfrac{1}{2}, \dfrac{9}{4}$

$f''(x) = [(2x-1)(8x-1) + (4x^2 - x - 18)(2)]$
$\quad\quad = 2(24x^2 - 12x - 35)$

$f''(x) = 0$ when $x = -0.98, 1.5$

When $x < -2$, $f'(x) < 0$ so f is decreasing
$\quad\quad\quad\quad\quad\quad f''(x) > 0$ so f is concave up

When $-2 < x < -0.98$,
$f'(x) > 0$ so f is increasing
$f''(x) > 0$ so f is concave up

When $-0.98 < x < 0.5$,
$f'(x) > 0$ so f is increasing
$f''(x) < 0$ so f is concave down

When $0.5 < x < 1.5$,
$f'(x) < 0$ so f is decreasing
$f''(x) < 0$ so f is concave down

When $1.5 < x < 2.25$,
$f'(x) < 0$ so f is decreasing
$f''(x) > 0$ so f is concave up
When $x > 2.25$,
$f'(x) > 0$ so f is increasing
$f''(x) > 0$ so f is concave up
$(-2, -125)$ is a relative minimum,
$(-0.98, -70.4)$ is an inflection point,
$(0.5, 0)$ is a relative maximum,
$(1.5, -26.2)$ is an inflection point, and $(2.25, -48.2)$ is a relative minimum.

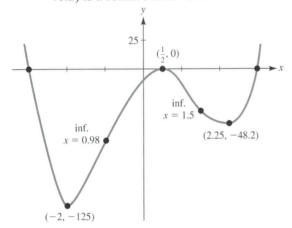

23. $f(x) = \dfrac{1}{2x+3}$

domain: $x \neq -\dfrac{3}{2}$

intercepts:

when $x = 0$, $f(0) = \dfrac{1}{3}$; point $\left(0, \dfrac{1}{3}\right)$

$f(x) \neq 0$ for any value of x
asymptotes:

$x = -\dfrac{3}{2}$ is a vertical asymptote

$y = 0$ is a horizontal asymptote

$f'(x) = -(2x+3)^{-2}(2) = \dfrac{-2}{(2x+3)^2}$

note that $f'(x) < 0$ for all x in domain

$f''(x) = -2[-2(2x+3)^{-3}(2)] = \dfrac{8}{(2x+3)^3}$

When $x < -1.5$,
$f'(x) < 0$ so f is decreasing
$f''(x) < 0$ so f is concave down

When $x > -1.5$,
$f'(x) < 0$ so f is decreasing
$f''(x) > 0$ so f is concave up

f is undefined for $x = -1.5$, so there are no
relative extrema or inflection points.

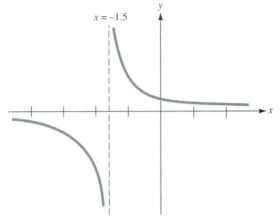

25. $f(x) = x - \dfrac{1}{x} = \dfrac{x^2 - 1}{x} = \dfrac{(x+1)(x-1)}{x}$

domain: $x \neq 0$
intercepts:
when $x = 0$, $f(0)$ undefined
$f(x) = 0$, $x = -1, 1$; points $(-1, 0)$, $(1, 0)$
asymptotes:
$x = 0$ is a vertical asymptote
no horizontal asymptote

$\left(\text{Note: } \lim_{x \to \pm\infty} \dfrac{x^2-1}{x} = \lim_{x \to \pm\infty} \dfrac{x - \frac{1}{x}}{1} = x, \text{ so}\right.$

$y = x$ is an oblique asymptote$)$

$f'(x) = 1 + \dfrac{1}{x^2} = \dfrac{x^2+1}{x}$

$f''(x) = -\dfrac{2}{x^3}$

When $x < 0$, $f'(x) > 0$ so f is increasing
$\qquad\qquad\qquad f''(x) > 0$ so f is concave up

When $x > 0$,
$f'(x) > 0$ so f is increasing
$f''(x) < 0$ so f is concave down
f is undefined for $x = 0$, so there are no

relative extrema or inflection points.

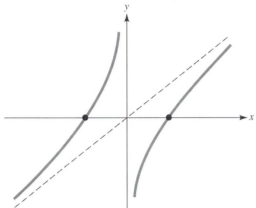

27. $f(x) = \dfrac{1}{x^2 - 9} = \dfrac{1}{(x+3)(x-3)}$

domain: $x \neq -3, 3$
intercepts:

when $x = 0$, $f(0) = -\dfrac{1}{9}$; point $\left(0, -\dfrac{1}{9}\right)$

$f(x) \neq 0$ for any value of x
asymptotes:
$x = -3$ and $x = 3$ are vertical asymptotes
$y = 0$ is a horizontal asymptote

$f'(x) = -(x^2 - 9)^2(2x) = \dfrac{-2x}{(x^2-9)^2}$

$f'(x) = 0$ when $x = 0$

$f''(x)$
$= \dfrac{(x^2-9)^2(-2) - (-2x)(2(x^2-9)(2x))}{(x^2-9)^4}$

$= \dfrac{6(x^2+3)}{(x^2-9)^3}$

When $x < -3$,
$f'(x) > 0$ so f is increasing
$f''(x) > 0$ so f is concave up
When $-3 < x < 0$,
$f'(x) > 0$ so f is increasing
$f''(x) < 0$ so f is concave down
When $0 < x < 3$,
$f'(x) < 0$ so f is decreasing
$f''(x) < 0$ so f is concave down
When $x > 3$, $f'(x) < 0$ so f is decreasing
$\qquad\qquad f''(x) > 0$ so f is concave up

$\left(0, -\dfrac{1}{9}\right)$ is a relative maximum. Since f is undefined for $x = -3, 3$, there are no other relative extrema or inflection points.

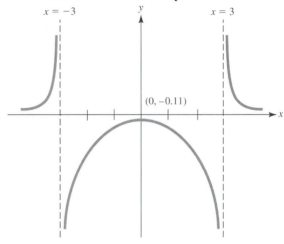

29. $f(x) = \dfrac{x^2 - 9}{x^2 + 1} = \dfrac{(x+3)(x-3)}{x^2 + 1}$

domain: all real numbers
intercepts:
when $x = 0$, $f(0) = -9$; point $(0, -9)$
$f(x) = 0$, $x = -3, 3$; points $(-3, 0)$, $(3, 0)$
asymptotes:
no vertical asymptotes
$y = 1$ is a horizontal asymptote

$f'(x) = \dfrac{(x^2+1)(2x) - (x^2-9)(2x)}{(x^2+1)^2}$

$\qquad = \dfrac{20x}{(x^2+1)^2}$

$f'(x) = 0$ when $x = 0$

$f''(x) = \dfrac{(x^2+1)^2(20) - (20x)[2(x^2+1)(2x)]}{(x^2+1)^4}$

$\qquad = \dfrac{20(-3x^2+1)}{(x^2+1)^3}$

$f''(x) = 0$ when $x = -\dfrac{1}{\sqrt{3}}, \dfrac{1}{\sqrt{3}}$

When $x < -\dfrac{1}{\sqrt{3}}$,

$f'(x) < 0$ so f is decreasing
$f''(x) < 0$ so f is concave down

When $-\dfrac{1}{\sqrt{3}} < x < 0,$

$f'(x) < 0$ so f is decreasing
$f''(x) > 0$ so f is concave up

When $0 < x < \dfrac{1}{\sqrt{3}},$

$f'(x) > 0$ so f is increasing
$f''(x) > 0$ so f is concave up

When $x > \dfrac{1}{\sqrt{3}},$

$f'(x) > 0$ so f is increasing
$f''(x) < 0$ so f is concave down

$(-0.58, -6.48)$ is an inflection point,
$(0, -9)$ is a relative minimum, and
$(0.58, -6.48)$ is an inflection point.

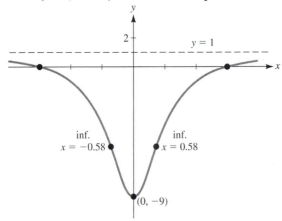

31. $f(x) = x^{3/2} = \sqrt{x^3}$

domain: $x \geq 0$
intercepts:
when $x = 0$, $f(0) = 0$; point $(0, 0)$
$f(x) = 0$, $x = 0$
asymptotes: no vertical or horizontal
asymptotes

$f'(x) = \dfrac{3}{2}x^{1/2} = \dfrac{3}{2}\sqrt{x}$

$f'(x) = 0$ when $x = 0$

$f''(x) = \dfrac{3}{4}x^{-1/2} = \dfrac{3}{4\sqrt{x}}$

When $x < 0$, f is undefined.
When $x > 0$, $f'(x) > 0$ so f is increasing
$\qquad\qquad\quad f''(x) > 0$ so f is concave up

$(0, 0)$ is a relative minimum.

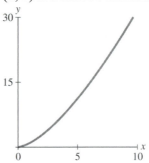

33. Answers will vary.

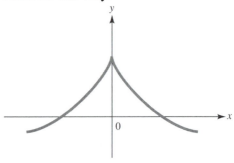

35. Answers will vary.

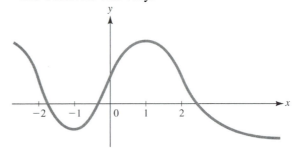

37. Answers will vary.

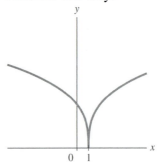

39. $f'(x) = x^3(x-2)^2$

(a) $f'(x) = 0$ when $x = 0, 2$

When

$x < 0$, $f'(x) < 0$ so f is decreasing
$0 < x < 2$, $f'(x) > 0$ so f is increasing
$x > 2$, $f'(x) > 0$ so f is increasing

(b) At $x = 0$, there is a relative minimum but there is no relative extrema at $x = 2$.

(c) $f''(x)$

$$= (x^3)[2(x-2)(1)] + (x-2)^2(3x^2)$$

$$= x^2(x-2)(5x-6)$$

$f''(x) = 0$ when $x = 0, \dfrac{6}{5}, 2$

When

$x < 0$, $f''(x) > 0$ so f is concave up
$0 < x < 1.2$, $f''(x) > 0$ so f is concave up
$1.2 < x < 2$, $f''(x) < 0$ so f is concave down
$x > 2$, $f''(x) > 0$ so f is concave up

Overall, f is concave up when $x < 0$, $0 < x < 1.2$, and when $x > 2$; f is concave down when $1.2 < x < 2$.

(d) At $x = 1.2$ and $x = 2$, there are inflection points.

41. $f'(x) = \dfrac{x+3}{(x-2)^2}$

(a) $f'(x) = 0$ when $x = -3$

$f'(x)$ is undefined when $x = 2$

When

$x < -3$, $f'(x) < 0$ so f is decreasing
$-3 < x < 2$, $f'(x) > 0$ so f is increasing
$x > 2$, $f'(x) > 0$ so f is increasing

(b) When $x = -3$, there is a relative minimum but there is no relative extrema at $x = 2$.

(c) $f''(x)$

$$= \frac{(x-2)^2(1) - (x+3)\cdot 2(x-2)(1)}{(x-2)^4}$$

$$= \frac{(x-2)^2[(x-2) - 2(x+3)]}{(x-2)^4}$$

$$= \frac{-x-8}{(x-2)^3}$$

$$= \frac{-(x+8)}{(x-2)^3}$$

$f''(x) = 0$ when $x = -8$

$f''(x)$ is undefined when $x = 2$

When

$x < -8$, $f''(x) < 0$ so f is concave down
$-8 < x < 2$, $f''(x) > 0$ so f is concave up
$x > 2$, $f''(x) < 0$ so f is concave down

(d) Since the concavity switches when $x = -8$, there is an inflection point when $x = -8$. The concavity switches when $x = 2$ as well, however f is undefined when $x = 2$.

43. To have a vertical asymptote of $x = 2$, the denominator must be zero for $x = 2$, so

$$5 + B(2) = 0 \rightarrow B = -\frac{5}{2}$$

To have a horizontal asymptote of $y = 4$

$$\lim_{x \to \pm\infty} \frac{Ax - 3}{5 - \frac{5}{2}x} = \lim_{x \to \pm\infty} \frac{A - \frac{3}{x}}{\frac{5}{x} - \frac{5}{2}}$$

$$= \frac{A}{-\frac{5}{2}}$$

$$= -\frac{2}{5}A$$

$$= 4,$$

so $A = -10$.

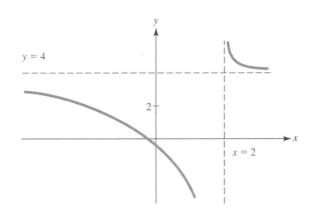

$y = 4$

$x = 2$

45. $C(x) = 3x^2 + x + 48$

$$A(x) = 3x + 1 + \frac{48}{x} = \frac{3x^2 + x + 48}{x}$$

(a) $x = 0$ is a vertical asymptote; there are no horizontal asymptotes.

(b) As $x \to \infty$, the graph of A approaches the line $y = 3x + 1$ asymptotically.

(c)

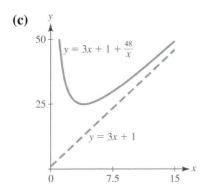

$y = 3x + 1 + \frac{48}{x}$

$y = 3x + 1$

47. $W(x) = \frac{200x}{100 - x}$

(a) domain: $0 \le x < 100$
intercepts:
when $x = 0$, $W(0) = 0$; point $(0, 0)$
$W(x) = 0$, $x = 0$
asymptotes: $x = 100$ is a vertical asymptote since $x \ge 0$, there is no horizontal asymptote

$$W'(x) = \frac{20,000}{(100 - x)^2}$$

$$W''(x) = \frac{40,000}{(100 - x)^3}$$

When $0 \le x < 100$,
$W'(x) > 0$ so W is increasing
$W''(x) > 0$ so W is concave up

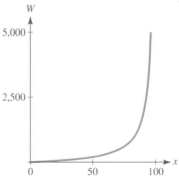

(b) $1500 = \frac{200x}{100 - x}$;
$150,000 - 1500x = 200x$; $x \approx 88.2\%$
will receive a new book, so
$100 - 88.2 = 11.8\%$ will not receive a new book.

49. $Q(x) = \frac{7x}{27 + x^2}$

(a) domain: $x \ge 0$
intercepts:
when $x = 0$, $Q(0) = 0$; point $(0, 0)$
$Q(x) = 0$, $x = 0$
asymptotes: the denominator is never zero, so there are no vertical asymptotes

$$\lim_{x \to \infty} Q(x) = \lim_{x \to \infty} \frac{\frac{7}{x}}{\frac{27}{x^2} + 1} = 0 \text{ so } y = 0 \text{ is}$$

a horizontal asymptote.

$$Q'(x) = \frac{(27 + x^2)(7) - (7x)(2x)}{(27 + x^2)^2}$$

$$= \frac{189 - 7x^2}{(27 + x^2)^2}$$

$$Q''(x) = \frac{-(189-7x^2)\cdot 2(27+x^2)(2x)}{(27+x^2)^4}$$

$$= \frac{2(27+x^2)[-7x(27+x^2)}{(27+x^2)^4}$$
$$\;\;\;\;\;\;\frac{-2x(189-7x^2)]}{(27+x^2)^4}$$

$$= \frac{2(-567x+7x^3)}{(27+x^2)^3}$$

$Q'(x)=0$ when $x=\sqrt{27}\approx 5.2$

$Q''(x)=0$ when $x=0,9$

When $0<x<5.2$,
$Q'(x)>0$ so Q is increasing
$Q''(x)<0$ so Q is concave down

When $5.2<x<9$,
$Q'(x)<0$ so Q is decreasing
$Q''(x)<0$ so Q is concave down

When $x>9$,
$Q'(x)<0$ so Q is decreasing
$Q''(x)>0$ so Q is concave up

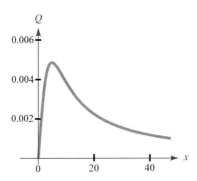

(b) Sales are maximized when $x=\sqrt{27}$, or a marketing expenditure of \$5,196. The corresponding maximum sales is

$$Q(5.196)=\frac{7(5.196)}{27+27}\approx 0.6736 \text{ or}$$

approximately 674 units.

51. $G(x)=\dfrac{1}{2,000}\left(\dfrac{800}{x}+5x\right)$

(a) total cost = cost driver + cost gas
cost driver = (#hrs)(pay/hr)
$$=\left(\frac{\#\text{mi}}{\text{mi/hr}}\right)(\text{pay/hr})$$

$$\text{cost gas}=(\#\text{mi})\left(\frac{\text{gal}}{\text{mi}}\right)\left(\frac{\text{cost}}{\text{gal}}\right)$$

$$C(x)=\left(\frac{400}{x}\right)(18)$$
$$+(400)\left[\frac{1}{2,000}\left(\frac{800}{x}+5x\right)\right](4.25)$$
$$=\frac{7,880}{x}+4.25x$$

domain: $30\le x\le 65$
intercepts: none in domain
asymptotes: none in domain

$$C'(x)=-\frac{7,880}{x^2}+4.25$$

$C'(x)=0$ when $x\approx 43$
When $30\le x<43$, $C'(x)<0$ so C is decreasing
When $43<x\le 65$, $C'(t)>0$ so C is increasing.

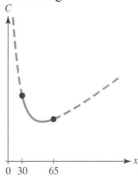

(b) When $x=43$, there is a minimum. So cost is minimized when the driver travels at 43 mph. The minimum cost is $C(43)\approx \$366.01$.

53. Answers will vary.

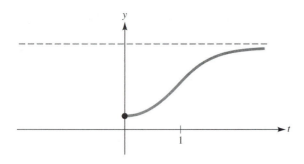

55. $f(n) = 3 + \dfrac{12}{n}$

(a) Since $\lim\limits_{n \to \infty} f(n) = \lim\limits_{n \to \infty}\left(3 + \dfrac{12}{n}\right) = 3$,

the y-values of the graph approach 3 as x gets large. Using this idea and a few points gives

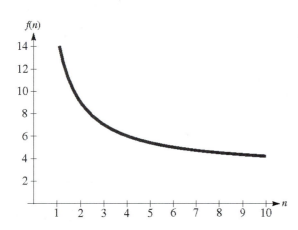

(b) Since n represents the number of trials, the positive integers are the practical values for n.

(c) $\lim\limits_{n \to \infty} f(n) = \lim\limits_{n \to \infty}\left(3 + \dfrac{12}{n}\right) = 3$ So, the rat's time decreases until it approaches 3 minutes.

57. $S(t) = \dfrac{100(t^2 - 3t + 25)}{t^2 + 7t + 25}$

(a) domain: $0 \le t \le 10$
intercepts:
when $t = 0$, $S(0) = 100$; point $(0, 100)$

$f(x) \ne 0$ for any x
asymptotes: there are no vertical asymptotes
$y = 100$ is a horizontal asymptote

$$S'(t) = \frac{100}{(t^2 + 7t + 25)^2}((t^2 + 7t + 25)(2t - 3)$$
$$- (t^2 - 3t + 25)(2t + 7))$$
$$= \frac{1000(t + 5)(t - 5)}{(t^2 + 7t + 25)^2}$$
$$= \frac{1000(t^2 - 25)}{(t^2 + 7t + 25)^2}$$

$S'(t) = 0$ when $t = 5$
When
$0 \le t < 5$, $S'(t) < 0$ so S is decreasing
$5 < 5 \le 10$, $S'(t) > 0$ so S is increasing

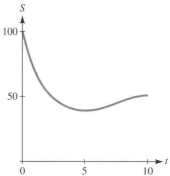

(b) When $t = 5$, there is a relative minimum, so her support is lowest when $t = 5$ and her minimum support level is $S(5) \approx 41.2\%$.

(c) When $t > 5$, $S'(t) > 0$ so $S'(10)$ is positive.

$$S''(t) = \frac{1000}{(t^2 + 7t + 25)^4}[(t^2 + 7t + 25)^2(2t)$$
$$- (t^2 - 25)(2(t^2 + 7t + 25)(2t + 7))]$$

When $t = 10$, $S''(10) < 0$ so S', or her approval rate, is decreasing.

59. $T(t) = -\dfrac{1}{36}t^3 + \dfrac{1}{8}t^2 + \dfrac{7}{3}t - 2$

(a) domain: $0 \le t \le 12$
intercepts: when $t = 0$, $T(0) = -2$;
point $(0, -2)$

x-intercepts too difficult to find

asymptotes: none

$$T'(t) = -\frac{1}{12}t^2 + \frac{1}{4}t + \frac{7}{3}$$

$T'(t) = 0$ when

$$t^2 - 3t - 28 = 0$$
$$(t + 4)(t - 7) = 0$$
$$t = 7 \text{ (deleting negative solution)}$$

$$T''(t) = -\frac{1}{6}t + \frac{1}{4}$$

$T''(t) = 0$ when

$$2t - 3 = 0$$
$$t = \frac{3}{2}$$

When $0 \le t < \frac{3}{2}$,

$T'(t) > 0$ so T is increasing
$T''(t) > 0$ so T is concave up

When $\frac{3}{2} < t < 7$,

$T'(t) > 0$ so T is increasing
$T''(t) < 0$ so T is concave down

When $7 < t \le 12$,
$T'(t) < 0$ so T is decreasing
$T''(t) < 0$ so T is concave down

Overall, T is increasing for $0 \le t < 7$; decreasing for $7 < t \le 12$; concave up for $0 \le t < \frac{3}{2}$; concave down for $\frac{3}{2} < t \le 12$. The critical value $t = 7$ corresponds to the point $(7, 10.9)$, which is an absolute maximum. When $t = \frac{3}{2}$, the corresponding point is $(1.5, 1.7)$,

which is an inflection point.

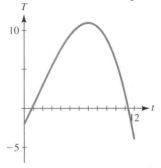

(b) The maximum occurs when $t = 7$, or at 1:00 P.M. The maximum temperature is approximately 10.9°C.

61. $f(x) = x^{2/3}(2x - 5)$

(a) $f'(x) = (x^{2/3})(2) + (2x - 5)\left(\frac{2}{3}x^{-1/3}\right)$

$$= 2x^{2/3} + \frac{2(2x - 5)}{3x^{1/3}}$$

$$= \frac{10(x - 1)}{3x^{1/3}}$$

$$= \frac{10}{3}x^{2/3} - \frac{10}{3}x^{-1/3}$$

$f'(x) = 0$ when $x = 1$.

When

$x < 0$, $f'(x) > 0$ so f is increasing
$0 < x < 1$, $f'(x) < 0$ so f is decreasing
$x > 1$, $f'(x) > 0$ so f is increasing

$(0, 0)$ is a relative maximum and
$(1, -3)$ is a relative minimum.
Since f is defined but f' is undefined for $x = 0$, there is a vertical tangent at $x = 0$.

(b) $f''(x) = \frac{20}{9}x^{-1/3} + \frac{10}{9}x^{-4/3}$

$$= \frac{10(2x + 1)}{9x^{4/3}}$$

$f''(x) = 0$ when $x = -\frac{1}{2}$

When

$x < -0.5$, $f''(x) < 0$ so f is concave down

$-0.5 < x < 0$, $f''(x) > 0$ so f is concave up

$x > 0$, $f''(x) > 0$ so f is concave up

$(-0.5, -3.8)$ is an inflection point.

(c) When $x = 0$, $f(0) = 0$, so y-intercept is 0. When $f(x) = 0$, $x = 0$, 2.5; so, x-intercepts are 0 and 2.5. There are no vertical or horizontal asymptotes.

(d)

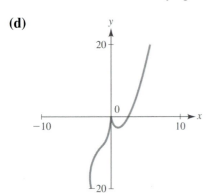

63. Let $f(x) = \dfrac{x-1}{x^2-1}$ and let $g(x) = \dfrac{x-1.01}{x^2-1}$.

(a) To use a graphing utility to sketch the graph of f,
Press $\boxed{y=}$ and input $(x-1)/(x^2-1)$ and press $\boxed{\text{Graph}}$.
At first appearance, the graph appears to be continuous at $x = 1$.
Use $\boxed{\text{2nd}}$ $\boxed{\text{calc}}$ and 1: value to evaluate $f(1)$. We see no y-value displayed for $x = 1$ which means $f(1)$ is undefined. From algebra, $f(x) = \dfrac{x-1}{(x+1)(x-1)}$.
We can cancel the common factor $x - 1$, which leaves a "hole" in the

graph of f at $x = 1$.

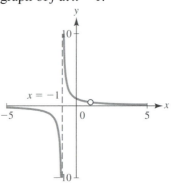

(b) To sketch a graph of g,
Press $\boxed{y=}$ and input $(x-1.01)/(x^2-1)$ and press $\boxed{\text{Graph}}$.
The graph of g appears to be the same as the graph for f. However, by tracing and zooming in at $x = 1$, we see the vertical asymptote appears at $x = 1$. In addition, using $\boxed{\text{2nd}}$ $\boxed{\text{calc}}$ and 1: value to evaluate $g(1)$ also produces an undefined y-value. The reason for this is not due to a "hole" in the graph for g but rather the vertical asymptote $x = 1$.

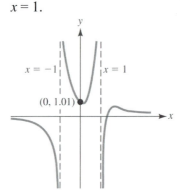

3.4 Optimization; Elasticity of Demand

1. $f(x) = x^2 + 4x + 5$, $-3 \le x \le 1$
$f'(x) = 2x + 4 = 2(x + 2)$
$f'(x) = 0$ when $x = -2$, which is in the interval $f(-2) = 1$, $f(-3) = 2$ and $f(1) = 10$.
So, $f(1) = 10$ is the absolute maximum and $f(-2) = 1$ is the absolute minimum.

3. $f(x) = \frac{1}{3}x^3 - 9x + 2, \ 0 \le x \le 2$

$f'(x) = x^2 - 9 = (x+3)(x-3)$

$f'(x) = 0$ when $x = -3$ and $x = 3$, which are not in the interval.

$f(0) = 2$, which is the absolute maximum and $f(2) = -\frac{40}{3}$, which is the absolute minimum.

5. $f(t) = 3t^5 - 5t^3, \ -2 \le t \le 0$

$f'(t) = 15t^4 - 15t^2 = 15t^2(t+1)(t-1)$

$f'(t) = 0$ when $t = -1$, $t = 0$ and $t = 1$, of which $t = -1$ and $t = 0$ are in the interval.

$f(-1) = 2, f(0) = 0, f(-2) = -56$

So, $f(-1) = 2$ is the absolute maximum and $f(-2) = -56$ is the absolute minimum.

7. $f(x) = (x^2 - 4)^5, \ -3 \le x \le 2$

$f'(x) = 5(x^2 - 4)^4(2x)$

$\quad = 10x(x+2)^4(x-2)^4$

$f'(x) = 0$ when $x = -2$, $x = 0$, and $x = 2$, all of which are in the interval.

$f(-2) = 0, f(0) = -1{,}024, f(2) = 0$ and $f(-3) = 3{,}125$

So, $f(-3) = 3{,}125$ is the absolute maximum and $f(0) = -1{,}024$ is the absolute minimum.

9. $g(x) = x + \frac{1}{x}, \ \frac{1}{2} \le x \le 3$

$g'(x) = 1 - \frac{1}{x^2} = \frac{x^2 - 1}{x^2} = \frac{(x+1)(x-1)}{x^2}$

$g'(x) = 0$ when $x = -1$ and $x = 1$, of which $x = 1$ is in the interval.

$g'(x)$ is undefined at $x = 0$, however, $x = 0$ is not in the interval

$g(1) = 2, \ g\left(\frac{1}{2}\right) = \frac{5}{2}, \ g(3) = \frac{10}{3}$

So, $g(3) = \frac{10}{3}$ is the absolute maximum and $g(1) = 2$ is the absolute minimum.

11. $f(u) = u + \frac{1}{u}, \ u > 0$

$f'(u) = 1 - \frac{1}{u^2} = \frac{u^2 - 1}{u^2} = \frac{(u+1)(u-1)}{u^2}$

$f'(u) = 0$ when $u = -1$ and $u = 1$, of which $u = 1$ is in the interval.

$f'(u)$ is undefined when $u = 0$, which is not in the interval

When $0 < x < 1$, $f'(x) < 0$ so f is decreasing

$\quad x > 1$, $f'(x) > 0$ so f is increasing

Since there are no endpoints, $f(1) = 2$ is the absolute minimum and there is no absolute maximum.

13. $f(x) = \frac{1}{x}, \ x > 0$

$f'(x) = -\frac{1}{x^2}$

$f'(x)$ is never zero and $f'(x)$ is undefined when $x = 0$, which is not in the domain. Also, there are no endpoints. So, there is no absolute maximum or absolute minimum.

15. $f(x) = \frac{1}{x+1}, \ x \ge 0$

$f'(x) = -(x+1)^{-2}(1) = -\frac{1}{(x+1)^2}$

$f'(x)$ is never zero and $f'(x)$ is undefined when $x = -1$, which is not in the domain. When $x > 0$, $f'(x) < 0$ so f is decreasing. So, $f(0) = 1$ is the absolute maximum and there is no absolute minimum.

17. $p(q) = 49 - q$ and $C(q) = \frac{1}{8}q^2 + 4q + 200$

(a) $R(q) = qp(q) = 49q - q^2$

$R'(q) = 49 - 2q$

$C'(q) = \frac{1}{4}q + 4$

The profit function is

$$P(q) = R(q) - C(q) = \frac{9}{8}q^2 + 45q - 200$$

$$P'(q) = -\frac{9}{4}q + 45$$

$P'(q) = 0$ when $q = 20$, so profit is maximized when 20 units are produced.

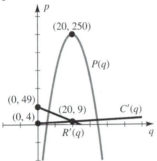

(b) $A(q) = \dfrac{C(q)}{q} = \dfrac{1}{8}q + 4 + \dfrac{200}{q}$

$$A'(q) = \frac{1}{8} - \frac{200}{q^2}$$

$A'(q) = 0$ when $q = 40$, so the average cost is minimized when 40 units are produced.

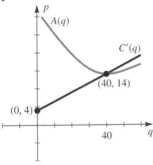

19. $p(q) = 180 - 2q$ and $C(q) = q^3 + 5q + 162$

(a) $R(q) = qp(q) = 180q - 2q^2$

$R'(q) = 180 - 4q$

$C'(q) = 3q^2 + 5$

The profit function is

$P(q) = R(q) - C(q)$

$\quad = -q^3 - 2q^2 + 175q - 162$

$P'(q) = -3q^2 - 4q + 175$

$P'(q) = 0$ when $q = 7$ (rejecting negative solution), so profit is maximized when 7 units are produced.

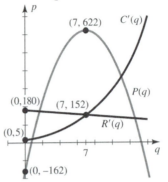

(b) $A(q) = \dfrac{C(q)}{q} = q^2 + 5 + \dfrac{162}{q}$

$$A'(q) = 2q - \frac{162}{q^2}$$

$A'(q) = 0$ when $q \approx 4.327$, so the average cost is minimized when 4.327 units are produced.

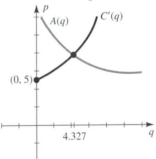

21. $p(q) = 1.0625 - 0.0025q$ and

$$C(q) = \frac{q^2 + 1}{q + 3}$$

(a) $R(q) = qp(q) = 1.0625q - 0.0025q^2$

$R'(q) = 1.0625 - 0.005q$

$$C'(q) = \frac{q^2 + 6q - 1}{(q + 3)^2}$$

The profit function is

$P(q) = R(q) - C(q)$

$$= 1.0625q - 0.0025q^2 - \frac{q^2 + 1}{q + 3}$$

$$= \frac{1}{q+3}[-0.0025q^3 + 0.055q^2$$

$$+ 3.1875q - 1]$$

$$P'(q) = \frac{1}{(q+3)^2}[(q+3)(-0.0075q^2$$

$$+ 0.11q + 3.1875)$$

$$+ 0.0025q^3 - 0.055q^2$$

$$- 3.1875q + 1]$$

$$= \frac{1}{(q+3)^2}[-0.005q^3$$

$$+ 0.0325q^2 + 0.33q$$

$$+ 10.5625]$$

Press $\boxed{y=}$ and input P, R', and C' for $y_1 =$, $y_2 =$, and $y_3 =$, respectively. Use window dimensions $[0, 45]5$ by $[0, 3]0.5$
Press $\boxed{\text{graph}}$
Use the maximum function under the calc menu to find the relative maximum of P occurs at $x = 17.3361$.

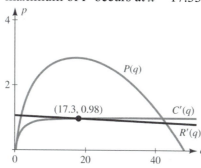

(b) $A(q) = \frac{C(q)}{q} = \frac{q^2 + 1}{q(q+3)}$

$$A'(q) = \frac{1}{(q^2 + 3q)^2}[2q(q^2 + 3q)$$

$$- (q^2 + 1)(2q + 3)]$$

$$= \frac{3q^2 - 2q - 3}{(q^2 + 3q)^2}$$

Press $\boxed{y=}$ and input A and C' for $y_1 =$ and $y_2 =$, respectively. Use window

dimensions of $[0, 6]0.5$ by $[0, 1.5]0.2$.
Press $\boxed{\text{graph}}$
Use the minimum function under the calc menu to find the relative minimum occurs at $q = 1.3874$.

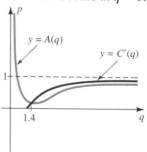

23. $D(p) = -1.3p + 10$

$$E(p) = -\frac{p}{D(p)} \cdot \frac{dD}{dp}$$

$$= \frac{-p}{-1.3p + 10}(-1.3)$$

$$E(4) = \frac{1.3(4)}{-1.3(4) + 10} = \frac{13}{12}$$

$|E(4)| > 1$, so the demand is elastic.

25. $D(p) = 200 - p^2$

$$E(p) = -\frac{p}{D(p)} \cdot \frac{dD}{dp}$$

$$= \frac{-p}{200 - p^2}(-2p)$$

$$= \frac{2p^2}{200 - p^2}$$

$$E(10) = \frac{2(10)^2}{200 - (10)^2} = 2$$

$|E(10)| > 1$, so the demand is elastic.

27. $D(p) = \dfrac{3,000}{p} - 100$

$E(p) = -\dfrac{p}{D(p)} \cdot \dfrac{dD}{dp}$

$= \dfrac{-p}{\dfrac{3,000}{p} - 100}\left(-\dfrac{3,000}{p^2}\right)$

$= \dfrac{p}{\dfrac{3,000 - 100p}{p}}\left(\dfrac{3,000}{p^2}\right)$

$= \dfrac{p^2}{100(30 - p)}\left(\dfrac{3,000}{p^2}\right)$

$= \dfrac{30}{30 - p}$

$E(10) = \dfrac{30}{30 - 10} = \dfrac{3}{2}$

$|E(10)| > 1$, so the demand is elastic.

29. Need to find the maximum absolute value of the slope of the graph. The slope is

$f'(x) = 4x - x^2$.

To maximize $|f'|$ on the interval

$-1 \le x \le 4$,

$f''(x) = 4 - 2x = 2(2 - x)$

$f''(x) = 0$ when $x = 2$

Now,

$|f'(2)| = |4| = 4$

$|f'(-1)| = |-5| = 5$

$|f'(4)| = |0| = 0$

So, slope is steepest when $x = -1$, and its value is $f'(-1) = -5$.

31. $P(q) = -2q^2 + 68q - 128$

(a) aver profit $A(q) = \dfrac{P(q)}{q}$

$A(q) = -2q + 68 - \dfrac{128}{q}$

marginal profit is P'

$P'(q) = -4q + 68$

(b) $A(q) = P'(q)$

$-2q + 68 - \dfrac{128}{q} = -4q + 68$

$-\dfrac{128}{q} = -2q$

$64 = q^2$

$q = 8$ units

(c) $A(q) = -2q + 68 - \dfrac{128}{q}$

$A'(q) = -2 + \dfrac{128}{q^2}$

$A'(q) = 0$ when

$0 = -2 + \dfrac{128}{q^2}$

$2 = \dfrac{128}{q^2}$

$q^2 = 64$

$q = 8$

When $0 \le q < 8$, $A'(q) > 0$ so A is increasing

$q > 8$, $A'(q) < 0$ so A is decreasing

So, A is a maximum when $q = 8$ units.

(d)

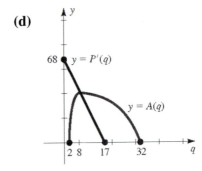

33. $q^2 + 3pq = 22$

(a) Using implicit differentiation,

$$2q\frac{dq}{dp} + (3p)\frac{dq}{dp} + (q)(3) = 0$$

$$\frac{dq}{dp} = \frac{-3q}{2q + 3p}$$

$$E(p) = -\frac{p}{q} \cdot \frac{dq}{dp}$$

$$D(p) = q$$

So, $E(p) = -\frac{p}{q}\left(\frac{-3q}{2q + 3p}\right) = \frac{3p}{2q + 3p}$.

(b) When $p = 3$, $q^2 + 9q = 22$, or $q = 2$ (rejecting negative root).

$$|E(3)| = \left|\frac{3 \cdot 3}{2 \cdot 2 + 3 \cdot 3}\right| = \left|\frac{9}{13}\right| = \frac{9}{13}$$

Since $|E(3)| < 1$, the demand is inelastic.

35. (a) When $q = 50$, $50 = 500 - 2p$, or $p = 225$. Further, when $q = 0$, $0 = 500 - 2p$, or $p = 250$. So, the range for price is $225 \le p \le 250$.

(b) $E(p) = -\frac{p}{q} \cdot \frac{dq}{dp} = \frac{-p}{500 - 2p}(-2)$

$$= \frac{2p}{500 - 2p} = \frac{p}{250 - p}$$

$$\left|\frac{p}{250 - p}\right| = 1 \text{ when } \frac{p}{250 - p} = \pm 1 \text{ or,}$$

when $p = 125$ and demand is of unit elasticity.

When $p < 125$, $|E_p| < 1$ and demand is inelastic.

When $p > 125$, $|E_p| > 1$ and demand is elastic.

(c) When the price is less than \$125, total revenue is increasing as price increases; when the price is \$125, total revenue is unaffected by a small change in price, when the price is more than \$125, total revenue is decreasing as price increases.

(d) If an unlimited number of prints is available, should charge \$125 each; if only 50 prints are available, should charge \$225, the value in the price interval which is closest to \$125.

37. The relationship between the number of Moppsy dolls and Floppsy dolls is given by $y = \frac{82 - 10x}{10 - x}$ with the relevant interval $0 \le x \le 8$.

Let C be the amount received from the sale of Floppsy doll. Then, $2C$ is the amount received from the sale of each Moppsy doll. The total revenue from the sale of both dolls is

$$R(x) = Cx + \frac{2C(82 - 10x)}{10 - x}$$

$$= C\left(\frac{164 - 10x - x^2}{10 - x}\right)$$

$$R'(x) = \frac{C}{(10 - x^2)}[(10 - x)(-10 - 2x)$$

$$- (164 - 10x - x^2)(-1)]$$

$$= C\left(\frac{x^2 - 20x + 64}{(10 - x)^2}\right)$$

$$= C\frac{(x - 16)(x - 4)}{(10 - x)^2}$$

$R'(x) = 0$ when $x = 4$ ($x = 16$ is not in the interval). Since $R(4) = 18C$, $R(0) = 16.4C$, and $R(8) = 10C$, revenue is maximized when 400 Floppsy and $\frac{82 - 10(4)}{10 - 4}$, or 700 Moppsy dolls are produced.

39. Let x be the number of hours worked after 8:00 A.M. before the coffee break. Then, $4 - x$ will be the number of hours worked after the break. The total number of units assembled will be

$$N(x) = f(x) + g(4-x)$$
$$= -x^3 + 6x^2 + 15x - \frac{1}{3}(4-x)^3$$
$$+ (4-x)^2 + 23(4-x)$$
$$N'(x) = -3x^2 + 12x + 15 - (4-x)^2(-1)$$
$$+ 2(4-x)(-1) + 23(-1)$$
$$N'(x) = -2x^2 + 6x = -2x(x-3)$$
$$N'(x) = 0 \text{ when } x = 0, x = 3$$

Testing these values along with the interval endpoints ($0 \le x \le 4$) gives
$N(0) = 86.67$, $N(3) = 95.67$; $N(4) = 92$
So, to assemble the maximum number of units, the break should be scheduled when $x = 3$. That is, at 11:00 A.M.

41. $q = b - ap$

(a) $E(p) = -\frac{p}{q} \cdot \frac{dq}{dp} = \frac{-p}{b-ap}(-a) = \frac{ap}{b-ap}$

(b) $|E(p)| = 1$ when $\left| \frac{ap}{b-ap} \right| = 1$, or when

$\frac{ap}{b-ap} = \pm 1$, or $p = \frac{b}{2a}$.

(c) $|E(p)| < 1$ when $p < \frac{b}{2a}$, so demand

is inelastic when $0 \le p < \frac{b}{2a}$.

$|E(p)| > 1$ when $p > \frac{b}{2a}$, so demand

is elastic when $\frac{b}{2a} < p \le \frac{b}{a}$.

43. $q = \frac{a}{p^m} = ap^{-m}$

The elasticity of demand is

$$E(p) = -\frac{p}{q} \cdot \frac{dq}{dp}$$
$$= -\frac{p}{\frac{a}{p^m}}(-amp^{-m-1})$$
$$= \frac{-p^{m+1}}{a}\left(\frac{-am}{p^{m+1}} \right)$$
$$= m$$

When
$0 < m < 1$, $|E(p)| < 1$ and demand is inelastic
$m = 1$, $|E(p)| = 1$ and demand is of unit elasticity
$m > 1$, $|E(p)| > 1$ and demand is elastic

45. $P(x) = 100(2x^3 - 45x^2 + 264x)$

(a) The period of time between 2000 and 2013 corresponds to the interval $2 \le x \le 15$.
$$P'(x) = 100(6x^2 - 90x + 264)$$
$$= 600(x-4)(x-11)$$
$$P'(x) = 0 \text{ when } x = 4 \text{ and } x = 11$$
$P(2) = 36,400$; $P(4) = 46,400$;
$P(11) = 12,100$; $P(15) = 58,500$
The maximum membership occurred when $x = 15$, or in the year 2013. The minimum membership occurred when $x = 11$, or in the year 2009.

(b) The maximum was
$P(15) = 58,500$ members and the minimum was
$P(11) = 12,100$ members.

47. $S(r) = c(R^2 - r^2)$, where c is a positive constant. The relevant interval is $0 \le r \le R$.
$$S'(r) = -2cr$$
$$S'(r) = 0 \text{ when } r = 0$$
(the left-hand endpoint of the interval)
With $S(0) = cR^2$ and $S(r) = 0$, the speed of the blood is greatest when $r = 0$, that is, at the central axis.

49. $F(p) = p^n(1-p)^{m-n}$

$$F'(p) = p^n(m-n)(1-p)^{m-n-1}(-1)$$
$$+ (1-p)^{m-n}(n)(p^{n-1})$$
$$= -p^n(m-n)(1-p)^{m-n-1}$$
$$+ p^{n-1}(n)(1-p)^{m-n}$$
$$= p^{n-1}(1-p)^{m-n-1}$$
$$[-(m-n)p + n(1-p)]$$
$$= p^{n-1}(1-p)^{m-n-1}$$
$$[-mp + np + n - np]$$

$F'(p) = 0$ when $p = 0$, 1, and $\dfrac{n}{m}$

$F(0) = 0$, $F(1) = 0$

Since n, m are positive and $m \geq n$, $\dfrac{n}{m}$ is in interval.

$$F\left(\frac{n}{m}\right) = \left(\frac{n}{m}\right)^n \left(1 - \frac{n}{m}\right)^{m-n}, \text{ and}$$

$F\left(\dfrac{n}{m}\right) > 0$, so, $p = \dfrac{n}{m}$ gives the absolute maximum.

51. $E(v) = \dfrac{1}{v}[0.074(v-35)^2 + 22]$

(a) $E'(v) = \left(\dfrac{1}{v}\right)[0.148(v-35)(1)]$

$$+ [0.074(v-35)^2 + 22]\left(\frac{-1}{v^2}\right)$$
$$= \frac{1}{v}[0.148v - 5.18$$
$$- \frac{1}{v}(0.074v^2 - 5.18v$$
$$+ 112.65)]$$
$$= \frac{1}{v}(0.148v - 5.18 - 0.074v$$
$$+ 5.18 - \frac{112.65}{v})$$
$$= \frac{1}{v}\left(0.074v - \frac{112.65}{v}\right)$$

So, $E'(v) = 0$ when

$$0.074v - \frac{112.65}{v} = 0$$
$$0.074v = \frac{112.65}{v}$$
$$v^2 \approx 1522.3$$
$$v \approx 39$$

$$E''(v) = \left(\frac{1}{v}\right)\left(0.074 + \frac{112.65}{v^2}\right)$$
$$+ \left(0.074v - \frac{112.65}{v}\right)\left(-\frac{1}{v^2}\right)$$

$E''(39) > 0$, so there is an absolute minimum when $v = 39$.

(b) Writing Exercise—Answers will vary.

53. $R(D) = D^2\left(\dfrac{C}{2} - \dfrac{D}{3}\right) = \dfrac{C}{2}D^2 - \dfrac{1}{3}D^3$

(a) To maximize $R'(D)$,

$$R'(D) = CD - D^2$$
$$R''(D) = C - 2D$$

$R''(D) = 0$ when $D = \dfrac{C}{2}$

$$R'''(D) = -2$$

$R'''\left(\dfrac{C}{2}\right)$ is negative, $D = \dfrac{C}{2}$ is a maximum for sensitivity. The sensitivity when $D = \dfrac{C}{2}$ is

$$R'\left(\frac{C}{2}\right) = C\left(\frac{C}{2}\right) - \left(\frac{C}{2}\right)^2 = \frac{C^2}{4}.$$

(b) The reaction when $D = \dfrac{C}{2}$ is

$$R\left(\frac{C}{2}\right) = \left(\frac{C}{2}\right)^2\left[\frac{C}{2} - \frac{\frac{C}{2}}{3}\right] = \frac{C^3}{12}.$$

55. (a) $P(x) = \dfrac{Ax}{B + x^m}$

$R(x) = 0$ when $x = \left(\dfrac{B}{m-1}\right)^{1/m}$.

$$R(x) = P'(x) = A\frac{(B + x^m) - mxx^{m-1}}{(B + x^m)^2}$$

$$= \frac{A[B + (1-m)x^m]}{(B + x^m)^2}$$

(b) $R'(x) = \dfrac{A}{(B + x^m)^4}[(B + x^m)^2[m(1-m)x^{m-1}] - [B + (1-m)x^m][2(B + x^m)(mx^{m-1})]]$

$$= \frac{A(B + x^m)nx^{m-1}}{(B + x^m)^3}[(B + x^m)(1-m) - 2(B + (1-m)x^m)]$$

$$= \frac{-Amx^{m-1}[B(1+m) + x^m(1-m)]}{(B + x^m)^3}$$

$R'(x) = 0$ when $x = \left[\dfrac{B(m+1)}{m-1}\right]^{1/m}$.

(c) Assuming $m > 1$, when $0 < x < \left[\dfrac{B(m+1)}{m-1}\right]^{1/m}$, $R'(x) > 0$ so R is increasing.

when $x > \left[\dfrac{B(m+1)}{m-1}\right]^{1/m}$, $R'(x) < 0$ so R is decreasing.

So there is a relative maximum when $x = \left[\dfrac{B(m+1)}{m-1}\right]^{1/m}$.

57. $I = \dfrac{E}{r + R}$, $P(r) = I^2 R = \dfrac{E^2 R}{(r + R)^2}$

$P'(R)$

$$= \frac{(r + R)^2(E^2) - (E^2 R)[2(r + R)]}{(r + R)^4}$$

$$= \frac{E^2(r + R)[(r + R) - 2R]}{(r + R)^4}$$

$$= \frac{E^2(r - R)}{(r + R)^3}$$

$P'(R) = 0$ when $R = r$

When $R = 0$, $P(0) = 0$

$0 < R < r$, $P'(R) > 0$ so P is increasing

$R > r$, $P'(R) < 0$ so P is decreasing

So, $R = r$ results in maximum power.

59. $F(v) = Av^2 + \dfrac{B}{v^2} = Av^2 + Bv^{-2}$

To find the minimum, need to find when $F'(v) = 0$.

$F'(v) = 2Av - 2Bv^{-3} = 2Av - \dfrac{2B}{v^3}$

$F'(v) = 0$ when

$$0 = 2Av - \frac{2B}{v^3}$$

$$0 = 2Av^4 - 2B$$

$$2Av^4 = 2B$$

$$v^4 = \frac{B}{A}, \text{ or } v = \sqrt[4]{\frac{B}{A}}$$

To show this corresponds to a minimum, use the second derivative test.

$$F''(v) = 2A + 6Bv^{-4} = 2A + \frac{6B}{v^4}$$

$$F''\left(\sqrt[4]{\frac{B}{A}}\right) = 2A + \frac{6B}{\left(\sqrt[4]{\dfrac{B}{A}}\right)^4}$$

$$= 2A + \frac{6B}{\dfrac{B}{A}} = 2A + 6A = 8A$$

Since $F''\left(\sqrt[4]{\dfrac{B}{A}}\right) > 0$, this is the minimum value.

61. Given $R'(p) = q(p)\big[-E(p) + 1\big]$.

When $E(p) > 1$,

$R'(p) = q(p)[\text{value} < 0]$ so $R'(p) < 0$ and R is decreasing.

When $E(p) < 1$,

$R'(p) = q(p)[\text{value} > 0]$ so $R'(p) > 0$ and R is increasing.

When $E(p) = 1$,

$R'(p) = q(p)(0)$ so R is maximized.

3.5   Additional Applied Optimization

1. Let x denote the number that exceeds its square, x^2, by the largest amount. Then, $f(x) = x - x^2$ is the function to be maximized.

$f'(x) = 1 - 2x$

$f'(x) = 0$ when $x = \dfrac{1}{2}$

$f''(x) = -2$, so $f''\left(\dfrac{1}{2}\right) < 0$

and there is a relative maximum when $x = \dfrac{1}{2}$. Further, since $f''(x) < 0$ for all x, it is the absolute maximum. So, $x = \dfrac{1}{2}$ is the desired number.

3. Let x be the first number and y be the second. Then, $P = xy$, or since $y = 50 - x$,

$P(x) = x(50 - x) = 50x - x^2$, which is the function to be maximized.

$P'(x) = 50 - 2x$

$P'(x) = 0$ when $x = 25$

$P''(x) = -2$, so $P''(25) < 0$

and there is a relative maximum when $x = 25$. Further, since $P''(x) < 0$ for all x in the domain $0 < x < 50$, it is the absolute maximum. So, $x = 25$ and $y = 50 - 25 = 25$ are the desired numbers.

5. Let x be the length of the field and y be the width. The amount of fencing is the perimeter of the field, or $P = 2x + 2y$.

Since the area is 3,600,

$\qquad A = xy$

$3,600 = xy$, or $y = \dfrac{3,600}{x}$

and $P(x) = 2x + 2\left(\dfrac{3,600}{x}\right) = 2x + \dfrac{7200}{x}$ which is the function to be minimized.

$P'(x) = 2 - \dfrac{7,200}{x^2}$

$P'(x) = 0$ when $x = 60$

$P''(x) = \dfrac{14,400}{x^3}$, or $P''(60) > 0$

and there is a relative minimum when $x = 60$. Further, since $P''(x) > 0$ for all x in the domain $x > 0$, it is the absolute maximum. So, the field should have a length of 60 meters and a width of 60 meters.

7. Let x be the length of the rectangle and y be the width. The area is $A = xy$.
Since the perimeter is fixed, let C represent its fixed value. Then,
$P = 2x + 2y$

$C = 2x + 2y$, so $y = \dfrac{C - 2x}{2}$ and $A(x) = x\left(\dfrac{C - 2x}{2}\right) = \dfrac{C}{2}x - x^2$ which is the function to be

maximized.

$A'(x) = \dfrac{C}{2} - 2x$

$A'(x) = 0$ when $x = \dfrac{C}{4}$

$A''(x) = -2$, so $A''\left(\dfrac{C}{4}\right) < 0$ and there is a relative maximum when $x = \dfrac{C}{4}$. Further, since $A''(x) < 0$

for all x in the domain $0 < x < \dfrac{C}{2}$, it is the absolute maximum. When $x = \dfrac{C}{4}$, $y = \dfrac{C}{4}$. So for any

given perimeter, a square is the rectangle having the maximum area.

9. Let x be the length of the rectangle and let y be the vertical distance above the rectangle along the side
of length 5. Then, $5 - y$ is the width of the rectangle. The area of the rectangle is $A = x(5 - y)$.

By similar triangles, $\dfrac{12}{5} = \dfrac{x}{y}$, or $y = \dfrac{5}{12}x$ and $A(x) = x\left(5 - \dfrac{5}{12}x\right) = 5x - \dfrac{5}{12}x^2$ which is the

function to be maximized.

$A'(x) = 5 - \dfrac{5}{6}x$

$A'(x) = 0$ when $x = 6$

$A''(x) = -\dfrac{5}{6}$, so $A''(6) < 0$

and there is a relative maximum when
$x = 6$. Further, since $A''(x) < 0$ for all x in the domain $0 < x < 12$, it is the absolute maximum. The
dimensions of the rectangle having the maximum area are

$x = 6$ and $y = \dfrac{5}{12}(6) = 2.5$.

11. Let x be the \$1.00 price increments above \$40.00. Then $40 + x$ will be the price per computer game,
$50 - 3x$ will be the number of units sold per month, and the profit will be
$P(x) = (50 - 3x)[(40 + x) - 25]$

$\qquad = 750 + 5x - 3x^2$
which is the function to be maximized.
$P'(x) = 5 - 6x$

$P'(x) = 0$ when $x = \dfrac{5}{6}$

$P''(x) = -6$, so $P''\left(\dfrac{5}{6}\right) < 0$ and there is a relative maximum when $x = \dfrac{5}{6}$. Further, since $P''(x) < 0$

for all x in the domain

$x \geq 0$, it is the absolute maximum. So, the selling price for maximum profit is $40 + \dfrac{5}{6} \approx \$41$.

13. Let x be the number of additional trees planted per acre. The number of oranges per tree will be $400 - 4x$ and the number of trees per acre $60 + x$. The yield per acre is

$$y(x) = \left(\frac{\text{\# of oranges}}{\text{tree}}\right)\left(\frac{\text{\# of trees}}{\text{acre}}\right)$$

$$= (400 - 4x)(60 + x)$$

$$= 24{,}000 + 160x - 4x^2$$

$y'(x) = 160 - 8x$

$y'(x) = 0$ when $x = 20$

$y''(x) = -8$, so $y''(20) < 0$ and there is a relative maximum when $x = 20$. Further, since $y''(x) < 0$ for all x in the domain

$x \geq 0$, it is the absolute maximum. So, the yield is maximized when there are $60 + 20 = 80$ trees per acre.

15. Profit = (#sold)(profit per card)
Let x be the number of 25 cent reductions in price. The profit per card will be
(selling price) − (cost to obtain)

$= (10 - 0.25x) - 5$

$= 5(1 - 0.05x)$

while the number of cards sold will be

$25 + 5x = 5(5 + x)$.

The total profit will be

$P(x) = 25(5 + x)(1 - 0.05x)$

$\qquad = 25(5 + 0.75x - 0.05x^2)$

$P'(x) = 18.75 - 2.5x$

$P'(x) = 0$ when $x = 7.5$

Seven and a half 25-cents reductions means a total reduction of $\$1.875$. That is, sell the cards for $10 - 1.875 = \$8.12$ or $\$8.13$ per card.

17. Let x be the distance down the opposite bank where the cable meets the bank. Then, the cost of the cable under the water is given by $C_w = 25\sqrt{x^2 + 1200^2}$.

The cost of the cable over land is $C_l = 20(1500 - x)$.

The total cost is given by

$C(x)$

$= 25(x^2 + 1{,}440{,}000)^{1/2} + 20(1500 - x)$

$C'(x)$

$= \dfrac{25}{2}(x^2 + 1{,}440{,}000)^{-1/2}(2x) + 20(-1)$

$= \dfrac{25x}{\sqrt{x^2 + 1{,}440{,}000}} - 20$

$C'(x) = 0$ when

$$0 = \frac{25x}{\sqrt{x^2 + 1,440,000}} - 20$$

$$20 = \frac{25x}{\sqrt{x^2 + 1,440,000}}$$

$$\sqrt{x^2 + 1,440,000} = \frac{5}{4}x$$

$$x^2 + 1,440,000 = \frac{25}{16}x^2$$

$$1,440,000 = \frac{9}{16}x^2$$

$$2,560,000 = x^2$$

$$x = 1600$$

Since the maximum value of x is 1500, disregard this answer and check the endpoints ($0 \le x \le$ 1500).

$C(0) = 60,000$

$C(1500) \approx 48,023$

So, the minimum cost occurs when the cable runs entirely under water.

19. Let $P(x)$ be the profit from the sale of the wine at time x in years.

$$\text{Profit} = \text{value} - \frac{\text{purchase}}{\text{cost}} - \frac{\text{storage}}{\text{cost}}$$

Let $V(x)$ be the value of the wine at time x, and let C be the purchase cost of the wine. Since the storage cost is $3x$,

$P(x) = V(x) - C - 3x$ which is the function to maximize and $P'(x) = V'(x) - 3$.

Since the rate of change of value is

$53 - 10x$,

$P'(x) = 50 - 10x$

$P'(x) = 0$ when $x = 5$

$P''(x) = -10$, so $P''(5) < 0$ and there is a relative maximum when $x = 5$. Further, since $P''(x) < 0$ for all x in the domain

$x \ge 0$, it is the absolute maximum. So, the wine should be sold 5 years from the time of purchase to maximize profit.

21. Let n denote the number of floors and $A(n)$ the corresponding average cost. Since the total cost is

$C(n) = 2n^2 + 500n + 600$ thousand dollars.

$$A(n) = \frac{C(n)}{n} = 2n + 500 + \frac{600}{n}$$

The relevant interval is $n > 0$.

$$A'(n) = 2 - \frac{600}{n^2} = \frac{2(n^2 - 300)}{n^2}$$

$A'(n) = 0$ when $n = \sqrt{300} \approx 17.32$.

When $\begin{array}{l} 0<n<17.32,\ A'(n)<0 \text{ so } A \text{ is decreasing} \\ n>17.32,\ A'(n)>0 \text{ so } A \text{ is increasing} \end{array}$

Since the number of floors must be an integer and $A(17) \approx 569.29$ and $A(18) \approx 569.33$, the average cost per floor is minimized when 17 floors are built.

23. (a) Let x be the number of bottles in each shipment. The costs include:

purchase cost $= (800)(20) = 16,000$

ordering cost $= \left(\dfrac{800}{x}\right)(10)$

ordering cost $= \left(\dfrac{x}{2}\right)(0.4)$

So, the total cost is $C(x) = 16,000 + \dfrac{8,000}{x} + 0.2x$ which is the function to be minimized.

$C'(x) = -\dfrac{8,000}{x^2} + 0.2$

$C'(x) = 0$ when $x = 200$

$C''(x) = \dfrac{16,000}{x^3}$, so $C''(200) > 0$ and there is a relative minimum when

$x = 200$. $C(200) = 16,080$,
$C(1) = 17,000.20$, $C(800) = 16,170$. So, the cost is minimized when 200 bottles are ordered in each shipment.

(b) The number of shipments is
$\dfrac{800}{200} = 4$ times a year, so the store orders every 3 months.

25. (a) Let x denote the number of machines used and $C(x)$ the corresponding total cost. Then
$C(x) = $ set up cost $+$ operating cost
$$= 20\left(\begin{array}{c}\text{number of} \\ \text{machines}\end{array}\right) + 15\left(\begin{array}{c}\text{number} \\ \text{of hours}\end{array}\right) \text{ Since each machine produces}$$
30 kickboards per hour, x machines produce $30x$ kickboards per hour and the number of hours required to produce 8,000 kickboards is $\dfrac{8,000}{30x}$.

So, $C(x) = 20x + 15\left(\dfrac{8,000}{30x}\right) = 20x + \dfrac{4,000}{x}$ $\quad C'(x) = 20 - \dfrac{4,000}{x^2}$

$C'(x) = 0$ when $x \approx 14$

Since the company owns 10 machines, the domain of C is
$1 \le x \le 10$. Further, $C(1) = 4,020$ and $C(10) = 600$, so cost is minimized when 10 machines are used.

(b) When 10 machines are used, the number of hours to produce the kickboards is $\dfrac{8,000}{30(10)}$ and the

supervisor would be paid $15\left(\dfrac{8,000}{300}\right) = \$400$.

(c) The cost of setting up 10 machines is $20(10) = \$200$.

27. Let $C(N)$ be the total cost of using N machines. Now, the setup cost of

N machines is aN and the operating cost of N machines is $\dfrac{b}{N}$. So, $C(N) = aN + \dfrac{b}{N}$ which is the

function to minimize.

$$C'(N) = a = \dfrac{b}{N^2}$$

$C'(N) = 0$ when $a = \dfrac{b}{N^2}$, or when $aN = \dfrac{b}{N}$ (setup cost = operating cost)

$C''(N) = \dfrac{2b}{N^3}$, which is positive for all N in the domain $N \geq 1$, so there is an absolute minimum when

setup cost equals operating cost.

29. (a) Let x be the number of units produced, $p(x)$ the price per unit, t the tax per unit, and $C(x)$ the total cost.

$$C(x) = \dfrac{7x^2}{8} + 5x + 100$$

Since $p(x) = 15 - \dfrac{3x}{8}$, the revenue is $R(x) = xp(x) = 15x - \dfrac{3x^2}{8}$.

Now profit is

$P(x) = \text{revenue} - \text{taxation} - \text{cost}$

$P(x) = 15x - \dfrac{3x^2}{8} - tx - \dfrac{7x^2}{8} - 5x - 100$ which is the function to be maximized.

$P'(x) = 15 - \dfrac{3x}{4} - t - \dfrac{7x}{4} - 5$

$\quad\ = -\dfrac{5}{2}x + 10 - t$

$P'(x) = 0$ when $x = \dfrac{2(10-t)}{5}$

$P''(x) = -\dfrac{5}{2}$, so $P''\left(\dfrac{2(10-t)}{5}\right) < 0$ and there is a relative maximum when $x = \dfrac{2}{5}(10-t)$.

Further, since $P''(x) < 0$ for all x in the domain $x > 0$, it is the absolute maximum.

(b) The government share is $G(x) = tx = \left(\dfrac{2}{5}\right)(10t - t^2)$ which is the function to be maximized.

$$G'(t) = \left(\dfrac{2}{5}\right)(10 - 2t)$$

$G'(t) = 0$ when $t = 5$

$G''(t) = -\dfrac{4}{5}$, so $G''(5) < 0$ and there is a relative maximum when $t = 5$. Further, since $G''(t) < 0$ for all t in the domain $t > 0$, it is the absolute maximum.

(c) From part (a), with $t = 0$,
$$x = \dfrac{2(10 - 0)}{5} = 4, \text{ and with } t = 5, \ x = \dfrac{2(10 - 5)}{5} = 2.$$

The price per unit for the two quantities produced is, respectively, $p(4) = 15 - \dfrac{3(4)}{8} = \$13.50$ and

$p(2) = 15 - \dfrac{3(2)}{8} = \$14.25.$

The difference between the two unit prices is $14.25 - 13.50$ or 75 cents, which represents the amount of tax passed on to the consumer. The monopolist will absorb \$4.25 of the tax.

(d) Writing Exercise—Answers will vary.

31. Let x be the number of miles from the house to plant A. Then, $18 - x$ is its distance from plant B, and $1 \le x \le 16$. Let $P(x)$ be the concentration off particulate matter at the house. Then,

$$P(x) = \dfrac{80}{x} + \dfrac{720}{18 - x} \text{ which is the function to minimize.}$$

$$P'(x) = -\dfrac{80}{x^2} + \dfrac{0 - (720)(-1)}{(18 - x)^2}$$

$$P'(x) = 0 \text{ when } \dfrac{80}{x^2} = \dfrac{720}{(180 - x)^2}$$

$2x^2 + 9x - 81 = 0$ or, $x = \dfrac{9}{2}$ (rejecting negative solution)

$P(4.5) = 0$, $P(1) \approx 122.4$, $P(16) = 365$, so the total pollution is minimized when the house is 4.5 miles from plant A.

33. Let x be the distance along the shoreline from A to P. Then, the distance from B to P is the hypotenuse of a right triangle, $d(B, P) = \sqrt{25 + x^2}$. The total distance along the shoreline from A to L is the leg of a right triangle, $d(A, L) = \sqrt{(13)^2 - (5)^2} = 12$.

So, the distance from P to L is
$d(P, L) = 12 - x$.
The path of the bird is from B to P, and then from P to L. If e is the energy per mile to fly over land (a constant), then the energy to fly this path is $E(x) = 2e\sqrt{25 + x^2} + e(12 - x)$ which is the function to be minimized.

$$E'(x) = e(25 + x^2)^{-1/2}(2x) - e$$

$$= \frac{2ex}{(25 + x^2)^{1/2}} - e$$

$$E'(x) = 0 \text{ when } \frac{2ex}{(25 + x^2)^{1/2}} = e$$

$$\frac{2x}{(25 + x^2)^{1/2}} = 1$$

$$2x = (25 + x^2)^{1/2}$$

$$4x^2 = 25 + x^2$$

$$\text{or, } x = \sqrt{\frac{25}{3}} = \frac{5\sqrt{3}}{3}$$

Since $0 \le x \le 12$, $E\left(\frac{5\sqrt{3}}{3}\right) \approx 20.7e$;

$E(0) = 22e$; $E(12) = 26e$

So, to minimize energy expended, the bird should fly to point P which is $\sqrt{\frac{25}{3}} \approx 2.9$ miles from point A.

35. $R(S) = \dfrac{cS}{a + S + bS^2}$

    (a) domain: using the quadratic formula, the denominator is never zero, so the practical domain is $[0, \infty)$
intercepts: when $S = 0$, $R(0) = 0$;
point $(0, 0)$ when $R(S) = 0$, $S = 0$
asymptotes: no vertical asymptotes (since denominator is never zero)

$$\lim_{S \to \infty} \frac{\frac{C}{S}}{\frac{a}{S^2} + \frac{1}{S} + b} = 0, \text{ so } y = 0 \text{ is a horizontal asymptote.}$$

$$R'(S)$$

$$= c\left[\frac{(a + S + bS^2)(1) - (S)(1 + 2bS)}{(a + S + bS^2)^2}\right]$$

$$= c\frac{a - bS^2}{(a + S + bS^2)^2}$$

$$R'(S) = 0 \text{ when } a - bS^2 = 0 \text{ or } S = \sqrt{\frac{a}{b}} \text{ (rejecting negative answer)}$$

When $0 \le S < \sqrt{\dfrac{a}{b}}$, $R'(S) > 0$, so R is increasing

$S > \sqrt{\dfrac{a}{b}}$, $R'(S) < 0$, so R is decreasing

So, there is a relative maximum (which is also the absolute maximum) when $S = \sqrt{\dfrac{a}{b}}$. (Note: the second derivative is too complex to use in sketching the graph.)

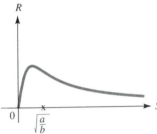

The lowest point is (0, 0). Since the graph starts concave down but then approaches the S axis asymptotically, there must be an inflection point. Since R approaches zero as S gets larger and larger, the growth rate, R', must also approach zero.

(b) Writing exercise—Answers will vary.

37. $E(v) = \dfrac{Cv^k}{v - v_w}$

(a) $E'(v) = \dfrac{(v - v_w)\left(kCv^{k-1}\right) - \left(Cv^k\right)(1)}{(v - v_w)^2}$

$\quad = \dfrac{Cv^{k-1}\left[k(v - v_w) - v\right]}{(v - v_w)^2}$

$E'(v) = 0$ when

$\quad 0 = Cv^{k-1}\left[k(v - v_w) - v\right]$

$\quad 0 = k(v - v_w) - v$

$\quad 0 = kv - kv_w - v$

$\quad kv_w = v(k - 1)$

$\quad v = \dfrac{kv_w}{k - 1}$

When $0 \le v < \dfrac{kv_w}{k - 1}$, $E' < 0$ so E is decreasing.

When $v > \dfrac{kv_w}{k - 1}$, $E' > 0$ so E is increasing.

So, E has a relative minimum when $v = \dfrac{kv_w}{k - 1}$.

(b) $F(k) = \dfrac{v_w k}{k-1}$, $k > 2$

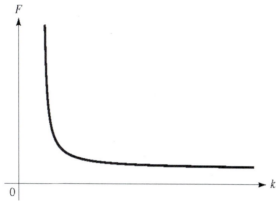

$F(k)$ is approximately v_w for large k.

39. Let x be the width (left to right) of the *printed* area and let y be the printed area's length (top to bottom). Then, the entire paper has a width of $x + 4$ and a length of $y + 8$. Need to minimize the area of the paper $A = (x + 4)(y + 8)$.

Since the printed area is 648 cm^2,

$$xy = 648$$
$$y = \frac{648}{x}$$

So, $A(x) = (x+4)\left(\dfrac{648}{x} + 8\right)$

$$= 648 + 8x + \frac{2592}{x} + 32$$

$$A'(x) = 8 - \frac{2592}{x^2}$$

$A'(x) = 0$ when $\quad 0 = 8 - \dfrac{2592}{x^2}$

$$8 = \frac{2592}{x^2}$$
$$x^2 = 324$$
$$x = 18$$

$$A''(x) = 0 + \frac{5184}{x^3}$$

Since $A''(18) > 0$, the absolute minimum occurs when $x = 18$. So, the paper should be $18 + 4 = 22$ cm wide and $\dfrac{648}{18} + 8 = 44$ cm long.

41. The amount of material is the amount for the circular top and bottom, and the amount for the curved side.

$$m = 2\pi r^2 + 2\pi rh$$

Since the volume is 6.89π,

$$V = \pi r^2 h$$

$$6.89\pi = \pi r^2 h, \text{ or } h = \frac{6.89}{r^2}$$

and $m(r) = 2\pi r^2 + 2\pi r \left(\frac{6.89}{r^2} \right)$

$$= 2\pi r^2 + \frac{13.78\pi}{r}$$

which is the function to be minimized.

$$m'(r) = 4\pi r - \frac{13.78\pi}{r^2}$$

$m'(r) = 0$ when $r \approx 1.51$

$m''(r) = 4\pi + \frac{27.56}{r^3}$, so $m''(1.51) > 0$ and there is a relative minimum when

$r = 1.51$. Further, since $m''(r) > 0$ for all r in the domain $r > 0$, it is an absolute minimum. So, the minimum material is when the can's radius is approximately 1.51 inches and its height is approximately

$\frac{6.89}{(1.51)^2} \approx 3.02$ inches. (These dimensions are not used due to packaging and

handling concerns.)

43. The cost of the material is the cost of the circular bottom and the cost of the curved side.

$$C = 3(\pi r^2) + 2(2\pi rh)$$

Since the volume is to be fixed, let K represent this fixed value.

$$V = \pi r^2 h$$

$$K\pi r^2 h, \text{ or } h = \frac{K}{\pi r^2}$$

and

$$C(r) = 3\pi r^2 + 4\pi r \left(\frac{K}{\pi r^2} \right) = 3\pi r^2 + \frac{4K}{r}$$

which is the function to be minimized.

$$C'(r) = 6\pi r - \frac{4K}{r^2}$$

$C'(r) = 0$ when $\frac{3\pi}{2} r^3 = K$

$$\frac{3\pi}{2} r^3 = \pi r^2 h$$

$$\text{or, } r = \frac{2}{3} h$$

$C''(r) = 6\pi + \frac{8K}{r^3}$, so $C'' \left(\frac{2}{3} h \right) > 0$ and there is a relative minimum when $r = \frac{2}{3} h$. Further, since

$C''(r) > 0$ for all r in the domain $r > 0$, it is an absolute minimum. So, a can with a fixed volume has its cost minimized whenever $r = \dfrac{2}{3}h$.

45. Let x be the length of the side of the square base and y be the height of the box. The volume of the box is $V = x^2 y$.

The cost of the four sides is $\begin{aligned}4(\text{cost per unit area})(\text{area}) &= 4(3)(xy)\\ &= 12xy\end{aligned}$

The cost of the bottom of the box is $(\text{cost per unit area})(\text{area}) = 4(x^2)$

Since there is 48 dollars available to build the box, $48 = 12xy + 4x^2$, or $y = \dfrac{48 - 4x^2}{12x} = \dfrac{4}{x} - \dfrac{x}{3}$ and

$V(x) = x^2\left(\dfrac{4}{x} - \dfrac{x}{3}\right) = 4x - \dfrac{1}{3}x^3$ which is the function to be maximized.

$V'(x) = 4 - x^2$

$V'(x) = 0$ when $x = 2$ (rejecting the negative solution)

$V''(x) = -2x$, so $V''(2) < 0$ and there is a relative maximum when $x = 2$. Further, since $V''(x) < 0$ for all x in the domain

$x > 0$, it is the absolute maximum. So the box has a maximum volume when its dimensions are 2 meters by 2 meters by $y = \dfrac{4}{2} - \dfrac{2}{3} = \dfrac{4}{3}$ meters.

47. Frank is right. In the cost function, $C(x) = 5\sqrt{(900)^2 + x^2} + 4(3,000 - x)$ note where the distance downstream appears. Since it is only part of the constant term in $C(x)$, it drops out when $C'(x)$. So, the critical value is always $x = 1,200$ (as long as the distance downstream is at least 1,200 meters).
When $0 \le x < 1,200$, $C'(x) < 0$ so C is decreasing
$\quad\quad x > 1,200$, $C'(x) > 0$ so C is increasing
So, the absolute minimum cost is always when the cable reaches the bank 1,200 meters downstream.

49. Let x be the distance down the paved road where the jeep reaches the road. Then, the time the jeep drives in sand is given by $t_s = \dfrac{d_s}{r_s} = \dfrac{\sqrt{x^2 + 32^2}}{48}$.

The time the jeep drives on the road is given by $t_r = \dfrac{d_r}{r_r} = \dfrac{16 - x}{80}$.

The total time is given by

$T(x) = \dfrac{1}{48}(x^2 + 1024)^{1/2} + \dfrac{1}{80}(16 - x)$

$T'(x) = \dfrac{1}{96}(x^2 + 1024)^{-1/2}(2x) + \dfrac{1}{80}(-1)$

$\quad\quad = \dfrac{x}{48\sqrt{x^2 + 1024}} - \dfrac{1}{80}$

$$0 = \frac{x}{48\sqrt{x^2 + 1024}} - \frac{1}{80}$$

$$\frac{1}{80} = \frac{x}{48\sqrt{x^2 + 1024}}$$

$$\sqrt{x^2 + 1024} = \frac{5}{3}x$$

$T'(x) = 0$ when $\quad x^2 + 1024 = \frac{25}{9}x^2$

$$1024 = \frac{16}{9}x^2$$

$$576 = x^2$$

$$x = 24$$

Since the maximum value of x is 16, disregard this answer and check the endpoints ($0 \le x \le 16$).

$T(0) \approx 0.867$ hr

$T(16) \approx 0.745$ hr

So, the minimum time to reach the power plant is 0.745 hour, or approximately 44.7 minutes (making the trip entirely in the sand). Since he has 50 minutes to deliver the ransom, he can make it in time.

51. Let S be the stiffness of the beam. Then, $S = kwh^3$, where k is a constant of proportionality. Since $w^2 + h^2 = 225$, or $h = \sqrt{225 - w^2}$, S can be expressed as a function of w, $S(w) = kw(225 - w^2)^{3/2}$ which is the function to be maximized.

$$S'(w) = k\left[w \cdot \frac{3}{2}(225 - w^2)^{1/2}(-2w) \right.$$

$$\left. + (225 - w^2)^{3/2}(1) \right]$$

$$= k(225 - w^2)^{1/2}$$

$$[-3w^2 + 225 - w^2]$$

$$= k(225 - w^2)^{1/2}(225 - 4w^2)$$

$S'(w) = 0$ when $w = \frac{15}{2}$ (rejecting the solution $w = 15$, which is not possible given the diameter)

When $0 < w < \frac{15}{2}$, $S'(w) > 0$ so C is increasing $\frac{15}{2} < w < 15$, $S'(x) < 0$ so S is decreasing. So, the

dimensions for maximum stiffness are $w = \frac{15}{2}$ inches and $y = \sqrt{225 - \left(\frac{15}{2}\right)^2} \approx 13.0$ inches.

53. The volume of the parcel is $V = x^2 y$.

The restriction given is

$4x + y = 108$(max), or $y = 108 - 4x$ and $V(x) = x^2(108 - 4x) = 108x^2 - 4x^3$ which is the function to be maximized.

$V'(x) = 216x - 12x^2 = 12x(18 - x)$

$V'(x) = 0$ when $x = 18$ (rejecting $x = 0$)

$V''(x) = 216 - 24x$, so $V''(18) < 0$ and there is a relative minimum when $x = 18$.

When $0 < x < 18$, $V'(x) > 0$ so V is increasing

$\quad\quad x > 18$, $V'(x) < 0$ so V is decreasing

So, the relative maximum is the absolute maximum. The maximum volume is

$108(18)^2 - 4(18)^3 = 11{,}664$ cubic inches.

Checkup for Chapter 3

1. Graph (a) is the graph of f, while graph (b) is the graph of f'; possible explanations include:

    (i) the degree of (a) is one larger than the degree of (b)

    (ii) the x-intercepts of (b) correspond to the relative extrema of (a)

2. (a) $f(x) = -4x^4 + 4x^3 + 5$

$\quad\quad f'(x) = -4x^3 + 12x^2 = -4x^2(x - 3)$

$\quad\quad f'(x) = 0$ when $x = 0, 3$

$\quad\quad$ When $x < 0$, $f'(x) > 0$ so f is increasing

$\quad\quad\quad\quad 0 < x < 3$, $f'(x) > 0$ so f is increasing

$\quad\quad\quad\quad x > 3$, $f'(x) < 0$ so f is decreasing

$\quad\quad$ There is no relative extrema when

$\quad\quad x = 0$, but when $x = 3$, f has a relative maximum.

(b) $f(t) = 2t^3 - 9t^2 + 12t + 5$

$\quad\quad f'(t) = 6t^2 - 18t + 12 = 6(t - 1)(t - 2)$

$\quad\quad f'(t) = 0$ when $t = 0, 3$

$\quad\quad$ When $t < 1$, $f'(t) > 0$ so f is increasing

$\quad\quad\quad\quad 1 < t < 2$, $f'(t) < 0$ so f is decreasing

$\quad\quad\quad\quad t > 2$, $f'(t) > 0$ so f is increasing

$\quad\quad$ When $t = 1$, f has a relative maximum, and when $t = 2$, f has a relative minimum.

(c) $g(t) = \dfrac{t}{t^2 + 9}$

$\quad\quad g'(t) = \dfrac{(t^2 + 9)(1) - (t)(2t)}{(t^2 + 9)^2}$

$\quad\quad\quad = \dfrac{(3 + t)(3 - t)}{(t^2 + 9)^2}$

$\quad\quad g'(t) = 0$ when $t = -3, 3$

$\quad\quad$ When $t < -3$, $g'(t) < 0$ so g is decreasing

$\quad\quad\quad\quad -3 < t < 3$, $g'(t) > 0$ so g is increasing

$\quad\quad\quad\quad t > 3$, $g'(t) < 0$ so g is decreasing

$\quad\quad$ When $t = -3$, g has a relative minimum, and when $t = 3$, g has a relative maximum.

(d) $g(x) = \dfrac{4-x}{x^2+9}$

$g'(x) = \dfrac{(x^2+9)(-1)-(4-x)(2x)}{(x^2+9)^2}$

$ = \dfrac{(x+1)(x-9)}{(x^2+9)^2}$

$g'(x) = 0$ when $x = -1, 9$

When $x < -1, g'(x) > 0$ so g is increasing

 $-1 < x < 9, g'(x) < 0$ so g is decreasing

 $x > 9, g'(x) > 0$ so g is increasing

When $x = -1$, g has a relative maximum, and when $x = 9$, g has a relative minimum.

3. (a) $f(x) = 3x^5 - 10x^4 + 2x - 5$

$f'(x) = 15x^4 - 40x^3 + 2$

$f''(x) = 60x^3 - 120x^2 = 60x^2(x-2)$

$f''(x) = 0$ when $x = 0, 2$

When $x < 0$, $f''(x) < 0$ so f is concave down

 $0 < x < 2$, $f''(x) < 0$ so f is concave down

 $x > 2$, $f''(x) > 0$ so f is concave up

There is an inflection point when $x = 2$.

(b) $f(x) = 3x^5 + 20x^4 - 50x^3$

$f'(x) = 15x^4 + 80x^3 - 150x^2$

$f''(x) = 60x^3 + 240x^2 - 300x$

$ = 60x(x+5)(x-1)$

$f''(x) = 0$ when $x = -5, 0, 1$

When $x < -5$, $f''(x) < 0$ so f is concave down

 $-5 < x < 0$, $f''(x) > 0$ so f is concave up

 $0 < x < 1$, $f''(x) < 0$ so f is concave down

 $x > 1$, $f''(x) > 0$ so f is concave up

There are inflection points when $x = -5, 0, 1$.

(c) $f(t) = \dfrac{t^2}{t-1}$

$f'(t) = \dfrac{(t-1)(2t)-(t^2)(1)}{(t-1)^2} = \dfrac{t^2-2t}{(t-1)^2}$

$$f''(t) = \frac{\begin{array}{c}(t-1)^2(2t-2)\\ -(t^2-2t)(2(t-1)\\ -(t^2-2t)(2(t-1)(1))\end{array}}{(t-1)^4}$$

$$f''(t) = \frac{2(t-1)^3 - 2t(t-2)(t-1)}{(t-1)^4}$$

$$f''(t) = \frac{2(t-1)[(t-1)^2 - t(t-2)]}{(t-1)^4}$$

$$f''(t) = \frac{2}{(t-1)^3}$$

$f''(t)$ is never zero, so there are no inflection points; $f''(t)$ is undefined for $t = 1$.

When $t < 1$, $f''(t) < 0$ so f is concave down
$\qquad t > 1$, $f''(t) > 0$ so f is concave up

(d) $g(t) = \dfrac{3t^2 + 5}{t^2 + 3}$

$$g'(t) = \frac{(t^2+3)(6t) - (3t^2+5)(2t)}{(t^2+3)^2}$$

$$= \frac{8}{(t^2+3)^2}$$

$$g''(t) = \frac{(t^2+3)^2(8) - (8t)(2(t^2+3)(2t))}{(t^2+3)^4}$$

$$= \frac{8(t^2+3)[(t^2+3) - 4t^2]}{(t^2+3)^4} \qquad g''(t) = 0 \text{ when } t = -1,\, 1$$

$$= \frac{24(1+t)(1-t)}{(t^2+3)^3}$$

$\qquad t < -1$, $g''(t) < 0$ so g is concave down

When $-1 < t < 1$, $g''(t) > 0$ so f is concave up
$\qquad t > 1$, $g''(t) < 0$ so g is concave down

There are inflection points when
$t = -1,\, 1$.

4. (a) $f(x) = \dfrac{2x-1}{x+3}$

$x + 3 = 0$ when $x = -3$, so there is a vertical asymptote of $x = -3$.

$$\lim_{x \to \pm\infty} \frac{2x-1}{x+3} = \lim_{x \to \pm\infty} \frac{2 - \frac{1}{x}}{1 + \frac{3}{x}} = \frac{2}{1} = 2 \text{ so there is a horizontal asymptote of}$$

$y = 2$.

(b) $f(x) = \dfrac{x}{x^2 - 1}$

$x^2 - 1 = (x+1)(x-1) = 0$ when

$x = -1, 1$; so, there are vertical asymptotes of $x = -1$ and $x = 1$.

$\displaystyle\lim_{x \to \infty} \frac{x}{x^2 - 1} = \lim_{x \to \infty} \frac{\frac{1}{x}}{1 - \frac{1}{x^2}} = \frac{0}{1} = 0$; so, there is a horizontal asymptote of

$y = 0$.

(c) $f(x) = \dfrac{x^2 + x - 1}{2x^2 + x - 3}$

$2x^2 + x - 3 = (2x+3)(x-1) = 0$ when $x = -\dfrac{3}{2}, 1$; so, there are vertical asymptotes of $x = -\dfrac{3}{2}$ and $x = 1$.

$\displaystyle\lim_{x \to \pm\infty} \frac{x^2 + x - 1}{2x^2 + x - 3} = \lim_{x \to \pm\infty} \frac{1 + \frac{1}{x} - \frac{1}{x^2}}{1 + \frac{1}{x} - \frac{3}{x^2}}$

$\phantom{\lim_{x \to \pm\infty}} = \dfrac{1}{2};$

so there is a horizontal asymptote of $y = \dfrac{1}{2}$.

(d) $f(x) = \dfrac{1}{x} - \dfrac{1}{\sqrt{x}} = \dfrac{\sqrt{x} - x}{x\sqrt{x}} = \dfrac{x^{1/2} - x}{x^{3/2}}$

$x^{3/2} = 0$ when $x = 0$; so, there is a vertical asymptote of $x = 0$.

$\displaystyle\lim_{x \to \pm\infty} \frac{x^{1/2} - x}{x^{3/2}} = \lim_{x \to \pm\infty} \frac{\frac{1}{x^{1/2}} - 1}{x^{1/2}} = 0$; so, there is a horizontal asymptote of

$y = 0$.

5. (a) $f(x) = 3x^4 - 4x^3$

When $x = 0$, $f(0) = 0$ so $(0, 0)$ is an intercept. When $f(x) = 0$, $3x^4 - 4x^3 = x^3(3x - 4) = 0$ so

$f(x) = 0$ when $x = 0, \dfrac{4}{3}$, and $\left(\dfrac{4}{3}, 0\right)$ is an intercept. There are no asymptotes.

$f'(x) = 12x^3 - 12x^2 = 12x^2(x - 1)$

$f'(x) = 0$ when $x = 0, 1$

$f''(x) = 36x^2 - 24x = 12x(3x - 2)$

$f''(x) = 0$ when $x = 0, \dfrac{2}{3}$

When $x < 0$, $f'(x) < 0$ so f is decreasing

$\phantom{When x < 0,}$ $f''(x) > 0$ so f is concave up

When $0 < x < \dfrac{2}{3}$, $f'(x) < 0$ so f is decreasing
$\qquad\qquad\qquad$ $f''(x) < 0$ so f is concave down

When $\dfrac{2}{3} < x < 1$, $f'(x) < 0$ so f is decreasing
$\qquad\qquad\qquad$ $f''(x) > 0$ so f is concave up

When $x > 1$, $f'(x) > 0$ so f is increasing
$\qquad\qquad\qquad$ $f''(x) > 0$ so f is concave up

There is a relative minimum when

$x = 1$, or $(1, -1)$. There are inflection points when $x = 0$, $\dfrac{2}{3}$, or $(0, 0)$ and $\left(\dfrac{2}{3}, -\dfrac{16}{27} \right)$.

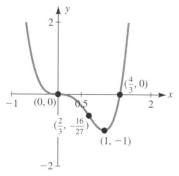

(b) $f(x) = x^4 - 3x^3 + 3x^2 + 1$

When $x = 0$, $f(0) = 1$, so $(0, 1)$ is an intercept. $f(x) = 0$ is too difficult to solve. There are no asymptotes.

$$f'(x) = 4x^3 - 9x^2 + 6x$$
$$\qquad = x(4x^2 - 9x + 6)$$
$$f'(x) = 0 \ \text{when} \ x = 0$$

$$f''(x) = 12x^2 - 18x + 6$$
$$\qquad = 6(2x - 1)(x - 1)$$

$$f''(x) = 0 \ \text{when} \ x = \dfrac{1}{2}, 1$$

When $x < 0$, $f'(x) < 0$ so f is decreasing
$\qquad\qquad\qquad$ $f''(x) > 0$ so f is concave up

When $0 < x < \dfrac{1}{2}$, $f'(x) > 0$ so f is increasing
$\qquad\qquad\qquad$ $f''(x) > 0$ so f is concave up

When $\dfrac{1}{2} < x < 1$, $f'(x) > 0$ so f is increasing
$\qquad\qquad\qquad$ $f''(x) < 0$ so f is concave down

When $x > 1$, $f'(x) > 0$ so f is increasing
$\qquad\qquad\qquad$ $f''(x) > 0$ so f is concave up

There is a relative minimum when

$x = 0$, or $(0, 1)$. There are inflection points when $x = \dfrac{1}{2}$, 1, or $\left(\dfrac{1}{2}, \dfrac{23}{16}\right)$ and $(1, 2)$.

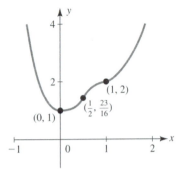

(c) $f(x) = \dfrac{x^2 + 2x + 1}{x^2}$

When $x = 0$, $f(0)$ is undefined.

When $f(x) = 0$, $x^2 + 2x + 1 = (x + 1)^2 = 0$, so

$f(x) = 0$ when $x = -1$, and $(-1, 0)$ is an intercept.

$x^2 = 0$ when $x = 0$, so there is a vertical asymptote of $x = 0$.

$$\lim_{x \to \pm\infty} \frac{x^2 + 2x + 1}{x^2} = \lim_{x \to \pm\infty} \frac{1 + \frac{2}{x} + \frac{1}{x^2}}{1}$$
$$= \frac{1}{1},$$

so there is a horizontal asymptote of $y = 1$.

Note: $\dfrac{x^2 + 2x + 1}{x^2} = 1$ when $x^2 + 2x + 1 = x^2$, $2x + 1 = 0$, or $x = -\dfrac{1}{2}$, so the graph will cross this

asymptote at $\left(-\dfrac{1}{2}, 1\right)$.

$$f'(x) = \frac{(x^2)(2x + 2) - (x^2 + 2x + 1)(2x)}{x^4}$$
$$= \frac{-2x^2 - 2x}{x^4}$$
$$= \frac{-2x(x + 1)}{x^4}$$
$$= \frac{-2(x + 1)}{x^3}$$

$f'(x) = 0$ when $x = -1$ and $f'(x)$ is undefined when $x = 0$.

$$f''(x) = \frac{(x^3)(-2) - (-2(x+1)(3x^2))}{x^6}$$

$$= \frac{2x^2[-x + 3(x+1)]}{x^6}$$

$$= \frac{2(2x+3)}{x^4}$$

$f''(x) = 0$ when $x = -\dfrac{3}{2}$ and $f''(x)$ is undefined when $x = 0$.

When $x < -\dfrac{3}{2}$, $f'(x) < 0$ so f is decreasing
$f''(x) < 0$ so f is concave down

When $-\dfrac{3}{2} < x < -1$, $f'(x) < 0$ so f is decreasing
$f''(x) > 0$ so f is concave up

When $-1 < x < 0$, $f'(x) > 0$ so f is increasing
$f''(x) > 0$ so f is concave up

When $x > 0$, $f'(x) < 0$ so f is decreasing
$f''(x) > 0$ so f is concave up

There is a relative minimum when

$x = -1$, or $(-1, 0)$. There is an inflection point when $x = -\dfrac{3}{2}$, or $\left(-\dfrac{3}{2}, \dfrac{1}{9}\right)$.

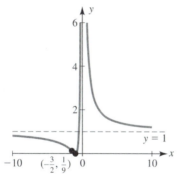

(d) $f(x) = \dfrac{1 - 2x}{(x-1)^2}$

When $x = 0$, $f(0) = 1$ so $(0, 1)$ is an intercept. When $f(x) = 0$, $1 - 2x = 0$, or $x = \dfrac{1}{2}$ so $\left(\dfrac{1}{2}, 0\right)$ is

an intercept.

$(x-1)^2 = 0$ when $x = 1$, so there is a vertical asymptote of $x = 1$.

$$\lim_{x \to \pm\infty} \frac{1 - 2x}{x^2 - 2x + 1} = \lim_{x \to \pm\infty} \frac{\frac{1}{x} - 2}{x - 2 + \frac{1}{x}}$$

$$= 0,$$

so there is a horizontal asymptote of
$y = 0$.

$$f'(x) = \frac{(x-1)^2(-2)}{- (1-2x)(2(x-1)(1))}\Big/(x-1)^4$$

$$= \frac{-2(x-1)[(x-1)+(1-2x)]}{(x-1)^4}$$

$$= \frac{2x}{(x-1)^3}$$

$f'(x) = 0$ when $x = 0$ and $f'(x)$ is undefined when $x = 1$.

$$f''(x) = \frac{(x-1)^3(2)}{-(2x)(3(x-1)^2(1))}\Big/(x-1)^6$$

$$= \frac{2(x-1)^2[(x-1)-3x]}{(x-1)^6}$$

$$= \frac{-2(1+2x)}{(x-1)^4}$$

$f''(x) = 0$ when $x = -\dfrac{1}{2}$ and $f''(x)$ is undefined when $x = 1$.

When $x < -\dfrac{1}{2}$, $f'(x) > 0$ so f is increasing
$\quad\quad\quad\quad\quad\ f''(x) > 0$ so f is concave up

When $-\dfrac{1}{2} < x < 0$, $f'(x) > 0$ so f is increasing
$\quad\quad\quad\quad\quad\quad\ f''(x) < 0$ so f is concave down

When $0 < x < 1$, $f'(x) < 0$ so f is decreasing
$\quad\quad\quad\quad\quad f''(x) < 0$ so f is concave down

When $x > 1$, $f'(x) > 0$ so f is increasing
$\quad\quad\quad\quad\ f''(x) < 0$ so f is concave down

There is a relative maximum when

$x = 0$, or $(0, 1)$. There is an inflection point when $x = -\dfrac{1}{2}$, or $\left(-\dfrac{1}{2}, \dfrac{8}{9}\right)$.

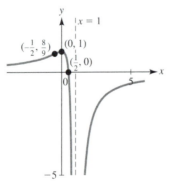

6. (a) Graph of f is increasing when $x < 0$ and $0 < x < 2$.

(b) Graph of f is decreasing when $x > 2$.

(c) Graph of f levels when $x = 0$ and
$x = 2$; from parts (a) and (b), $x = 0$ is not a relative extremum and $x = 2$ is a relative maximum.

(d) Graph of f is concave down when
$x < 0$ and $x > 1$.

(e) Graph of f is concave up when
$0 < x < 1$; from parts (d) and (e), there are inflection points when $x = 0$ and
$x = 1$.

(f) Graph of f goes through points
$(-1, 0)$, $(4, 0)$, $(0, 1)$, $(1, 2)$ and $(2, 3)$.

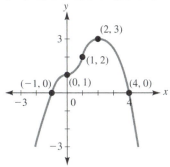

7. (a) $f(x) = x^3 - 3x^2 - 9x + 1;\ -2 \leq x \leq 4$

$f'(x) = 3x^2 - 6x - 9 = 3(x+1)(x-3)$

$f'(x) = 0$ when $x = -1$, $x = 3$ both in interval

$f(-2) = (-2)^3 - 3(-2)^2 - 9(-2) + 1$
$\qquad = -1$

$f(-1) = (-1)^3 - 3(-1)^2 - 9(-1) + 1 = 6$

$f(3) = (3)^3 - 3(3)^2 - 9(3) + 1 = -26$

$f(4) = (4)^3 - 3(4)^2 - 9(4) + 1 = -19$

absolute max $= 6$
absolute min $= -26$

(b) $g(t) = -4t^3 + 9t^2 + 12t - 5;\ -1 \leq t \leq 4$

$g'(t) = -12t^2 + 18t + 12$
$\qquad = -6(2t+1)(t-2)$

$g'(t) = 0$ when $t = -\dfrac{1}{2}$, $t = 2$ both in interval

$g(-1) = -4(-1)^3 + 9(-1)^2 + 12(-1) - 5$
$\qquad = -4$

$$g\left(-\frac{1}{2}\right)$$

$$=-4\left(-\frac{1}{2}\right)^3+9\left(-\frac{1}{2}\right)^2+12\left(-\frac{1}{2}\right)-5$$

$$=-\frac{33}{4}$$

$$g(2)=-4(2)^3+9(2)^2+12(2)-5=23$$

$$g(4)=-4(4)^3+9(4)^2+12(4)-5$$
$$=-69$$

absolute max $= 23$

absolute min $= -69$

(c) $h(u)=8\sqrt{u}-u+3;\ 0\le u\le 25$

$$=8u^{1/2}-u+3$$

$$h'(u)=4u^{-1/2}-1=\frac{4}{\sqrt{u}}-1$$

$h'(u)=0$ when $u=16$ in interval

$$h(0)=8\sqrt{0}-0+3=3$$

$$h(16)=8\sqrt{16}-16+3=19$$

$$h(25)=8\sqrt{25}-25+3=18$$

absolute max $= 19$

absolute min $= 3$

8. $f(t)=-t^3+7t^2+200t$ is the number of letters the clerk can sort in t hours. The clerk's rate of output is $R(t)=f'(t)=-3t^2+14t+200$ letters per hour. The relevant interval is $0\le t\le 4$.

$R'(t)=f''(t)=-6t+14$

$R'(t)=0$ when $t=\dfrac{7}{3}$

$R\left(\dfrac{7}{3}\right)=216.33,\ R(0)=200,$ and

$R(4)=208$

So, the rate of output is greatest when $t=\dfrac{7}{3}$ hours; that is, after 2 hours and 20 minutes, at 8:20 A.M.

9. Profit = revenue − costs

= (#sold)(selling price)

− (#sold)(cost per unit)

$P(x)=20(180-x)x-20(180-x)90$

$=20(180-x)(x-90)$

and the relevant domain is $x \geq 90$

$$P'(x) = 20[(180 - x)(1) + (x - 90)(-1)]$$
$$= 20(270 - 2x)$$
$$= 40(135 - x)$$

$P'(x) = 0$ when $x = 135$

When $90 \leq x < 135$, $P'(x) > 0$ so P is increasing

$\quad\quad x > 135$, $P'(x) < 0$ so P is decreasing

So, when the selling price is \$135 per unit, the profit is maximized.

10. $C(t) = \dfrac{0.05t}{t^2 + 27}$

(a) The relevant domain of the function is $t \geq 0$. When $t = 0$, $C(0) = 0$ so $(0, 0)$ is an intercept. When $C(t) = 0$, $t = 0$.

$t^2 + 27$ is never zero, so there are no vertical asymptotes.

$$\lim_{t \to \pm\infty} \frac{0.05t}{t^2 + 27} = \lim_{t \to \pm\infty} \frac{0.05}{t + \frac{27}{t}} = 0, \text{ so there is a horizontal asymptote of}$$

$y = 0$.

$$C'(t) = \frac{(t^2 + 27)(0.05) - (0.05t)(2t)}{(t^2 + 27)^2}$$

$$= \frac{1.35 - 0.05t^2}{(t^2 + 27)^2}$$

$C'(t) = 0$ when $t = \sqrt{27}$

$$C''(t) = \frac{1}{(t^2 + 27)^4}[(t^2 + 27)^2(-0.1t)$$

$$- (1.35 - 0.05t^2)(2(t^2 + 27)(2t))]$$

$$= \frac{1}{(t^2 + 27)^4}(t(t^2 + 27)[-0.1(t^2 + 27) \quad C''(t) = 0 \text{ when } t = 9$$

$$- 4(1.35 - 0.05t^2)])$$

$$= \frac{t(0.1t^2 - 8.1)}{(t^2 + 27)^3}$$

When $0 < t < \sqrt{27}$, $C'(x) > 0$ so C is increasing

$\quad\quad\quad\quad\quad\quad\quad\quad C''(x) < 0$ so C is concave down

When $\sqrt{27} < t < 9$, $C'(x) < 0$ so C is decreasing

$\quad\quad\quad\quad\quad\quad\quad\quad C''(x) < 0$ so C is concave down

When $t > 9$, $C'(x) < 0$ so C is decreasing

$\quad\quad\quad\quad\quad\quad C''(x) > 0$ so C is concave up

There is an absolute maximum when $t = \sqrt{27}$, or approximately $(5.20, 0.005)$. There is an inflection point when $t = 9$, or approximately

(9, 0.004).

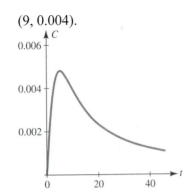

(b) $C'(t) < 0$ when $t = \sqrt{27}$, so C is decreasing when $t > \sqrt{27}$. The rate of decrease is maximized when $C''(t) = 0$ for $t > \sqrt{27}$, or when $t = 9$.

(c) $\displaystyle\lim_{t \to +\infty} \frac{0.05t}{t^2 + 27} = 0$, so the concentration tends to zero in the long run.

11. $P(t) = \dfrac{15t^2 + 10}{t^3 + 6}$

The relevant domain is $t \geq 0$.

(a) When $t = 0$, $P(0) = \dfrac{10}{6}$ or

1.667 million bacteria.

(b) $P'(t) = \dfrac{(t^3 + 6)(30t) - (15t^2 + 10)(3t^2)}{(t^3 + 6)^2}$

$ = \dfrac{-15t(t^3 + 2t - 12)}{(t^3 + 6)^2}$

$P(t) = 0$ when $t = 0, 2$

When $t = 0$, $P(0) = \dfrac{10}{6}$.

When $\begin{array}{l} 0 < t < 2,\ P'(t) > 0 \text{ so } P \text{ is increasing} \\ t > 2,\ P'(t) < 0 \text{ so } P \text{ is decreasing} \end{array}$

So, when $t = 2$, the bacteria population is maximized and the maximum population is 5 million.

(c) $\displaystyle\lim_{x \to \infty} \frac{15t^2 + 10}{t^3 + 6} = \lim_{x \to \infty} \frac{15 + \frac{10}{t^2}}{t + \frac{6}{t^2}} = 0$, so in the long run, the bacteria population dies out.

Use all of the above information to graph P, noting also that $P(t)$ is never zero, so there are no other intercepts. $t^3 + 6$ is never zero, so there are no vertical asymptotes. $y = 0$ is the horizontal asymptote and $P(2) = 5$.

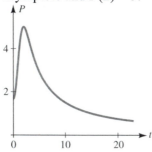

Review Exercises

1. $f(x) = -2x^3 + 3x^2 + 12x - 5$

When $x = 0, f(0) = -5$ so $(0, -5)$ is an intercept. $f(x) = 0$ is too difficult to solve. There are no asymptotes.

$f'(x) = -6x^2 + 6x + 12 = -6(x+1)(x-2)$

$f'(x) = 0$ when $x = -1, 2$

$f''(x) = -12x + 6 = -6(2x-1)$

$f''(x) = 0$ when $x = \dfrac{1}{2}$.

When $x < -1$, $f'(x) < 0$ so f is decreasing
$f''(x) > 0$ so f is concave up

When $-1 < x < \dfrac{1}{2}$,

$f'(x) > 0$ so f is increasing
$f''(x) > 0$ so f is concave up

When $\dfrac{1}{2} < x < 2$, $f'(x) > 0$ so f is increasing
$f''(x) < 0$ so f is concave down

When $x > 2$, $f'(x) < 0$ so f is decreasing
$f''(x) < 0$ so f is concave down

Overall f is decreasing when $x < -1$ and $x > 2$

f is increasing when $-1 < x < 2$

f is concave down when $x > \dfrac{1}{2}$

f is concave up when $x < \dfrac{1}{2}$.

There is a relative minimum when $x = -1$, or $(-1, -12)$, and a relative maximum when $x = 2$, or

(2, 15). There is an inflection point when $x = \dfrac{1}{2}$, or $\left(\dfrac{1}{2}, \dfrac{3}{2}\right)$.

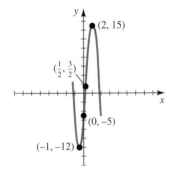

3. $f(x) = 3x^3 - 4x^2 - 12x + 17$

When $x = 0$, $f(0) = 17$ so (0, 17) is an intercept. $f(x) = 0$ is too difficult to solve. There are no asymptotes.

$f'(x) = 9x^2 - 8x - 12$

$f'(x) = 0$ when $x \approx -0.79, 1.68$

$f''(x) = 18x - 8 = 2(9x - 4)$

$f''(x) = 0$ when $x = \dfrac{4}{9}$

When $x < -0.79$, $f'(x) > 0$ so f is increasing
$\qquad\qquad$ $f''(x) < 0$ so f is concave down

When $-0.79 < x < \dfrac{4}{9}$, $f'(x) < 0$ so f is decreasing
$\qquad\qquad$ $f''(x) < 0$ so f is concave down

When $\dfrac{4}{9} < x < 1.68$, $f'(x) < 0$ so f is decreasing
$\qquad\qquad$ $f''(x) > 0$ so f is concave up

When $x > 1.68$, $f'(x) > 0$ so f is increasing
$\qquad\qquad$ $f''(x) > 0$ so f is concave up

Overall, f is decreasing when $-0.79 < x < 1.68$. f is increasing when $x < -0.79$ and $x > 1.68$. f is concave down when $x < \dfrac{4}{9}$. f is concave up when $x < \dfrac{4}{9}$.

There is a relative maximum when $x = -0.79$, or $(-0.79, 22.51)$, and a relative minimum when $x = 1.68$, or $(1.68, -0.23)$. There is an inflection point when $x = \dfrac{4}{9}$, or $(0.44, 11.1)$.

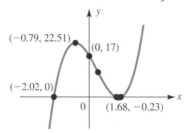

5. $f(t) = 3t^5 - 20t^3$

When $t = 0$, $f(0) = 0$ so $(0, 0)$ is an intercept. When $f(t) = 3t^5 - 20t^3 = t^3(3t^2 - 20) = 0$ so

$t = 0, \pm\sqrt{\dfrac{20}{3}}$ and $\left(\pm\sqrt{\dfrac{20}{3}}, 0\right)$ are intercepts.

There are no asymptotes.

$f'(t) = 15t^4 - 60t^2 = 15t^2(t+2)(t-2)$

$f'(t) = 0$ when $t = -2, 0, 2$

$f''(t) = 60t^3 - 120t = 60t(t^2 - 2)$

$f''(t) = 0$ when $t = -\sqrt{2}, 0, \sqrt{2}$

When $t < -2$, $f'(x) > 0$ so f is increasing
$\qquad\qquad f''(x) < 0$ so f is concave down

When $-2 < t < -\sqrt{2}$, $f'(x) < 0$ so f is decreasing
$\qquad\qquad f''(x) < 0$ so f is concave down

When $-\sqrt{2} < t < 0$, $f'(x) < 0$ so f is decreasing
$\qquad\qquad f''(x) > 0$ so f is concave up

When $0 < t < \sqrt{2}$, $f'(x) < 0$ so f is decreasing
$\qquad\qquad f''(x) < 0$ so f is concave down

When $\sqrt{2} < t < 2$, $f'(x) < 0$ so f is decreasing
$\qquad\qquad f''(x) > 0$ so f is concave up

When $t > 2$, $f'(x) > 0$ so f is increasing
$\qquad\qquad f''(x) > 0$ so f is concave up

Overall, f is decreasing when $-2 < t < 2$. f is increasing when $t < -2$ and $t > 2$. f is concave down when $t < -\sqrt{2}$ and $0 < t < \sqrt{2}$. f is concave up when $-\sqrt{2} < t < 0$ and $t > \sqrt{2}$.
There is a relative maximum when $t = -2$, or $(-2, 64)$, and a relative minimum when $t = 2$, or

(2, 64). There are inflection points when $t = -\sqrt{2}, \sqrt{2}$, or $(-1.4, 39.6)$ and $(1.4, -39.6)$.

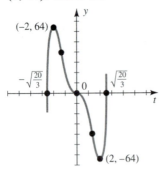

7. $g(t) = \dfrac{t^2}{t+1}$

When $t = 0$, $g(0) = 0$ so $(0, 0)$ is an intercept. When $g(t) = 0$, $t = 0$.

$t + 1 = 0$ when $t = -1$, so there is a vertical asymptote of $t = -1$.

$$\lim_{t \to \pm\infty} \frac{t^2}{t+1} = \lim_{t \to \pm\infty} \frac{t}{1 + \frac{1}{t}} = \pm\infty, \text{ so there are no horizontal asymptotes.}$$

Note: $y = t - 1$ is an oblique asymptote.

$$g'(t) = \frac{(t+1)(2t) - (t^2)(1)}{(t+1)^2}$$

$$= \frac{t^2 + 2t}{(t+1)^2}$$

$$= \frac{t(t+2)}{(t+1)^2}$$

$g'(t) = 0$ when $t = -2, 0$ and $g'(t)$ is undefined when $t = -1$.

$g''(t)$

$$= \frac{(t+1)^2(2t+2) - (t^2 + 2t)(2(t+1)(1))}{(t+1)^4}$$

$$= \frac{2(t+1)[(t+1)^2 - (t^2 + 2t)]}{(t+1)^4}$$

$$= \frac{2}{(t+1)^3}$$

$g''(t)$ is never zero and $g''(t)$ is undefined when $t = -1$.

When $t < -2$, $g'(x) > 0$ so g is increasing

$g''(x) < 0$ so g is concave down

When $-2 < t < -1$, $g'(x) < 0$ so g is decreasing

$g''(x) < 0$ so g is concave down

When $-1 < t < 0$, $g'(x) < 0$ so g is decreasing

$g''(x) > 0$ so g is concave up

When $t > 0$, $g'(x) > 0$ so g is increasing
$$g''(x) > 0 \text{ so } g \text{ is concave up}$$

Overall, g is decreasing when $-2 < t < -1$ and $-1 < t < 0$. g is increasing when $t < -2$ and $t > 0$. g is concave down when $t < -1$. g is concave up when $t > -1$.

There is a relative maximum when $t = -2$, or $(-2, -4)$, and a relative minimum when $t = 0$, or $(0, 0)$. There are no inflection points.

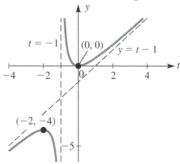

9. $F(x) = 2x + \dfrac{8}{x} + 2 = \dfrac{2x^2 + 2x + 8}{x}$

When $x = 0$, $F(0)$ is undefined.

$F(x) = 0$, $2(x^2 + x + 4) = 0$, which has no solution. Denominator is zero when $x = 0$, so there is a vertical asymptote of $x = 0$.

$$\lim_{x \to \pm\infty} \frac{2x^2 + 2x + 8}{x} = \lim_{x \to \pm\infty} \frac{2x + 2 + \frac{8}{x}}{1}$$
$$= \pm\infty$$

so there are no horizontal asymptotes. Note: $y = 2x + 2$ is an oblique asymptote.

$$F'(x) = 2 - \frac{8}{x^2}$$

$F'(x) = 0$ when $x = -2, 2$ and $F'(x)$ is undefined when $x = 0$.

$$F''(x) = \frac{16}{x^3}$$

$F''(x)$ is never zero and $F''(x)$ is undefined when $x = 0$.

When $x < -2$, $F'(x) > 0$ so F is increasing
$$F''(x) < 0 \text{ so } F \text{ is concave down}$$

When $-2 < x < 0$, $F'(x) < 0$ so F is decreasing
$$F''(x) < 0 \text{ so } F \text{ is concave down}$$

When $0 < x < 2$, $F'(x) < 0$ so F is decreasing
$$F''(x) > 0 \text{ so } F \text{ is concave up}$$

When $x > 2$, $F'(x) > 0$ so F is increasing
$$F''(x) > 0 \text{ so } F \text{ is concave up}$$

Overall, F is decreasing when $-2 < x < 0$ and $0 < x < 2$, F is increasing when $x < -2$ and $x > 2$, F is concave down when $x < 0$, F is concave up when $x > 0$.

There is a relative maximum when $x = -2$, or $(-2, -6)$, and a relative minimum when $x = 2$, or

(2, 10). There are no inflection points.

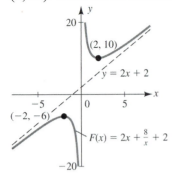

11. Graph (b) is the graph of f, and graph (a) is the graph of f'. Possible reasons include:

   (i) The degree of graph (b) is one greater than the degree of graph (a).

   (ii) Graph (a) is always positive, and graph (b) is always increasing.

13. $f'(x) = x^3 (2x-3)^2 (x+1)^5 (x-7)$

   $f'(x) = 0$ when $x = -1, 0, \dfrac{3}{2}, 7$

   When $x < -1$, $f'(x) < 0$ so f is decreasing

   $\quad -1 < x < 0$, $f'(x) > 0$ so f is increasing

   $\quad 0 < x < \dfrac{3}{2}$, $f'(x) < 0$ so f is decreasing

   $\quad \dfrac{3}{2} < x < 7$, $f'(x) < 0$ so f is decreasing

   $\quad x > 7$, $f'(x) > 0$ so f is increasing.

   There is a relative minimum when $x = -1$ and $x = 7$. There is a relative maximum when $x = 0$. There is no relative extremum when $x = \dfrac{3}{2}$.

15. $F'(x) = \dfrac{x(x-2)^2}{x^4 + 1}$

   $f'(x) = 0$, when $x = 0, 2$

   When $x < 0$, $f'(x) < 0$ so f is decreasing

   $\quad 0 < x < 2$, $f'(x) > 0$ so f is increasing

   $\quad x > 2$, $f'(x) < 0$ so f is increasing

   There is a relative minimum when $x = 0$, but there is no relative extrema when $x = 2$.

17. (a) $f'(x) > 0$ so f is increasing when
   $x < 0$ and $x > 5$.

   (b) $f'(x) < 0$ so f is decreasing when
   $0 < x < 5$.

(c) $f''(x) > 0$ so f is concave up when
$-6 < x < -3$ and $x > 2$.

(d) $f''(x) < 0$ so f is concave down when $x < -6$ and $-3 < x < 2$.

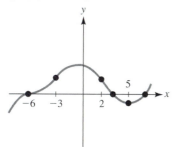

Note: since there are no points given, graphs can shift in y-direction, although not in x-direction.

19. (a) $f'(x) > 0$ so f is increasing when
$1 < x < 2$.

(b) $f'(x) < 0$ so f is decreasing when
$x < 1$ and $x > 2$.

(c) $f''(x) > 0$ so f is concave up when
$x < 2$ and $x > 2$.

(d) $f'(1) = 0$, so graph levels when $x = 0$.

$f'(2)$ is undefined, so graph has a vertical asymptote, hole or vertical tangent when $x = 2$.

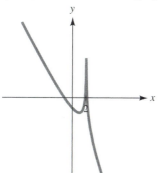

Note: since there are no points given, graphs can shift in y-direction, although not in x-direction.

21. $f(x) = -2x^3 + 3x^2 + 12x - 5$

$f'(x) = -6x^2 + 6x + 12 = -6(x+1)(x-2)$

$f'(x) = 0$ when $x = -1, 2$

$f''(x) = -12x + 6$

$f''(-1) = 18 > 0$, so there is a relative minimum when $x = -1$, or $(-1, -12)$; $f''(2) = -18 < 0$, so there is a relative maximum when $x = 2$, or $(2, 15)$.

23. $f(x) = \dfrac{x^2}{x+1}$

$f'(x) = \dfrac{(x+1)(2x) - (x^2)(1)}{(x+1)^2} = \dfrac{x(x+2)}{(x+1)^2}$

$f'(x) = 0$ when $x = -2, 0$

$f''(x)$

$= \dfrac{(x+1)^2(2x+2) - (x^2+2x)(2(x+1)(1))}{(x+1)^2}$

$f''(-2) = -2 < 0$, so there is a relative maximum when $x = -2$, or $(-2, -4)$; $f''(0) = 2 > 0$, so there is a relative minimum when $x = 0$, or $(0, 0)$.

25. $f(x) = -2x^3 + 3x^2 + 12x - 5$

$f'(x) = -6x^2 + 6x + 12$

$f'(x) = -6(x+1)(x-2)$

$f'(x) = 0$ when $x = -1, 2$, both of which are in the interval $-3 \le x \le 3$.

$f(-1) = -12, f(2) = 15, f(-3) = 40, f(3) = 4$.

So, $f(-3) = 40$ is the absolute maximum and $f(-1) = -12$ the absolute minimum.

27. $g(s) = \dfrac{s^2}{s+1}$

$g'(s) = \dfrac{(s+1)(2s) - (s^2)(1)}{(s+1)^2}$

$g'(s) = \dfrac{s(s+2)}{(s+1)^2}$

$g'(s) = 0$ when $s = -2, 0$, of which only

$s = 0$ is in the interval $-\dfrac{1}{2} \le s \le 1$.

$g\left(-\dfrac{1}{2}\right) = \dfrac{1}{2}$, $g(0) = 0$, and $g(1) = \dfrac{1}{2}$

So, $g\left(-\dfrac{1}{2}\right) = g(1) = \dfrac{1}{2}$ is the absolute maximum and $g(0) = 0$ the absolute minimum.

29. $f'(x) = x(x-1)^2$

(a) $f'(x) = 0$ when $x = 0, 1$

When $x < 0$, $f'(x) < 0$ so f is decreasing

$0 < x < 1$, $f'(x) > 0$ so f is increasing

$x > 1$, $f'(x) > 0$ so f is increasing

(b) $f''(x) = x[2(x-1)(1)] + (x-1)^2(1)$
$\qquad = (3x-1)(x-1)$

$f''(x) = 0$ when $x = \dfrac{1}{3}, 1$

When $x < \dfrac{1}{3}$, $f''(x) > 0$ so f is concave up

$\qquad \dfrac{1}{3} < x < 1$, $f''(x) < 0$ so f is concave down

$\qquad x > 1$, $f''(x) > 0$ so f is concave up

(c) There is a relative minimum when

$x = 0$ and there are inflection points when $x = \dfrac{1}{3}$ and $x = 1$.

(d)

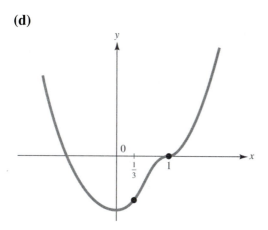

Note: Since there are no points given, graph can shift in y-direction, although not in x-direction.

31. Profit = revenue $-$ costs
$\qquad$ = (#sold)(selling price)
$\qquad\qquad$ $-$ (#sold)(cost per unit)

$P(x) = 100(20 - x)x - 100(20 - x)5$
$\qquad = 100(20 - x)(x - 5)$

$P'(x) = 100[(20 - x)(1) + (x - 5)(-1)]$
$\qquad = 100(25 - 2x)$

$P'(x) = 0$ when $x = 12.50$

When $5 \le x < 12.5$, $P'(x) > 0$ so P is increasing
$\qquad x > 12.5$, $P'(x) < 0$ so P is decreasing

So, when the price is \$12.50 per unit, the profit is maximized.

33. Let r denote the radius, h the height, C the (fixed) cost (in cents), and V the volume of the container.
$V = \pi r^2 h$

$C = \text{cost of bottom} + \text{cost of side}$

$\quad = 3(\text{area of bottom}) + 2(\text{area of side})$

or $C = 3\pi r^2 + 4\pi rh$

Solving for h, $h = \dfrac{C - 3\pi r^2}{4\pi r}$ and $V(r) = \pi r^2 \left(\dfrac{C - 3\pi r^2}{4\pi r} \right) = \dfrac{rC}{4} - \dfrac{3\pi r^3}{4}$

$V'(r) = \dfrac{C}{4} - \dfrac{9\pi r^2}{4}$

$V'(r) = 0$ when $\dfrac{C}{4} = \dfrac{9\pi r^2}{4}$, or $C = 9\pi r^2$. So, $h = \dfrac{9\pi r^2 - 3\pi r^2}{4\pi r}$, or $h = \dfrac{3r}{2}$.

$V''(r) = -\dfrac{9\pi r}{2}$

$V''\left(\dfrac{3}{2}r\right) < 0$, so there is a relative maximum when $h = 1.5r$. Further, $V''(r) < 0$ for all r, so the

volume is maximized when the height is 1.5 times the radius of the cylindrical container.

35. Let x be the width of the pasture and let y be its length. The area of the enclosed pasture is $A = xy$.

(a) Since there are 320 feet of fencing to use in enclosing the pasture,
$$2(x + y) = 320$$
$$y = 160 - x$$

So,

$A(x) = x(160 - x) = 160x - x^2$

$A'(x) = 160 - 2x$

$A'(x) = 0$ when $x = 80$

When $0 < x < 80$, $A'(x) > 0$ so A is
      increasing
      $80 < x < 160$, $A'(x) < 0$ so A is
      decreasing

So, to maximize the area, the dimensions are width = 80 feet and
length = 160 − 80 = 80 feet.

(b) Since there are 320 feet of fencing to use in enclosing the pasture and fencing is only needed on
three sides (choosing the width as the side opposite the wall),

$x + 2y = 320$

$\quad y = 160 - \dfrac{1}{2}x$

Now,

$A(x) = x\left(160 - \dfrac{1}{2}x\right) = 160x - \dfrac{1}{2}x^2$

$A'(x) = 160 - x$

$A'(x) = 0$ when $x = 160$

When $0 < x < 160$, $A'(x) > 0$ so A is
    increasing
    $160 < x < 320$, $A'(x) < 0$ so A is
    decreasing

So, to maximize the area, the dimensions are
width (side opposite the wall)

$= 160$ feet and length $= 160 - \dfrac{1}{2}(160) = 80$ feet.

37. Let Q be the point on the opposite bank straight across from the starting point. With $QP = x$, the distance walked along the bank is $1 - x$. The distance across the water is given by the Pythagorean theorem to be $\sqrt{1 + x^2}$. The time t is

t = time in the water + time on the land

$= \dfrac{\text{distance in the water}}{\text{speed in the water}}$

$\qquad + \dfrac{\text{distance on the land}}{\text{speed on the land}}$

$= \dfrac{1}{4}(1 + x^2)^{1/2} + \dfrac{1}{5}(1 - x)$

The relevant interval is $0 \le x \le 1$ and $t'(x) = \dfrac{x}{4\sqrt{1 + x^2}} - \dfrac{1}{5}$

$t'(x) = 0$ when $\dfrac{x}{4\sqrt{1 + x^2}} = \dfrac{1}{5}$

$\qquad\qquad\qquad 5x = 4\sqrt{1 + x^2}$

$\qquad\qquad\qquad 25x^2 = 16 + 16x^2$, or $x = \pm\dfrac{4}{3}$

Neither of these critical values is in the interval $0 \le x \le 1$. So, the absolute minimum must occur at an endpoint.

$t(0) = 0.45$, $t(1) = \dfrac{\sqrt{2}}{4} \approx 0.354$

The minimum time is when $x = 1$. That is, when you row all the way to town.

39. Let x denote the number of machines used and $C(x)$ the corresponding cost of producing the 400,000 medals. Then $C(x) =$ set-up cost + operating cost

$$= 80 \text{ (number of machines)}$$
$$+ 5.76 \text{ (number of hours)}$$

Each machine can produce 200 medals per hour, so x machines can produce $200x$ medals per hour, and it will take

$\dfrac{400,000}{200x}$ hours to produce the

19. $4^{2x-1} = 16$

$(2^2)^{2x-1} = (2)^4$

$2^{2(2x-1)} = 2^4$

$2^{4x-2} = 2^4$

By the equality rule of exponential functions, $4x - 2 = 4$, or $x = \dfrac{3}{2}$.

21. $2^{3-x} = 4^x$

$2^{3-x} = (2^2)^x$

$2^{3-x} = 2^{2x}$

By the equality rule of exponential functions, $3 - x = 2x$, or $x = 1$.

23. $(2.14)^{x-1} = (2.14)^{1-x}$

By the equality rule of exponential functions, $x - 1 = 1 - x$, or $x = 1$.

25. $10^{x^2-1} = 10^3$

By the equality rule of exponential functions, $x^2 - 1 = 3$, or $x^2 = 4$.

$x = -2, x = 2$

27. $\left(\dfrac{1}{8}\right)^{x-1} = 2^{3-2x^2}$

Since $\dfrac{1}{8} = 2^{-3}$,

$(2^{-3})^{x-1} = 2^{3-2x^2}$

$2^{-3(x-1)} = 2^{3-2x^2}$

By the equality rule of exponential functions, $-3(x-1) = 3 - 2x^2$

$2^2 - 3x = 0$, or $x(2x - 3) = 0$

$x = 0, \; x = \dfrac{3}{2}$

29. $y = 3^{1-x}$

Press $\boxed{y =}$ and input $3 \wedge (1 - x)$ for $y_1 =$.

Use window dimensions $[-5, 5]1$ by $[-1, 9]1$.

Press $\boxed{\text{graph}}$

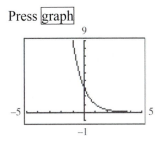

31. $y = 4 - e^{-x}$

Press $\boxed{y =}$ and input $4 - e \wedge (-x)$.

Use window dimensions $[-5, 5]1$ by $[-5, 5]1$.

Press $\boxed{\text{graph}}$

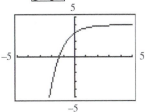

33. $y = Cb^x$

$12 = Cb^2$ and $24 = Cb^3$

Solving the first equation for C, $C = \dfrac{12}{b^2}$

Substituting into second equation,

$24 = \left(\dfrac{12}{b^2}\right)b^3$

$24 = 12b$

$2 = b$

$C = \dfrac{12}{(2)^2} = 3$

35.   The future value after t years, if P dollars is invested at an annual interest rate r and interest is compounded k times per year, will be $B(t) = P\left(1 + \dfrac{r}{k}\right)^{kt}$ dollars continuously, will be $B(t) = Pe^{rt}$ dollars. When $P = \$1{,}000$, $r = 0.07$, $t = 10$, and

(a) $k = 1$

$$B(10) = 1{,}000\left(1 + \frac{0.07}{1}\right)^{1 \cdot 10}$$

$$\approx \$1{,}967.15$$

(b) $k = 4$

$$B(10) = 1{,}000\left(1 + \frac{0.07}{4}\right)^{4 \cdot 10}$$

$$\approx \$2{,}001.60$$

(c) $k = 12$

$$B(10) = 1{,}000\left(1 + \frac{0.07}{12}\right)^{12 \cdot 10}$$

$$\approx \$2{,}009.66$$

(d) compounded continuously

$$B(10) = 1{,}000e^{0.07(10)} \approx \$2{,}013.75$$

37. The present value when $t = 5$, $r = 0.07$ and $B = \$10{,}000$ is

(a) compounded annually,

$$P = 10{,}000\left(1 + \frac{0.07}{1}\right)^{-1(5)}$$

$$\approx \$7{,}129.86$$

(b) compounded quarterly,

$$P = 10{,}000\left(1 + \frac{0.07}{4}\right)^{-4(5)}$$

$$\approx \$7{,}068.25$$

(c) compounded daily,

$$P = 10{,}000\left(1 + \frac{0.07}{365}\right)^{-365(5)}$$

$$\approx \$7{,}047.12$$

(d) compounded continuously,

$$P = 10{,}000e^{-0.07(5)}$$

$$\approx \$7{,}046.88$$

39. When interest is compounded quarterly at an annual rate of 6%,

$$r_e = \left(1 + \frac{r}{k}\right)^k - 1$$

$$= \left(1 + \frac{0.06}{4}\right)^4 - 1$$

$$\approx 0.614, \text{ or } 6.14\%$$

41. When interest is compounded continuously at an annual rate of 5%,

$$r_e = e^r - 1 = e^{0.05} - 1 \approx 0.0513, \text{ or } 5.13\%$$

43. The amount she must invest now, when $T = 4$, $r = 0.05$, $B = \$5{,}000$ and interest is compounded continuously is

$$P = 5{,}000e^{-0.05(4)}$$

$$\approx \$4{,}093.65$$

45. $p = 300e^{-0.02x}$

(a) When $x = 100$, $p = 300e^{-0.02(100)}$

$$p \approx 40.60058$$

The market price is $40.60.

(b) revenue = (#sold)(selling price)

$$R(x) = xp$$

$$R(100) = 100(300e^{-0.02(100)})$$

$$R(100) \approx 4{,}060.058$$

The corresponding revenue is $4,060.

(c) When $x = 50$,

$$R(50) = 50(300e^{-0.02(5)})$$

$$R(50) \approx 5{,}518.1916$$

The corresponding revenue is $5,518.

$$R(100) - R(50) = 4{,}060 - 5{,}518$$

$$= -1{,}458$$

When 100 units are produced, the revenue is $1,458 less than when 50 units are produced.

47. (a) $r_e = \left(1 + \dfrac{r}{k}\right)^k - 1$

$$= \left(1 + \dfrac{0.079}{2}\right)^2 - 1$$

$$\approx 0.0806$$

(b) $r_e = \left(1 + \dfrac{0.078}{4}\right)^4 - 1 \approx 0.0803$

(c) $r_e = \left(1 + \dfrac{0.077}{12}\right)^{12} - 1 \approx 0.0798$

(d) $r_e = e^r - 1 = e^{0.0765} - 1 \approx 0.0795$
From lowest to highest, *d, c, b, a*.

49. The value of $500 in five years, at an annual inflation rate of 4%, will be

$$B(t) = P\left(1 + \dfrac{r}{k}\right)^{kt}$$

$$B(5) = 500\left(1 + \dfrac{0.04}{1}\right)^{1(5)} \approx 608.326$$

To break even, he should sell the stamp for $608.33.

51. $f(t) = e^{-0.2t}$

(a) The fraction of toasters still working after 3 years is
$$f(3) = e^{-0.2(3)} \approx 0.5488.$$

(b) The fraction which fails during the first year is $f(0) - f(1) = e^0 - e^{-0.2(1)}$
$$\approx 1 - 0.8187$$
$$= 0.1813$$

(c) The fraction which fails during the third year is
$$f(2) - f(3) = e^{-0.2(2)} - e^{-0.2(3)}$$
$$\approx 0.6703 - 0.5488$$
$$= 0.1215$$

53. Investing $24 at 7% compounded continuously for 364 years would yield

$$B(t) = Pe^{rt}$$
$$B(364) = 24e^{0.07(364)}$$
$$\approx 2.7928 \times 10^{12}$$
$$\approx 2{,}792.8 \text{ billion dollars}$$
$$\underline{-\ \ 25.2}$$
$$2{,}767.6$$

Investing the money would have resulted in the better deal for the sellers by $2,767.6 billion dollars.

55. (a) Since the GDP increases at a rate of 2.7% per year, the pattern of growth is
$$500 \to 1.027(500)$$
$$\to 1.027\left[1.027(500)\right]$$
$$\to 1.027\left[(1.027)(1.027)(500)\right]$$

it can be modeled by the function
$$\text{GDP}(t) = 500(1.027)^t \text{ billion dollars}$$

(b) In 2015, when $t = 10$,
$$G(10) = 500(1.027)^{10} = 652.64 \text{ or}$$
approximately $652.6 billion.

57. $M = \dfrac{Ai}{1 - (1+i)^{-n}}$

When $A = 150{,}000$, $i = \dfrac{0.09}{12} = 0.0075$,

$n = 360$,
$$M = \dfrac{150{,}000(0.0075)}{1 - (1.0075)^{-360}} = \$1{,}206.93 \text{ for the}$$

monthly payment.

59. (a) The potential buyer is offering to pay you, $1000 + 160(36) = \$6{,}760$. Using the amortization formula, monthly payments would be
$$M = \dfrac{5{,}000\left(\frac{0.12}{12}\right)}{1 - \left(1 + \frac{0.12}{12}\right)^{-12 \cdot 3}} \approx \$166.07.$$

This way, you would receive $\$1{,}000 + (166.07)(36) = \$6{,}978.52$. This is $\$6{,}978.52 - \$6{,}760 = \$218.52$ more than the potential buyer is offering.

(b) Writing Exercise—Answers will vary.

61. $D(x) = 12e^{-0.07x}$

(a) At the center of the city, the density is $D(0) = 12$, or 12,000 people per square mile.

(b) Ten miles from the center, the density is
$$D(10) = 12e^{-0.07(10)}$$
$$= 12e^{-0.7}$$
$$\approx 5.959,$$
or 5,959 people per square mile.

63. $P(t) = 50e^{0.02t}$

(a) For the current population, $t = 0$ so $P(0) = 50e^0 = 50$ so the current population is 50 million.

(b) When $t = 30$,
$P(30) = 50e^{0.02(30)} \approx 91.11$ so the population will be approximately 91.11 million.

65. $C(t) = 3 \cdot 2^{-0.75t}$

(a) When $t = 0$, $C(0) = 3 \cdot 2^{-0.75(0)}$
$$= 3 \cdot 2^0$$
$$= 3 \cdot 1$$
$$= 3 \text{ mg/ml}$$
When $t = 1$,
$C(1) = 3 \cdot 2^{-0.75(1)} \approx 1.7838$ mg/ml.

(b) average rate of change $= \dfrac{C(t_2) - C(t_1)}{t_2 - t_1}$
$$= \dfrac{C(2) - C(1)}{2 - 1}$$
$C(2) = 3 \cdot 2^{-0.75(2)} \approx 1.0607$
average rate of change
$$= \dfrac{1.0607 - 1.7838}{2 - 1}$$
$$= -0.7231 \text{ mg/ml per hour}$$

67. $P(t) = A \cdot 2^{0.001t}$

(a) Since $P(10) = 10,000$,
$$10,000 = A \cdot 2^{0.001(10)}$$
$$\dfrac{10,000}{2^{0.01}} = A$$
$$A \approx 9,931$$

(b) When $t = 0$, $P(0) = 9,931 \cdot 2^{0.01(0)}$
$$= 9,931 \cdot 2^0$$
$$= 9,931 \text{ bacteria}$$
When $t = 20$, $P(20) = 9,931 \cdot 2^{0.01(20)}$
$$\approx 10,070 \text{ bacteria}$$
Since time is measured in minutes, one hour corresponds to $t = 60$ and
$$P(60) = 9,931 \cdot 2^{0.001(60)}$$
$$\approx 10,353 \text{ bacteria}$$

(c) Keeping t measured in minutes, the average rate of change is
$$\dfrac{P(120) - P(60)}{120 - 60}.$$
$$P(120) = 9,931 \cdot 2^{0.001(120)}$$
$$\approx 10,792 \text{ bacteria}$$
average rate of change
$$= \dfrac{10,792 - 10,353}{60}$$
$$\approx 7.32 \text{ bacteria per minute}$$

69. $C(t) = 0.065(1 + e^{-0.025t})$

(a) When $t = 0$,
$$C(0) = 0.065(1 + e^0) = 0.13 \text{ g/cm}^3.$$

(b) When $t = 20$,
$$C(20) = 0.065(1 + e^{-0.025(20)})$$
$$\approx 0.1044 \text{ g/cm}^3$$
Since t is measured in minutes, one hour corresponds to $t = 60$ and
$$C(60) = 0.065(1 + e^{-0.025(60)})$$
$$\approx 0.0795 \text{ g/cm}^3$$

(c) average rate of change $= \dfrac{C(1) - C(0)}{1 - 0}$

$C(1) = 0.065(1 + e^{-0.025(1)}) \approx 0.1284$

average rate of change

$= \dfrac{0.1284 - 0.13}{1 - 0}$

≈ -0.0016 g/cm^3 per minute

(d) As $t \to \infty$,

$\displaystyle\lim_{t \to \infty} C(t) = \lim_{t \to \infty} 0.065(1 + e^{-0.025t}).$

Since $\displaystyle\lim_{x \to \infty} e^{-x} = 0,$

$\displaystyle\lim_{t \to \infty} C(t) = 0.065(1 + 0) = 0.065$ g/cm^3

(e) Press $\boxed{y=}$ and input

$0.065(1 + e \, \char`\^ \, (-0.025t))$

for $y_1 =$.

Use window dimensions [0, 180]60 by

[0, 0.2].05

Press $\boxed{\text{graph}}$

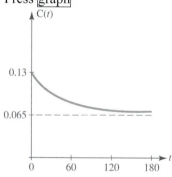

71. $I = I_0 e^{-kx}$

When $x = 3$ meters, $I = 0.1I_0$. So

$0.1I_0 = I_0 e^{-k \cdot 3}$, or

$0.1 = e^{-3k}$

When $x = 1$ meter,

$I = I_0 e^{-k \cdot 1}$

$= I_0 (e^{-3k})^{1/3}$

$= I_0 (0.1)^{1/3}$

$\approx 0.46 I_0$

73. $Q(t) = Q_0 e^{-0.0001t}$.

When $t = 5{,}000$, $Q(5{,}000) = 200$. So,

$200 = Q_0 e^{-0.0001(5{,}000)}$

$Q_0 = \dfrac{200}{e^{-0.5}} \approx 329.7$ grams.

75. $f(x) = \dfrac{1}{2}\left(\dfrac{1}{4}\right)^x$

x	-2.2	-1.5	0	1.5	2.3
$f(x)$	10.5561	4	0.5	0.0625	0.02062

Press $\boxed{y=}$ and input .5(.25 $\char`\^$ x) for $y_1 =$.
Press $\boxed{\text{2nd}}$ $\boxed{\text{TBLSET}}$ and enter ask
independent with auto dependent. Press
$\boxed{\text{2nd}}$ $\boxed{\text{table}}$ and enter $x = -2.2, -1.5, 0, 1.5,$
and 2.3. The output values are displayed
automatically.

77. To use a calculator to evaluate $\left(1 + \dfrac{1}{n}\right)^n$

for $n = -1{,}000, -2{,}000 \ldots -50{,}000$, press
$\boxed{y=}$ and input $(1 + (1 \div x)) \char`\^ x$ for $y_1 =$.
Press $\boxed{\text{2nd}}$ $\boxed{\text{TBLSET}}$ and input
TblStart $= -1{,}000$, ΔTbl $= -1{,}000$ and
auto independent with auto dependent.
Press 2nd table.
Following are some values from this table:

x	y_1
$-1{,}000$	2.7196
$-2{,}000$	2.719
$-3{,}000$	2.7187
$-4{,}000$	2.7186
$-5{,}000$	2.7186
$\vdots$	$\vdots$
$-48{,}000$	2.7183
$-49{,}000$	2.7183
$-50{,}000$	2.7183

As n decreases without bound, $\left(1+\dfrac{1}{n}\right)^n$

approaches $e \approx 2.71828$.

79. To use a calculator to estimate

$$\lim_{x\to\infty}\left(2-\frac{5}{2n}\right)^{n/3}.$$

Press $\boxed{y=}$ and input
$(2-(5\div(2x)))\,\hat{}\,(x\div 3)$ for $y_1 =$.
Press $\boxed{\text{2nd}}$ $\boxed{\text{TBLSET}}$ and input
Tblstart $= 10$ and ΔTbl $= 10$. Use auto
independent with auto dependent. Press $\boxed{\text{2nd}}$
$\boxed{\text{table}}$. The following are a few values from
the table:

x	y_1
10	6.4584
20	66.071
$\vdots$	$\vdots$
100	7.12×10^9

These values suggest that

$$\lim_{n\to+\infty}\left(2-\frac{5}{2n}\right)^{n/3} = +\infty.$$

4.2   Logarithmic Functions

1. Using the TI-84 Plus, press LN, the number,
a right parenthesis, and then ENTER. When
combined with powers of e, press LN, 2nd,
e^x, the power of e, right parenthesis, and
then ENTER.
So, $\ln 1 = 0$, $\ln 2 \approx 0.693$, $\ln e = 1$,

$\ln 5 \approx 1.609$, $\ln\left(\dfrac{1}{5}\right) \approx -1.609$, $\ln e^2 = 2$.

Since $\ln x$ has a domain of $x > 0$, $\ln 0$ and
$\ln(-2)$ yield ERR: DOMAIN.

3. Since $\ln x$ and e^x are inverse operations,
$\ln e^3 = 3$.

5. Since e^x and $\ln x$ are inverse operations,
$e^{\ln 5} = 5$.

7. $\begin{aligned}[t] e^{3\ln 2 - 2\ln 5} &= e^{\ln 2^3 - \ln 5^2} \\ &= e^{\ln 8 - \ln 25} \\ &= e^{\ln\left(\frac{8}{25}\right)} \\ &= \frac{8}{25} \end{aligned}$

9. $\begin{aligned}[t] \log_3 270 &= \log_3 27 + \log_3 10 \\ &= \log_3 27 + \log_3 5 + \log_3 2 \\ &= \log_3 3^3 + \log_3 5 + \log_3 2 \end{aligned}$

Since $\log_3 x$ and 3^x are inverse
operations,

$$= 3 + \log_3 5 + \log_3 2$$

11. $\begin{aligned}[t] \log_3 100 &= \log_3 (10)^2 \\ &= 2\log_3 10 \\ &= 2\log_3 (2\cdot 5) \\ &= 2(\log_3 2 + \log_3 5) \\ &= 2\log_3 2 + 2\log_3 5 \end{aligned}$

13. $\begin{aligned}[t] \log_2(x^4 y^3) &= \log_2 x^4 + \log_2 y^3 \\ &= 4\log_2 x + 3\log_2 y \end{aligned}$

15. $\begin{aligned}[t] \ln\sqrt[3]{x^2 - x} &= \ln(x^2 - x)^{1/3} \\ &= \frac{1}{3}\ln(x^2 - x) \\ &= \frac{1}{3}\ln[x(x-1)] \\ &= \frac{1}{3}[\ln x + \ln(x-1)] \\ &= \frac{1}{3}\ln x + \frac{1}{3}\ln(x-1) \end{aligned}$

17. $\ln\left[\dfrac{x^2(3-x)^{2/3}}{\sqrt{x^2+x+1}}\right]$

$= \ln[x^2(3-x)^{2/3}] - \ln\sqrt{x^2+x+1}$

$= \ln x^2 + \ln(3-x)^{2/3} - \ln(x^2+x+1)^{1/2}$

$= 2\ln x + \dfrac{2}{3}\ln(3-x) - \dfrac{1}{2}\ln(x^2+x+1)$

19. $\ln(x^3 e^{-x^2}) = \ln x^3 + \ln e^{-x^2} = 3\ln x - x^2$

21. $4^x = 53$

Taking the natural log of both sides gives

$\ln 4^x = \ln 53$.

Using a rule of logarithms gives

$x \ln 4 = \ln 53$

$$x = \frac{\ln 53}{\ln 4} \approx 2.864$$

23. $\log_3(2x - 1) = 2$

Rewriting in exponential form gives

$2x - 1 = 3^2$ or $x = 5$.

25. $2 = e^{0.06x}$

Taking the natural log of both sides gives

$\ln 2 = 0.06x$, or $x = \dfrac{\ln 2}{0.06} \approx 11.552$.

27. $3 = 2 + 5e^{-4x}$

$1 = 5e^{-4x}$

$\dfrac{1}{5} = e^{-4x}$

Taking the natural log of both sides gives

$\ln \dfrac{1}{5} = -4x$, or $x = \dfrac{\ln\left(\frac{1}{5}\right)}{-4}$.

Since $\ln \dfrac{1}{5} = \ln 1 - \ln 5 = 0 - \ln 5 = -\ln 5$,

$x = \dfrac{-\ln 5}{-4} = \dfrac{\ln 5}{4} \approx 0.402$.

29. $-\ln x = \dfrac{t}{50} + C$

$\ln x = \dfrac{-t}{50} - C$

$e^{\ln x} = e^{(-t/50)-C}$, or $x = e^{(-t/50)-C}$

31. $\ln x = \dfrac{1}{3}(\ln 16 + 2\ln 2)$

$= \dfrac{1}{3}(\ln 16 + \ln 4)$

$= \dfrac{1}{3}\ln(16 \cdot 4)$

$= \ln 64^{1/3}$

So, $\ln x = \ln 4$

$e^{\ln x} = e^{\ln 4}$

or $x = 4$.

33. $3^x = e^2$

Taking the natural log or both sides gives

$\ln 3^x = \ln e^2$. So, $x \ln 3 = 2$.

$x = \dfrac{2}{\ln 3} \approx 1.820$

35. $\dfrac{25e^{0.1x}}{e^{0.1x} + 3} = 10$

$25e^{0.1x} = 10(e^{0.1x} + 3)$

$25e^{0.1x} = 10e^{0.1x} + 30$

$15e^{0.1x} = 30$

$e^{0.1x} = 2$

$\ln e^{0.1x} = \ln 2$

$0.1x = \ln 2$

$x = \dfrac{\ln 2}{0.1} = 10\ln 2 \approx 6.9315$

37. $\log_2 x = 5$

Rewriting in exponential form

$x = 2^5$

$\ln x = \ln 2^5$

$\ln x = 5\ln 2 \approx 3.4657$

39. $\log_5(2x) = 7$

Rewriting in exponential form,

$2x = 5^7$

$x = \dfrac{5^7}{2}$

$\ln x = \ln\left(\dfrac{5^7}{2}\right)$

$= \ln 5^7 - \ln 2$

$= 7\ln 5 - \ln 2 \approx 10.5729$

41. $\ln \dfrac{1}{\sqrt{ab^3}} = \ln 1 - \ln\left(\sqrt{ab^3}\right)$

$= 0 - \ln(ab^3)^{1/2}$

$= -\dfrac{1}{2}\ln(ab^3)$

$= -\dfrac{1}{2}[\ln a + \ln b^3]$

$= -\dfrac{1}{2}\ln a - \dfrac{1}{2}\ln b^3$

$= -\dfrac{1}{2}\ln a - \dfrac{3}{2}\ln b$

Since $\ln a = 2$ and $\ln b = 3$,

$= -\dfrac{1}{2}(2) - \dfrac{3}{2}(3)$

$= -\dfrac{11}{2}$

43. $B(t) = Pe^{rt}$

After a certain time, the investment will have grown to $B(t) = 2P$ at the interest rate of 0.06. So, $2P = Pe^{0.06t}$

$2 = e^{0.06t}$

$\ln 2 = \ln e^{0.06t}$

$\ln 2 = 0.06t$

and $t = \dfrac{\ln 2}{0.06} = 11.55$ years.

45. $B(t) = Pe^{rt}$

Since money doubles in 13 years,

$2P = B(13) = Pe^{13t}$

$2 = e^{13r}$

$\ln 2 = \ln e^{13r}$

$\ln 2 = 13r$

and $r = \dfrac{\ln 2}{13} = 0.0533$. The annual interest rate is 5.33%.

47. $B(t) = Pe^{rt}$

Since money doubles in 12 years,

$2P = B(12) = Pe^{12r}$

$2 = e^{12r}$

$\ln 2 = 12r$

and $r = \dfrac{\ln 2}{12} \approx 0.05776$.

To find t when money triples,

$3P = B(t) = Pe^{0.05776t}$

$3 = e^{0.5776t}$

$\ln 3 = 0.05776t$

$t = \dfrac{\ln 3}{0.05776} \approx 19.02$ years

49. At 6% compounded annually, the effective interest rate is

$\left(1 + \dfrac{r}{k}\right)^k - 1 = \left(1 + \dfrac{0.06}{1}\right)^1 - 1 = 0.06.$

At r% compounded continuously, the effective interest rate is $e^r - 1$. Setting the two effective rates equal to each other yields

$e^r - 1 = 0.06,$

$e^r = 1.06,$

$r = \ln 1.06 = 0.0583$ or 5.83%.

51. $Q(t) = 500 - Ae^{-kt}$

When $t = 0$, $Q(0) = 300$ and

$300 = 500 - Ae^{-k(0)}$

$300 = 500 - A$, or $A = 200$

So, $Q(t) = 500 - 200e^{-kt}$

When $t = 6$ months, $Q(6) = 410$ and

$$410 = 500 - 200e^{-k(6)}$$
$$200e^{-6k} = 90$$
$$e^{-6k} = \frac{9}{20}$$
$$\ln e^{-6k} = \ln \frac{9}{20}$$
$$-6k = \ln \frac{9}{20}$$
$$k = \frac{\ln \frac{9}{20}}{-6}$$

So, $Q(t) = 500 - 200e^{\frac{\ln 0.45}{6}t}$.
When $t = 12$ months,
$$Q(12) = 500 - 200e^{\frac{\ln 0.45}{6}(12)} = 459.5 \text{ units.}$$

53. $S(x) = \ln(x + 2)$
$D(x) = 10 - \ln(x + 1)$

(a) When $x = 10$,
$D(10) = 10 - \ln(10 + 1) \approx \$7.60$.

(b) When $x = 100$,
$S(100) = \ln(100 + 2) \approx \$4.62$.

(c)
$$\ln(x + 2) = 10 - \ln(x + 1)$$
$$\ln(x + 2) + \ln(x + 1) = 10$$
$$\ln(x^2 + 3x + 2) = 10$$
$$e^{\ln(x^2 + 3x + 2)} = e^{10}$$
$$x^2 + 3x + 2 = e^{10}$$
$$x^2 + 3x + (2 - e^{10}) = 0$$
$$x_e = \frac{-3 + \sqrt{(3)^2 - 4(1)(2 - e^{10})}}{2(1)}$$
$$x_e \approx 147 \text{ units}$$
$$P_e = \ln(147 + 2) \approx \$5.00$$

55. $A(t) = 300 \ln(t + 3)$

(a) When $t = 0$,
$A(0) = 300 \ln 3 \approx \$329.6 \text{ million}$

(b) $A(0)$ is the initial amount invested, so need to find t when
$$A(t) = 2(329.6) = 659.2$$

$$659.2 = 300 \ln(t + 3)$$
$$2.197 \approx \ln(t + 3)$$
$$e^{2.197} \approx e^{\ln(t+3)}$$
$$8.998 \approx t + 3$$
$$t \approx 5.998$$
or approximately 6 years.

(c) In millions of dollars, one billion is 1,000 million. So,
$$1,000 = 300 \ln(t + 3)$$
$$\frac{10}{3} = \ln(t + 3)$$
$$e^{10/3} = e^{\ln(t+3)}$$
$$e^{10/3} = t + 3$$
$$t \approx 25.03$$
or approximately 25 years.

57. $C(t) = 0.4(2 - 0.13e^{-0.02t})$

(a) After 20 seconds, the drug concentration is
$$C(20) = 0.4(2 - 0.13e^{-0.02(20)})$$
$$\approx 0.765 \text{ g/cm}^3$$
After 60 seconds, it is
$$C(60) = 0.4(2 - 0.13e^{-0.02(60)})$$
$$\approx 0.784 \text{ g/cm}^3$$

(b) To find the time for the given concentration,
$$0.75 = 0.4(2 - 0.13e^{-0.02t})$$
$$1.875 = 2 - 0.13e^{-0.02t}$$
$$0.13e^{-0.02t} = 0.125$$
$$e^{-0.02t} \approx 0.9615$$
$$-0.02t \approx \ln 0.9615$$
$$t \approx \frac{\ln 0.9615}{-0.02}$$
$$\approx 1.96 \text{ seconds}$$

59. (a)

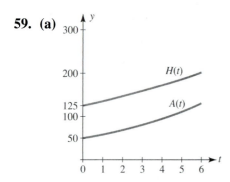

(b) $H = 125e^{0.08t}$

$$\frac{H}{125} = e^{0.08t}$$

$$\ln \frac{H}{125} = \ln e^{0.08t}$$

$$\ln \frac{H}{125} = 0.08t$$

$$t = \frac{\ln \frac{H}{125}}{0.08}$$

Now, $A(t) = 50e^{0.16t}$

So, $A(H) = 50e^{0.16(\ln(H/125)/0.08)}$

$$= 50e^{2\ln(H/125)}$$

$$= 50e^{\ln(H/125)^2}$$

$$= 50 \frac{H^2}{15,625}$$

$$= \frac{2H^2}{625}$$

61. The decay function is of the form
$R(t) = R_0 e^{-kt}$. From the text page 306, the
half-life of ^{14}C is 5,730 years, so

$$\frac{1}{2}R_0 = R_0 e^{-k(5,730)}$$

$$\frac{1}{2} = e^{-5,730k}$$

$$\ln \frac{1}{2} = \ln e^{-5,730k}$$

$$\ln \frac{1}{2} = -5,730k$$

$$\frac{\ln \frac{1}{2}}{-5,730} = k \text{ or } k = \frac{\ln 2}{5,730}$$

When 28% of the original amount

remains,

$$0.28R_0 = R_0 e^{-kt}$$

$$0.28 = e^{-kt}$$

$$\ln 0.28 = \ln e^{-kt}$$

$$\ln 0.28 = -kt$$

$$\frac{\ln 0.28}{-k} = t$$

Substituting k from above,

$$t = \frac{\ln 0.28}{-\left(\frac{\ln 2}{5,730}\right)}$$

$$= \frac{-5,730\ln 0.28}{\ln 2}$$

$$\approx 10,523 \text{ years.}$$

63. The decay function is of the form
$R(t) = R_0 e^{-kt}$. From the text page 306, the
half-life of ^{14}C is 5,730 years, so

$$\frac{1}{2}R_0 = R_0 e^{-k(5,730)}$$

$$\frac{1}{2} = e^{-5,730k}$$

$$\ln \frac{1}{2} = \ln e^{-5,730k}$$

$$\ln \frac{1}{2} = -5,730k$$

$$\frac{\ln \frac{1}{2}}{-5,730} = k \text{ or } k = \frac{\ln 2}{5,730}$$

When 99.7% of the original amount
remains,

$$0.997R_0 = R_0 e^{-kt}$$

$$0.997 = e^{-kt}$$

$$\ln 0.997 = \ln e^{-kt}$$

$$\ln 0.997 = -kt$$

$$\frac{\ln 0.997}{-k} = t$$

Substituting k from above,

$$t = \frac{\ln 0.997}{-\left(\frac{\ln 2}{5,730}\right)}$$

$$= \frac{-5,730\ln 0.997}{\ln 2}$$

$$\approx 24.8 \text{ years}$$

So, the painting in question was painted only 24.8 years ago. If the painting was actually $2003 - 1640 = 363$ years old, and p represents the percentage of ^{14}C currently present, $pR_0 = R_0 e^{-\left(\frac{\ln 2}{5,730}\right)(363)}$

$$p = e^{-\left(\frac{\ln 2}{5,730}\right)(363)}$$
$$p \approx 0.957, \text{ or } 95.7\%$$

65. $P(t) = 51 + 100\ln(t+3)$

(a) When $t = 0$, $P(0) = 51 + 100\ln(0+3)$
≈ 160.86 thousand,
or 160,860 people

(b) To find t when
$P(t) = 2(160.86) = 321.72$ solve
$321.72 = 51 + 100\ln(t+3)$
$270.72 = 100\ln(t+3)$
$2.7072 = \ln(t+3)$
$e^{2.7072} = e^{\ln(t+3)}$
$e^{2.7072} = t+3$
$\qquad t = e^{2.7072} - 3 \approx 12$ years

(c) average rate of growth $= \dfrac{P(10) - P(0)}{10 - 0}$
$P(10) = 51 + 100\ln(10+3) \approx 307.49$
average rate of growth
$= \dfrac{307.49 - 160.86}{10 - 0}$
≈ 14.66 thousand, or 14,660 people per year

67. Intensity function is of the form
$I(t) = I_0 e^{-kt}$. When $t = 20.9$ hours,
$I(20.9) = \frac{1}{2}I_0$. So, $\frac{1}{2}I_0 = I_0 e^{-k(20.9)}$
$$\frac{1}{2} = e^{-20.9k}$$
$$\ln\frac{1}{2} = \ln e^{-20.9k}$$
$$\ln\frac{1}{2} = -20.9k$$
$$\frac{\ln\frac{1}{2}}{-20.9} = k \text{ or } k = \frac{\ln 2}{20.9}$$

(a) When $t = 24$ hours,
$I(24) = I_0 e^{-\left(\frac{\ln 2}{20.9}\right)(24)} \approx I_0 \cdot 0.451.$
So approximately 45.1% of the original amount should be detected.

(b) $I(25) = I_0 e^{\left(\frac{\ln 0.5}{20.9}\right)(25)} \approx I_0 \cdot 0.436$
A total of 43.6% should remain in the entire body, and
$43.6\% - 41.3\% = 2.3\%$ remains outside of the thyroid gland.

69. $p(t) = 0.89[0.01 + 0.99(0.85)^t]$

(a) When $t = 0$,
$P(0) = 0.89[0.01 + 0.99(0.85)^0]$
$= 0.89[0.01 + 0.99]$
$= 0.89$

(b) To find t when $p(t) = 0.5$,
$0.5 = 0.89[0.01 + 0.99(0.85)^t]$
$0.5618 \approx 0.01 + 0.99(0.85)^t$
$0.5518 \approx 0.99(0.85)^t$
$0.5574 \approx (0.85)^t$
$\ln 0.5574 \approx \ln(0.85)^t$
$\ln 0.5574 \approx t\ln(0.85)$
$\qquad t \approx \dfrac{\ln 0.5574}{\ln 0.85} \approx 3.6$ seconds

(c) Press $\boxed{y=}$ and input
$0.89(0.01 + 0.99(0.85 \wedge t))$ for $y_1 =$.

Use window dimensions [0, 20]4 by [0, 1].25. Press graph

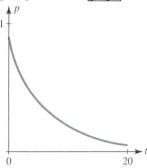

71. The decay function is of the form $Q(t) = Q_0 e^{-kt}$. Since the half-life is 1,690 years, $Q(1690) = \frac{1}{2}Q_0$ and

$$\frac{1}{2}Q_0 = Q_0 e^{-k(1,690)}$$

$$\frac{1}{2} = e^{-1,690k}$$

$$\ln\frac{1}{2} = \ln e^{-1,690k}$$

$$\ln\frac{1}{2} = -1,690k$$

$$\frac{\ln\frac{1}{2}}{-1,690} = k \text{ or } k = \frac{\ln 2}{1,690}$$

The initial amount, $Q_0 = 50$ grams, will reduce to 5 grams when

$$5 = 50 e^{-kt}$$

$$\frac{1}{10} = e^{-kt}$$

$$\ln\frac{1}{10} = \ln e^{-kt}$$

$$\ln\frac{1}{10} = -kt$$

$$\frac{\ln\frac{1}{10}}{-k} = t \text{ or } t = \frac{\ln 10}{k}$$

Substituting k from above,

$$t = \frac{\ln 10}{\left(\frac{\ln 2}{1,690}\right)} = \frac{1,690\ln 10}{\ln 2} \approx 5,614 \text{ years.}$$

73. $f(t) = 70 + Ae^{-kt}$
When $t = 0$, $f(0) = 212$.

So, $212 = 70 + Ae^{-k(0)}$
$212 = 70 + A$, or $A = 142$

and $f(t) = 70 + 142 e^{-kt}$.
Now, let t_i be the ideal drinking temperature. Then,

$t_i + 15 = f(2) = 70 + 142 e^{-k(2)}$ or,

$t_i = 55 + 142 e^{-2k}$.

Also, $t_i = f(4) = 70 + 142 e^{-k(4)}$ so,

$70 + 142 e^{-4k} = 55 + 142 e^{-2k}$.

$142 e^{-4k} - 142 e^{-2k} + 15 = 0$

Letting $u = e^{-2k}$, $142u^2 - 142u + 15 = 0$.
Using the quadratic formula,

$$u = \frac{142 \pm \sqrt{(-142)^2 - (4)(142)(15)}}{2(142)} \text{ so,}$$

$u \approx 0.1200445$ or $u \approx 0.8799555$ or,

$e^{-2k} \approx 0.1200445$ or 0.8799555

$t_i = 55 + 142 e^{-2k}$ So,

$t_i = 55 + 142(0.1200445)$

$\approx 72.05° \text{ F}$

or

$t_i = 55 + 142(0.8799555)$

$\approx 179.95° \text{ F}$

Since 72 degrees is approximately room temperature, the realistic answer is 179.95 degrees.

75. $T = T_a + (98.6 - T_a)(0.97)^t$
When $T = 40°F$ and $T_a = 10°F$,

$$40 = 10 + (98.6 - 10)(0.97)^t$$

$$\frac{30}{88.6} = (0.97)^t$$

$$\ln\frac{30}{88.6} = \ln(0.97)^t$$

$$\ln\frac{30}{88.6} = t\ln(0.97)$$

$$\text{so, } t = \frac{\ln\frac{30}{88.6}}{\ln(0.97)} \approx 35.55 \text{ hr}$$

This means the murder occurred around 1:27 A.M. on Wednesday. Blohardt was in jail at this time, so Scélérat must have committed the murder.

77. $R = \dfrac{2}{3}\log_{10}\left(\dfrac{E}{10^{4.4}}\right)$

(a) When $E = 5.96 \times 10^{16}$,

$$R = \frac{2}{3}\log_{10}\left(\frac{5.96 \times 10^{16}}{10^{4.4}}\right) \approx 8.25.$$

(b) When $R = 9.0$,

$$9.0 = \frac{2}{3}\log_{10}\left(\frac{E}{10^{4.4}}\right)$$

$$13.5 = \log_{10}\left(\frac{E}{10^{4.4}}\right)$$

$$10^{13.5} = 10^{\log_{10}\left(\frac{E}{10^{4.4}}\right)}$$

$$10^{13.5} = \frac{E}{10^{4.4}}$$

$$E = (10^{13.5})(10^{4.4}) = 10^{17.9} \text{ joules}$$

79. The midpoint of the segment joining the points (a, b) and (b, a) is $\left(\dfrac{a+b}{2}, \dfrac{b+a}{2}\right)$.
This point is on the line $y = x$; the slope of the line joining the points is
$\dfrac{a-b}{b-a} = \dfrac{-(b-a)}{b-a} = -1$. So the line is perpendicular to the line $y = x$, which has slope = 1. Now, using the midpoint found above, the distance from (a, b) to the line
$y = x$ is $\sqrt{\left(a - \dfrac{a+b}{2}\right)^2 + \left(b - \dfrac{a+b}{2}\right)^2}$.
Similarly, the distance from (b, a) to the
line $y = x$ is $\sqrt{\left(b - \dfrac{a+b}{2}\right)^2 + \left(a - \dfrac{a+b}{2}\right)^2}$
which is the same distance. So, the reflection of the point (a, b) in the line $y = x$ is (b, a).

81. $y = Cx^k$

$\ln y = \ln(Cx^k)$

$\quad = \ln C + \ln x^k$

$\quad = \ln C + k\ln x$

$\ln y = k\ln x + \ln c$ is of the form

$Y = mX + b$

So, $\ln y$ is a linear function of $\ln x$.

83. $x = \ln(3.42 \times 10^{-8.1})$
Input $\ln(3.42 * 10 \wedge -8.1)$ and see that the output is approximately -17.4213. So $x \approx -17.4213$. Note: Do not input $\ln(3.42 \boxed{\text{2nd}} \boxed{\text{EE}} - 8.1)$ as this results in an error.

85. $e^{0.113x} + 4.72 = 7.031 - x$

$x + e^{0.113x} - 2.311 = 0$

Press $\boxed{y =}$ and input

$x + e \wedge (0.113x) - 2.311$ for $y_1 =$

Press $\boxed{\text{graph}}$

Press $\boxed{\text{2nd}} \boxed{\text{calc}}$ and use the zero function to find $x \approx 1.1697$.

87. (a) $(\log_a b)(\log_b a) = \left(\dfrac{\ln b}{\ln a}\right)\left(\dfrac{\ln a}{\ln b}\right) = 1$

(b) $\dfrac{\log_b x}{\log_b a} = \dfrac{\frac{\ln x}{\ln b}}{\frac{\ln a}{\ln b}}$

$\qquad = \dfrac{\ln x}{\ln b} \cdot \dfrac{\ln b}{\ln a}$

$\qquad = \dfrac{\ln x}{\ln a}$

$\qquad = \log_a x$

4.3 Differentiation of Logarithmic and Exponential Functions

1. $f(x) = e^{5x}$

$$f'(x) = e^{5x}\frac{d}{dx}(5x) = 5e^{5x}$$

3. $f(x) = xe^x$

$$f'(x) = x\frac{d}{dx}e^x + e^x\frac{d}{dx}x$$

$$= x\left[e^x\frac{d}{dx}x\right] + e^x \cdot 1$$

$$= xe^x + e^x$$

$$= e^x(x+1)$$

5. $f(x) = 30 + 10e^{-0.05x}$

$$f'(x) = 0 + 10e^{-0.05x}\frac{d}{dx}(-0.05x)$$

$$= -0.5e^{-0.05x}$$

7. $f(x) = (x^2 + 3x + 5)e^{6x}$

$$f'(x) = (x^2 + 3x + 5)\frac{d}{dx}e^{6x}$$

$$+ e^{6x}\frac{d}{dx}(x^2 + 3x + 5)$$

$$= (x^2 + 3x + 5)\left[e^{6x}\frac{d}{dx}6x\right]$$

$$+ e^{6x}(2x + 3)$$

$$= 6(x^2 + 3x + 5)(e^{6x}) + (2x+3)e^{6x}$$

$$= e^{6x}[(6x^2 + 18x + 30) + (2x+3)]$$

$$= (6x^2 + 20x + 33)e^{6x}$$

9. $f(x) = (1 - 3e^x)^2$

$$f'(x) = 2(1 - 3e^x)\frac{d}{dx}(1 - 3e^x)$$

$$= 2(1 - 3e^x)(0 - 3e^x)$$

$$= -6e^x(1 - 3e^x)$$

11. $f(x) = e^{\sqrt{3x}}$

$$f'(x) = e^{\sqrt{3x}}\frac{d}{dx}\left(\sqrt{3x}\right)$$

$$= e^{\sqrt{3x}}\left(\sqrt{3}\frac{d}{dx}\sqrt{x}\right)$$

$$= \sqrt{3}e^{\sqrt{3x}}\left(\frac{1}{2}x^{-1/2}\right)$$

$$= \frac{\sqrt{3}}{2\sqrt{x}}e^{\sqrt{3x}}$$

$$= \frac{3}{2\sqrt{3x}}e^{\sqrt{3x}}$$

13. $f(x) = \ln x^3 = 3\ln x$

$$f'(x) = 3\left(\frac{1}{x}\right) = \frac{3}{x}$$

15. $f(x) = x^2\ln x$

$$f'(x) = x^2\frac{d}{dx}(\ln x) + \ln x\frac{d}{dx}(x^2)$$

$$= x^2 \cdot \frac{1}{x} + 2x\ln x$$

$$= x + 2x\ln x$$

$$= x(1 + 2\ln x)$$

17. $f(x) = \sqrt[3]{e^{2x}} = e^{2x/3}$

$$f'(x) = e^{2x/3}\frac{d}{dx}\left(\frac{2x}{3}\right) = \frac{2}{3}e^{2x/3}$$

19. $f(x) = \ln\left(\frac{x+1}{x-1}\right)$

$$f'(x) = \frac{1}{\left(\frac{x+1}{x-1}\right)}\frac{d}{dx}\left(\frac{x+1}{x-1}\right)$$

$$= \frac{x-1}{x+1}\left[\frac{(x-1)(1) - (x+1)(1)}{(x-1)^2}\right]$$

$$= \frac{-2}{(x+1)(x-1)}$$

21. $f(x) = e^{-2x} + x^3$

$$f'(x) = e^{-2x}\frac{d}{dx}(-2x) + 3x^2$$
$$= -2e^{-2x} + 3x^2$$

23. $g(s) = (e^s + s + 1)(2e^{-s} + s)$

$$g'(s) = (e^s + s + 1)\left(2e^{-s}\frac{d}{ds}(-s) + 1\right)$$
$$+ (2e^{-s} + s)(e^s + 1)$$
$$= (e^s + s + 1)(-2e^{-s} + 1)$$
$$+ (2e^{-s} + s)(e^s + 1)$$
$$= -2e^0 - 2se^{-s} - 2e^{-s} + e^s + s + 1$$
$$+ 2e^0 + se^s + 2e^{-s} + s$$
$$= 1 + 2s + e^s + se^s - 2se^{-s}$$

25. $h(t) = \dfrac{e^t + t}{\ln t}$

$$h'(t) = \frac{(\ln t)\frac{d}{dt}(e^t + t) - (e^t + t)\frac{d}{dt}(\ln t)}{(\ln t)^2}$$
$$= \frac{(\ln t)(e^t + 1) - (e^t + t)\left(\frac{1}{t}\right)}{(\ln t)^2}$$
$$= \frac{t(\ln t)(e^t + 1) - e^t - t}{t(\ln t)^2}$$

27. $f(x) = \dfrac{e^x + e^{-x}}{2} = \dfrac{1}{2}(e^x + e^{-x})$

$$f'(x) = \frac{1}{2}(e^x - e^{-x})$$

29. $f(t) = \sqrt{\ln t + t} = (\ln t + t)^{1/2}$

$$f'(t) = \frac{1}{2}(\ln t + t)^{-1/2}\frac{d}{dt}(\ln t + t)$$
$$= \frac{\frac{1}{t} + 1}{2(\ln t + t)^{1/2}}$$
$$= \frac{1 + t}{2t\sqrt{\ln t + t}}$$

31. $f(x) = \ln(e^{-x} + x)$

$$f'(x) = \frac{1}{e^{-x} + x}\frac{d}{dx}(e^{-x} + x)$$
$$= \frac{-e^{-x} + 1}{e^{-x} + x}$$

33. $g(u) = \ln\left(u + \sqrt{u^2 + 1}\right)$

$$= \ln(u + (u^2 + 1)^{1/2})$$
$$g'(u) = \frac{1}{u + (u^2 + 1)^{1/2}}\frac{d}{du}(u + (u^2 + 1)^{1/2})$$
$$= \frac{1 + \frac{1}{2}(u^2 + 1)^{-1/2}(2u)}{u + (u^2 + 1)^{1/2}}$$
$$= \frac{1 + u(u^2 + 1)^{-1/2}}{u + (u^2 + 1)^{1/2}} \cdot \frac{u - (u^2 + 1)^{1/2}}{u - (u^2 + 1)^{1/2}}$$
$$= \frac{u + u^2(u^2 + 1)^{-1/2} - (u^2 + 1)^{1/2} - u}{u^2 - (u^2 + 1)}$$
$$= \frac{\frac{u^2}{(u^2 + 1)^{1/2}} - (u^2 + 1)^{1/2}}{-1}$$
$$= \frac{-u^2}{(u^2 + 1)^{1/2}} + (u^2 + 1)^{1/2}\frac{(u^2 + 1)^{1/2}}{(u^2 + 1)^{1/2}}$$
$$= \frac{-u^2 + u^2 + 1}{(u^2 + 1)^{1/2}}$$
$$= \frac{1}{(u^2 + 1)^{1/2}}$$

35. $f(x) = \dfrac{2^x}{x}$

$$f'(x) = \frac{x\frac{d}{dx}(2^x) - (2^x)(1)}{x^2}$$
$$= \frac{x(\ln 2)2^x - 2^x}{x^2}$$
$$= \frac{2^x(x\ln 2 - 1)}{x^2}$$

37. $f(x) = x \log_{10} x$

$$f'(x) = (x)\frac{d}{dx}(\log_{10} x) + (\log_{10} x)(1)$$

$$= x \cdot \frac{1}{\ln 10} \cdot \frac{1}{x} + \log_{10} x$$

$$= \frac{1}{\ln 10} + \log_{10} x$$

$$= \frac{1}{\ln 10} + \frac{\ln x}{\ln 10}$$

$$= \frac{1 + \ln x}{\ln 10}$$

39. $f(x) = e^{1-x}; \; 0 \le x \le 1$

$$f'(x) = (e^{1-x})(-1) = -e^{1-x}$$

So, $f'(x) = 0$ when

$$-e^{1-x} = 0$$

$$e^{1-x} = 0 \to \text{no solution}$$

$$f(0) = e^{1-0} = e \approx 2.718$$

$$f(1) = e^{1-1} = e^0 = 1$$

abs max $= e$; abs min $= 1$

41. $f(x) = (3x-1)e^{-x}; \; 0 \le x \le 2$

$$f'(x) = (3x-1)(e^{-x})(-1) + (e^{-x})(3)$$

$$= e^{-x}[-1(3x-1) + 3]$$

$$= e^{-x}(4 - 3x)$$

So, $f'(x) = 0$ when

$$e^{-x} = 0 \to \text{no solution}$$

$$4 - 3x = 0 \to x = \frac{4}{3}$$

$$f(0) = [3(0)-1]e^{-0} = -1$$

$$f\left(\frac{4}{3}\right) = \left[3\left(\frac{4}{3}\right) - 1\right]e^{-4/3}$$

$$= 3e^{-4/3}$$

$$\approx 0.791$$

$$f(2) = [3(2)-1]e^{-2} = 5e^{-2} \approx 0.677$$

abs max $= 3e^{-4/3}$; abs min $= -1$

43. $g(t) = t^{3/2}e^{-2t}; \; 0 \le t \le 1$

$$g'(t) = (t^{3/2})(e^{-2t} \cdot -2) + (e^{-2t})\left(\frac{3}{2}t^{1/2}\right)$$

$$= t^{1/2}e^{-2t}\left[-2t + \frac{3}{2}\right]$$

So, $g'(t) = 0$ when

$$t^{1/2} = 0 \to t = 0$$

$$e^{-2t} = 0 \to \text{no solution}$$

$$-2t + \frac{3}{2} = 0 \to t = \frac{3}{4}$$

$$g\left(\frac{3}{4}\right) = \left(\frac{3}{4}\right)^{3/2}(e^{-2\left(\frac{3}{4}\right)})$$

$$= \frac{3\sqrt{3}}{8}e^{-3/2}$$

$$\approx 0.1449$$

$$g(0) = 0; \; g(1) = e^{-2} \approx 0.1353$$

abs max $= \frac{3\sqrt{3}}{8}e^{-3/2}$; abs min $= 0$

45. $f(x) = \frac{\ln(x+1)}{x+1}, \; 0 \le x \le 2$

$$f'(x) = \frac{(x+1) \cdot \frac{1}{x+1} - \ln(x+1) \cdot 1}{(x+1)^2}$$

$$= \frac{1 - \ln(x+1)}{(x+1)^2}$$

So, $f'(x) = 0$ when

$$1 - \ln(x+1) = 0$$

$$1 = \ln(x+1)$$

$$e^1 = e^{\ln(x+1)}$$

$$e = x+1, \text{ or } x = e-1$$

$$f(e-1) = \frac{\ln(e-1+1)}{(e-1)+1} = \frac{1}{e} \approx 0.3679$$

$$f(0) = \frac{\ln(0+1)}{(0+1)^2} = 0$$

$$f(2) = \frac{\ln(2+1)}{2+1} \approx 0.3662$$

abs max $= \frac{1}{e}$; abs min $= 0$

47. $f(x) = xe^{-x}$; $x = 0$

$f'(x) = (x)(e^{-x} \cdot -1) + (e^{-x})(1)$

$\qquad = e^{-x}(1-x)$

So, $m = f'(0) = e^0(1-0) = 1$.

Also, $f(0) = 0$, so point $(0, 0)$ is on tangent line and $y - 0 = 1(x - 0)$, or $y = x$.

49. $f(x) = \dfrac{e^{2x}}{x^2}$, $x = 1$

$f'(x) = \dfrac{(x^2)(e^{2x} \cdot 2) - (e^{2x})(2x)}{x^4}$

$\qquad = \dfrac{2xe^{2x}(x-1)}{x^4}$

$\qquad = \dfrac{2e^{2x}(x-1)}{x^3}$

so, $m = f'(1) = \dfrac{2e^2(1-1)}{1^3} = 0$.

Since the slope of the line tangent is zero, the tangent line is horizontal and of the form $y = b$. Since $f(1) = e^2$, the tangent line is $y = e^2$.

51. $f(x) = x^2 \ln \sqrt{x}$; $x = 1$

$f(x) = x^2 \ln x^{1/2} = \dfrac{1}{2} x^2 \ln x$

$f'(x) = \left(\dfrac{1}{2}x^2\right)\left(\dfrac{1}{x}\right) + (\ln x)(x)$

$\qquad = \dfrac{x}{2} + x \ln x$

So, $m = f'(1) = \dfrac{1}{2} + \ln 1 = \dfrac{1}{2}$. Also,

$f(1) = 1 \ln 1 = 0$, so the point $(1, 0)$ is on

tangent line and $y - 0 = \dfrac{1}{2}(x-1)$, or

$y = \dfrac{1}{2}x - \dfrac{1}{2}$.

53. $f(x) = e^{2x} + 2e^{-x}$

$f'(x) = e^{2x} \cdot 2 + 2e^{-x} \cdot -1 = 2e^{2x} - 2e^{-x}$

$f''(x) = 2e^{2x} \cdot 2 - 2e^{-x} \cdot -1 = 4e^{2x} + 2e^{-x}$

55. $f(t) = t^2 \ln t$

$f'(t) = (t^2)\left(\dfrac{1}{t}\right) + (\ln t)(2t)$

$\qquad = t(1 + 2\ln t)$

$f''(t) = (t)\left(2 \cdot \dfrac{1}{t}\right) + (1 + 2\ln t)(1)$

$\qquad = 2 + 1 + 2\ln t$

$\qquad = 3 + 2\ln t$

57. $f(x) = (2x+3)^2(x - 5x^2)^{1/2}$

$\ln f(x) = \ln[(2x+3)^2(x-5x^2)^{1/2}]$

$\qquad = \ln(2x+3)^2 + \ln(x-5x^2)^{1/2}$

$\qquad = 2\ln(2x+3) + \dfrac{1}{2}\ln(x - 5x^2)$

Differentiating,

$\dfrac{f'(x)}{f(x)}$

$= 2\left(\dfrac{1}{2x+3}\right)(2) + \dfrac{1}{2}\left(\dfrac{1}{x-5x^2}\right)(1-10x)$

$= \dfrac{4}{2x+3} + \dfrac{1-10x}{2(x-5x^2)}$

Multiplying both sides by $f(x)$,

$f'(x) = (2x+3)^2(x-5x^2)^{1/2}$

$\qquad \left[\dfrac{4}{2x+3} + \dfrac{1-10x}{2(x-5x^2)}\right]$

59. $f(x) = \dfrac{(x+2)^5}{\sqrt[6]{3x-5}}$

$\ln f(x) = \ln\left[\dfrac{(x+2)^5}{(3x-5)^{1/6}}\right]$

$\qquad = \ln(x+2)^5 - \ln(3x-5)^{1/6}$

$\qquad = 5\ln(x+2) - \dfrac{1}{6}\ln(3x-5)$

Differentiating, $\dfrac{f'(x)}{f(x)} = \dfrac{5}{x+2} - \dfrac{3}{6(3x-5)}$.

Multiplying both sides by $f(x)$

$$f'(x) = \frac{(x+2)^5}{(3x-5)^{1/6}}\left[\frac{5}{x+2} - \frac{1}{2(3x-5)}\right]$$

61. $f(x) = (x+1)^3(6-x)^2\sqrt[3]{2x+1}$

$\ln f(x) = \ln[(x+1)^3(6-x)^2(2x+1)^{1/3}]$

$= \ln(x+1)^3 + \ln(6-x)^2 + \ln(2x+1)^{1/3}$

$= 3\ln(x+1) + 2\ln(6-x) + \frac{1}{3}\ln(2x+1)$

Differentiating,

$\frac{f'(x)}{f(x)} = \frac{3}{x+1} + \frac{-2}{6-x} + \frac{2}{3(2x+1)}$

Multiplying both sides by $f(x)$,

$f'(x) = (x+1)^3(6-x)^2(2x+1)^{1/3}$

$\qquad \cdot \left[\frac{3}{x+1} - \frac{2}{6-x} + \frac{2}{3(2x+1)}\right]$

63. $f(x) = 5^{x^2}$

$\ln f(x) = \ln 5^{x^2} = x^2\ln 5$

Differentiating, $\frac{f'(x)}{f(x)} = (\ln 5)2x$

Multiplying both sides by $f(x)$,

$f'(x) = (2\ln 5)\cdot x \cdot 5^{x^2}$

65. $D(p) = 3{,}000e^{-0.04p}$

(a) $E(p)$

$= \frac{p}{q}\cdot\frac{dq}{dp}$

$= \frac{p}{3{,}000e^{-0.04p}}(3{,}000e^{-0.04p}\cdot -0.04)$

$= -0.04p$

$|E(p)| = |-0.04p| = 0.04p$

Demand is of unit elasticity when
$0.04p = 1$, or $p = 25$. Demand is
elastic when $0.04p > 1$, or $p > 25$.
Demand is inelastic when $0.04p < 1$,
or $p < 25$.

(b) $E(15) = -0.04(15) = -0.60$, so a 2%
increase in price results in a $(-0.60)(2) =$
-1.2, or 1.2% decrease in demand.

(c) $R(p) = p\cdot 3000e^{-0.04p}$

$R'(p) = (p)(3000e^{-0.04p}\cdot -0.04)$

$\qquad\qquad + (3000e^{-0.04p})(1)$

$= 3000e^{-0.04p}(-0.04p+1)$

So $R'(p) = 0$ when $p = 25$.

67. $D(p) = 5000(p+11)e^{-0.1p}$

(a) $E(p) = \frac{p}{q}\cdot\frac{dq}{dp}$

$\frac{dq}{dp} = \frac{dD}{dp}$

$= 5000[(p+11)(e^{-0.1p}\cdot -0.1)$

$\qquad\qquad + (e^{-0.1p})(1)]$

$= 5000e^{0.1p}[-0.1(p+11)+1]$

$= 5000e^{-0.1p}(-0.1p-0.1)$

$= -500e^{-0.1p}(p+1)$

So,

$E(p) = \dfrac{p}{5000(p+11)e^{-0.1p}}$

$\qquad\qquad \cdot -500e^{-0.1p}(p+1)$

$= \dfrac{-p(p+1)}{10(p+11)}$

$|E(p)| = \left|\dfrac{-p(p+1)}{10(p+11)}\right| = \dfrac{p(p+1)}{10(p+11)}$

Demand is of unit elasticity when

$\dfrac{p(p+1)}{10(p+11)} = 1$

$p^2 + p = 10p + 110$

$p^2 - 9p - 110 = 0$

$p = \dfrac{9 \pm \sqrt{(-9)^2 - (4)(1)(-110)}}{2(1)} \approx 15.91$

(rejecting the negative price)
Demand is elastic when

$\dfrac{p(p+1)}{10(p+11)} > 1$ or $p > 15.91$.

Demand is inelastic when

$\dfrac{p(p+1)}{10(p+11)} < 1$ or $p < 15.91$.

(b) $E(15) = \dfrac{-15(15+1)}{10(15+11)} \approx -0.923$, so a 2%

increase in price results in a $(-0.923)(2) \approx -1.85$, or 1.85% decrease in demand.

(c) $R(p) = p \cdot 5000(p+11)e^{-0.1p}$

$R'(p) = 5000[(p^2+11p)(e^{-0.1p} \cdot -0.1)$
$\qquad\qquad + (e^{-0.1p})(2p+11)]$

$\quad = 5000e^{-0.1p}[-0.1(p^2+11p)$
$\qquad\qquad + (2p+11)]$

$\quad = 5000e^{-0.1p}(-0.1p^2 + 0.9p + 11)$

$R'(p) = 0$ when $-0.1p^2 + 0.9p + 11 = 0$

$p = \dfrac{-0.9 \pm \sqrt{(0.9)^2 - (4)(0.1)(11)}}{2(-0.1)}$

$\quad \approx 15.91$ (rejecting the negative price)

69. $C(x) = e^{0.2x}$

(a) $C'(x) = 0.2e^{0.2x}$

(b) $A(x) = \dfrac{e^{0.2x}}{x}$

Marginal cost equals average cost

when $0.2e^{0.2x} = \dfrac{e^{0.2x}}{x}$

$0.2 = \dfrac{1}{x}$, or $x = 5$ units.

71. $C(x) = 12x^{1/2}e^{x/10}$

(a) $C'(x) = 12\left[(x^{1/2})\left(e^{x/10} \cdot \dfrac{1}{10} \right) \right.$
$\qquad\qquad \left. + (e^{x/10})\left(\dfrac{1}{2}x^{-1/2} \right) \right]$

$\quad = 6x^{-1/2}e^{x/10}\left(\dfrac{1}{5}x + 1 \right)$

$\dfrac{6e^{x/10}\left(\dfrac{1}{5}x+1 \right)}{\sqrt{x}}$

(b) $A(x) = \dfrac{12x^{1/2}e^{x/10}}{x} = \dfrac{12e^{x/10}}{x^{1/2}}$

Marginal cost equals average cost

when $\dfrac{6e^{x/10}\left(\dfrac{1}{5}x+1 \right)}{x^{1/2}} = \dfrac{12e^{x/10}}{x^{1/2}}$

$\dfrac{1}{5}x + 1 = 2$ or $x = 5$ units.

73. (a) $Q(t) = 20{,}000e^{-0.4t}$

The rate of depreciation after t years is

$Q'(t) = 20{,}000e^{-0.4t}(-0.4)$
$\qquad = -8{,}000e^{-0.4t}$

So, the rate after 5 years is

$Q'(5) = -8{,}000e^{-2}$
$\qquad = -\$1{,}082.68$ per year.

(b) The percentage rate of change t years from now will be

$100\left[\dfrac{Q'(t)}{Q(t)} \right] = 100\left(\dfrac{-8{,}000e^{-0.4t}}{20{,}000e^{-0.4t}} \right)$
$\qquad\qquad = -40\%$ per year

which is a constant, independent of time.

75. $f(x) = 20 - 15e^{-0.2x}$

(a) When a change in x is made, the corresponding change in f can be approximated by $\Delta f \approx f'(x)\Delta x$. Here, $x = 10$ thousand initially, $\Delta x = 1$ thousand, and

$f'(x) = -15e^{-0.2x}(-0.2) = 3e^{-0.2x}$.

So, $\Delta f \approx f'(10) \cdot 1 = 3e^{-0.2(10)} \approx 0.406.$

An increase of one thousand additional complimentary copies will increase sales by 0.406 thousand, or 406 copies.

(b) The actual change in sales is
$$\Delta f = f(11) - f(10)$$
$$= (20 - 15e^{-2.2}) - (20 - 15e^{-2})$$
$$= 0.368, \text{ or } 368 \text{ copies}$$

77. $R = E + T$

When $t = t_0$, $R = 11 + 8 = 9$

Now, $\dfrac{E'(t_0)}{E(t_0)} = 0.09$, $\dfrac{T'(t_0)}{T(t_0)} = -0.02$

or, $E'(t_0) = 0.09E(t_0)$ and

$T'(t_0) = -0.02T(t_0)$. Using logarithmic differentiation,

$$\ln R = \ln(E + T)$$
$$\frac{R'}{R} = \frac{E' + T'}{E + T}$$
$$\frac{R'(t_0)}{R(t_0)} = \frac{0.09(11) - 0.02(8)}{11 + 8}$$
$$= 0.0437$$

So, 4.37% is the relative rate of growth of revenue when $t = t_0$.

79. $\dfrac{-(\ln q)'}{(\ln p)'}$

The derivative of $\ln q$, with respect to p is $\dfrac{1}{q} \cdot \dfrac{dq}{dp}$. The derivative of $\ln p$ is $\dfrac{1}{p}$. So,

$$\frac{-(\ln q)'}{(\ln p)'} = \frac{-\dfrac{1}{q} \cdot \dfrac{dq}{dp}}{\dfrac{1}{p}} = -\frac{p}{q} \cdot \frac{dq}{dp} = E(p)$$

81. $F(t) = B + (1 - B)e^{-kt}$

(a) $F'(t) = 0 + (1 - B)e^{-kt}(-k)$
$$= -k(1 - B)e^{-kt}$$

$F'(t)$ represents the rate at which recall is changing. That is, the rate at which you are forgetting material.

(b) $F - B = B + (1 - B)e^{-kt} - B$
$$= (1 - B)e^{-kt}$$

since $F'(t) = -k(1 - B)e^{-kt}$,
$$F'(t) = -k(F - B)$$

That is, $F'(t)$ is proportional to

$F - B$. This means that the rate you forget material is proportional to the fraction remaining that will be forgotten.

(c)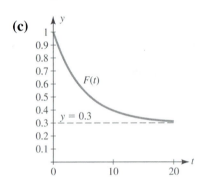

83. (a) $P(t) = 50e^{0.02t}$

The rate of change of the population t years from now will be

$P'(t) = 50e^{0.02t}(0.02) = e^{0.02t}.$

So, the rate of change 10 years from now will be

$P'(10) = e^{0.2} = 1.22$ million per year.

(b) The percentage rate of change t years from now will be

$$100\left[\frac{P'(t)}{P(t)}\right] = 100\left(\frac{e^{0.02t}}{50e^{0.02t}}\right)$$
$$= \frac{100}{50}$$
$$= 2\% \text{ per year}$$

which is a constant, independent of time.

85. $N(t) = \dfrac{600}{1 + 3e^{-0.02t}}$

(a) $N'(t) = \dfrac{0 - (600)(3e^{-0.02t} \cdot -0.02)}{(1 + 3e^{-0.02t})^2}$

$= \dfrac{36e^{-0.02t}}{(1 + 3e^{-0.02t})^2}$ individuals

per year

$N'(t)$ is never zero. $N'(t) > 0$ for all values of t, so the population is always increasing.

(b) Using logarithmic differentiation,
$\ln N'(t)$

$= \ln\left[\dfrac{36e^{-0.02t}}{(1 + 3e^{-0.02t})^2}\right]$

$= \ln 36 + \ln e^{0.02t} - \ln(1 + 3e^{-0.02t})^2$

$= \ln 36 - 0.02t - 2\ln(1 + 3e^{-0.02t})$

$\dfrac{N''(t)}{N'(t)} = -0.02 - 2\dfrac{-0.06e^{-0.02t}}{1 + 3e^{-0.02t}}$

$N''(t) = \left[\dfrac{-0.02(1 + 3e^{-0.02t}) + 0.12e^{-0.02t}}{1 + 3e^{-0.02t}}\right]N'(t)$

$N''(t) = \left[\dfrac{-0.02 + 0.06e^{-0.02t}}{1 + 3e^{-0.02t}}\right]\left[\dfrac{36e^{-0.02t}}{(1 + 3e^{-0.02t})^2}\right]$

$N''(t) = (-0.02 + 0.06e^{-0.02t})\left[\dfrac{36e^{-0.02t}}{(1 + 3e^{-0.02t})^3}\right]$

So $N''(t) = 0$ when

$-0.02 + 0.06e^{-0.02t} = 0$

$e^{-0.02t} = \dfrac{1}{3}$

$\ln e^{-0.02t} = \ln\dfrac{1}{3}$

$-0.02t = \ln\dfrac{1}{3}$

$t = \dfrac{\ln\frac{1}{3}}{-0.02}$

$= \dfrac{-\ln\frac{1}{3}}{0.02}$

$= 50\ln 3$

When
$0 < t < 50\ln 3$, $N''(t) > 0$, so $N'(t)$ is increasing
$t > 50\ln 3$, $N''(t) < 0$, so $N'(t)$ is decreasing

(c) As $t \to \infty$, $e^{-0.02t} \to 0$, so

$\lim\limits_{t \to \infty} \dfrac{600}{1 + 3e^{-0.02t}} = 600$

So, in the long run, the number of individuals approaches 600.

87. $P_1(t) = \dfrac{21}{1 + 25e^{-0.3t}}$, $P_2(t) = \dfrac{20}{1 + 17e^{-0.6t}}$

(a) $P_1'(t) = \dfrac{0 - (21)(25e^{-0.3t} \cdot -0.3)}{(1 + 25e^{-0.3t})^2}$

$P_1'(10) = \dfrac{157.5e^{-3}}{(1 + 25e^{-3})^2}$

≈ 1.556 cm per day

$P_2'(t) = \dfrac{0 - (20)(17e^{-0.6t} \cdot -0.6)}{(1 + 17e^{-0.6t})^2}$

$= \dfrac{204e^{-0.6t}}{(1 + 17e^{-0.6t})^2}$

Using logarithmic differentiation,
$\ln P_2'(t)$

$= \ln\left[\dfrac{204e^{-0.6t}}{(1 + 17e^{-0.6t})^2}\right]$

$= \ln 204 + \ln e^{0.6t} - \ln(1 + 17e^{-0.6t})^2$

$= \ln 204 - 0.6t - 2\ln(1 + 17e^{-0.6t})$

$\dfrac{P_2''(t)}{P_2'(t)} = -0.6 - 2\left(\dfrac{-10.2e^{-0.6t}}{1 + 17e^{-0.6t}}\right)$

$P_2''(t) = \left[\dfrac{-0.6(1 + 17e^{-0.6t}) + 20.4e^{-0.6t}}{1 + 17e^{-0.6t}}\right]P_2'(t)$

$= \left[\dfrac{-0.6 + 10.2e^{-0.6t}}{1 + 17e^{-0.6t}}\right]\left[\dfrac{204e^{-0.6t}}{(1 + 17e^{-0.6t})^2}\right]$

$= (-0.6 + 10.2e^{-0.6t})\left[\dfrac{204e^{-0.6t}}{(1 + 17e^{-0.6t})^3}\right]$

Since $-0.6 + 10.2e^{-0.6(10)} < 0$,

$P_2''(10) < 0$ so P_2' is decreasing. In other words, the rate of growth of the second plant is decreasing.

(b) $\dfrac{21}{1+25e^{-0.3t}} = \dfrac{20}{1+17e^{-0.6t}}$

$\dfrac{21}{1+25e^{-0.3t}} - \dfrac{20}{1+17e^{-0.6t}} = 0$

Press $\boxed{y=}$ and input
$21/(1 + 25e \wedge (-.3x)) - 20/(1 + 17e \wedge (-.6x))$ for $y_1 =$.

Use window dimensions [0, 30]5 by [−10, 10]1

Press $\boxed{\text{graph}}$.

Use the zero function under the calc menu to find that the plants have the same height at approximately 20.71 days.

$P_1(20.71) \approx 20$ cm

$P_1'(20.71) \approx 0.286$ cm/day and

$P_2'(20.71) \approx 0.000818$ cm/day, so P_1 is growing at a faster rate when they have the same height.

89. Show $\dfrac{d}{dx}(b^x) = (\ln b)b^x$

(a) Rewrite $b^x = e^{x \ln b}$

Note: $e^{x \ln b} = e^{\ln b^x} = b^x$

$\dfrac{d}{dx}(e^{x \ln b}) = (e^{x \ln b})(\ln b)$

Note: In b is a constant, so

$\dfrac{d}{dx} x \ln b = \ln b.$

So, $\dfrac{d}{dx}(b^x) = (e^{x \ln b})(\ln b)$

$= (b^x)(\ln b)$

$= (\ln b)b^x$

(b) $y = b^x$

$\ln y = \ln b^x$

$\ln y = x \ln b$

Now, take the derivative of both sides

$\dfrac{y'}{y} = \ln b$

$y' = (\ln b)y$

$\dfrac{d}{dx}(b^x) = (\ln b)b^x$

91. To use a numerical differentiation utility to find $f'(c)$, where $c = 0.65$ and

$f(x) = \ln\left[\dfrac{\sqrt[3]{x+1}}{(1+3x)^4}\right].$

Press $\boxed{y=}$ and input f for $y_1 =$.

Press $\boxed{\text{2nd}}$ $\boxed{\text{calc}}$ and use dy/dx option.

Enter $x = 0.65$ and display shows

$dy/dx = -3.866.$

So, $f'(0.65) = -3.866.$

The slope of the line tangent at $x = 0.65$ is

$m = -3.866.$

The point on the tangent line is

$f(0.65) = -4.16.$

$y + 4.16 = -3.866(x - 0.65)$

$y = -3.866x - 1.6475$

Press $\boxed{y=}$ and enter this line for $y_2 =$.

Use window dimensions [0, 3]1 by [−10, 2]1.

Press $\boxed{\text{graph}}$.

An easier method is to

Press $\boxed{y=}$ and input f for $y_1 =$.

Use window dimensions [−5, 5].5 by [−5, 5].5.

Press $\boxed{\text{graph}}$.

Use the tangent function under the draw menu and enter $x = 0.65$.

The tangent line is drawn and the equation

is displayed at the bottom of the screen.

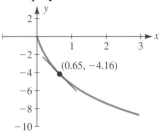

(0.65, −4.16)

93. $Q(t) = Q_0 \dfrac{e^{kt}}{t}$ so $Q'(t) = Q_0 \dfrac{kte^{kt} - e^{kt}}{t^2}$

The percentage rate of change is

$$100 \frac{Q'(t)}{Q(t)} = 100 Q_0 \frac{kte^{kt} - e^{kt}}{t^2} \left(\frac{t}{Q_0 e^{kt}} \right)$$

$$= 100 \frac{kt - 1}{t}$$

4.4 Additional Applications; Exponential Models

1. When $x = 1, f(1) = 0$ which eliminates f_1, f_3, and f_4. As $x \to \infty, f(x) \to 0$. Now,

$$\lim_{x \to \infty} x \ln x^5 = \infty$$

$$\lim_{x \to \infty} \frac{5 \ln x}{x} = \lim_{x \to \infty} \frac{\frac{5}{x}}{1} = 0$$

$$\lim_{x \to \infty} (x-1)e^{-2x} = \lim_{x \to \infty} \frac{x-1}{e^{2x}}$$

$$= \lim_{x \to \infty} \frac{1}{2e^{2x}}$$

$$= 0$$

which eliminates f_2.

$$f_5'(x) = \frac{(x)\left(5 \cdot \frac{1}{x}\right) - (5 \ln x)(1)}{x^2} = \frac{5 - 5 \ln x}{x^2}$$

$f_5'(x) = 0$ when $5 - 5 \ln x = 0$

$$5 = 5 \ln x$$
$$1 = \ln x$$
$$e^1 = e^{\ln x}, \text{ or } x = e$$

$$f_6'(x) = (x-1)(-2e^{-2x}) + (e^{-2x})(1)$$

$$= e^{-2x}[-2(x-1) + 1]$$

$$= e^{-2x}(3 - 2x)$$

$f_6'(x) = 0$ when $3 - 2x = 0$, or $x = \dfrac{3}{2}$

which eliminates f_6. So, this is the graph of f_5.

3. As $x \to \infty, f(x) \to 2$.

$$\lim_{x \to \infty} 2 - e^{-2x} = 2$$

$$\lim_{x \to \infty} x \ln x^5 = \infty$$

$$\lim_{x \to \infty} \frac{2}{1 - e^{-x}} = \frac{2}{1 - 0} = 2$$

$$\lim_{x \to \infty} \frac{2}{1 + e^{-x}} = \frac{2}{1 + 0} = 2$$

$$\lim_{x \to \infty} \frac{\ln x^5}{x} = 0$$

$$\lim_{x \to \infty} (x-1)e^{-2x} = 0$$

which eliminates f_2, f_5, and f_6.

As $x \to 0^+, f(x) \to -\infty$.

$$\lim_{x \to 0^+} 2 - e^{-2x} = 2 - 1 = 1$$

$$\lim_{x \to 0^+} \frac{2}{1 + e^{-x}} = \frac{2}{1 + 1} = 2$$

which eliminates f_1 and f_4. So, this is the graph of f_3.

5. $f(t) = 2 + e^t$

When $x = 0, f(0) = 3$ so $(0, 3)$ is an intercept. When $f(t) = 0$, $2 + e^t = 0$,

$e^t = -2$ has no solution.

$\lim_{x \to -\infty} 2 + e^t = 2$, so $y = 2$ is a horizontal asymptote. $\lim_{x \to -\infty} 2 + e^t = +\infty$

$f'(t) = e^t$, so there are no critical values. $f'(t) > 0$ for all values of t, so f is always increasing. Since $f''(t) = e^t$ as well,

$f''(t) > 0$ and f is always concave up.

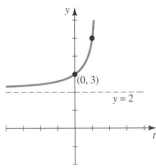

7. $g(x) = 2 - 3e^x$

When $x = 0$, $g(0) = -1$ so $(0, -1)$ is an intercept. When $f(x) = 0$, $0 = 2 - 3e^x$,

$$3e^x = 2, \ e^x = \frac{2}{3}$$

$$\ln e^x = \ln \frac{2}{3}, \ \text{or} \ x = \ln \frac{2}{3}.$$

So $\left(\ln \frac{2}{3}, 0 \right)$ is an intercept.

$\lim\limits_{x \to -\infty} 2 - 3e^x = 2$, so $y = 2$ is a horizontal asymptote.

$\lim\limits_{x \to +\infty} 2 - 3e^x = -\infty$. $g'(x) = -3e^x$ so there are no critical values. $g'(x) < 0$ for all values of x, so g is always decreasing. Since $g''(x) = -3e^x$ as well, $g''(x) < 0$ and g is always concave down.

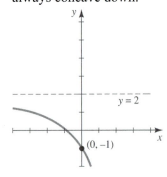

9. $f(x) = \dfrac{2}{1 + 3e^{-2x}}$

When $x = 0$, $f(0) = \dfrac{1}{2}$, so $\left(0, \dfrac{1}{2} \right)$ is an intercept. $f(x) = 0$ has no solution.

$\lim\limits_{x \to -\infty} \dfrac{2}{1 + 3e^{-2x}} = 0$, so $y = 0$ is a horizontal asymptote.

$\lim\limits_{x \to +\infty} \dfrac{2}{1 + 3e^{-2x}} = 2$, so $y = 2$ is a horizontal asymptote.

$f'(x) = \dfrac{0 - (2)(-6e^{-2x})}{(1 + 3e^{-2x})^2}$, so $f'(x) = 0$

when $12e^{-2x} = 0$.

Since $12e^{-2x}$ is never zero, there are no critical values. $f'(x) > 0$ for all values of x, so f is always increasing. Using logarithmic differentiation,

$$\ln f'(x) = \ln \left[\frac{12e^{-2x}}{(1 + 3e^{-2x})^2} \right]$$

$$= \ln 12 + \ln e^{-2x} - \ln(1 + 3e^{-2x})^2$$

$$= \ln 12 - 2x - 2\ln(1 + 3e^{-2x})$$

$$\frac{f''(x)}{f'(x)} = -2 - 2 \cdot \frac{-6e^{-2x}}{1 + 3e^{-2x}}$$

$$f''(x) = \left[\frac{-2(1 + 3e^{-2x}) + 12e^{-2x}}{1 + 3e^{-2x}} \right] f'(x)$$

$$= \left[\frac{-2 + 6e^{-2x}}{1 + 3e^{-2x}} \right] \left[\frac{12e^{-2x}}{(1 + 3e^{-2x})^2} \right]$$

$$= (-2 + 6e^{-2x}) \left[\frac{12e^{-2x}}{(1 + 3e^{-2x})^3} \right]$$

So $f''(x) = 0$ when $-2 + 6e^{-2x} = 0$;

$$e^{-2x} = \frac{1}{3}$$

$$\ln e^{-2x} = \ln \frac{1}{3}$$

$$-2x = \ln \frac{1}{3}$$

$$x = \frac{\ln \frac{1}{3}}{-2}, \ \text{or} \ x = \frac{\ln 3}{2} \approx 0.549.$$

When
$0 < x < 0.549$, $f''(x) > 0$, so f is concave up
$x > 0.549$, $f''(x) < 0$, so f is concave down
Since the concavity changes at $x = 0.549$,

the point (0.549, 1) is an inflection point.

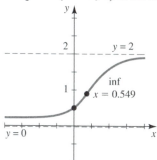

11. $f(x) = xe^x$

When $x = 0$, $f(0) = 0$, so $(0, 0)$ is an intercept.

$f(x) = 0$ when $x = 0$.

$\lim\limits_{x \to -\infty} xe^x = 0$ so $y = 0$ is a horizontal asymptote.

$\lim\limits_{x \to +\infty} xe^x = +\infty$.

$f'(x) = (x)(e^x) + (e^x)(1) = e^x(x+1)$, so $f'(x) = 0$ when $x = -1$.

When $x < -1$, $f'(x) < 0$, so f is decreasing

When $x > -1$, $f'(x) > 0$, so f is increasing.

The point $\left(-1, -\dfrac{1}{e}\right)$ is a relative minimum.

$f''(x) = (e^x)(1) + (x+1)(e^x) = e^x(x+2)$ so $f''(x) = 0$ when $x < -2$.

When $x < -2$, $f''(x) < 0$, so f is concave down. When $x > -2$, $f''(x) > 0$, so f is concave up.

Since the concavity changes at $x = -2$, the point $\left(-2, -\dfrac{2}{e^2}\right)$ is an inflection point.

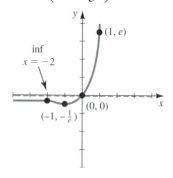

13. $f(x) = xe^{2-x}$

When $x = 0$, $f(0) = 0$, so $(0, 0)$ is an intercept.

$f(x) = 0$ when $x = 0$.

$\lim\limits_{x \to -\infty} xe^{2-x} = -\infty$

$\lim\limits_{x \to +\infty} xe^{2-x} = \lim\limits_{x \to +\infty} \dfrac{x}{e^{x-2}} = \lim\limits_{x \to +\infty} \dfrac{1}{e^x} = 0$,

so $y = 0$ is a horizontal asymptote.

$f'(x) = (x)(-e^{2-x}) + (e^{2-x})(1)$
$\qquad = e^{2-x}(1-x)$,

so $f'(x) = 0$ when $x = 1$.

When $x < 1$, $f'(x) > 0$, so f is increasing

$x > 1$, $f'(x) < 0$, so f is decreasing.

The point $(1, e)$ is a relative maximum.

$f''(x) = (e^{2-x})(-1) + (1-x)(-e^{2-x})$
$\qquad = e^{2-x}(x-2)$

So $f''(x) = 0$ when $x = 2$.

When

$x < 2$, $f''(x) < 0$ so f is concave down

$x > 2$, $f''(x) > 0$ so f is concave up

Since the concavity changes at $x = 2$, the point $(2, 2)$ is an inflection point.

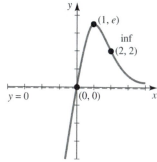

15. $f(x) = x^2 e^{-x}$

When $x = 0$, $f(0) = 0$ so $(0, 0)$ is an intercept. $f(x) = 0$ when $x = 0$.

$\lim\limits_{x \to -\infty} x^2 e^{-x} = +\infty$

$$\lim_{x \to +\infty} x^2 e^{-x} = \lim_{x \to +\infty} \frac{x^2}{e^x}$$

$$= \lim_{x \to +\infty} \frac{2x}{e^x}$$

$$= \lim_{x \to +\infty} \frac{2}{e^x}$$

$$= 0,$$

so $y = 0$ is a horizontal asymptote.

$$f'(x) = (x^2)(-e^{-x}) + (e^{-x})(2x)$$

$$= e^{-x}(2x - x^2)$$

$$= e^{-x}(2 - x)x$$

So, $f'(x) = 0$ when $x = 0, 2$.

When $x < 0$, $f'(x) < 0$, so f is decreasing.
When $0 < x < 2$, $f'(x) > 0$, so f is increasing. When $x > 2$, $f'(x) < 0$, so f is decreasing.
The point $(0, 0)$ is a relative minimum and the point $\left(2, \dfrac{4}{e^2}\right)$ is a relative maximum.

$$f''(x) = (e^{-x})(2 - 2x) + (2x - x^2)(-e^{-x})$$

$$= e^{-x}[(2 - 2x) - (2x - x^2)]$$

$$= e^{-x}(x^2 - 4x + 2)$$

So $f''(x) = 0$ when $x = 2 \pm \sqrt{2}$.

When $x < 2 - \sqrt{2}$, $f''(x) > 0$ so f is concave up. When $2 - \sqrt{2} < x < 2 + \sqrt{2}$, $f''(x) < 0$ so f is concave down.
$x > 2 + \sqrt{2}$, $f''(x) > 0$, so f is concave up.
The points $(0.59, 0.19)$ and $(3.41, 0.38)$ are inflection points.

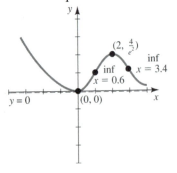

17. $f(x) = \dfrac{6}{1 + e^{-x}}$

When $x = 0$, $f(0) = 3$, so $(0, 3)$ is an intercept. $f(x) = 0$ has no solution.

$$\lim_{x \to -\infty} \frac{6}{1 + e^{-x}} = 0, \text{ so } y = 0 \text{ is a horizontal}$$

asymptote. $\displaystyle\lim_{x \to +\infty} \frac{6}{1 + e^{-x}} = 6$, so $y = 6$ is a horizontal asumptote.

$$f'(x) = \frac{0 - (6)(-e^{-x})}{(1 + e^{-x})^2} \text{ so } f'(x) = 0 \text{ when}$$

$$6e^{-x} = 0$$

Since $6e^{-x}$ is never zero, there are no critical values. $f'(x) > 0$ for all values of x, so f is always increasing. Using logarithmic differentiation.

$$\ln f'(x) = \ln \left[\frac{6e^{-x}}{(1 + e^{-x})^2} \right]$$

$$= \ln 6 + \ln e^{-x} - \ln(1 + e^{-x})^2$$

$$= \ln 6 - x - 2\ln(1 + e^{-x})$$

$$\frac{f''(x)}{f'(x)} = -1 - 2 \cdot \frac{-e^{-x}}{1 + e^{-x}}$$

$$f''(x) = \left[\frac{-(1 + e^{-x}) + 2e^{-x}}{1 + e^{-x}} \right] f'(x)$$

$$= \left[\frac{-1 + e^{-x}}{1 + e^{-x}} \right] \left[\frac{6e^{-x}}{(1 + e^{-x})^2} \right]$$

$$= (-1 + e^{-x}) \left[\frac{6e^{-x}}{(1 + e^{-x})^3} \right]$$

So $f''(x) = 0$ when $-1 + e^{-x} = 0$, or when $x = 0$.
When
$x < 0$, $f''(x) > 0$, so f is concave up
$x > 0$, $f''(x) < 0$, so f is concave down

The point (0, 3) is an inflection point.

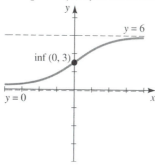

The point (e, 1) is an inflection point.

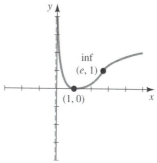

19. $f(x) = (\ln x)^2$, $x > 0$

When $f(x) = 0$, $(\ln x)^2 = 0$; $\ln x = 0$, or $x = 1$. So $(1, 0)$ is an intercept.

$\lim\limits_{x \to 0^+} (\ln x)^2 = \infty$ so $x = 0$ is a vertical asymptote.

$\lim\limits_{x \to +\infty} (\ln x)^2 = +\infty$

$f'(x) = 2(\ln x)\left(\dfrac{1}{x}\right)$ so $f'(x) = 0$ when $x = 1$.

When
$0 < x < 1$, $f'(x) < 0$, so f is decreasing
$x > 1$, $f'(x) > 0$, so f is increasing
The point $(1, 0)$ is a relative minimum.

$f''(x) = \left(\dfrac{2}{x}\right)\left(\dfrac{1}{x}\right) + (\ln x)\left(-\dfrac{2}{x^2}\right)$

$= \dfrac{2}{x^2}(1 - \ln x)$

So $f''(x) = 0$ when $1 - \ln x = 0$

$$1 = \ln x$$
$$e^1 = e^{\ln x}, \text{ or } x = e.$$

When
$0 < x < e$, $f''(x) > 0$, so f is concave up
$x > e$, $f''(x) < 0$, so f is concave down

21. $N(t) = N_0 e^{kt}$

Let $t = 0$ correspond to the year 2005. Then, $N_0 = 4$ billion. Further, the year 2010 corresponds to $t = 5$ and $N(5) = 12$ billion.

$$N(t) = 4e^{kt}$$
$$12 = 4e^{k \cdot 5}$$
$$3 = e^{5k}$$
$$\ln 3 = \ln e^{5k}$$
$$\ln 3 = 5k$$
$$k = \frac{\ln 3}{5} \approx 0.2197$$

So, $N(t) = 4e^{0.2197t}$. The year 2015 corresponds to $t = 10$ and $N(10) = 4e^{0.2197(10)} \approx 35.99$ billion. Sales should be approximately 36 billion.

23. $f(t) = 1 - e^{-0.03t}$

(a) When $t = 0$, $f(0) = 0$, and $f(t) = 0$, when $t = 0$.

$\lim\limits_{t \to \infty} 1 - e^{-0.03t} = 1$, so $y = 1$ is a horizontal asymptote.

$f'(t) = 0.03e^{-0.03t}$

$f'(t) > 0$ for all values of t, so f is always increasing.

$f''(t) = -0.09e^{-0.03t}$

$f''(t) < 0$ for all values of t, so f is

always concave down.

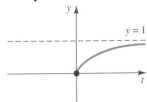

(b) The fraction of tankers that sink in fewer than 10 days is
$$f(10) = 1 - e^{-0.3}.$$
The fraction of tankers that remain afloat for at least 10 days is therefore
$$1 - f(10) = e^{-0.3} = 0.7408.$$

(c) The fraction of tankers that can be expected to sink between the 15th and 20th days is $f(20) - f(15)$
$$= (1 - e^{-0.6}) - (1 - e^{-0.45})$$
$$= -e^{-0.6} + e^{-0.45}$$
$$= -0.5488 + 0.6373$$
$$= 0.0888.$$

25. $Q(t) = 40 - Ae^{-kt}$
When $t = 0$, $Q(0) = 20$, so
$$20 = 40 - Ae^{-k \cdot 0}, \text{ or } A = 20.$$
Now, $Q(t) = 40 - 20e^{-kt}$.
When $t = 1$, $Q(1) = 30$, so
$$30 = 40 - 20e^{-k \cdot 1}$$
$$20e^{-k} = 10; \quad e^{-k} = \frac{1}{2}; \quad -k = \ln\frac{1}{2}$$
$$k = -\ln\frac{1}{2} = \ln 2$$
Now, $Q(t) = 40 - 20e^{-(\ln 2)t}$.
When $t = 3$, $Q(3) = 40 - 20e^{-(\ln 2)(3)}$
$$= 37.5 \text{ units per day.}$$

27. $f(x) = 15 - 20e^{-0.3x}$

(a) $\Delta f \approx f'(x)$, where x is the current number of complimentary copies.
$$f'(x) = 6e^{-0.3x}$$
$$\Delta f \approx f'(9) = 6e^{-0.3(9)} \approx 0.403$$

So, approximately 403 additional copies will be sold.

(b) $\Delta f = f(x_2) - f(x_1) = f(10) - f(9)$
$$f(10) = 15 - 20e^{-0.3(10)} \approx 14.004$$
$$f(9) = 15 - 20e^{-0.3(9)} \approx 13.656$$
$\Delta f = 0.348$, or 348 additional copies. The approximation is off by
55 copies, or $\dfrac{55}{348} \approx 16\%$.

29. $p(t) = \dfrac{Ce^{kt}}{1 + Ce^{kt}}$

Since $p(0) = \dfrac{1}{200}$,
$$\frac{1}{200} = \frac{Ce^{k \cdot 0}}{1 + Ce^{k \cdot 0}}$$
$$\frac{1}{200} = \frac{C}{1 + C}$$
$$1 + C = 200C, \text{ or } C = \frac{1}{199}$$

So, $p(t) = \dfrac{\frac{1}{199}e^{kt}}{1 + \frac{1}{199}e^{kt}} = \dfrac{e^{kt}}{199 + e^{kt}}$.

Since $p(4) = \dfrac{1}{100}$,
$$\frac{1}{100} = \frac{e^{k \cdot 4}}{199 + e^{k \cdot 4}}$$
$$199 + e^{4k} = 100e^{4k}$$
$$\frac{199}{99} = e^{4k}$$
$$\ln\frac{199}{99} = 4k$$
$$k = \frac{\ln\frac{199}{99}}{4} \approx 0.1745$$

So, $p(t) = \dfrac{e^{0.1745t}}{199 + e^{0.1745t}}$.

Using logarithmic differentiation to find the rate of change,

$$\ln p(t) = \ln\left[\frac{e^{0.1745t}}{199 + e^{0.1745t}}\right]$$

$$= 0.1745t - \ln(199 + e^{0.1745t})$$

$$\frac{p'(t)}{p(t)} = 0.1745 - \frac{0.1745e^{0.1745t}}{199 + e^{0.1745t}}$$

$$= \frac{(0.1745)(199)}{199 + e^{0.1745t}}$$

$$= \frac{34.7255}{199 + e^{0.1745t}}$$

So, $p'(t) = \dfrac{34.7255}{199 + e^{0.1745t}}\left(\dfrac{e^{0.1745t}}{199 + e^{0.1745t}}\right)$

$$= \frac{34.7255e^{0.1745t}}{(199 + e^{0.1745t})^2}.$$

To maximize this rate, use logarithmic differentiation again.

$\ln p'(t)$

$$= \ln\left[\frac{34.7255e^{0.1745t}}{(199 + e^{0.1745t})^2}\right]$$

$$= \ln 34.7255 + 0.1745t - 2\ln(199 + e^{0.1745t})$$

$$\frac{p''(t)}{p'(t)} = 0 + 0.1745 - 2\left(\frac{0.1745e^{0.1745t}}{199 + e^{0.1745t}}\right)$$

$$= \frac{(0.1745)(199) - 0.1745e^{0.1745t}}{199 + e^{0.1745t}}$$

$$= \frac{34.7255 - 0.1745e^{0.1745t}}{199 + e^{0.1745t}}$$

and,

$$p''(t) = \left[\frac{34.7255 - 0.1745e^{0.1745t}}{199 + e^{0.1745t}}\right]$$

$$\left[\frac{34.7255e^{0.1745t}}{(199 + e^{0.1745t})^2}\right]$$

$p''(t) = 0$ when

$$34.7255 - 0.1745e^{0.1745t} = 0$$

$$\frac{34.7255}{0.1745} = e^{0.1745t}$$

$\ln 199 = 0.1745t$, or

$$t = \frac{\ln 199}{0.1745} \approx 30.33 \text{ weeks}$$

$p(30.33) \approx 0.5$, so roughly half of the trading volume is due to day trading.

31. $V(t) = 8,000e^{\sqrt{t}}$

The prevailing interest rate of 6% is the same as the percentage rate of change of V, so $6 = 100\dfrac{V'(t)}{V(t)}$.

Now, $V'(t) = 8,000e^{\sqrt{t}}\left(\dfrac{1}{2}t^{-1/2}\right)$

So, $100\dfrac{V'(t)}{V(t)} = 100\dfrac{8,000e^{\sqrt{t}}\left(\frac{1}{2}t^{-1/2}\right)}{8,000e^{\sqrt{t}}} = \dfrac{50}{t^{1/2}}$

and $6 = \dfrac{50}{t^{1/2}}$, $t^{1/2} = \dfrac{25}{3}$, or

$$t = \left(\frac{25}{3}\right)^2 \approx 69.44$$

When $0 < t < \dfrac{625}{9}$, the percentage rate of growth $100\dfrac{V'(t)}{V(t)} > 6\%$. When $t > \dfrac{625}{9}$, the percentage rate of growth is $100\dfrac{V'(t)}{V(t)} < 6\%$. So, the land should be sold approximately 69.44 years from now.

33. Since the current value of the stamp is $1,200 and its value increases linearly at a rate of $200 per year, its value t years from now is $V(t) = 1,200 + 200t$.

Now, interest is compounded continuously, so its present value is

$$P(t) = (200t + 1,200)e^{-0.08t}$$

Need to find t to maximize this value.

$$P'(t) = (200t + 1,200)\left(-0.08e^{-0.08t}\right)$$

$$+ e^{-0.08t}(200)$$

$$= -16te^{-0.08t} - 96e^{-0.08t} + 200e^{-0.08t}$$

$$= e^{-0.08t}(-16t + 104)$$

$P'(t) = 0$ when $-16t + 104 = 0$, or $t = 6.5$ years.

$$P''(t) = \left(e^{-0.08t}\right)(-16)$$
$$+ \left(-16t + 104\right)\left(-0.08e^{-0.08t}\right)$$
$$= -16e^{-0.08t} + 1.28te^{-0.08t} - 8.32e^{-0.08t}$$
$$= e^{-0.08t}\left(-24.32 + 1.28t\right)$$

When $t = 6.5$,

$P''(6.5) \approx 0.595(-16)$, which is negative.

So, the present value is maximized and the collection should be sold after 6.5 years.

35. $V(t) = V_0\left(1 - \dfrac{2}{L}\right)^t$

(a) When $L = 8$,

$$V(t) = 875\left(1 - \frac{2}{8}\right)^t = 875(0.75)^t.$$

When $t = 5$,

$$V(5) = 875(0.75)^5 \approx \$207.64.$$

The annual rate of depreciation is the derivative, and logarithmic differentiation must be used.

$$\ln V = \ln[875(0.75)^t]$$
$$= \ln 875 + \ln(0.75)^t$$
$$= \ln 875 + t\ln 0.75$$

Differentiating,

$$V(t) = V_0\left(1 - \frac{2}{L}\right)^t$$
$$V'(t) = V_0\left(1 - \frac{2}{L}\right)^t \ln\left(1 - \frac{2}{L}\right)$$

(b) In general, the percentage rate of change is

$$100\frac{V'(t)}{V(t)} = 100\frac{\ln\left(1 - \frac{2}{L}\right)V(t)}{V(t)}$$
$$= 100\ln\left(1 - \frac{2}{L}\right).$$

37. $N(t) = 2(1 - e^{-.037t})$

To graph this function and see what happens as $t \to \infty$, press $\boxed{y =}$ and input N for $y_1 =$. Use window dimensions of $[0, 200]10$ by $[0, 3]1$.
Press $\boxed{\text{graph}}$.
The value of N approaches the maximum of 2 million viewers.

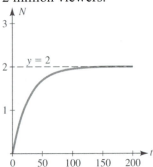

39. $P(t) = P_0 e^{kt}$

Let $t = 0$ correspond to the year 1997. Then, $P_0 = 60$ million. Further, the year 2002 corresponds to $t = 5$ and $P(t) = 90$ million.

$$P(t) = 60e^{kt}$$
$$90 = 60e^{k \cdot 5}$$
$$1.5 = e^{5k}$$
$$\ln 1.5 = \ln e^{5k}$$
$$\ln 1.5 = 5k$$
$$k = \frac{\ln 1.5}{5} \approx 0.0811$$

So, $P(t) = 60e^{0.0811t}$. The year 2012 corresponds to $t = 15$ and

$$P(15) = 60e^{0.0811(15)} \approx 202.52 \text{ million}.$$

The population should be approximately 202.5 million.

41. $L(t) = \dfrac{\ln(t + 1)}{t + 1}$

(a) $L'(t) = \dfrac{(t+1)\left(\frac{1}{t+1}\cdot 1\right) - \ln(t+1)(1)}{(t+1)^2}$

$= \dfrac{1 - \ln(t+1)}{(t+1)^2}$

So, $L'(t) = 0$ when

$1 - \ln(t+1) = 0$

$\ln(t+1) = 0$

$e^{\ln(t+1)} = e^1$

$t + 1 = e$, or $t = e - 1$.

When $0 \le t < e - 1$, $L'(t) > 0$, so L is increasing. When $e - 1 < t \le 5$, $L'(t) < 0$, so L is decreasing.

$L(e-1) = \dfrac{\ln(e-1+1)}{e-1+1}$

$= \dfrac{\ln e}{e}$

$= \dfrac{1}{e} \approx 0.368$

$L(0) = \dfrac{\ln(1)}{1} = 0$

$L(1) = \dfrac{\ln 2}{2} \approx 0.347$

So, at the age $e - 1 \approx 1.7$ years of age, a child's learning capacity is the greatest.

(b) Need to maximize the rate of learning, or maximize the first derivative.

$L''(t) = \dfrac{(t+1)^2\left(\frac{-1}{t+1}\right) - [1 - \ln(t+1)][2(t+1)(1)]}{(t+1)^4}$

$= \dfrac{-(t+1) - 2(t+1)[1 - \ln(t+1)]}{(t+1)^4}$

$= \dfrac{(t+1)[-1 - 2(1 - \ln(t+1))]}{(t+1)^4}$

$= \dfrac{-1 - 2[1 - \ln(t+1)]}{(t+1)^3}$

So $L''(t) = 0$ when

$-1 - 2[1 - \ln(t+1)] = 0$, or

$-2[1 - \ln(t+1)] = 1$

$1 - \ln(t+1) = -\dfrac{1}{2}$

$1 + \dfrac{1}{2} = \ln(t+1)$

$e^{1.5} = e^{\ln(t+1)}$

$e^{1.5} = t + 1$, or $t = e^{1.5} - 1$

When $0 \le t < e^{1.5} - 1$, $L''(t) < 0$, so $L'(t)$ is decreasing. When $e^{1.5} - 1 < t \le 5$, $L''(t) > 0$, so $L'(t)$ is increasing.

$L'(e^{1.5} - 1) = \dfrac{1 - \ln(e^{1.5} - 1 + 1)}{(e^{1.5} - 1 + 1)^2}$

$= \dfrac{1 - \ln e^{1.5}}{(e^{1.5})^2}$

$= \dfrac{1 - 1.5}{e^3} \approx -0.025$

$L'(0) = \dfrac{1 - \ln(0+1)}{(0+1)^2} = 0$

$L'(5) = \dfrac{1 - \ln(5+1)}{(5+1)^2} \approx -0.022$

So, a child's learning capability is increasing most rapidly at birth.

43. $Q(t) = Q_0 e^{-0.0015t}$

(a) The percentage rate is

$100\dfrac{Q'(t)}{Q(t)} = 100\dfrac{-0.0015 Q_0 e^{-0.0015t}}{Q_0 e^{-0.0015t}}$

$= -0.15\%$ per year

(b) When 10% is depleted, 90% remains, so $0.9 Q_0 = Q_0 e^{-0.0015t}$

$0.9 = e^{-0.0015t}$

$\ln 0.9 = -0.0015t$, or

$t = \dfrac{\ln 0.9}{-0.0015} \approx 70.24$ years

The percentage rate of change is constant, so the rate at this time is 0.15%.

45. $N(t) = \dfrac{B}{1 + Ce^{-kt}}$

(a) When $t = 0$, $N(0) = 0.1B$ so

$$0.1B = \dfrac{B}{1 + Ce^{-k \cdot 0}}$$

$$0.1 = \dfrac{1}{1 + C}$$

$$1 + C = \dfrac{1}{0.1}$$

$$C = 9$$

When $t = 2$, $N(2) = 0.25B$, so

$$0.25B = \dfrac{B}{1 + 9e^{-k(2)}}$$

$$0.25 = \dfrac{1}{1 + 9e^{-2k}}$$

$$1 + 9e^{-2k} = \dfrac{1}{0.25}$$

$$9e^{-2k} = 3$$

$$e^{-2k} = \dfrac{1}{3}$$

$$\ln e^{-2k} = \ln \dfrac{1}{3}$$

$$-2k = \ln \dfrac{1}{3},$$

or, $k = \dfrac{\ln \frac{1}{3}}{-2} = \dfrac{-\ln \frac{1}{3}}{2} = \dfrac{\ln 3}{2}$

(b) $N(t) = \dfrac{B}{1 + 9^{-\left(\frac{\ln 3}{2}\right)t}}$

$$0.5B = \dfrac{B}{1 + 9e^{(-t/2)(\ln 3)}}$$

$$1 + 9e^{(-t/2)(\ln 3)} = \dfrac{1}{0.5}$$

$$9e^{(-t/2)(\ln 3)} = 1$$

$$e^{\ln 3^{-t/2}} = \dfrac{1}{9}$$

$$3^{-t/2} = \dfrac{1}{9}$$

$$\ln 3^{-t/2} = \ln \dfrac{1}{9}$$

$$-\dfrac{t}{2} \ln 3 = \ln \dfrac{1}{9}$$

$$-\dfrac{t}{2} = \dfrac{\ln \frac{1}{9}}{\ln 3},$$

or $t = \dfrac{-2 \ln \frac{1}{9}}{\ln 3} = \dfrac{2 \ln 9}{\ln 3} = 4$ hours

(c) Need to maximize the rate at which news is spreading (maximize the first derivative).

$$N(t) = \dfrac{B}{1 + 9e^{-(\ln 3/2)t}}$$

To use the result from page 331, consider $\dfrac{\ln 3}{2} = B\left(\dfrac{\ln 3}{2B}\right) = Bk$.

Then, $N''(t) = 0$ when

$$t = \dfrac{\ln 9}{\ln \frac{3}{2}} = \dfrac{2 \ln 9}{\ln 3} = \dfrac{\ln 81}{\ln 3} = \log_3 81 = 4.$$

So, the news is spreading most rapidly after 4 hours.

47. $f(t) = \dfrac{2}{1 + 3e^{-0.8t}}$

(a) When $t = 0$, $f(0) = \dfrac{1}{2}$, so $\left(0, \dfrac{1}{2}\right)$ is an intercept. $f(t) = 0$ has no solution.

$$\lim_{t \to \infty} \dfrac{2}{1 + 3e^{-0.8t}} = 2, \text{ so } y = 2 \text{ is a}$$

horizontal asymptote.

$$f'(t) = \frac{0 - (2)(-2.4e^{-0.8t})}{(1+3e^{-0.8t})^2}$$

$$= \frac{4.8e^{-0.8}}{(1+3e^{-0.8t})^2}$$

$f'(t) > 0$ for all values of t, so f is always increasing. Using logarithmic differentiation,

$\ln f'(t)$

$$= \ln\left[\frac{4.8e^{-0.8t}}{(1+3e^{-0.8t})^2}\right]$$

$$= \ln 4.8 + \ln e^{-0.8t} - \ln(1+3e^{-0.8t})^2$$

$$= \ln 4.8 - 0.8t - 2\ln(1+3e^{-0.8t})$$

$$\frac{f''(t)}{f'(t)} = -0.8 - 2 \cdot \frac{-2.4e^{-0.8t}}{1+3e^{-0.8t}}$$

$f''(t)$

$$= \left[\frac{-0.8(1+3e^{-0.8t}) + 4.8e^{-0.8t}}{1+3e^{-0.8t}}\right] f'(t)$$

$$= \left[\frac{-0.8 + 2.4e^{-0.8t}}{1+3e^{-0.8t}}\right]\left[\frac{4.8e^{-0.8t}}{(1+3e^{-0.8t})^2}\right]$$

$$= (-0.8 + 2.4e^{-0.8t})\left[\frac{4.8e^{-0.8t}}{(1+3e^{-0.8t})^3}\right]$$

$f''(t) = 0$ when

$$-0.8 + 2.4e^{-0.8t} = 0$$

$$e^{-0.8t} = \frac{1}{3}$$

$$\ln e^{-0.8t} = \ln\frac{1}{3}$$

$$-0.8t = \ln\frac{1}{3},$$

or $t = \dfrac{\ln\frac{1}{3}}{-0.8} = \dfrac{-\ln\frac{1}{3}}{0.8} = \dfrac{\ln 3}{0.8} = \dfrac{5\ln 3}{4}$

When

$0 < t < \dfrac{5\ln 3}{4}$, $f''(t) > 0$, so f is concave up

$t > \dfrac{5\ln 3}{4}$, $f''(t) < 0$, so f is concave down

The point $(1.37, 1)$ is an inflection point.

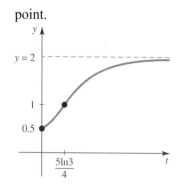

(b) $f(0) = 0.5$ thousand people, or 500 people.

(c) $f(3) = \dfrac{2}{1+3(0.907)} = 1.572$, so 1,572 people have caught the disease.

(d) $\displaystyle\lim_{t\to\infty} \frac{2}{1+3e^{-0.8t}} = 2$, so in the long run, approximately 2,000 people will contract the disease.

49. $p(x) = e^{-0.2x}$; $f(x) = 5x^{0.9}$

The per capita rate of increase function is

$$R(x) = \frac{\ln[e^{-0.2x}(5x^{0.9})]}{x}$$

$$= -0.2 + [\ln 5 + 0.9\ln x]\frac{1}{x}$$

So,

$R'(x)$

$$= 0 + [\ln 5 + 0.9\ln x]\left(\frac{-1}{x^2}\right) + \left(\frac{1}{x}\right)\left(0 + \frac{0.9}{x}\right)$$

$$= \frac{-\ln 5 - 0.9\ln x + 0.9}{x^2}$$

So, $R'(x) = 0$ when

$$0 = -\ln 5 - 0.9\ln x + 0.9$$

$$0.9\ln x = -\ln 5 + 0.9$$

$$\ln x = \frac{-\ln 5 + 0.9}{0.9}$$

$$e^{\ln x} = e^{(-\ln 5 + 0.9)/0.9}$$

$$x = e^{(-\ln 5 + 0.9)/0.9} \approx 0.45$$

Since

$R''(x)$

$$= \frac{(x^2)\left(\frac{-0.9}{x}\right) + [-\ln 5 - 0.9\ln x + 0.9](2x)}{x^4}$$

$$= \frac{x[-0.9 + 2(-\ln 5 - 0.9\ln x + 0.9)]}{x^4}$$

$$= \frac{-2\ln 5 - 1.8\ln x + 1.8}{x^3}$$

and $R''(0.45) < 0$, so $x = 0.45$ corresponds to the absolute maximum. The ideal reproductive age is 0.45 years.

51. $E(C) = C\left(aR + \dfrac{b}{C}\right)^2$

(a) $E'(C) = (C)\left[2\left(aR + \dfrac{b}{C}\right)\left(\dfrac{-b}{C^2}\right)\right]$

$$\qquad\qquad + \left(aR + \frac{b}{C}\right)^2 \quad (1)$$

$$= \frac{-2b}{C}\left(aR + \frac{b}{C}\right) + \left(aR + \frac{b}{C}\right)^2$$

$$= \left(aR + \frac{b}{C}\right)\left[\frac{-2b}{C} + aR + \frac{b}{C}\right]$$

$$= \left(aR + \frac{b}{C}\right)\left(aR - \frac{b}{C}\right)$$

So $E'(C) = 0$ when $aR - \dfrac{b}{C} = 0$

(rejecting the negative solution)

$$aR = \frac{b}{C}$$

$$C = \frac{b}{aR}$$

$E''(C)$

$$= \left(aR + \frac{b}{C}\right)\left(\frac{b}{C^2}\right) + \left(aR - \frac{b}{C}\right)\left(\frac{-b}{C^2}\right)$$

$$= \frac{b}{C^2}\left[\left(aR + \frac{b}{C}\right) - \left(aR - \frac{b}{C}\right)\right]$$

$$= \frac{b}{C^2}\left(\frac{2b}{C}\right)$$

$$= \frac{2b^2}{C^3}$$

When $C = \dfrac{b}{aR}$, $E''\left(\dfrac{b}{aR}\right) = \dfrac{2b^2}{\left(\dfrac{b}{aR}\right)^2}$.

Since a, b, R are all positive,

$E''\left(\dfrac{b}{aR}\right) > 0$. So, the absolute minimum

occurs when $C = \dfrac{b}{aR}$.

(b) $E(C) = mCe^{k/C}$

Use logarithmic differentiation,

$\ln E(C) = \ln mCe^{k/C}$

$\ln E(C) = \ln m + \ln C + \ln e^{k/C}$

$\ln E(C) = \ln m + \ln C + \dfrac{k}{C}$

$$\frac{E'(C)}{E(C)} = \frac{1}{C} - \frac{k}{C^2}$$

or, $E'(C) = \dfrac{C - k}{C^2} E(C)$

$$= \frac{C - k}{C^2}(mCe^{k/C})$$

$$= \frac{(C - k)m}{C}e^{k/C}$$

So, $E'(C) = 0$ when $C - k = 0$, or $C = k$. We want the same value of C for a minimum in both models, so

$k = \dfrac{b}{aR}$.

From the first model, the minimum

value is $E\left(\dfrac{b}{aR}\right) = \dfrac{b}{aR}\left(aR + b \cdot \dfrac{aR}{b}\right)^2$

$$= \frac{b}{aR}(2aR)^2$$

$$= 4abR.$$

For the second model to have this

same minimum, $E(k) = mke^{k/k}$

$4abR = m\left(\dfrac{b}{aR}\right)e$, so $m = 4a^2R^2e^{-1}$.

53. (a) Assuming continuous growth, the situation can be modeled by a function of the form $Q(t) = Q_0e^{kt}$.

Let $t = 0$ be the year 1947. Since $r = 0.06$ and $Q_0 = 1{,}139$,

$$Q(t) = 1{,}139 e^{0.06t}.$$

In the year 1954, $t = 7$ and

$$Q(7) = 1{,}139 e^{0.06(7)}$$
$$\approx 1{,}733 \text{ staff members}$$

(b) Let the original size of the staff be Q_0 and double the staff be $2Q_0$. Then,

$$2Q_0 = Q_0 e^{0.06t}$$
$$2 = e^{0.06t}$$
$$\ln 2 = \ln e^{0.06t}$$
$$\ln 2 = 0.06t, \text{ or } t = \frac{\ln 2}{0.06} = 11.55$$

So, any size staff doubles in approximately 11.55 years.

(c) Writing Exercise—Answers will vary.

55. $w(t) = \dfrac{10}{1 + 15 e^{-0.05t}}; \ p(t) = e^{-0.01t}$

(a) total weight
= (weight per fish)(number of fish)
= (weight per fish)
   [(beginning number of fish)
   (proportion remaining)]

$$E(t) = \left(\frac{10}{1 + 15 e^{-0.05t}} \right)(1{,}000 e^{-0.01t})$$

$$E(t) = 10{,}000 \frac{e^{-0.01t}}{1 + 15 e^{-0.05t}}$$

(b) Using logarithmic differentiation,

$\ln E(t)$

$$= \ln \left[10{,}000 \frac{e^{-0.01t}}{1 + 15 e^{-0.05t}} \right]$$

$$= \ln 10{,}000 + \ln e^{-0.01t}$$
$$\qquad - \ln(1 + 15 e^{-0.05t})$$

$$= \ln 10{,}000 - 0.01t - \ln(1 + 15 e^{-0.05t})$$

$$\frac{E'(t)}{E(t)} = -0.01 - \frac{-0.75 e^{-0.05t}}{1 + 15 e^{-0.05t}}$$

$$E'(t) = \left[\frac{-0.01(1 + 15 e^{-0.05t})}{+ 0.75 e^{-0.05t}} \right] E(t)$$

$$= \left[\frac{-0.01 + 0.6 e^{-0.05t}}{1 + 15 e^{-0.05t}} \right]$$
$$\left[10{,}000 \frac{e^{-0.01t}}{1 + 15 e^{-0.05t}} \right]$$

$$= [-0.01 + 0.6 e^{-0.05t}]$$
$$\left[10{,}000 \frac{e^{-0.01t}}{(1 + 15 e^{-0.05t})^2} \right]$$

So, $E(t) = 0$ when

$$-0.01 + 0.6 e^{-0.05t} = 0.$$

$$0.6 e^{-0.05t} = 0.01$$

$$e^{-0.05t} = \frac{1}{60}$$

$$\ln e^{-0.05t} = \ln \frac{1}{60}$$

$$-0.05t = \ln \frac{1}{60}, \text{ or }$$

$$t = \frac{\ln \frac{1}{60}}{-0.05} = \frac{-\ln \frac{1}{60}}{0.05} = \frac{\ln 60}{0.05} \approx 81.9$$

For the domain $t \geq 0$, when

$$0 \leq t < \frac{\ln 60}{0.05}, \ E'(t) > 0, \text{ so } E \text{ is}$$

increasing. When $t > \dfrac{\ln 60}{0.05}, \ E'(t) < 0,$

so E is decreasing. So, the relative maximum is also the absolute maximum.

When $t \approx 81.9$ days, the yield is the maximum, namely

$$E(81.9) = 10{,}000 \frac{e^{-0.01(81.9)}}{1 + 15 e^{-0.05(81.9)}}$$
$$\approx 3{,}527 \text{ pounds}$$

(c)

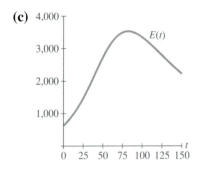

57. $Q(t) = (A)(1 - e^{-kt})$

(a) As t gets very large,

$$\lim_{t \to \infty} Q(t) = \lim_{t \to \infty} A(1 - e^{-kt})$$

Since e^{-kt} approaches zero,

$$= A(1 - 0) = A$$

Also, $Q(0) = A(1 - 1) = 0$. So, the graph of Q looks like

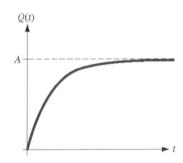

(b) Q approaches the value A, so given enough time, a person can recall nearly all of the relevant facts in his/her memory.

59. $f(t) = \dfrac{A}{1 + Ce^{-kt}}$

The epidemic is spreading most rapidly when the rate of change, or derivative, is maximized.

$$f'(t) = \frac{0 - (A)(-kCe^{-kt})}{(1 + Ce^{-kt})^2}$$

$$= \frac{kACe^{-kt}}{(1 + Ce^{-kt})^2}$$

$$= kAC \frac{e^{-kt}}{(1 + Ce^{-kt})^2}$$

The possible min/max of f' are the zeros of f''.

$$f''(t) = kAC\left[\frac{(1 + Ce^{-kt})^2(-ke^{-kt})}{(1 + Ce^{-kt})^4}\right.$$

$$\left. - \frac{(e^{-kt})[2(1 + Ce^{-kt})(-kCe^{-kt})]}{(1 + Ce^{-kt})^4}\right]$$

$$= -k^2 ACe^{-kt}(1 + Ce^{-kt})$$

$$\left[\frac{1 + Ce^{-kt} - 2Ce^{-kt}}{(1 + Ce^{-kt})^4}\right]$$

So, $f''(t) = 0$ when

$$1 - Ce^{-kt} = 0$$

$$1 = Ce^{-kt}$$

$$\frac{1}{C} = e^{-kt}$$

$$\ln\frac{1}{C} = \ln e^{-kt}$$

$$\ln\frac{1}{C} = -kt, \text{ or}$$

$$t = \frac{\ln\frac{1}{C}}{-k} = \frac{-\ln\frac{1}{C}}{k} = \frac{\ln C}{k}$$

Checking with f''' shows this value of t corresponds to the absolute maximum. The absolute maximum is

$$f\left(\frac{\ln C}{k}\right) = \frac{A}{1 + Ce^{-k(\ln C/k)}}$$

$$= \frac{A}{1 + Ce^{-\ln C}}$$

$$= \frac{A}{1 + Ce^{\ln(1/C)}}$$

$$= \frac{A}{1 + C \cdot \frac{1}{C}}$$

$$= \frac{A}{2}$$

So the epidemic is spreading most rapidly when half of those susceptible are infected.

61. $y(t) = \dfrac{c}{b-a}(a^{-at} - e^{-bt})$

(a) $y'(t) = \dfrac{c}{b-a}(-ae^{-at} + be^{-bt})$

So, $y'(t) = 0$ when

$$-ae^{-at} + be^{-bt} = 0$$
$$be^{-bt} = ae^{-at}$$
$$\frac{e^{-bt}}{e^{-at}} = \frac{a}{b}$$
$$e^{-bt+at} = \frac{a}{b}$$
$$\ln e^{-bt+at} = \ln\frac{a}{b}$$
$$(a-b)t = \ln\frac{a}{b}$$
$$t = \frac{\ln\frac{a}{b}}{a-b} = \frac{\ln\frac{b}{a}}{b-a}$$
$$y''(t) = \frac{c}{b-a}(a^2 e^{-at} - b^2 e^{-bt})$$
$$y''\left(\frac{\ln\frac{a}{b}}{a-b}\right) < 0, \text{ so the maximum}$$

occurs when $t = \dfrac{\ln\frac{a}{b}}{a-b}$.

In the long run,

$$\lim_{t\to+\infty} \frac{c}{b-a}(e^{-at} - e^{-kt}) = \frac{c}{b-a}(0-0)$$
$$= 0.$$

So, the concentration approaches zero.

(b)

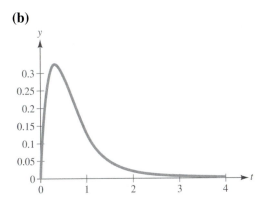

(c) Writing Exercise—Answers will vary.

63. $N(t) = 500(0.03)^{(0.4)^t}$

(a) When $t = 0$, $N(0) = 500(0.03)^{(0.4)^0}$
$$= 15 \text{ employees.}$$

When $t = 5$, $N(5) = 500(0.03)^{(0.4)^5}$
$$\approx 482 \text{ employees.}$$

$$300 = 500(0.03)^{(0.4)^t}$$
$$\frac{3}{5} = (0.03)^{(0.4)^t}$$
$$\ln 0.6 = \ln(0.03)^{(0.4)^t}$$
$$\ln 0.6 = (0.4)^t \ln(0.03)$$
$$\frac{\ln 0.6}{\ln 0.03} = (0.4)^t$$
$$0.145677 \approx (0.4)^t$$
$$\ln 0.145677 \approx \ln(0.4)^t$$
$$\ln 0.145677 \approx t\ln(0.4), \text{ so}$$
$$t \approx \frac{\ln 0.145677}{\ln 0.4}$$
$$\approx 2.10 \text{ years}$$

Since $\lim\limits_{t\to+\infty} (0.4)^t = 0$,

$$\lim_{t\to+\infty} 500(0.3)^{(0.4)^t} = 500 \text{ employees.}$$

(b) To sketch the graphs of N and

$F(t) = 500(0.03)^{-(0.4)^{-t}}$ on the same graph,

Press $\boxed{y=}$ and input N for $Y_1 =$.

Use window dimensions of $[-6, 6]2$ by $[0, 100]100$

Press $\boxed{\text{graph}}$.

Press $\boxed{y=}$ and input F for $Y_2 =$.

Press $\boxed{\text{graph}}$.

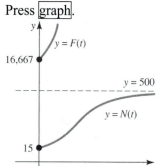

Writing Exercise—Answers will vary.

65. $p(x) = Ax^s c^{-sx/r}$

(a) $P'(x) = A\left[(x^s)\left(\dfrac{-s}{r} e^{-sx/r} \right) \right.$

$\left. + (e^{-sx/r})(sx^{s-1}) \right]$

$= sAx^{s-1}e^{-sx/r}\left[\dfrac{-x}{r} + 1 \right]$

So $p'(x) = 0$ when $\dfrac{-x}{r} + 1 = 0$, or

$x = r$. When

$0 \le x < r$, $p'(x) > 0$, so p is increasing

$x > r$, $p'(x) < 0$, so p is decreasing

Since the domain of p is $p \ge 0$, this means the absolute maximum occurs when $x = r$.

(b) Rewrite $p'(x)$ as

$p'(x) = sAe^{-sx/r}\left(\dfrac{-x^s}{r} + x^{s-1} \right)$

Then

$p''(x)$

$= sA\left[(e^{-sx/r})\left(\dfrac{-sx^{s-1}}{r} + (s-1)x^{s-2} \right) \right.$

$\left. + \left(\dfrac{-x^s}{r} + x^{s-1} \right)\left(\dfrac{-s}{r}e^{-sx/r} \right) \right]$

$= sAx^{s-2}e^{-sx/r}\left[\dfrac{-sx}{r} + s - 1 \right.$

$+ \left(\dfrac{-x^2}{r} + x \right)\dfrac{-s}{r} \right]$

$= sAx^{s-2}e^{-sx/r}\left(\dfrac{-s}{r}x + s - 1 \right.$

$\left. + \dfrac{s}{r^2}x^2 - \dfrac{s}{r}x \right)$

$= sAx^{s-2}e^{-sx/r}\left[\dfrac{s}{r^2}x^2 - \dfrac{2s}{r}x \right.$

$\left. + (s-1) \right]$

$= \dfrac{1}{r^2}sAe^{s-2}e^{-sx/r}[sx^2 - 2rsx$

$+ r^2(s-1)]$

Using the quadratic formula,

$x = \dfrac{2rs \pm \sqrt{(2rs)^2 - (4)(s)r^2(s-1)}}{2(s)}$

$x = \dfrac{2rs \pm 2r\sqrt{s^2 - s(s-1)}}{2s}$

$x = \dfrac{rs \pm r\sqrt{s}}{s} = \dfrac{r}{s}\left(s \pm \sqrt{s} \right)$

So, there are two possible inflection points. (Checking with $p''(x)$ shows that they both are inflection points.)

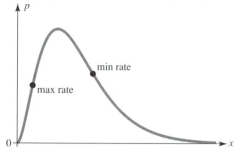

(c) When $0 < s < 1$, $s - \sqrt{s} < 0$, so $x < 0$. Since the practical domain is $x > 0$, this value is rejected and there is only one inflection point.

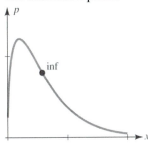

67. $T(t) = -5 + Ae^{-kt}$

   (a) When $t = 0$, $T(0) = 80$, so

$$80 = -5 + Ae^0, \text{ or } A = 85.$$

      When $t = 20$, $T(20) = 25$, so

$$25 = -5 + 85e^{-k \cdot 20}$$
$$30 = 85e^{-20k}$$
$$\frac{6}{17} = e^{-20k}$$
$$\ln \frac{6}{17} = \ln e^{-20k}$$
$$\ln \frac{6}{17} = -20k, \text{ or}$$
$$k = \frac{\ln \frac{6}{17}}{-20} = \frac{-\ln \frac{6}{17}}{20} = \frac{\ln \frac{17}{6}}{20}$$

   (b) $T(t) = -5 + 85e^{-0.052t}$

      When $t = 0$, $T(0) = 80$, so $(0, 80)$ is an intercept. When $T(t) = 0$,

$$0 = -5 + 85e^{-0.052t}$$
$$5 = 85e^{-0.052t}$$
$$\frac{1}{17} = e^{-0.052t}$$
$$\ln \frac{1}{17} = \ln e^{-0.052t}$$
$$\ln \frac{1}{17} = -0.052t, \text{ so}$$
$$t = \frac{\ln \frac{1}{17}}{-0.052} = \frac{-\ln \frac{1}{17}}{0.052} = \frac{\ln 17}{0.052} \approx 54.5$$

      So, $(54.5, 0)$ is an intercept.

$$\lim_{t \to +\infty} -5 + 85e^{-0.052t} = -5, \text{ so } y = -5$$

      is a horizontal asymptote.

$$T'(t) = -4.42e^{-0.052t}$$

$T'(t) < 0$ for all values of t, so T is always decreasing.

$$T''(t) = 0.23e^{-0.052t}$$

$T''(t) > 0$ for all values of t, so T is always concave up.

As $t \to +\infty$, the temperature approaches $-5°$C.

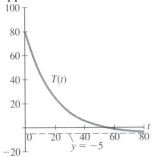

   (c) $T(30) = -5 + 85e^{-0.052(30)} \approx 12.8°$C

   (d) The temperature will be $0°$C after approximately 54.5 minutes (see part b).

69. $f(x) = \frac{1}{\sigma\sqrt{2\pi}} e^{-(x-\mu)^2/2\sigma^2}$

   (a) Noting that σ, $\sqrt{2\pi}$ and μ, are all constants, and that

$$-\frac{(x-\mu)^2}{2\sigma^2} = -\frac{1}{2\sigma^2}(x^2 - 2\mu x + \mu^2)$$
$$= -\frac{1}{2\sigma^2}x^2 + \frac{\mu}{\sigma^2}x - \frac{\mu^2}{2\sigma^2}$$

$$f'(x)$$
$$= \frac{1}{\sigma\sqrt{2\pi}}$$
$$\left[\left(e^{-(x-\mu)^2/2\sigma^2} \right) \left(-\frac{1}{\sigma^2}x + \frac{\mu}{\sigma^2} \right) \right]$$

So, $f'(x) = 0$ when

$$\frac{1}{\sigma\sqrt{2\pi}} = 0 \to \text{no solution}$$
$$e^{-(x-\mu)^2/2\sigma^2} = 0 \to \text{no solution}$$

$$-\frac{1}{\sigma^2}x + \frac{\mu}{\sigma^2} = 0$$

$$-x + \mu = 0$$

$$x = \mu$$

$$f''(x)$$

$$= \frac{1}{\sigma\sqrt{2\pi}}\left[\left(e^{-(x-\mu)^2/2\sigma^2}\right)\left(-\frac{1}{\sigma^2}\right)\right.$$

$$+ \left(-\frac{1}{\sigma^2}x + \frac{\mu}{\sigma^2}\right)\left(e^{-(x-\mu)^2/2\sigma^2}\right)$$

$$\left. \left(-\frac{1}{\sigma^2}x + \frac{\mu}{\sigma^2}\right)\right]$$

$$= \frac{1}{\sigma\sqrt{2\pi}}e^{-(x-\mu)^2/2\sigma^2}$$

$$\cdot \left[-\frac{1}{\sigma^2} + \left(\frac{1}{\sigma^4}x^2 - \frac{2\mu}{\sigma^4} + \frac{\mu^2}{\sigma^4}\right)\right]$$

So, $f''(x) = 0$ when

$$\frac{1}{\sigma\sqrt{2\pi}} = 0 \rightarrow \text{no solution}$$

$$e^{-(x-\mu)^2/2\sigma^2} = 0 \rightarrow \text{no solution}$$

$$\frac{1}{\sigma^4}x^2 - \frac{2\mu}{\sigma^4}x + \frac{\mu^2}{\sigma^4} - \frac{1}{\sigma^2} = 0$$

$$\frac{1}{\sigma^4}(x^2 - 2\mu x + \mu^2 - \sigma^2) = 0$$

$$x^2 - 2\mu x + (\mu^2 - \sigma^2) = 0$$

$$(x - (\mu - \sigma))(x - (\mu + \sigma)) = 0$$

$$x - (\mu - \sigma) = 0, \text{ or}$$

$$x = \mu - \sigma$$

$$x - (\mu + \sigma) = 0, \text{ or}$$

$$x = \mu + \sigma$$

So there are inflection points at $x = \mu - \sigma$ and $x = \mu + \sigma$. To test the critical value $x = \mu$ from the first derivative, note that

$$f''(\mu) = \frac{1}{\sigma\sqrt{2\pi}}e^{-(\mu-\mu)^2/2\sigma^2}$$

$$\cdot \left(\frac{1}{\sigma^4}\mu^2 - \frac{2\mu}{\sigma^4}\mu + \frac{\mu^2}{\sigma^4} - \frac{1}{\sigma^2}\right)$$

$$= \frac{1}{\sigma\sqrt{2\pi}} \cdot e^0\left(-\frac{1}{\sigma^2}\right) < 0$$

The function is concave down and there is an absolute max at $x = \mu$.

(b) $f(\mu + c) = \frac{1}{\sigma\sqrt{2\pi}}e^{-[(\mu+c)-\mu]^2/2\sigma^2}$

$$f(\mu - c) = \frac{1}{\sigma\sqrt{2\pi}}e^{-[(\mu-c)-\mu]^2/2\sigma^2}$$

$$-[(\mu+c)-\mu]^2 = -c^2$$

$$-[(\mu-c)-\mu]^2 = -c^2$$

So, $f(\mu + c) = f(\mu - c)$. This means that the graph of f is symmetric about the line $\mu = c$.

Checkup for Chapter 4

1. (a) $\dfrac{(3^{-2})(9^2)}{(27)^{2/3}} = \dfrac{\left(\frac{1}{3^2}\right)(9^2)}{\left(\sqrt[3]{27}\right)^2} = \dfrac{\left(\frac{1}{9}\right)(81)}{(3)^2} = 1$

(b) $\sqrt[3]{(25)^{1.5}\left(\dfrac{8}{27}\right)} = \sqrt[3]{(25)^{1.5}}\sqrt[3]{\dfrac{8}{27}}$

$$= [(25)^{1.5}]^{1/3}\dfrac{\sqrt[3]{8}}{\sqrt[3]{27}}$$

$$= (25)^{0.5}\left(\dfrac{2}{3}\right)$$

$$= \sqrt{25}\left(\dfrac{2}{3}\right)$$

$$= \dfrac{10}{3}$$

(c) $\log_2 4 + \log_4 16^{-1}$

$\log_2 4 = a$ if and only if $2^a = 4$, so $a = 2$.

$\log_4 16^{-1} = \log_4\left(\dfrac{1}{16}\right)$.

Now, $\log_4\left(\dfrac{1}{16}\right) = b$ if and only if

$4^b = \dfrac{1}{16}$, so $b = -2$

$\log_2 4 + \log_4 16^{-1} = 2 - 2 = 0$

(d) $\left(\dfrac{8}{27}\right)^{-2/3}\left(\dfrac{16}{81}\right)^{3/2} = \left(\dfrac{27}{8}\right)^{2/3}\left(\dfrac{16}{81}\right)^{3/2}$

$\qquad\qquad = \left(\sqrt[3]{\dfrac{27}{8}}\right)^{2}\left(\sqrt{\dfrac{16}{81}}\right)^{3}$

$\qquad\qquad = \left(\dfrac{3}{2}\right)^{2}\left(\dfrac{4}{9}\right)^{3}$

$\qquad\qquad = \left(\dfrac{9}{4}\right)\left(\dfrac{64}{729}\right)$

$\qquad\qquad = \dfrac{16}{81}$

2. (a) $(9x^4 y^2)^{3/2} = 9^{3/2}(x^4)^{3/2}(y^2)^{3/2}$

$\qquad\qquad = \left(\sqrt{9}\right)^{3}(x^6)(y^3)$

$\qquad\qquad = 27x^6 y^3$

(b) $(3x^2 y^{4/3})^{-1/2}$

$\qquad = \left(\dfrac{1}{3x^2 y^{4/3}}\right)^{1/2}$

$\qquad = \dfrac{(1)^{1/2}}{(3)^{1/2}(x^2)^{1/2}(y^{4/3})^{1/2}}$

$\qquad = \dfrac{\sqrt{1}}{\left(\sqrt{3}\right)(x)(y^{2/3})}$

$\qquad = \dfrac{1}{\sqrt{3}xy^{2/3}}$

(c) $\left(\dfrac{y}{x}\right)^{3/2}\left(\dfrac{x^{2/3}}{y^{1/6}}\right)^{2}$

$\qquad = \left(\dfrac{y^{3/2}}{x^{3/2}}\right)\left(\dfrac{x^{4/3}}{y^{1/3}}\right)$

$\qquad = (x^{4/3-3/2})(y^{3/2-1/3})$

$\qquad = x^{-1/6}y^{7/6}$

$\qquad = \dfrac{y^{7/6}}{x^{1/6}}$

(d) $\left(\dfrac{x^{0.2}y^{-1.2}}{x^{1.5}y^{0.4}}\right)^{5}$

$\qquad = [(x^{0.2-1.5})(y^{-1.2-0.4})]^5$

$\qquad = (x^{-1.3}y^{-1.6})^5$

$\qquad = (x^{-1.3})^5(y^{-1.6})^5$

$\qquad = x^{-6.5}y^{-8}$

$\qquad = \dfrac{1}{x^{6.5}y^8}$

3. (a) $4^{2x-x^2} = \dfrac{1}{64}$

$\qquad 4^{2x-x^2} = 4^{-3}$

$\qquad \text{So, } 2x - x^2 = -3$

$\qquad\qquad 0 = x^2 - 2x + 3$

$\qquad\qquad 0 = (x-3)(x+1)$

$\qquad\qquad x = 3, -1$

(b) $e^{1/x} = 4$

$\qquad \ln e^{1/x} = \ln 4$

$\qquad \dfrac{1}{x} = \ln 4$

$\qquad x = \dfrac{1}{\ln 4}$

(c) $\log_4 x^2 = 2$ if and only if $4^2 = x^2$,

$\qquad$ so, $x = \pm 4$.

(d) $\dfrac{25}{1+2e^{-0.5t}} = 3$

$\qquad \dfrac{25}{3} = 1 + 2e^{-0.5t}$

$\qquad \dfrac{22}{3} = 2e^{-0.5t}$

$\qquad \dfrac{11}{3} = e^{-0.5t}$

$\qquad \ln\dfrac{11}{3} = \ln e^{-0.5t}$

$\qquad \ln\dfrac{11}{3} = -0.5t, \text{ or}$

$\qquad t = \dfrac{\ln\frac{11}{3}}{-0.5} = -2\ln\dfrac{11}{3} = 2\ln\dfrac{3}{11}$

4. (a) $y = \dfrac{e^x}{x^2 - 3x}$

$$\frac{dy}{dx} = \frac{(x^2 - 3x)(e^x \cdot 1) - (e^x)(2x - 3)}{(x^2 - 3x)^2}$$

$$= \frac{e^x[(x^2 - 3x) - (2x - 3)]}{(x^2 - 3x)^2}$$

$$= \frac{e^x(x^2 - 5x + 3)}{(x^2 - 3x)^2}$$

(b) $y = \ln(x^3 + 2x^2 - 3x)$

$$\frac{dy}{dx} = \frac{1}{x^3 + 2x^2 - 3x}(3x^2 + 4x - 3)$$

$$= \frac{3x^2 + 4x - 3}{x^3 + 2x^2 - 3x}$$

(c) $y = x^3 \ln x$

$$\frac{dy}{dx} = (x^3)\left(\frac{1}{x} \cdot 1\right) + (\ln x)(3x^2)$$

$$= x^2 + 3x^2 \ln x$$

$$= x^2(1 + 3\ln x)$$

(d) $y = \dfrac{e^{-2x}(2x - 1)^3}{1 - x^2}$

Using logarithmic differentiation,
$\ln y$

$$= \ln\left[\frac{e^{-2x}(2x - 1)^3}{1 - x^2}\right]$$

$$= \ln e^{-2x} + \ln(2x - 1)^3 - \ln(1 - x^2)$$

$$= -2x + 3\ln(2x - 1) - \ln(1 - x^2)$$

$$\frac{y'}{y} = -2 + 3 \cdot \frac{2}{2x - 1} - \frac{-2x}{1 - x^2}$$

$$y' = \left(-2 + \frac{6}{2x - 1} + \frac{2x}{1 - x^2}\right)y$$

$$= \left(-2 + \frac{6}{2x - 1} + \frac{2x}{1 - x^2}\right)\left[\frac{e^{-2x}(2x - 1)^3}{1 - x^2}\right]$$

$$= \left(-1 + \frac{3}{2x - 1} + \frac{x}{1 - x^2}\right)\left[\frac{2e^{-2x}(2x - 1)^3}{1 - x^2}\right]$$

5. (a) $y = x^2 e^{-x}$

When $x = 0$, $y = 0$ so $(0, 0)$ is an intercept. When $y = 0$, $x = 0$.

Also, $\displaystyle\lim_{x \to -\infty} x^2 e^{-x} = +\infty$

$$\lim_{x \to +\infty} x^2 e^{-x} = \lim_{x \to +\infty} \frac{x^2}{e^x}$$

$$= \lim_{x \to +\infty} \frac{2x}{e^x}$$

$$= \lim_{x \to +\infty} \frac{2}{e^x}$$

$$= 0$$

so $y = 0$ is a horizontal asymptote.

$$y' = (x^2)(-e^{-x}) + (e^{-x})(2x)$$

$$= xe^{-x}(2 - x)$$

so $y' = 0$ when $x = 0, 2$.

Rewriting, $y' = e^{-x}(2x - x^2)$, so

$$y'' = (e^{-x})(2 - 2x) + (2x - x^2)(-e^{-x})$$

$$= e^{-x}[(2 - 2x) - (2x - x^2)]$$

$$= e^{-x}(2 - 4x + x^2)$$

So, $y'' = 0$ when $2 - 4x + x^2 = 0$.
Using the quadratic formula,
$x = 2 \pm \sqrt{2}$.

When $x < 0$,
$y' < 0$, so y is decreasing
$y'' > 0$, so y is concave up

When $0 < x < 2 - \sqrt{2}$,
$y' > 0$, so y is increasing
$y'' > 0$, so y is concave up

When $2 - \sqrt{2} < x < 2$,
$y' > 0$, so y is increasing
$y'' < 0$, so y is concave down

When
$2 < x < 2 + \sqrt{2}$,
$y' < 0$, so y is decreasing
$y'' < 0$, so y is concave down

When $x > 2 + \sqrt{2}$,
$y' < 0$, so y is decreasing
$y'' > 0$, so y is concave up

Overall, y is increasing when

$0 < x < 2$; y is decreasing when $x < 0$ and
$x > 2$; y is concave up when $x < 2 - \sqrt{2}$
and $x > 2 + \sqrt{2}$;
y is concave down when
$2 - \sqrt{2} < x < 2 + \sqrt{2}$.
The point $(0, 0)$ is a relative minimum,
the point $\left(2, \dfrac{4}{e^2}\right)$ is a relative

maximum, and the points
$(0.59, 0.19)$, $(3.41, 0.38)$ are inflection
points.

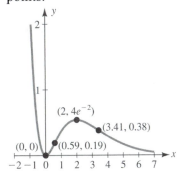

(2, $4e^{-2}$)
(3.41, 0.38)
(0, 0) (0.59, 0.19)

(b) $y = \dfrac{\ln \sqrt{x}}{x^2} = \dfrac{\ln x^{1/2}}{x^2} = \dfrac{\frac{1}{2}\ln x}{x^2} = \dfrac{\ln x}{2x^2}$

Note that the domain of y is $x > 0$, and
$x = 0$ is a vertical asymptote.
When $y = 0$, $x = 1$ so $(1, 0)$ is an
intercept.

$$\lim_{x \to \infty} \frac{\ln x}{2x^2} = \lim_{x \to \infty} \frac{\frac{1}{x}}{4x} = \lim_{x \to \infty} \frac{1}{4x^2} = 0 \text{ so}$$

$y = 0$ is a horizontal asymptote.

$$y' = \frac{(2x^2)\left(\frac{1}{x}\right) - (\ln x)(4x)}{(2x^2)^2}$$

$$= \frac{2x(1 - 2\ln x)}{4x^4}$$

$$= \frac{1 - 2\ln x}{2x^3}$$

So $y' = 0$ when $1 - 2\ln x = 0$

$$1 = 2\ln x$$

$$\frac{1}{2} = \ln x$$

$$e^{1/2} = e^{\ln x}, \text{ or}$$

$$x = e^{1/2}$$

$$y'' = \frac{(2x^3)\left(-2 \cdot \frac{1}{x}\right) - (1 - 2\ln x)(6x^2)}{(2x^3)^2}$$

$$= \frac{2x^2[(-2 - 3)(1 - 2\ln x)]}{4x^6}$$

$$= \frac{-2 - 3 + 6\ln x}{4x^6}$$

$$= \frac{-5 + 6\ln x}{4x^6}$$

So, $y'' = 0$ when $-5 + 6\ln x = 0$

$$6\ln x = 5$$

$$\ln x = \frac{5}{6}$$

$$e^{\ln x} = e^{5/6}$$

$$x = e^{5/6}$$

When $0 < x < e^{1/2}$,
$y' > 0$, so y is increasing
$y'' < 0$, so y is concave down

When $e^{1/2} < x < e^{5/6}$,
$y' < 0$, so y is decreasing
$y'' < 0$, so y is concave down

When $x > e^{5/6}$,
$y' < 0$, so y is decreasing
$y'' > 0$, so y is concave up
Overall, y is increasing when
$0 < x < e^{1/2}$; y is decreasing when
$x > e^{1/2}$; y is concave up when
$x > e^{5/6}$; y is concave down when
$0 < x < e^{5/6}$.

The point $\left(e^{1/2}, \dfrac{1}{4e}\right)$ is a relative

maximum and the point

$\left(e^{5/6}, \dfrac{5}{12e^{5/3}}\right)$ is an inflection point.

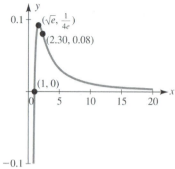

(c) $y = \ln\left(\sqrt{x} - x\right)^2 = 2\ln(x^{1/2} - x)$

Note that the domain of y is $x > 0$ and $x \neq 1$. When $y = 0$, $\ln(x^{1/2} - x)^2 = 0$

$$x^{1/2} - x = \pm 1$$

$0 = x - x^{1/2} + 1$ has no solution.

$0 = x - x^{1/2} - 1$ is solved by letting $u = x^{1/2}$, so $0 = u^2 - u - 1$.

$u = \dfrac{1 \pm \sqrt{1+4}}{2} \approx 1.62$ (rejecting the negative solution)

$x^{1/2} \approx 1.62$ so $x \approx 2.6$

So, $(2.6, 0)$ is an intercept. Since y is undefined when $x = 1$, there is vertical asymptote at $x = 1$. Similarly, there is a vertical asymptote at $x = 0$. Since

$$\lim_{x \to +\infty} \left(\sqrt{x} - x\right)^2 = +\infty,$$

$$\lim_{x \to +\infty} \ln\left(\sqrt{x} - x\right)^2 = +\infty$$

$$y' = 2 \cdot \frac{1}{x^{1/2} - x}\left(\frac{1}{2}x^{-1/2} - 1\right)$$

$$= \frac{\frac{1}{x^{1/2}} - 2}{x^{1/2} - x} \cdot \frac{x^{1/2}}{x^{1/2}}$$

$$= \frac{1 - 2x^{1/2}}{x - x^{3/2}}$$

So $y' = 0$ when $1 - 2x^{1/2} = 0$

$$1 = 2x^{1/2}$$

$$\frac{1}{2} = x^{1/2}, \text{ or}$$

$$x = \frac{1}{4}$$

$$y'' = \frac{(x - x^{3/2})(-x^{-1/2}) \\ - (1 - 2x^{1/2})\left(1 - \frac{3}{2}x^{1/2}\right)}{(x - x^{3/2})^2}$$

$$= \frac{-x^{1/2} + x - (1 - 2x^{1/2} - \frac{3}{2}x^{1/2} + 3x)}{(x - x^{3/2})^2}$$

$$= \frac{-2x + \frac{5}{2}x^{1/2} - 1}{(x - x^{3/2})^2}$$

So $y'' = 0$ when $-2x + \dfrac{5}{2}x^{1/2} - 1 = 0$.

To solve, let $u = x^{1/2}$, so

$-2u^2 + \dfrac{5}{2}u - 1 = 0$. Using the quadratic formula, there are no solutions.

When $0 < x < \dfrac{1}{4}$,

$y' > 0$, so y is increasing
$y'' < 0$, so y is concave down

When $\dfrac{1}{4} < x < 1$,

$y' < 0$, so y is decreasing
$y'' < 0$, so y is concave down

When $x > 1$,

$y' > 0$, so y is increasing
$y'' < 0$, so y is concave down

Overall, y is increasing when

$0 < x < \dfrac{1}{4}$ and $x > 1$; y is decreasing

when $\dfrac{1}{4} < x < 1$; y is concave down

when $0 < x < 1$ and $x > 1$.

The point $\left(\dfrac{1}{4}, \ln\dfrac{1}{16}\right)$ is a relative

maximum and there are no inflection points.

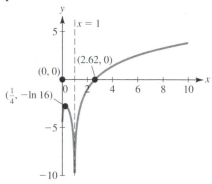

(d) $y = \dfrac{4}{1+e^{-x}}$

When $x = 0$, $y = 2$ so $(0, 2)$ is an intercept. $y = 0$ has no solution.

$\displaystyle\lim_{x \to -\infty} \dfrac{4}{1+e^{-x}} = 0$ so $y = 0$ is a horizontal asymptote.

$\displaystyle\lim_{x \to +\infty} \dfrac{4}{1+e^{-x}} = 4$ so $y = 4$ is a horizontal asymptote.

$y' = \dfrac{0 - (4)(-e^{-x})}{(1+e^{-x})^2} = \dfrac{4e^{-x}}{(1+e^{-x})^2}$

So y' is never zero. Further, $y' > 0$ for all values of x, so y is always increasing. Using logarithmic differentiation,

$\ln y' = \ln\left[\dfrac{4e^{-x}}{(1+e^{-x})^2}\right]$

$\qquad = \ln 4 + \ln e^{-x} - \ln(1+e^{-x})^2$

$\qquad = \ln 4 - x - 2\ln(1+e^{-x})$

$\dfrac{y''}{y'} = -1 - 2 \cdot \dfrac{-e^{-x}}{1+e^{-x}} = -1 + \dfrac{2e^{-x}}{1+e^{-x}}$

$y'' = \left[\dfrac{-(1+e^{-x}) + 2e^{-x}}{1+e^{-x}}\right]y'$

$\quad = \left[\dfrac{-1+e^{-x}}{1+e^{-x}}\right]\left[\dfrac{4e^{-x}}{(1+e^{-x})^2}\right]$

$\quad = (-1+e^{-x})\left[\dfrac{4e^{-x}}{(1+e^{-x})^3}\right]$

So, $y'' = 0$ when $-1+e^{-x} = 0$

$e^{-x} = 1$

$-x = \ln 1$

or $x = 0$

When

$x < 0$, $y'' > 0$ so y is concave up

$x > 0$, $y'' < 0$ so y is concave down

The point $(0, 2)$ is an inflection point.

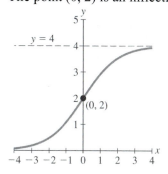

6. In general, $B(t) = Pe^{rt}$.

Here, $B(t) = 2000e^{0.05(t)}$.

When $t = 3$, $B(3) = 2000e^{0.05(3)}$

$= 2000e^{0.15}$

$\approx \$2{,}323.67$

For a balance of $\$3{,}000$,

$3000 = 2000e^{0.05t}$

$\dfrac{3}{2} = e^{0.05t}$

$\ln\dfrac{3}{2} = \ln e^{0.05t}$

$\ln\dfrac{3}{2} = 0.05t$, or

$t = \dfrac{\ln\left(\frac{3}{2}\right)}{0.05} \approx 8.1$ years

7. (a) $8,000 = (P)\left(1 + \dfrac{0.0625}{2}\right)^{2(10)}$

$P = \dfrac{8,000}{(1.03125)^{20}} = \$4,323.25$

(b) $8,000 = Pe^{0.0625(10)}$

$P = 8,000e^{-0.625} = \$4,282.09$

8. $p = \dfrac{\ln(t+1)}{t+1} + 5$

Note that the domain is $t > -1$.

(a) $p'(t) = \dfrac{(t+1)\left(\frac{1}{t+1} \cdot 1\right) - [\ln(t+1)](1)}{(t+1)^2} + 0$

$= \dfrac{1 - \ln(t+1)}{(t+1)^2}$

So, $p'(t) = 0$ when

$1 - \ln(t+1) = 0$

$1 = \ln(t+1)$

$e^1 = e^{\ln(t+1)}$

$e = t+1$, or

$t = e - 1$

When

$-1 < t < e-1$, $p' > 0$ so p is increasing

$t > e-1$, $p' < 0$ so p is decreasing

(b) The price is decreasing most rapidly when the first derivative is maximized.

$p'' = \dfrac{(t+1)^2\left(\frac{-1}{t+1} \cdot 1\right) - [1 - \ln(t+1)][2(t+1)(1)]}{(t+1)^4}$

$= \dfrac{(t+1)[-1 - 2(1 - \ln(t+1))]}{(t+1)^4}$

$= \dfrac{-3 + 2\ln(t+1)}{(t+1)^3}$

So, $p'' = 0$ when

$-3 + 2\ln(t+1) = 0$

$2\ln(t+1) = 3$

$\ln(t+1) = \dfrac{3}{2}$

$e^{\ln(t+1)} = e^{3/2}$

$t + 1 = e^{3/2}$, or

$t = e^{3/2} - 1$

$p''' = \dfrac{(t+1)^3\left(\frac{2}{t+1}\right) - [-3 + 2\ln(t+1)][3(t+1)^2]}{(t+1)^6}$

$= \dfrac{(t+1)^2[2 - 3(-3 + 2\ln(t+1))]}{(t+1)^6}$

$= \dfrac{11 - 6\ln(t+1)}{(t+1)^6}$

So, when $t = e^{3/2} - 1$,

$p''' = \dfrac{11 - 6\ln(e^{3/2} - 1 + 1)}{(e^{3/2} - 1 + 1)^6} = \dfrac{11 - 6e^{3/2}}{e^9}$

Since $p''' < 0$, $t = e^{3/2} - 1$ is a maximum.

(c) $\displaystyle\lim_{t \to \infty} \dfrac{\ln(t+1)}{t+1} + 5$

$= \displaystyle\lim_{t \to \infty} \dfrac{\ln(t+1)}{t+1} + \lim_{t \to \infty} 5$

$= \displaystyle\lim_{t \to \infty} \dfrac{\frac{1}{t+1} \cdot 1}{1} + 5$

$= \displaystyle\lim_{t \to \infty} \dfrac{1}{t+1} + 5$

$= 0 + 5$

$= 5$

So, in the long run, the price approaches $500.

9. $D = q(p) = 1,000(p+2)e^{-p}$

(a) $q'(p)$

$= 1,000[(p+2)(-e^{-p}) + (e^{-p})(1)]$

$= -1,000e^{-p}[(p+2) - 1]$

$= -1,000e^{-p}(p+1)$

So, $q'(p) = 0$ when $p = -1$.

For the practical domain $p \geq 0$,
$q'(p) < 0$ so q decreases.

(b) $R = pq = 1,000p(p+2)e^{-p}$

Rewriting R as $1,000(p^2 + 2p)e^{-p}$,

$R'(p) = 1,000[(p^2 + 2p)(-e^{-p})$
$\qquad\qquad + (e^{-p})(2p+2)]$
$\quad = -1,000e^{-p}[(p^2 + 2p)$
$\qquad\qquad\qquad - (2p+2)]$
$\quad = -1,000e^{-p}(p^2 - 2)$

So $R'(p) = 0$ when $p = \sqrt{2}$.
When
$0 \leq x < \sqrt{2}$, $R'(q) > 0$, so R is
increasing
$x > \sqrt{2}$, $R'(q) < 0$ so R is decreasing
$R''(p)$
$= -1,000[(e^{-p})(2p) + (p^2 - 2)(-e^{-p})]$
$= 1,000e^{-p}[p^2 - 2p - 2]$

So $R''(\sqrt{2}) < 0$ and the maximum

revenue occurs when the price is
approximately \$141.42. The
maximum revenue is

$R(\sqrt{2}) = 1,000(\sqrt{2})(\sqrt{2}+2)e^{-\sqrt{2}}$
$\qquad \approx 1,173.8714$ hundred or
$\qquad\qquad$ \$117,387.14

10. $R(t) = R_0 e^{-kt}$

Since the half-life is 5,730 years,

$\frac{1}{2}R_0 = R_0 e^{-k(5,730)}$

$\ln\frac{1}{2} = \ln e^{-5,730k}$

$\ln\frac{1}{2} = -5,730k$, so

$k = \dfrac{\ln\left(\frac{1}{2}\right)}{-5,730} = \dfrac{-\ln\left(\frac{1}{2}\right)}{5,730} = \dfrac{\ln 2}{5,730}$

So, $R(t) = R_0 e^{-(\ln 2/5,730)t}$

When 45% remains,

$0.45R_0 = R_0 e^{-(\ln 2/5,730)t}$

$\ln 0.45 = \ln e^{-(\ln 2/5,730)t}$

$\ln 0.45 = -\left(\ln\dfrac{2}{5,730}\right)t$

$t = \dfrac{-5,730\ln 0.45}{\ln 2}$
$\quad \approx 6,601$ years old

11. $N(T) = 10,000(8+t)e^{-0.1t}$

(a) When $t = 0$,
$N(0) = 10,000(8)e^0 = 80,000$ bacteria

(b) $N'(t) = 10,000[(8+t)(-0.1e^{-0.1t})$
$\qquad\qquad + (e^{-0.1t})(1)]$
$\quad = 10,000e^{-0.1t}[-0.1(8+t)+1]$
$\quad = 10,000e^{-0.1t}(0.2 - 0.1t)$
So, $N'(t) = 0$ when
$0.2 - 0.1t = 0$
$0.2 = 0.1t$, or
$t = 2$
$N''(t) = 10,000[(e^{-0.1t})(-0.1)$
$\qquad\qquad + (0.2-0.1t)(-0.1e^{-0.1t})]$
$\quad = 10,000e^{-0.1t}$
$\qquad [-0.1 - 0.1(0.2-0.1t)]$
$\quad = 10,000e^{-0.1t}(-0.12 + 0.01t)$
When $t = 2$, $N''(2) < 0$, so the
maximum occurs when $t = 2$ and is
$N(2) = 10,000(8+2)e^{-0.1(2)}$
$\quad \approx 81,873$ bacteria

(c) $\lim\limits_{t\to\infty} 10,000(8+t)e^{-0.1t}$
$= 10,000\lim\limits_{t\to\infty}\dfrac{8+t}{e^{0.1t}}$
$= 10,000\lim\limits_{t\to\infty}\dfrac{1}{0.1e^{0.1t}}$
$= 10,000(0)$
$= 0$
So, the bacterial colony dies off in the
long run.

Review Problems

1. $f(x) = 5^x$

When $x = 0$, $f(0) = 1$, so $(0, 1)$ is an intercept. $f(x) = 0$ has no solution.

$\lim\limits_{t \to -\infty} 5^x = \lim\limits_{t \to -\infty} \dfrac{1}{5^{-x}} = 0$ so $y = 0$ is a

horizontal asymptote.

$\lim\limits_{t \to +\infty} 5^x = +\infty$ so $f(x)$ increases without

bound as x increases.

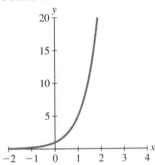

3. $f(x) = \ln x^2$

Note that the domain of f is $x \neq 0$, so $x = 0$ is a vertical asymptote.

When $f(x) = 0$, $x = \pm 1$ so $(-1, 0)$ and $(1, 0)$ are intercepts.

$\lim\limits_{x \to -\infty} \ln x^2 = \lim\limits_{x \to \infty} x^2 = +\infty$, so f increases

without bound as x decreases and as x increases.

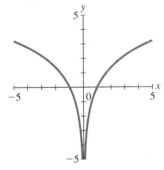

5. (a) $f(x) = Ae^{-kx}$

Since $f(0) = 10$, $10 = Ae^0$, or $A = 10$

and $f(x) = 10e^{-kx}$.

Since $f(1) = 25$,

$25 = 10e^{-k(1)}$

$\dfrac{5}{2} = e^{-k}$

$\ln \dfrac{5}{2} = \ln e^{-k}$

$k = -\ln \dfrac{5}{2}$

So, $f(t) = 10e^{-\left(-\ln \frac{5}{2}\right)t} = 10e^{\left(\ln \frac{5}{2}\right)t}$

$f(4) = 10e^{-4\ln(5/2)}$

$\qquad = 10e^{\ln(5/20)^4}$

$\qquad = 10\left(\dfrac{5}{2}\right)^4$

$\qquad = \dfrac{3125}{8}$

(b) $f(x) = Ae^{kx}$

Since $f(1) = 3$, $3 = Ae^{k(1)}$, or

$A = 3e^{-k}$.

Since $f(2) = 10$, $10 = Ae^{k(2)}$, or

$A = 10e^{-2k}$.

So, $3e^{-k} = 10e^{-2k}$

$\dfrac{3}{10} = e^{-k}$

and $A = 3\left(\dfrac{3}{10}\right) = \dfrac{9}{10}$.

$f(3) = \dfrac{9}{10}e^{k(3)}$

$\qquad = \dfrac{9}{10}(e^{-k})^{-3}$

$\qquad = \dfrac{9}{10}\left(\dfrac{3}{10}\right)^{-3}$

$\qquad = \dfrac{9}{10}\left(\dfrac{1000}{27}\right)$

$\qquad = \dfrac{100}{3}$

(c) $f(x) = 30 + Ae^{-kx}$

Since $f(0) = 50$, $50 = 30 + Ae^0$, or

$A = 20$ and $f(x) = 30 + 20e^{-kx}$.

Since $f(3) = 40$, $40 = 30 + 20e^{-k(3)}$

$$10 = 20e^{-3k}$$

$$\frac{1}{2} = e^{-3k}$$

$$f(9) = 30 + 20e^{-k(9)}$$

$$= 30 + 20(e^{-3k})^3$$

$$= 30 + 20\left(\frac{1}{2}\right)^3$$

$$= 30 + \frac{5}{2}$$

$$= \frac{65}{2}$$

(d) $f(t) = \dfrac{6}{1 + Ae^{-kt}}$

Since $f(0) = 3$, $3 = \dfrac{6}{1 + Ae^0}$, or $A = 1$.

Now, $f(t) = \dfrac{6}{1 + e^{-kt}}$. Since $f(5) = 2$,

$2 = \dfrac{6}{1 + e^{-k(5)}}$, $1 + e^{-5k} = 3$, $e^{-5k} = 2$.

So, $f(10) = \dfrac{6}{1 + e^{-k(10)}}$

$$= \dfrac{6}{1 + (e^{-5k})^2}$$

$$= \dfrac{6}{1 + (2)^2}$$

$$= \dfrac{6}{5}$$

7. $8 = 2e^{0.04x}$

$$e^{0.04x} = 4$$

$$0.04x = \ln 4$$

$$x = 25\ln 4$$

9. $4\ln x = 8$, $\ln x = 2$, or $x = e^2 \approx 7.389$.

11. $\log_9(4x - 1) = 2$ if and only if

$$4x - 1 = 9^2$$

$$4x = 82, \text{ or } x = \frac{41}{2}$$

13. $e^{2x} + e^x - 2 = 0$

Letting $u = e^x$, $\quad u^2 + u - 2 = 0$

$$(u + 2)(u - 1) = 0 \text{ or,}$$

$$u = -2, 1.$$

If $u = -2$, $e^x = -2$ and there is no solution. If $u = 1$, $e^x = 1$, so $x = 0$.

15. $y = x^2 e^{-x}$

$$\frac{dy}{dx} = (x^2)(-e^{-x}) + (e^{-x})(2x)$$

$$= xe^{-x}(-x + 2)$$

17. $y = x\ln x^2 = 2x\ln x$

$$\frac{dy}{dx} = (2x)\left(\frac{1}{x}\right) + (\ln x)(2) = 2(1 + \ln x)$$

19. $y = \log_3(x^2) = \dfrac{\ln(x^2)}{\ln 3} = \dfrac{2}{\ln 3}\ln x$

$$\frac{dy}{dx} = \frac{2}{\ln 3} \cdot \frac{1}{x} = \frac{2}{x\ln 3}$$

21. $y = \dfrac{e^{-x} + e^x}{1 + e^{-2x}}$

$$\frac{dy}{dx} = \frac{(1 + e^{-2x})(-e^{-x} + e^x) - (e^{-x} + e^x)(-2e^{-2x})}{(1 + e^{-2x})^2}$$

$$= \frac{\begin{array}{c}-e^{-x} - e^{-3x} + e^x + e^{-x}\\ + 2e^{-3x} + 2e^{-x}\end{array}}{(1 + e^{-2x})^2}$$

$$= \frac{e^{-3x} + 2e^{-x} + e^x}{(1 + e^{-2x})^2}$$

$$= \frac{(e^{-2x} + 1)(e^{-x} + e^x)}{(1 + e^{-2x})^2}$$

$$= \frac{e^{-x} + e^x}{1 + e^{-2x}}$$

$$= \frac{e^{-x} + e^x}{1 + e^{-2x}} \cdot \frac{e^{-x}}{e^{-x}}$$

$$= e^x$$

23. $y = \ln(e^{-2x} + e^{-x})$

$$\frac{dy}{dx} = \frac{1}{e^{-2x} + e^{-x}}(-2e^{-2x} - e^{-x})$$

$$= \frac{-e^{-x}(2e^{-x} + 1)}{e^{-x}(e^{-x} + 1)}$$

$$= -\frac{2e^{-x} + 1}{e^{-x} + 1}$$

25. $y = \dfrac{e^{-x}}{x + \ln x}$

$$\frac{dy}{dx} = \frac{(x + \ln x)(-e^{-x}) - (e^{-x})\left(1 + \frac{1}{x}\right)}{(x + \ln x)^2}$$

$$= \frac{-xe^{-x} - e^{-x}\ln x - e^{-x} - \frac{e^{-x}}{x}}{(x + \ln x)^2} \cdot \frac{x}{x}$$

$$= \frac{-x^2 e^{-x} - xe^{-x}\ln x - xe^{-x} - e^{-x}}{x(x + \ln x)^2}$$

$$= \frac{-e^{-x}(x^2 + x\ln x + x + 1)}{x(x + \ln x)^2}$$

27. $ye^{x - x^2} = x + y$

$$(y)[(e^{x - x^2})(1 - 2x)] + (e^{x - x^2})\frac{dy}{dx} = 1 + \frac{dy}{dx}$$

$$(e^{x - x^2})\frac{dy}{dx} - \frac{dy}{dx} = 1 - y(e^{x - x^2})(1 - 2x)$$

$$(e^{x - x^2} - 1)\frac{dy}{dx} = 1 - y(e^{x - x^2})(1 - 2x)$$

$$\frac{dy}{dx} = \frac{1 - y(e^{x - x^2})(1 - 2x)}{e^{x - x^2} - 1}$$

$$= \frac{1 + y(e^{x - x^2})(2x - 1)}{e^{x - x^2} - 1}$$

29. $y = \dfrac{(x^2 + e^{2x})^3 e^{-2x}}{(1 + x - x^2)^{2/3}}$

Using logarithmic differentiation,

$\ln y$

$$= \ln\left[\frac{(x^2 + e^{2x})^3 e^{-2x}}{(1 + x - x^2)^{2/3}}\right]$$

$$= \ln(x^2 + e^{2x})^3 + \ln e^{-2x} - \ln(1 + x - x^2)^{2/3}$$

$$= 3\ln(x^2 + e^{2x}) - 2x - \frac{2}{3}\ln(1 + x - x^2)$$

$$\frac{y'}{y} = 3 \cdot \frac{2x + 2e^{2x}}{x^2 + e^{2x}} - 2 - \frac{2}{3} \cdot \frac{1 - 2x}{1 + x - x^2}$$

$$y' = \left[\frac{6(x + e^{2x})}{x^2 + e^{2x}} - 2 - \frac{2(1 - 2x)}{3(1 + x - x^2)}\right]$$

$$\left[\frac{(x^2 + e^{2x})^3 e^{-2x}}{(1 + x - x^2)^{2/3}}\right]$$

$$= \left[\frac{3(x + e^{2x})}{x^2 + e^{2x}} - 1 - \frac{(1 - 2x)}{3(1 + x - x^2)}\right]$$

$$\left[\frac{2(x^2 + e^{2x})^3 e^{-2x}}{(1 + x - x^2)^{2/3}}\right]$$

31. $f(x) = e^x - e^{-x}$

When $x = 0$, $f(0) = 0$ so $(0, 0)$ is an intercept.
When $f(x) = 0$, $x = 0$.

$\displaystyle\lim_{x \to -\infty} e^x - e^{-x} = -\infty$ so f decreases without bound as x decreases.

$\displaystyle\lim_{x \to +\infty} e^x - e^{-x} = +\infty$ so f increases without bound as x increases.

$f'(x) = e^x + e^{-x}$

$f'(x)$ is never zero; further, $f'(x) > 0$ for all values of x, so f is always increasing.

$f''(x) = e^x - e^{-x}$

So, $f''(x) = 0$ when $x = 0$.

When
$x < 0$, $f''(x) < 0$ so f is concave down
$x > 0$, $f''(x) > 0$ so f is concave up

The point (0, 0) is an inflection point.

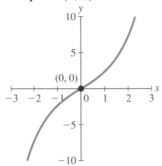

33. $f(t) = t + e^{-t}$

When $t = 0$, $f(0) = 1$ so (0, 1) is an intercept.
$f(t) = 0$ has no solution.

$$\lim_{t \to -\infty} \left(t + e^{-t}\right) = +\infty \text{ (since } e^{-t} \text{ increases}$$

more rapidly than t decreases).

$$\lim_{t \to +\infty} \left(t + e^{-t}\right) = t, \text{ so } y = t \text{ is an oblique}$$

asymptote.

$$f'(t) = 1 - e^{-t}$$

So $f'(t) = 0$ when $1 - e^{-t} = 0$

$$1 = e^{-t}$$
$$\ln 1 = -t, \text{ or } t = 0.$$

When $t < 0$, $f'(t) < 0$ so f is decreasing
　　$t > 0$, $f'(t) > 0$ so f is increasing

The point (0, 1) is a relative minimum.

$$f''(t) = e^{-t}$$

So, $f''(t)$ is never zero; further $f''(t) > 0$
for all values of t, so f is always concave
up.

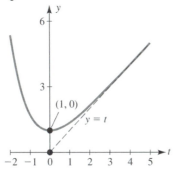

35. $F(u) = u^2 + 2\ln(u + 2)$

Note that the domain is $u > -2$, so $u = -2$ is
a vertical asymptote.

When $u = 0$, $F(0) = 2\ln 2$ so $(0, 2\ln 2)$ is an
intercept.
$F(u) = 0$ is too difficult to solve.

$$\lim_{x \to +\infty} u^2 + 2\ln(u + 2) = +\infty \text{ so } F \text{ increases}$$

without bound as u increases.

$$F'(u) = 2u + 2 \cdot \frac{1}{u + 2} \cdot 1$$
$$= 2\left(u + \frac{1}{u + 2}\right)$$
$$= 2\frac{u^2 + 2u + 1}{u + 2}$$
$$= 2\frac{(u + 1)^2}{u + 2}$$

So, $F'(u) = 0$ when $u = -1$.

When $-2 < u < -1$, $F'(u) > 0$ so F increases
　　$u > -1$, $F'(u) > 0$ so F increases

$F''(u)$
$$= 2\left[\frac{(u + 2)2(u + 1) - (u + 1)^2(1)}{(u + 2)^2}\right]$$
$$= 2(u + 1)\left[\frac{2(u + 2) - (u + 1)}{(u + 2)^2}\right]$$
$$= 2(u + 1)\left[\frac{u + 3}{(u + 2)^2}\right]$$

So, $F''(u) = 0$ when $u = -1$ (rejecting
$u = -3$ since it is not in the domain of F).
When
$-2 < u < -1$, $F''(u) < 0$ so F is concave
down
　　$u > -1$, $F''(u) > 0$ so F is concave up

The point (−1, 1) is an inflection point.

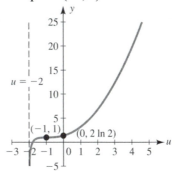

37. $G(x) = \ln(e^{-2x} + e^{-x})$

When $x = 0$, $G(0) = \ln 2$ so $(0, \ln 2)$ is an intercept. When $G(x) = 0$,

$\ln(e^{-2x} + e^{-x}) = 0$; $e^{-2x} + e^{-x} = 1$;

$e^{-2x} + e^{-x} - 1 = 0$

Letting $u = e^{-x}$, $u^2 + u - 1 = 0$

$$u = \frac{-1 \pm \sqrt{1 + (4)(1)(1)}}{2(1)} = \frac{-1 \pm \sqrt{5}}{2}$$

So, $e^{-x} = \frac{-1 \pm \sqrt{5}}{2}$

$$\ln e^{-x} = \ln\left(\frac{-1 + \sqrt{5}}{2}\right) \text{ (rejecting}$$

negative value)

$$-x = \ln\left(\frac{-1 + \sqrt{5}}{2}\right), \text{ or}$$

$$x = -\ln\left(\frac{-1 + \sqrt{5}}{2}\right) \approx 0.48$$

So, $(0.48, 0)$ is an intercept.

$\lim\limits_{x \to -\infty} \ln(e^{-2x} + e^{-x}) = +\infty$ so G increases without bound as x decreases.

$\lim\limits_{x \to +\infty} \ln(e^{-2x} + e^{-x}) = \lim\limits_{x \to 0^+} \ln x = -\infty$ so G decreases without bound as x increases.

$$G'(x) = \frac{1}{e^{-2x} + e^{-x}}(-2e^{-2x} - e^{-x})$$

So, $G'(x) = 0$ when

$-2e^{-2x} - e^{-x} = 0$

$-e^{-x}(2e^{-x} + 1) = 0$

$2e^{-x} + 1 = 0$ (since e^{-x} is never zero)

$e^{-x} = -\dfrac{1}{2}$ has no solution.

$G'(x)$ is never zero; further, $G'(x) < 0$ for all x so G is always decreasing.

$G''(x)$

$$= \left[\frac{(e^{-2x} + e^{-x})(4e^{-2x} + e^{-x})}{(e^{-2x} + e^{-x})^2}\right.$$

$$\left. - \frac{(-2e^{-2x} - e^{-x})(-2e^{-2x} - e^{-x})}{(e^{-2x} + e^{-x})^2}\right]$$

$$= \frac{\begin{array}{c}4e^{-4x} + 5e^{-3x} + e^{-2x}\\ -(4e^{-4x} + 4e^{-3x} + e^{-2x})\end{array}}{(e^{-2x} + e^{-x})^2}$$

$$= \frac{e^{-3x}}{(e^{-2x} + e^{-x})^2}$$

Since e^{-3x} is never zero, $G''(x)$ is never zero; further $G''(x) > 0$ for all x so G is always concave up.

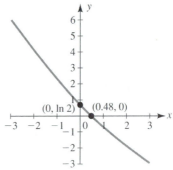

39. $f(x) = \ln(4x - x^2)$, $1 \le x \le 3$

$$f'(x) = \frac{4 - 2x}{4x - x^2}$$

So, $f'(x) = 0$ when $4 - 2x = 0$, or $x = 2$.

$f(2) = \ln 4$; $f(1) = \ln 3$; $f(3) = \ln 3$

The function's largest value is $\ln 4$ and its smallest value is $\ln 3$.

41. $h(t) = (e^{-t} + e^t)^5$, $-1 \le t \le 1$

$h'(t) = 5(e^{-t} + e^t)^4(-e^{-t} + e^t)$

So, $h'(t) = 0$ when

$-e^{-t} + e^t = 0$ (since $e^{-t} + e^t$ is never

$e^{-t}(-1 + e^{2t}) = 0$ zero)

$e^{2t} = 1$, or $t = 0$

$$h(0) = 32; \; h(-1) = \left(e + \frac{1}{e}\right)^5 \approx 280,$$

$$h(1) = \left(e + \frac{1}{e}\right)^5$$

So, the function's largest value is $\left(e + \frac{1}{e}\right)^5$
and its smallest value is 32.

43. $y = x \ln x^2$, $x = 1$

When $x = 1$, $y = \ln 1 = 0$ so point $(1, 0)$ is on the tangent line.

$$y' = (x)\left(\frac{2x}{x^2}\right) + (\ln x^2) = 2 + \ln x^2$$

slope $= y' = 2 + \ln(1)^2 = 2$

So, the equation of the tangent line is
$y - 0 = 2(x - 1)$, or $y = 2x - 2$.

45. $y = x^3 e^{2-x}$, $x = 2$

When $x = 2$, $y = 8$ so point $(2, 8)$ is on the tangent line.

$$y' = (x^3)(-e^{2-x}) + (e^{2-x})(3x^2)$$

slope $= y' = (2)^3(-e^0) + (e^0)(3 \cdot 4) = 4$

So, the equation of the tangent line is
$y - 8 = 4(x - 2)$, or $y = 4x$.

47. $f(x) = e^{kx}$

Since $f(3) = 2$, $2 = e^{3k}$. Now, $f(9) = e^{9k}$.
Using the facts that $(e^{3k})^3 = e^{9k}$,

$$f(9) = (e^{3k})^3 = (2)^3 = 8.$$

49. Since the money doubles in 15 years,

$$B(15) = P\left(1 + \frac{r}{4}\right)^{4 \cdot 15} = P\left(1 + \frac{r}{4}\right)^{60} = 2P.$$

Now, $B(30) = P\left(1 + \frac{r}{4}\right)^{4 \cdot 30}$

$$= \left[P\left(1 + \frac{r}{4}\right)^{60}\right]^2$$

$$= (2P)^2$$

$$= 4P$$

So the money quadruples in 30 years.

51. Since the decay is exponential and 500 grams were present initially,

$$Q(t) = 500e^{-kt}$$

Also, $Q(50) = 500e^{-50k} = 400$, so

$$e^{-50k} = \frac{4}{5}.$$

Now, $Q(200) = 500e^{-200k}$

$$= 500(e^{-50k})^4$$

$$= 500\left(\frac{4}{5}\right)^4$$

$$= 204.8 \text{ grams}$$

53. Since the growth is exponential,

$P(t) = P_0 e^{kt}$ where the initial number of bacteria is $P_0 = 5,000$. Also,

$P(10) = 5,000e^{10k} = 8,000$, so $e^{10k} = \frac{8}{5}$.

Now, $P(30) = 5000(e^{30k})$

$$= 5000(e^{10k})^3$$

$$= 5000\left(\frac{8}{5}\right)^3$$

$$= 20,480 \text{ bacteria}$$

55. $Q(x) = 50 - 40e^{-0.1x}$

(a) When $x = 0$, $Q(0) = 10$ so $(0, 10)$ is an intercept.
$Q(x) = 0$ when

$$50 - 40e^{-0.1x} = 0$$

$$50 = 40e^{-0.1x}$$

$$\frac{5}{4} = e^{-0.1x}$$

$$\ln\frac{5}{4} = -0.1x, \text{ or}$$

$$x = \frac{\ln\frac{5}{4}}{-0.1}$$

Since the relevant domain is $x \geq 0$, this intercept will not be on the graph.

$\lim\limits_{x \to \infty} 50 - 40e^{-0.1x} = 50$, so $y = 50$ is a
horizontal asymptote.

$Q'(x) = 4e^{-0.1x}$

Now, $Q'(x)$ is never zero. Further, $Q'(x) > 0$ for all x so Q is always increasing.

$Q''(x) = -0.4e^{-0.1x}$ which is never zero. Further, $Q''(x) < 0$ for all x so Q is always concave down.

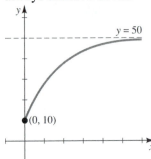

(b) When no money is spent on advertising,

$Q(0) = 50 - 40e^0 = 10$.

So, 10,000 units will be sold.

(c) If 8 thousand dollars are spent on advertising,

$Q(8) = 50 - 40e^{-0.1(8)} \approx 32.027$.

So, approximately 32,027 units will be sold.

(d) For sales of 35 thousand units,

$$35 = 50 - 40e^{-0.1x}$$

$$40e^{-0.1x} = 15$$

$$e^{-0.1x} = \frac{3}{8}$$

$$-0.1x = \ln \frac{3}{8}$$

$$x = \frac{\ln \frac{3}{8}}{-0.1} = 10\ln \frac{8}{3} \approx 9.81$$

So, approximately \$9,810 dollars must be spent on advertising.

(e) $\lim\limits_{x \to \infty} 50 - 40e^{-0.1x} = 50$

So, approximately (just less than) 50,000 units is the optimal sales projection.

57. $B(t) = P\left(1 + \dfrac{r}{k}\right)^{kt}$

(a) Compounded quarterly, with $P = 2,000$, $B(t) = 5,000$ and $r = 0.08$,

$$5,000 = 2,000\left(1 + \frac{0.08}{4}\right)^{4t}$$

$$2.5 = \left(1 + \frac{0.08}{4}\right)^{4t}$$

$$\ln 2.5 = \ln\left(1 + \frac{0.08}{4}\right)^{4t}$$

$$\ln 2.5 = 4t \cdot \ln\left(1 + \frac{0.08}{4}\right)$$

$$t = \frac{\ln 2.5}{4\ln\left(1 + \frac{0.08}{4}\right)} \approx 11.57 \text{ years}$$

(b) Using the same values but compounded continuously,

$$5,000 = 2,000e^{0.08t}$$

$$2.5 = e^{0.08t}$$

$$\ln 2.5 = \ln e^{0.08t}$$

$$\ln 2.5 = 0.08t$$

$$t = \frac{\ln 2.5}{0.08} \approx 11.45 \text{ years}$$

59. The present value of \$10,000 payable after 10 years.

(a) at 7% compounded monthly is

$$P = B\left(1 + \frac{r}{k}\right)^{-kt}$$

$$= 10,000\left(1 + \frac{0.07}{12}\right)^{-12(10)}$$

$$\approx \$4,975.96$$

(b) at 6% compounded continuously is

$$P = Be^{-rt}$$

$$= 10,000e^{-0.06(10)}$$

$$\approx \$5,488.12.$$

61. When interest is compounded quarterly, the effective rate is

$$\left(1+\frac{.0825}{4}\right)^4 - 1 \approx 0.08509, \text{ or } 8.51\%.$$

When interest is compounded continuously, the effective rate is

$$e^{.082} - 1 \approx 0.08546, \text{ or } 8.55\%.$$

So, 8.20% compounded continuously has the greater effective interest rate.

63. $P(t) = \dfrac{30}{1 + 2e^{-0.05t}}$

(a) When $t = 0$, $P(0) = 10$ so $(0, 10)$ is an intercept. $P(t) = 0$ has no solution.

$$\lim_{x \to \infty} \frac{30}{1 + 2e^{-0.05t}} = 30, \text{ so } y = 30 \text{ is a}$$

horizontal asymptote.

$$P'(t) = \frac{0 - (30)(-0.1e^{-0.05t})}{(1 + 2e^{-0.05t})^2}$$

$$= \frac{3e^{-0.05t}}{(1 + 2e^{-0.05t})^2}$$

Since $3e^{-0.05t}$ is never zero, $P'(t)$ is never zero. Further, $P'(t) > 0$ for all t, so P is always increasing. Using logarithmic differentiation,

$$\ln P'(t)$$

$$= \ln\left[\frac{3e^{-0.05t}}{(1 + 2e^{-0.05t})^2}\right]$$

$$= \ln 3 + \ln e^{-0.05t} - \ln(1 + 2e^{-0.05t})^2$$

$$= \ln 3 - 0.05t - 2\ln(1 + 2e^{-0.05t})$$

$$\frac{P''(t)}{P'(t)} = -0.05 - 2 \cdot \frac{-0.1e^{-0.05t}}{1 + 2e^{-0.05t}}$$

$$P''(t)$$

$$= \left[\frac{-0.05(1 + 2e^{-0.05t}) + 0.2e^{-0.05t}}{1 + 2e^{-0.05t}}\right]P'(t)$$

$$= \left[\frac{-0.05 + 0.1e^{-0.05t}}{1 + 2e^{-0.05t}}\right]\left[\frac{3e^{-0.05t}}{(1 + 2e^{-0.05t})^2}\right]$$

$$= (-0.05 + 0.1e^{-0.05t})\left[\frac{3e^{-0.05t}}{(1 + 2e^{-0.05t})^3}\right]$$

So $P''(t) = 0$ when

$$-0.05 + 0.1e^{-0.05t} = 0$$

$$e^{-0.05t} = 0.5$$

$$-0.05t = 0.5, \text{ or}$$

$$t = \frac{\ln 0.5}{-0.05}$$

$$= 20\ln 2 \approx 13.9$$

When $0 < t < 13.9$, $P''(t) > 0$, so P is concave up. When $t > 13.9$, $P''(t) < 0$ so P is concave down. The point $(13.9, 15.0)$ is an inflection point.

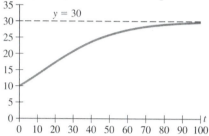

(b) The current population is

$$P(0) = \frac{30}{1 + 2e^0} = 10, \text{ or}$$

10,000,000 people.

(c) The population in 20 years will be

$$P(20) = \frac{30}{1 + 2e^{-0.05(20)}}$$

$$= \frac{30}{1 + 2e^{-1}}$$

$$= 17.2835,$$

or 17,283,500 people.

(d) $\displaystyle\lim_{x \to \infty} \frac{30}{1 + 2e^{-0.05t}} = 30$

So, the population approaches 30,000,000 in the long run.

65. (a) The rate of change of the carbon monoxide level t years from now is $Q'(t) = 0.12e^{0.03t}$. The rate two years from now is

$$Q'(2) = 0.12e^{0.03(2)} = 0.13 \text{ parts per million per year.}$$

(b) The percentage rate of change of the carbon monoxide level t years from now is $100\left[\dfrac{Q'(t)}{Q(t)}\right] = 100\left(\dfrac{0.12e^{0.03t}}{4e^{0.03t}}\right)$

$$= 3\% \text{ per year,}$$

which is constant (independent of time).

67. $V(t) = 2{,}000e^{\sqrt{2t}}$

The percentage rate of change of the value of the asset is

$100\dfrac{V'(t)}{V(t)}$

$= 100\dfrac{2000e^{\sqrt{2t}}\left[\frac{1}{2}(2t)^{-1/2}(2)\right]}{2{,}000e^{\sqrt{2t}}}$

$= 100\dfrac{1}{\sqrt{2t}}$

which will equal the prevailing interest rate when $\dfrac{1}{\sqrt{2t}} = 0.05$

$$\sqrt{2t} = \dfrac{1}{0.05} = 20$$
$$2t = 400$$
$$t = 200 \text{ years}$$

When $0 < t < 200$, the percentage rate is more than the prevailing rate. When $t > 200$, the prevailing rate is greater, so, it's best to sell the asset after 200 years.

69. $Q(t) = Q_0 e^{-kt}$

(a) When $t = \lambda$, $Q(\lambda) = \dfrac{1}{2}Q_0$, so

$\dfrac{1}{2}Q_0 = Q_0 e^{-k(\lambda)}$

$\dfrac{1}{2} = e^{-k(\lambda)}$

$\ln\dfrac{1}{2} = -k\lambda, \; k = \dfrac{\ln\frac{1}{2}}{-\lambda} = \dfrac{\ln 2}{\lambda}$

So, $Q(t) = Q_0 e^{-\left(\frac{\ln 2}{\lambda}\right)t}$.

(b) $Q_0 e^{-\left(\frac{\ln 2}{\lambda}\right)t} = Q_0(0.5)^{kt}$

$e^{-\left(\frac{\ln 2}{\lambda}\right)t} = (0.5)^{kt}$

$-\dfrac{\ln 2}{\lambda}t = kt\ln 0.5$

$k = \dfrac{-\ln 2}{\lambda \ln 0.5} = \dfrac{\ln\frac{1}{2}}{\lambda \ln\frac{1}{2}} = \dfrac{1}{\lambda}$

71. $R(t) = R_0 e^{-\left(\frac{\ln 2}{5{,}730}\right)t}$

Since the Bronze age began about 5,000 years ago, the maximum percentage is

$\dfrac{R(5{,}000)}{R_0} = \dfrac{R_0 e^{-(\ln 2/5{,}730)(5{,}000)}}{R_0}$

$\approx 0.5462, \text{ or } 54.62\%.$

73. $T(t) = 35e^{-0.32t}$

$27 = 35e^{-0.32t}$ or $t = 0.811$ min.
Rescuers have about 49 seconds before the girl loses consciousness.

$\dfrac{dT}{dt} = -35(0.32)e^{-0.32t}$

So, when $t = 0.811$,

$\dfrac{dT}{dt} = (-35)(0.32)(e^{-0.32(0.811)}) \approx -8.64$

So, the girl's temperature is dropping at a rate of $8.64°C$ per minute.

75. $C(t) = Ate^{-kt}$

(a) $C'(t) = A[(t)(-ke^{-kt}) + (e^{-kt})(1)]$

$= Ae^{-kt}(-kt + 1)$

So, $C'(t) = 0$ when $-kt + 1 = 0$, or

$t = \dfrac{1}{k}$.

When

$0 < t < \dfrac{1}{k}$, $C'(t) > 0$, so C is increasing

$t > \dfrac{1}{k}$, $C'(t) < 0$, so C is decreasing

So, the maximum occurs when $t = \dfrac{1}{k}$.

Since the maximum occurs after

2 hours, $2 = \dfrac{1}{k}$, or $k = \dfrac{1}{2}$. The maximum is 10, so $10 = A(2)e^{-\frac{1}{2}(2)}$, or $A = 5e$.

(b) To find when the concentration falls to 1 microgram/ml,
$$C(t) = 5ete^{-0.5t}$$
$$5ete^{-0.5t} = 1$$
$$5ete^{-0.5t} - 1 = 0$$
Press $\boxed{y =}$ and input $5e \wedge (1)xe \wedge (-.5x) - 1$ for $y_1 =$.
Use window dimensions of $[-5, 20]2$ by $[-10, 10]1$
Press $\boxed{\text{graph}}$.
Press $\boxed{\text{2nd}}$ $\boxed{\text{calc}}$ and use the zero function to find $t \approx 9.78$ hours.

77. $P(t) = \dfrac{40}{1 + Ce^{-kt}}$

Let $t = 0$ in the year 1960. Then,
$$P(0) = \dfrac{40}{1 + Ce^{-k(0)}}$$
$$3 = \dfrac{40}{1 + C}$$
$$1 + C = \dfrac{40}{3}$$
$$C = \dfrac{37}{3}$$
and $P(t) = \dfrac{40}{1 + \frac{37}{3}e^{-kt}}$.

In the year 1975, $t = 15$ and $P(15) = 4$ billion, so

$$4 = \dfrac{40}{1 + \frac{37}{3}e^{-k(15)}}$$
$$1 + \dfrac{37}{3}e^{-15k} = 10$$
$$e^{-15k} = \dfrac{27}{37}$$
$$\ln e^{-15k} = \ln\dfrac{27}{37}$$
$$-15k = \ln\dfrac{27}{37}$$
$$k = \dfrac{\ln\frac{27}{37}}{-15} \approx 0.0210$$

The predicted population in 2010, when $t = 50$, would be
$$P(50) = \dfrac{40}{1 + \frac{37}{3}e^{-0.0210(50)}}$$
$$\approx 7.52 \text{ billion people.}$$

79. $\text{pH} = -\log_{10}[H_3O^+]$
For milk and lime, $\text{pH}m = 3\text{pH}_1$.
For lime and orange, $\text{pH}_l = \dfrac{1}{2}\text{pH}_0$.
If $\text{pH}_0 = 3.2$, $\text{pH}_l = \dfrac{1}{2}(3.2) = 1.6$.
Then, $1.6 = -\log_{10}[H_3O^+]_l$
$$-1.6 = \log_{10}[H_3O^+]_l$$
$$10^{-1.6} = 10^{\log_{10}[H_3O^+]_l} \text{ or }$$
$$[H_3O^+]_l = 10^{-1.6} \approx 0.0251$$

81. (a) $D(t)$
$$= (D_0 - 0.00046)e^{-0.162t} + 0.00046$$
With $D_0 = 0.008$,
$$D(10) = (0.008 - 0.00046)e^{-0.162(10)} + 0.00046$$
$$= 0.00195, \text{ or } 1.95 \text{ deaths per } 1,000 \text{ women.}$$
$D(25) = 0.000590$, or 0.59 deaths per 1,000 women.

(b) When $t = 0$, $D(0) = 0.008$ so $(0, 0.008)$ is an intercept.
When $D(t) = 0$,

$0.00754e^{-0.162t} + 0.00046 = 0$

$e^{-0.162t} = -0.061008$, which has no

solution.

$D'(t) = -0.00122e^{-0.162t}$

So $D'(t)$ is never zero. Further,

$D'(t) < 0$ for all t, so D is always

decreasing.

$D''(t) = 0.000198e^{-0.162t}$

$D''(t)$ is never zero. Further,

$D''(t) > 0$ for all t, so D is always

concave up.

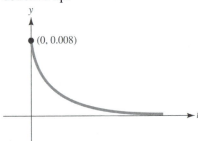

83. $R(t) = R_0 e^{-\left(\frac{\ln 2}{5,730}\right)t}$

(a) $R(3.8 \times 10^6) = R_0 e^{-\left(\frac{\ln 2}{5,730}\right)(3.8 \times 10^6)}$

$= R_0 e^{-459.7}$

Note: different calculators evaluate

$e^{-459.7}$ differently; as a result, you

may get 0 or you may get $\dfrac{1}{(2.3)^{200}}$.

In either case, $R(3.8 \times 10^6) \approx 0$.

Since $\lim\limits_{t \to +\infty} e^{-t} = 0$, we can't

distinguish ages for large values of t.

(b) Writing Exercise—Answers will vary.

85. $P(t) = \dfrac{202.31}{1 + e^{3.938 - 0.314t}}$

(a) To use this formula to compute the
population of US for the years 1790,
1800, 1830, 1860, 1880, 1900, 1920,
1940, 1960, 1980, 1990, and 2000,

Press $\boxed{y =}$ and input $P(t)$ for $y_1 =$.
Press $\boxed{2nd}$ $\boxed{tblset}$ and use Tblstart $= 0$,
ΔTbl $= 1$, auto independent and auto
dependent.
Press $\boxed{2nd}$ $\boxed{table}$.
Given below are the parts of the table
corresponding to the years above.

Year	t	Population (in millions)
1790	0	3.8671
1800	1	5.2566
1830	4	12.957
1860	7	30.207
1880	9	50.071
1900	11	77.142
1920	13	108.43
1940	15	138.37
1960	17	162.29
1980	19	178.78
1990	20	184.57
2000	21	189.03

(b) Press $\boxed{y =}$ and input $P(t)$ for $y_1 =$.
Use window dimensions $[0, 28]4$ by
$[0, 200]25$
Press $\boxed{graph}$.
The rate population is growing is given
by $P'(t) = \dfrac{63.52534e^{3.938 - 0.314t}}{(1 + e^{3.938 - 0.314t})^2}$.
Press $\boxed{y =}$ and input $P'(t)$ for $y_2 =$.
Deselect $y_1 =$ so only $P'(t)$ is active.
Use window dimensions $[0, 28]4$ by
$[0, 20]2$
Use the maximum function under the
calc menu to find that the maximum of
$P'(t)$ occurs at $x \approx 12.5$. So, the
population is growing most rapidly

when $t = 12.5$ or in 1915.

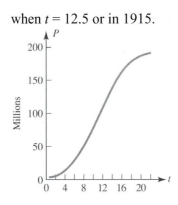

(c) Writing Exercise—Answers will vary.

87. To draw graphs of $y = \sqrt{3^x}$, $y = \sqrt{3^{-x}}$

and $y = 3^{-x}$ on the same axes,

Press $\boxed{y =}$ and input $\sqrt{} (3 \wedge x)$ for $y_1 =$.

$\sqrt{} (3 \wedge (-x))$ for $y_2 =$, and $3 \wedge (-x)$ for

$y_3 =$.

Use window dimensions $[-4, 4]1$ by
$[-2, 6]1$.

Press $\boxed{\text{graph}}$.

The graph of $y = \sqrt{3^{-x}}$ is a reflection of the

graph of $y = \sqrt{3^x}$ is across the y-axis.

The graph of $y = \sqrt{3^{-x}}$ is the graph of

$y = 3^{-x}$ vertically compressed. Similarly,

the graph of $y = \sqrt{3^x}$ is vertically

compressed in addition to being reflected
across the y-axis.

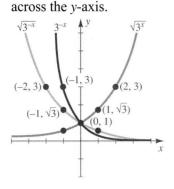

89. Using the conversion formula for
logarithms, we will change all logarithms
to natural logarithms:

$$\log_5(x + 5) - \log_2 x - \log_{10}(x^2 + 2x)^2 = 0$$

$$\frac{\ln(x + 5)}{\ln 5} - \frac{\ln(x)}{\ln 2} - \frac{\ln(x^2 + 2x)^2}{\ln 10} = 0$$

Press $\boxed{y =}$ and input $\ln(x + 5)/\ln(5) -$

$\ln(x)/\ln(2) - \ln((x^2 + 2x)^2)/\ln(10)$ for

$y_1 =$.

Press $\boxed{\text{graph}}$.

Use the zero function under the calc menu to
find that $x \approx 1.066$ is a root. There is no

other real root because x^2 increases much
more rapidly than any other argument,
making $y_1 =$ monotonically decreasing.

91. To make a table for $\left(\sqrt{n}\right)^{\sqrt{n+1}}$ and

$\left(\sqrt{n+1}\right)^{\sqrt{n}}$ with $n = 8, 9, 12, 20, 25, 31,$

$37, 38, 43, 50, 100,$ and $1,000$, press $\boxed{y =}$

and input $\sqrt{n} \wedge \sqrt{n+1}$ for $y_1 =$ and

$\sqrt{n+1} \wedge \sqrt{n}$ for $y_2 =$.

Press $\boxed{\text{2nd}} \boxed{\text{tblset}}$ and use ask independent
and auto dependent.

Press $\boxed{\text{2nd}} \boxed{\text{table}}$ and input each value of n
given.

n	$\left(\sqrt{n}\right)^{\sqrt{n+1}}$	$\left(\sqrt{n+2}\right)^{\sqrt{n}}$
8	22.63	22.36
9	32.27	31.62
12	88.21	85.00
20	957.27	904.84
25	3,665	3,447
31	16,528	15,494
37	68,159	63,786
38	85,679	80,166
43	261,578	244,579
50	1.17×10^6	1.09×10^6
1000	1.1×10^{10}	1.1×10^{10}
1000	2.9×10^{47}	2.8×10^{47}

$$\left(\sqrt{n}\right)^{\sqrt{n+1}} \geq \left(\sqrt{n+1}\right)^{\sqrt{n}}$$

This inequality holds for all $n \geq 8$. To confirm, since $(n+1)^{\sqrt{n}} \leq (n+1)^{\sqrt{n+1}}$

$$\lim_{n \to \infty} \frac{(n+1)^{\sqrt{n}}}{n^{\sqrt{n+1}}} \leq \lim_{n \to \infty} \frac{(n+1)^{\sqrt{n+1}}}{n^{\sqrt{n+1}}}$$

$$\leq \lim_{n \to \infty} \left(\frac{n+1}{n}\right)^{\sqrt{n+1}}$$

$$\leq \lim_{n \to \infty} e^{\ln\left(\frac{n+1}{n}\right)^{\sqrt{n+1}}}$$

$$\leq \lim_{n \to \infty} e^{\sqrt{n+1} \ln\left(\frac{n+1}{n}\right)}$$

$$\leq \lim_{n \to \infty} e^{[(\ln(n+1)/n)(n+1)^{-1/2}]}$$

$$\leq e^{\lim_{n \to \infty} \frac{\ln\left(\frac{n+1}{n}\right)}{(n+1)^{-1/2}}}$$

Using l'Hôpital's rule,

$$\leq e^{\lim_{n \to \infty} \frac{\left(\frac{n}{n+1}\right)\frac{n(1)-(n+1)(1)}{n^2}}{-\frac{1}{2}(n+1)^{-3/2}(1)}}$$

$$\leq e^{\lim_{n \to \infty} \frac{-\frac{1}{n}}{-\frac{1}{2}(n+1)^{-3/2}(n+1)}}$$

$$\leq e^{\lim_{n \to \infty} \frac{2(n+1)^{1/2}}{n}}$$

Using l'Hôpital's rule again,

$$\leq e^{\lim_{n \to \infty} \frac{1}{(n+1)^{1/2}}}$$

$$\leq e^0$$

$$\leq 1$$

Since the ratio of $(n+1)^{\sqrt{n}}$ to $n^{\sqrt{n+1}}$ is less than or equal to one,

$$(n+1)^{\sqrt{n}} \leq n^{\sqrt{n+1}}.$$

Chapter 5

Integration

5.1 Indefinite Integration and Differential Equations

1. $I = \int -3\,dx = -3x + C$

3. $I = \int x^5\,dx = \dfrac{x^6}{6} + C$

5. $I = \int \dfrac{1}{x^2}\,dx = \int x^{-2}\,dx = -x^{-1} + C$

$\qquad = -\dfrac{1}{x} + C$

7. $I = \int \dfrac{2}{\sqrt{t}}\,dt$

$\qquad = 2\int t^{-1/2}\,dt$

$\qquad = 2\dfrac{t^{1/2}}{\frac{1}{2}} + C$

$\qquad = 4t^{1/2} + C$

$\qquad = 4\sqrt{t} + C$

9. $I = \int u^{-2/5}\,du = \dfrac{u^{3/5}}{\frac{3}{5}} + C = \dfrac{5}{3}u^{3/5} + C$

11. $I = \int \left(3t^2 - \sqrt{5}t + 2\right)dt$

$\qquad = 3\int t^2\,dt - \sqrt{5}\int t^{1/2}\,dt + 2\int dt$

$\qquad = 3\left(\dfrac{t^3}{3}\right) - \sqrt{5}\left(\dfrac{t^{3/2}}{\frac{3}{2}}\right) + 2t + C$

$\qquad = t^3 - \dfrac{2\sqrt{5}}{3}t^{3/2} + 2t + C$

13. $I = \int \left(3\sqrt{y} - 2y^{-3}\right)dy$

$\qquad = 3\int y^{1/2}\,dy - 2\int y^{-3}\,dy$

$\qquad = 3\dfrac{y^{3/2}}{\frac{3}{2}} - 2\dfrac{y^{-2}}{-2} + C$

$\qquad = 2y^{3/2} + y^{-2} + C$

$\qquad = 2y^{3/2} + \dfrac{1}{y^2} + C$

15. $I = \int \left(\dfrac{e^x}{2} + x\sqrt{x}\right)dx$

$\qquad = \dfrac{1}{2}\int e^x\,dx + \int x^{3/2}\,dx$

$\qquad = \dfrac{1}{2}e^x + \dfrac{x^{5/2}}{\frac{5}{2}} + C$

$\qquad = \dfrac{e^x}{2} + \dfrac{2}{5}x^{5/2} + C$

17. $I = \int u^{1.1}\left(\dfrac{1}{3u} - 1\right)du$

$\qquad = \int \left(\dfrac{u^{1.1}}{3u} - u^{1.1}\right)du$

$\qquad = \int \left(\dfrac{u^{0.1}}{3} - u^{1.1}\right)du$

$\qquad = \dfrac{1}{3}\int u^{0.1}\,du - \int u^{1.1}\,du$

$\qquad = \dfrac{1}{3}\cdot\dfrac{u^{1.1}}{1.1} - \dfrac{u^{2.1}}{2.1} + C$

$\qquad = \dfrac{u^{1.1}}{3.3} - \dfrac{u^{2.1}}{2.1} + C$

19. $I = \int \dfrac{x^2 + 2x + 1}{x^2}\, dx$

$= \int \left(1 + \dfrac{2}{x} + \dfrac{1}{x^2}\right) dx$

$= \int dx + 2\int \dfrac{1}{x}\, dx + \int x^{-2}\, dx$

$= x + 2\ln|x| + \dfrac{x^{-1}}{-1} + C$

$= x + 2\ln|x| - \dfrac{1}{x} + C$

$= x + \ln x^2 - \dfrac{1}{x} + C$

21. $I = \int (x^3 - 2x^2)\left(\dfrac{1}{x} - 5\right) dx$

$= \int (x^2 - 2x - 5x^3 + 10x^2)\, dx$

$= \int (-5x^3 + 11x^2 - 2x)\, dx$

$= -5\int x^3\, dx + 11\int x^2\, dx - 2\int x\, dx$

$= -\dfrac{5x^4}{4} + \dfrac{11x^3}{3} - \dfrac{2x^2}{2} + C$

$= -\dfrac{5}{4}x^4 + \dfrac{11}{3}x^3 - x^2 + C$

23. $I = \int \sqrt{t}\,(t^2 - 1)\, dt$

$= \int (t^{5/2} - t^{1/2})\, dt$

$= \int t^{5/2}\, dt - \int t^{1/2}\, dt$

$= \dfrac{2t^{7/2}}{7} - \dfrac{2t^{3/2}}{3} + C$

$= \dfrac{2}{7}t^{7/2} - \dfrac{2}{3}t^{3/2} + C$

25. $I = \int (e^t + 1)^2\, dt$

$= \int (e^{2t} + 2e^t + 1)\, dt$

$= \int e^{2t}\, dt + 2\int e^t\, dt + \int dt$

$= \dfrac{1}{2}e^{2t} + 2e^t + t + C$

27. $I = \int \left(\dfrac{1}{3y} - \dfrac{5}{\sqrt{y}} + e^{-y/2}\right) dy$

$= \dfrac{1}{3}\int \dfrac{1}{y}\, dy - 5\int y^{-1/2}\, dy + \int e^{-\frac{1}{2}y}\, dy$

$= \dfrac{1}{3}\ln|y| - 5\dfrac{y^{1/2}}{\frac{1}{2}} + \dfrac{1}{-\frac{1}{2}}e^{-\frac{1}{2}y} + C$

$= \dfrac{1}{3}\ln|y| - 10\sqrt{y} - 2e^{-y/2} + C$

29. $I = \int t^{-1/2}(t^2 - t + 2)\, dt$

$= \int (t^{3/2} - t^{1/2} + 2t^{-1/2})\, dt$

$= \int t^{3/2}\, dt - \int t^{1/2}\, dt + 2\int t^{-1/2}\, dt$

$= \dfrac{t^{5/2}}{\frac{5}{2}} - \dfrac{t^{3/2}}{\frac{3}{2}} + 2\dfrac{t^{1/2}}{\frac{1}{2}} + C$

$= \dfrac{2}{5}t^{5/2} - \dfrac{2}{3}t^{3/2} + 4t^{1/2} + C$

31. $\dfrac{dy}{dx} = 3x - 2$

$\int \dfrac{dy}{dx}\, dx = \int (3x - 2)\, dx$

$\int \dfrac{dy}{dx}\, dx = 3\int x\, dx - 2\int dx$

$y = 3\dfrac{x^2}{2} - 2x + C$

$y = \dfrac{3}{2}x^2 - 2x + C$

Since $y = 2$ when $x = -1$,

$2 = \dfrac{3}{2}(-1)^2 - 2(-1) + C$

$2 = \dfrac{3}{2} + 2 + C$, or

$C = -\dfrac{3}{2}$

So, $y = \dfrac{3}{2}x^2 - 2x - \dfrac{3}{2}$.

33.
$$\frac{dy}{dx} = \frac{2}{x} - \frac{1}{x^2}$$

$$\int \frac{dy}{dx}\,dx = \int\left(\frac{2}{x} - \frac{1}{x^2}\right)dx$$

$$\int \frac{dy}{dx}\,dx = 2\int\frac{1}{x}\,dx - \int x^{-2}\,dx$$

$$y = 2\ln|x| - \frac{x^{-1}}{-1} + C$$

$$= \ln x^2 + \frac{1}{x} + C$$

Since $y = -1$ when $x = 1$,

$$-1 = \ln 1 + \frac{1}{1} + C$$

$$-1 = 0 + 1 + C, \text{ or}$$
$$C = -2$$

So, $y = \ln x^2 + \dfrac{1}{x} - 2.$

35.
$$f'(x) = 4x + 1$$
$$\int f'(x)\,dx = \int(4x+1)\,dx$$
$$\int f'(x)\,dx = 4\int x\,dx + \int dx$$
$$f(x) = 4\frac{x^2}{2} + x + C$$
$$= 2x^2 + x + C$$

Since the function goes through the point (1, 2),

$$2 = 2(1)^2 + 1 + C, \text{ or}$$
$$C = -1$$

So, $f(x) = 2x^2 + x - 1.$

37. $f'(x) = -x(x+1) = -x^2 - x;\ (-1, 5)$

$$\int f'(x)\,dx = \int(-x^2 - x)\,dx$$
$$= \int -x^2\,dx - \int x\,dx$$
$$= -\int x^2\,dx - \int x\,dx$$
$$f(x) = -\frac{x^3}{3} - \frac{x^2}{2} + C$$

Since the function goes through the point (−1, 5),

$$5 = -\frac{(-1)^3}{3} - \frac{(-1)^2}{2} + C$$

$$5 = \frac{1}{3} - \frac{1}{2} + C, \text{ or}$$

$$C = \frac{31}{6}$$

So, $f(x) = -\dfrac{x^3}{3} - \dfrac{x^2}{2} + \dfrac{31}{6}.$

39.
$$f'(x) = x^3 - \frac{2}{x^2} + 2$$

$$\int f'(x)\,dx = \int\left(x^3 - \frac{2}{x^2} + 2\right)dx$$

$$\int f'(x)\,dx = \int x^3\,dx - 2\int x^{-2}\,dx + 2\int dx$$

$$f(x) = \frac{x^4}{4} - 2\frac{x^{-1}}{-1} + 2x + C$$

$$= \frac{1}{4}x^4 + \frac{2}{x} + 2x + C$$

Since the function goes through the point (1, 3),

$$3 = \frac{1}{4}(1)^4 + \frac{2}{1} + 2(1) + C, \text{ or}$$

$$C = -\frac{5}{4}$$

So, $f(x) = \dfrac{1}{4}x^4 + \dfrac{2}{x} + 2x - \dfrac{5}{4}.$

41.
$$f'(x) = e^{-x} + x^2$$
$$\int f'(x)\,dx = \int(e^{-x} + x^2)\,dx$$
$$\int f'(x)\,dx = \int e^{-x}\,dx + \int x^2\,dx$$
$$f(x) = \frac{1}{-1}e^{-x} + \frac{x^3}{3} + C$$
$$= -e^{-x} + \frac{1}{3}x^3 + C$$

Since the function goes through the point (0, 4),

$$4 = -e^0 + \frac{1}{3}(0) + C, \text{ or}$$

$$C = 5$$

So, $f(x) = -e^{-x} + \dfrac{1}{3}x^3 + 5.$

43. $\dfrac{dy}{dx} = -2y$; $y = 3$ when $x = 0$.

Cross multiplying gives $dy = -2y\,dx$.

Multiplying both sides by $-\dfrac{1}{2y}$ gives

$-\dfrac{1}{2y}\,dy = dx$.

Integrating both sides,

$$\int -\dfrac{1}{2y}\,dy = \int dx$$

$$-\dfrac{1}{2}\int \dfrac{1}{y}\,dy = \int dx$$

$$-\dfrac{1}{2}\ln y + C_1 = x + C_2$$

$$-\dfrac{1}{2}\ln y = x + C_2 - C_1$$

$$-\dfrac{1}{2}\ln y = x + C_3$$

$$\ln y = -2x - 2C_3$$

$$\ln y = -2x - C_4$$

Solving for y,

$$e^{\ln y} = e^{-2x - C_4}$$

$$y = e^{-2x} \cdot e^{-C_4}$$

$$y = Ce^{-2x}$$

Since $y = 3$ when $x = 0$,

$3 = Ce^0$, or $C = 3$.

So, $y = 3e^{-2x}$.

45. $\dfrac{dy}{dx} = e^{x+y}$; $y = 0$ when $x = 0$

Cross multiplying,

$dy = e^{x+y}\,dx$

Since $e^{x+y} = e^x \cdot e^y$

$dy = e^x \cdot e^y\,dx$

Multiplying both sides by e^{-y} gives

$e^{-y}\,dy = e^x\,dx$

Integrating both sides,

$-e^{-y} + C_1 = e^x + C_2$

$-e^{-y} = e^x + C_3$

$e^{-y} = -e^x - C_3$

$e^{-y} = -e^x - C$

Since $y = 0$ when $x = 0$,

$e^0 = -e^0 + C$, or $C = 2$.

So, $e^{-y} = 2 - e^x$.

47. $C(q) = \int C'(q)\,dq$

$$= \int (3q^2 - 24q + 48)\,dq$$

$$= 3\int q^2\,dq - 24\int q\,dq + 48\int dq$$

$$= 3\dfrac{q^3}{3} - 24\dfrac{q^2}{2} + 48q + C$$

$$= q^3 - 12q^2 + 48q + C$$

Since the cost is $5,000 for producing 10 units,

$5000 = (10)^3 - 12(10)^2 + 48(10) + C$, or

$\quad C = 4720$

So, $C(q) = q^3 - 12q^2 + 48q + 4720$.

When 30 units are produced, the cost is

$C(30) = (30)^3 - 12(30)^2 + 48(30) + 4720$

$\qquad = \$22,360$.

49. $\qquad R'(q) = 100q^{-1/2}$

$$\int R'(q)\,dq = \int 100q^{-1/2}\,dq = 100\int q^{-1/2}\,dq$$

$$R(q) = 100 \cdot 2q^{1/2} + R(0)$$

$$= 200q^{1/2} + R(0)$$

$$C'(q) = 0.4q\,dq$$

$$\int C'(q)\,dq = \int 0.4q\,dq = 0.4\int q\,dq$$

$$C(q) = 0.4 \cdot \dfrac{q^2}{2} + C(0) = 0.2q^2 + C(0)$$

Now, profit = revenues − costs, so

$P(q) = R(q) - C(q)$

$\qquad = 200q^{1/2} + R(0) - 0.2q^2 - C(0)$

When $q = 16$, $P(16) = 520$, so

$$520 = 200\sqrt{16} + R(0) - 0.2(16)^2$$
$$- C(0)$$
$$520 = 800 - 51.2 + R(0) - C(0)$$
$$-228.8 = R(0) - C(0)$$

This makes the profit function

$$P(q) = 200\sqrt{q} - 0.2q^2 - 228.8.$$

When $q = 25$,

$$P(25) = 200\sqrt{25} - 0.2(25)^2 - 228.8$$
$$= \$646.20.$$

51.
$$N(t) = \int N'(t)\,dt$$
$$= \int (154t^{2/3} + 37)\,dt$$
$$= 154\int t^{2/3}\,dt + 37\int dt$$
$$= 154\frac{t^{5/3}}{\frac{5}{3}} + 37t + C$$
$$= \frac{462}{5}t^{5/3} + 37t + C$$

Since there are no subscribers when $t = 0$, $C = 0$.

So, $N(t) = \dfrac{462}{5}t^{5/3} + 37t.$

Eight months from now, the number of subscribers will be

$$N(8) = \frac{462}{5}(8)^{5/3} + 37(8)$$
$$\approx 3,253 \text{ subscribers.}$$

53. $R'(q) = 100 - 2q$

   (a) Since $P'(q) = R'(q)$,
$$P(q) = \int R'(q)\,dq$$
$$= \int (100 - 2q)\,dq$$
$$= 100\int dq - 2\int q\,dq$$
$$= 100q - 2\frac{q^2}{2} + C$$
$$= 100q - q^2 + C$$

Since the profit is \$700 when 10 units are produced,

$$700 = 100(10) - (10)^2 + C, \text{ or}$$
$$C = -200$$

So, $P(q) = 100q - q^2 - 200$.

   (b) Since $R'(q) = P'(q)$, to maximize P,
$R'(q) = 0$ when $100 - 2q = 0$, or
$q = 50$.
Further, $R''(q) = -2$, so $R''(50) < 0$
and the maximum profit occurs when
$q = 50$. The maximum profit is

$$P(50) = 100(50) - (50)^2 - 200$$
$$= \$2,300.$$

55.
$$c(x) = \int c'(x)\,dx$$
$$= \int \left(0.9 + 0.3\sqrt{x}\right)dx$$
$$= 0.9\int dx + 0.3\int x^{1/2}\,dx$$
$$= 0.9x + 0.3\frac{x^{3/2}}{\frac{3}{2}} + C$$
$$= 0.9x + 0.2x^{3/2} + C$$

Since the consumption is 10 billion when $x = 0$, $10 = 0.9(0) + 0.2(0) + C$, or
$$C = 10$$

So, $c(x) = 0.9x + 0.2x^{3/2} + 10$.

57. Rate revenue changes
= (# barrels)(rate selling price changes)
$$\frac{dR}{dt} = 400(98 + 0.04t), \text{ where } t \text{ is in months.}$$

$$\text{Revenue} = \int_0^{24} 400(98 + 0.04t)\,dt$$
$$= 400(98t + 0.02t^2)\Big|_0^{24}$$
$$= 400[(98(24) + 0.2(24)^2) - 0]$$
$$\approx \$986,880$$

59. Let $P(t)$ be the population of the town
t months from now. Since $\dfrac{dP}{dt} = 4 + 5t^{2/3}$,

then, $P(t) = \int \dfrac{dP}{dt}\,dt$

$\quad = \int (4 + 5t^{2/3})\,dt$

$\quad = 4\int dt + 5\int t^{2/3}\,dt$

$\quad = 4t + 5\dfrac{t^{5/3}}{\frac{5}{3}} + C$

$\quad = 4t + 3t^{5/3} + C$

Since the population is 10,000 when $t = 0$,

$10,000 = 4(0) + 3(0) + C$, or

$\quad C = 10,000$

So, $P(t) = 4t + 3t^{5/3} + 10,000$.

When $t = 8$, $P(8) = 4(8) + 3(8)^{5/3} + 10,000$

$\qquad = 10,128$ people.

61. $\dfrac{dP}{dt} = 200e^{0.1t} + 150e^{-0.03t}$

$P(t) = \int \dfrac{dP}{dt}\,dt$

$\quad = \int \left(200e^{0.1t} + 150e^{-0.03t} \right) dt$

$\quad = \int 200e^{0.1t}\,dt + \int 150e^{-0.03t}\,dt$

$\quad = 200\int e^{0.1t}\,dt + 150\int e^{-0.03t}\,dt$

$\quad = 200 \cdot \dfrac{1}{01}e^{0.1t} + C_1 + 150$

$\qquad \cdot \dfrac{1}{-0.03}e^{-0.03t} + C_2$

$\quad = 2000e^{0.1t} - 5000e^{-0.03t} + C_3$

$\quad = 2000e^{0.1t} - 5000e^{-0.03t} + C$

When $t = 0$, $P(0) = 200,000$ so,

$200,000 = 2000e^0 - 5000e^0 + C$

or, $C = 203,000$.

When $t = 12$,

$P(12) = 2000e^{0.1(12)} - 5000e^{-0.03(12)}$

$\qquad + 203,000$

$\quad \approx 6640 - 3488 + 203,000$

The population will be approximately 206,152 bacteria.

63. $f'(x) = 0.1(10 + 12x - 0.6x^2)$

(a) To maximize the rate of learning,

$f''(x) = 0.1(12 - 1.2x)$.

So, $f''(x) = 0$ when $12 - 1.2x = 0$, or $x = 10$.

Further, $f'''(x) = 0.1(-1.2) = -0.12$ so $f'''(10) < 0$ and the absolute maximum occurs when $x = 10$. The maximum rate is

$f'(10) = 0.1[10 + 12(10) - 0.6(10)^2]$

$\qquad = 7$ items per minute.

(b) $f(x) = \int f'(x)\,dx$

$\quad = \int [0.1(10 + 12x - 0.6x^2)]dx$

$\quad = \int (1 + 1.2x - 0.06x^2)\,dx$

$\quad = \int dx + 1.2\int x\,dx - 0.06\int x^2\,dx$

$\quad = x + 1.2\dfrac{x^2}{2} - 0.06\dfrac{x^3}{3} + C$

$\quad = x + 0.6x^2 - 0.02x^3 + C$

Since no items are memorized when $x = 0$, $C = 0$.

So, $f(x) = x + 0.6x^2 - 0.02x^3$.

(c) $f'(x) = 0.1(10 + 12x - 0.6x^2)$

$\qquad = 1 + 1.2x - 0.06x^2$

So, $f'(x) = 0$ when

$x = \dfrac{-1.2 \pm \sqrt{(1.2)^2 - 4(-0.06)(1)}}{2(-0.06)}$ or,

$x \approx 20.8$ (rejecting the negative solution)

$f''(20.8) < 0$, so the absolute maximum is

$f(20.8)$

$\quad = (20.8) + 0.6(20.8)^2 - 0.02(20.8)^3$

$\quad \approx 100$ items

65. $M'(t) = 0.4t - 0.005t^2$

65. (a) $M(t) = \int M'(t)\,dt$

$$= \int (0.4t - 0.005t^2)\,dt$$

$$= 0.4\int t\,dt - 0.005\int t^2\,dt$$

$$= 0.4\frac{t^2}{2} - 0.005\frac{t^3}{3} + C$$

$$= 0.2t^2 - \frac{0.005}{3}t^3 + C$$

Since $M(t) = 0$ when $t = 0$, $C = 0$.

So, $M(t) = 0.2t^2 - \frac{0.005}{3}t^3$.

In ten minutes, Rob can memorize

$$M(10) = 0.2(10)^2 - \frac{0.005}{3}(10)^3$$

$$= 18\frac{1}{3} \text{ items.}$$

(b) $M(20) - M(10)$

$$= \left[0.2(20)^2 - \frac{0.005}{3}(20)^3\right] - 18\frac{1}{3}$$

$$= 66\frac{2}{3} - 18\frac{1}{3}$$

$$= 48\frac{1}{3} \text{ items}$$

67. $V'(t) = 0.15 - 0.09e^{0.006t}$

(a) $V(t) = \int V'(t)\,dt$

$$= \int (0.15 - 0.09e^{0.006t})\,dt$$

$$= 0.15t - 0.09 \cdot \frac{1}{0.006}e^{0.006t} + C$$

$$= 0.15t - 15e^{0.006t} + C$$

When $t = 0$, $V(0) = 30$ so,

$$30 = 0.15(0) - 15e^0 + C, \text{ or } C = 45$$

So, $V(t) = 0.15t - 15e^{0.006t} + 45$.

(b) $V(60) = 0.15(60) - 15e^{0.006(60)} + 45$

$$\approx 32.5 \text{ cm}^3$$

$V(120)$

$$= 0.15(120) - 15e^{0.006(120)} + 45$$

$$\approx 32.18 \text{ cm}^3$$

(c) $V(90) = 0.15(90) - 15e^{0006(90)} + 45$

$$\approx 32.8 \text{ cm}^3$$

So, unfortunately, the procedure does not succeed.

69. $v'(r) = -ar$

$$v(r) = \int v'(r)\,dr$$

$$= \int -ar\,dr$$

$$= -a\int r\,dr$$

$$= -a\frac{r^2}{2} + C$$

$$= -\frac{a}{2}r^2 + C$$

Since $v(R) = 0$, $0 = -\frac{a}{2}(R)^2 + C$, or

$$C = \frac{aR^2}{2}$$

So, $v(r) = -\frac{a}{2}r^2 + \frac{aR^2}{2} = \frac{a}{2}(R^2 - r^2)$.

71. $T'(t) = 7e^{-0.35t}$

(a) $T(t) = \int T'(t)\,dt$

$$= \int 7e^{-0.35t}\,dt$$

$$= 7\int e^{-0.35t}\,dt$$

$$= 7 \cdot \frac{1}{-0.35}e^{-0.35t} + C$$

$$= -20e^{-0.35t} + C$$

Since the temperature was $-4°C$ when $t = 0$,

$$-4 = -20e^0 + C, \text{ or}$$

$$C = 16$$

So, $T(t) = -20e^{-0.35t} + 16$.

(b) After two hours,
$$T(2) = -20e^{-0.35(2)} + 16 \approx 6.07°C.$$

(c) For the temperature to reach 10°C,
$$10 = -20e^{-0.35t} + 16$$
$$6 = 20e^{-0.35t}$$
$$\frac{3}{10} = e^{-0.35t}$$
$$\ln\frac{3}{10} = \ln e^{-0.35t}$$
$$\ln\frac{3}{10} = -0.35t, \text{ or}$$
$$t = \frac{\ln\frac{3}{10}}{-0.35} = \frac{-20}{7}\ln\frac{3}{10}$$
$$= \frac{20}{7}\ln\frac{10}{3} \approx 3.44 \text{ hours.}$$

73. In the 0.7 seconds it takes for our spy to react, the car travels $(88)(0.7) = 61.6$ feet. Once he reacts, the speed of the car will be zero when
$$88 + \int -28\,dt = 0$$
$$88 - 28t = 0$$
$$t = \frac{22}{7} \text{ seconds}$$

During this time, the car travels an additional
$$\int_0^{22/7}(88 - 28t)dt = 88t - 14t^2\Big|_0^{22/7}$$
$$= 88\left(\frac{22}{7}\right) - 14\left(\frac{22}{7}\right)^2$$
$$\approx 138.29 \text{ feet}$$

So, the car travels $61.6 + 138.29 = 199.89$ feet. If the camel remains in the road during the entire $\frac{22}{7} + 0.7 = 3.84$ seconds, the camel will be hit.

75. $a(t) = -23$

(a) Since acceleration is the derivative of velocity, $v(t) = \int -23\,dt = -23t + C$.

The velocity when the brakes are applied is 67 ft/sec, so

$67 = -23(0) + C$, or $C = 67$ and
$v(t) = -23t + 67$.
Since velocity is the derivative of distance,
$$s(t) = \int v(t)\,dt$$
$$= \int (-23t + 67)dt$$
$$= -23\int t\,dt + 67\int dt$$
$$= -23\frac{t^2}{2} + 67t + C$$
$$= -\frac{23}{2}t^2 + 67t + C$$

Since the distance is to be measured from the point the brakes are applied,
$$s(0) = 0 \text{ and } 0 = -\frac{23}{2}(0) + 67(0) + C,$$
or $C = 0$. So, $s(t) = -\frac{23}{2}t^2 + 67t.$

(b) To use the graphing utility to sketch graphs of $v(t)$ and $s(t)$ on same screen, Press $\boxed{y=}$ and input $v(t)$ for $y_1 =$ and input $s(t)$ for $y_2 =$.
Use window dimensions [0, 5]1 by [0, 200]10.
Press $\boxed{\text{graph}}$.

(c) The car comes to a complete stop when $v(t) = 0$.
Press $\boxed{\text{trace}}$ and verify that the cross-hairs on the line $y_1 = -23t + 67$.
Move along line until it appears to be at the t-intercept.
Use the zoom-in function under the zoom menu to find that the velocity $= 0$ when $t \approx 2.9$ seconds.

To find how far the car travels in 2.9 seconds, go back to the original graphing screen. Use the value function under the calc menu and input 2.9 for x and press enter. Use the $\uparrow$ arrow to verify that $y_2 = -\dfrac{23}{2}t^2 + 67t$ is displayed. The car travels 97.6 feet in 2.9 seconds.

To find how fast the car travels when $s = 45$ feet, trace along the parabola $s(t)$ and use the zoom-in function to find that it takes approximately 0.77 seconds and 5.05 seconds to travel 45 feet. Next, go back to the original graphing screen and use the value function under the calc menu. Input $x = 0.77$ and verify $y_1 = -23t + 67$ is displayed. The car is traveling 49.2 feet/sec when it has traveled 45 feet.

Repeat this process with $x = 5.05$ to find the velocity at 5.05 is 49.15 (decelerating).

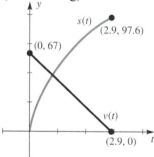

77.
$$\int b^x\,dx = \int e^{x\ln b}\,dx$$
$$= \int e^{(\ln b)x}\,dx$$
$$= \frac{1}{\ln b}e^{x\ln b} + C$$
$$= \frac{1}{\ln b}b^x + C$$

5.2 Integration by Substitution

1. **(a)** $u = 3x + 4$

   (b) $u = 3 - x$

   (c) $u = 2 - t^2$

   (d) $u = 2 + t^2$

3. Let $u = 2x + 6$. Then $du = 2\,dx$ or $dx = \dfrac{du}{2}$. So
$$\int (2x+6)^5\,dx = \frac{1}{2}\int u^5\,du = \frac{(2x+6)^6}{12} + C.$$

5. Let $u = 4x - 1$. Then $du = 4\,dx$ or $dx = \dfrac{du}{4}$.

   So $\int \sqrt{4x-1}\,dx = \dfrac{1}{4}\int u^{1/2}\,du$
$$= \frac{1}{4}\frac{2u^{3/2}}{3} + C$$
$$= \frac{(4x-1)^{3/2}}{6} + C.$$

7. Let $u = 1 - x$. Then $du = -dx$ or $dx = -du$. So $\int e^{1-x}\,dx = -\int e^u\,du = -e^{1-x} + C.$

9. Let $u = x^2$. Then $\dfrac{du}{dx} = 2x$ or $\dfrac{1}{2}du = x\,dx$.
$$\int xe^{x^2}\,dx = \int e^{x^2}\cdot x\,dx$$
$$= \int e^u \cdot \frac{1}{2}\,du$$
$$= \frac{1}{2}\int e^u\,du$$
$$= \frac{1}{2}e^{x^2} + C.$$

11. Let $u = t^2 + 1$. Then $\dfrac{du}{dt} = 2t$ or

$$\frac{1}{2} du = t\, dt.$$

$$\int t(t^2 + 1)^5 dt = \int (t^2 + 1)^5 t\, dt$$

$$= \int u^5 \cdot \frac{1}{2} du$$

$$= \frac{1}{2} \int u^5 du$$

$$= \frac{(t^2 + 1)^6}{12} + C.$$

13. Let $u = x^3 + 1$. Then $\dfrac{du}{dx} = 3x^2$ or

$$\frac{1}{3} du = x^2 dx.$$

$$\int x^2 (x^3 + 1)^{3/4} dx = \int (x^3 + 1)^{3/4} x^2 dx$$

$$= \int u^{3/4} \cdot \frac{1}{3} du$$

$$= \frac{1}{3} \int u^{3/4} du$$

$$= \frac{4(x^3 + 1)^{7/4}}{21} + C$$

15. Let $u = y^5 + 1$. Then $\dfrac{du}{dy} = 5y^4$, or

$$\frac{1}{5} du = y^4 dy.$$

$$\int \frac{2y^4}{y^5 + 1} dy = 2 \int \frac{1}{y^5 + 1} y^4 dy$$

$$= 2 \int \frac{1}{u} \cdot \frac{1}{5} du$$

$$= \frac{2}{5} \int \frac{1}{u} du$$

$$= \frac{2}{5} \ln \left| y^5 + 1 \right| + C$$

17. Let $u = x^2 + 2x + 5$. Then

$$\frac{du}{dx} = 2x + 2 = 2(x + 1), \text{ or}$$

$$\frac{1}{2} du = (x + 1) dx.$$

$$\int (x + 1)(x^2 + 2x + 5)^{12} dx$$

$$= \int (x^2 + 2x + 5)^{12} (x + 1) dx$$

$$= \int u^{12} \cdot \frac{1}{2} du$$

$$= \frac{1}{2} \int u^{12} du$$

$$= \frac{(x^2 + 2x + 5)^{13}}{26} + C$$

19. Let $u = x^5 + 5x^4 + 10x + 12$. Then

$$\frac{du}{dx} = 5x^4 + 20x^3 + 10$$

$$= 5(x^4 + 4x^3 + 2),$$

or

$$\frac{1}{5} du = (x^4 + 4x^3 + 2) dx$$

$$\int \frac{3x^4 + 12x^3 + 6}{x^5 + 5x^4 + 10x + 12} dx$$

$$= \int \frac{3(x^4 + 4x^3 + 2)}{x^5 + 5x^4 + 10x + 12} dx$$

$$= 3 \int \frac{1}{x^5 + 5x^4 + 10x + 12}(x^4 + 4x^3 + 2) dx$$

$$= 3 \int \frac{1}{u} \cdot \frac{1}{5} du$$

$$= \frac{3}{5} \int \frac{1}{u} du$$

$$= \frac{3}{5} \ln \left| x^5 + 5x^4 + 10x + 12 \right| + C$$

21. Let $t = u^2 - 2u + 6$. Then

$$\frac{dt}{du} = 2u - 2 = 2(u - 1), \text{ or } \frac{1}{2} dt = (u - 1) du.$$

$$\int \frac{3u - 3}{(u^2 - 2u + 6)^2} du$$

$$= \int \frac{3(u - 1)}{(u^2 - 2u + 6)^2} du$$

$$= 3 \int \frac{1}{(u^2 - 2u + 6)^2} (u - 1) du$$

$$= 3 \int \frac{1}{t^2} \cdot \frac{1}{2} dt$$

$$= \frac{3}{2} \int t^{-2} dt$$

$$= \frac{-3}{2(u^2 - 2u + 6)} + C$$

23. Let $u = \ln 5x$. Then $\dfrac{du}{dx} = \dfrac{1}{5x} \cdot 5 = \dfrac{1}{x}$, or

$du = \dfrac{1}{x} dx$.

$$\int \frac{\ln 5x}{x} dx = \int \ln 5x \cdot \frac{1}{x} dx$$

$$= \int u \, du$$

$$= \frac{(\ln 5x)^2}{2} + C$$

25. Let $u = \ln x$. Then $\dfrac{du}{dx} = \dfrac{1}{x}$, or $du = \dfrac{1}{x} dx$.

$$\int \frac{1}{x(\ln x)^2} dx = \int \frac{1}{(\ln x)^2} \cdot \frac{1}{x} dx$$

$$= \int \frac{1}{u^2} du$$

$$= -\frac{1}{\ln x} + C$$

27. Let $u = x^2 + 1$. Then $\dfrac{du}{dx} = 2x$, or

$\dfrac{1}{2} du = x \, dx$.

$$\int \frac{2x \ln(x^2 + 1)}{x^2 + 1} dx = 2 \int \frac{\ln(x^2 + 1)}{x^2 + 1} \cdot x \, dx$$

$$= 2 \int \frac{\ln u}{u} \cdot \frac{1}{2} du$$

$$= \int \frac{\ln u}{u} du$$

Substitution must be used a second time.

Let $t = \ln u$. Then $\dfrac{dt}{du} = \dfrac{1}{u}$, or $dt = \dfrac{1}{u} du$.

$$\int \frac{\ln u}{u} du = \int \ln u \cdot \frac{1}{u} du$$

$$= \int t \, dt$$

$$= \frac{t^2}{2} + C$$

$$= \frac{(\ln u)^2}{2} + C$$

$$= \frac{[\ln(x^2 + 1)]^2}{2} + C$$

29. Let $u = e^x - e^{-x}$. Then $\dfrac{du}{dx} = e^x + e^{-x}$, or

$du = (e^x + e^{-x}) dx$.

$$\int \frac{e^x + e^{-x}}{e^x - e^{-x}} dx = \int \frac{1}{e^x - e^{-x}} (e^x + e^{-x}) dx$$

$$= \int \frac{1}{u} du$$

$$= \ln \left| e^x - e^{-x} \right| + C$$

31. Let $u = 2x + 1$. Then $\dfrac{du}{dx} = 2$, or

$\dfrac{1}{2} du = dx$. Further, $x = \dfrac{u - 1}{2}$.

$$\int \frac{x}{2x + 1} dx = \frac{1}{4} \int \frac{u - 1}{u} du$$

$$= \frac{1}{4} \int \left(1 - \frac{1}{u} \right) du$$

$$= \frac{1}{4} \int du - \frac{1}{4} \int \frac{1}{u} du$$

$$= \frac{1}{4} u - \frac{1}{4} \ln |u| + C$$

$$= \frac{1}{4} (2x + 1) - \frac{1}{4} \ln |2x + 1| + C$$

This can also be written as

$$= \frac{1}{2}x + \frac{1}{4} - \frac{1}{4}\ln|2x+1| + C$$

$$= \frac{1}{2}x - \frac{1}{4}\ln|2x+1| + C,$$

where the $\frac{1}{4}$ has been added to the constant C. (In mathematics, the same C is often used for the original constant and for the constant after it is changed.)

33. Let $u = 2x + 1$. Then $\frac{du}{dx} = 2$, or

$\frac{1}{2}du = dx$. Further, $x = \frac{u-1}{2}$.

$$\int x\sqrt{2x-1}\,dx$$

$$= \frac{1}{4}\int (u-1)u^{1/2}\,du$$

$$= \frac{1}{4}\int (u^{3/2} - u^{1/2})\,du$$

$$= \frac{1}{4}\left(\frac{2}{5}(2x+1)^{5/2} - \frac{2}{3}(2x+1)^{3/2} \right) + C$$

$$= \frac{1}{10}(2x+1)^{5/2} - \frac{1}{6}(2x+1)^{3/2} + C$$

35. Let $u = \sqrt{x} + 1$. Then

$$\frac{du}{dx} = \frac{1}{2}x^{-1/2} = \frac{1}{2x^{1/2}}, \text{ or } 2\,du = \frac{1}{\sqrt{x}}\,dx.$$

$$\int \frac{1}{\sqrt{x}(\sqrt{x}+1)}\,dx = \int \frac{1}{\sqrt{x}+1} \cdot \frac{1}{\sqrt{x}}\,dx$$

$$= 2\int \frac{1}{u}\,du$$

$$= 2\ln\left|\sqrt{x}+1\right| + C$$

$$= 2\ln\left(\sqrt{x}+1\right) + C$$

37. $y = \int \frac{dy}{dx}\,dx = \int (3-2x)^2\,dx$

Let $u = 3 - 2x$. Then, $\frac{du}{dx} = -2$, or

$$-\frac{1}{2}du = dx.$$

$$y = \int (3-2x)^2\,dx = \frac{1}{2}\int u^2\,du$$

$$y = -\frac{1}{2}\cdot\frac{u^3}{3} + C = -\frac{1}{6}(3-2x)^3 + C$$

Since $y = 0$ when $x = 0$,

$$0 = -\frac{1}{6}(3-2(0))^3 + C$$

$$0 = -\frac{9}{2} + C, \text{ or } C = \frac{9}{2}$$

So, $y = -\frac{1}{6}(3-2x)^3 + \frac{9}{2}$.

39. $y = \int \frac{dy}{dx}\,dx = \int \frac{1}{x+1}\,dx$

Let $u = x + 1$. Then $\frac{du}{dx} = 1$, or $du = dx$.

$$\int \frac{1}{x+1}\,dx = \int \frac{1}{u}\,du = \ln|x+1| + C$$

Since $y = 1$ when $x = 0$,

$1 = \ln|0+1| + C$, or $C = 1$.

So, $y = \ln|x+1| + 1$.

41. $y = \int \frac{dy}{dx}\,dx = \int \frac{x+2}{x^2+4x+5}\,dx$

Let $u = x^2 + 4x + 5$. Then

$$\frac{du}{dx} = 2x + 4 = 2(x+2), \text{ or } \frac{1}{2}du = (x+2)dx.$$

$$\int \frac{x+2}{x^2+4x+5}\,dx = \int \frac{1}{x^2+4x+5}(x+2)dx$$

$$= \frac{1}{2}\int \frac{1}{u}\,du$$

$$= \frac{1}{2}\ln\left|x^2+4x+5\right| + C$$

Since $y = 3$ when $x = -1$,

$$3 = \frac{1}{2}\ln\left|(-1)^2 + 4(-1) + 5\right| + C \text{ or,}$$

$$C = 3 - \frac{1}{2}\ln 2.$$

So, $y = \frac{1}{2}\ln\left|x^2+4x+5\right| + 3 - \frac{1}{2}\ln 2$.

43. $f(x) = \int f'(x)dx = \int (1-2x)^{3/2} dx$

Let $u = 1 - 2x$. Then $\dfrac{du}{dx} = -2$, or

$-\dfrac{1}{2} du = dx$.

$\int (1-2x)^{3/2} dx = -\dfrac{1}{2} \int u^{3/2} du$

$= -\dfrac{1}{2} \left[\dfrac{2}{5} (1-2x)^{5/2} \right] + C$

$= -\dfrac{1}{5} (1-2x)^{5/2} + C$

Since the function goes through the point

$(0, 0)$, $0 = -\dfrac{1}{5}[1 - 2(0)]^{5/2} + C$, or

$C = \dfrac{1}{5}$.

So, $f(x) = -\dfrac{1}{5}(1-2x)^{5/2} + \dfrac{1}{5}$.

45. $f(x) = \int f'(x)dx = \int xe^{-4-x^2} dx$

Let $u = 4 - x^2$. Then $\dfrac{du}{dx} = -2x\,dx$, or

$-\dfrac{1}{2} du = x\,dx$.

$\int xe^{4-x^2} dx = \int e^{4-x^2} \cdot x\,dx$

$= -\dfrac{1}{2} \int e^u du$

$= -\dfrac{1}{2} e^{4-x^2} + C$

Since $y = 1$ when $x = -2$,

$1 = -\dfrac{1}{2} e^{4-(-2)^2} + C$, or

$C = \dfrac{3}{2}$.

So, $f(x) = -\dfrac{1}{2} e^{4-x^2} + \dfrac{3}{2}$.

47. $\dfrac{dy}{dx} = \dfrac{2-y}{(x+1)^2}$

Cross-multiplying gives

$(x+1)^2\, dy = (2-y)\, dx$

Multiplying both sides by

$\dfrac{1}{(x+1)^2 (2-y)}$ gives

$\dfrac{1}{2-y}\, dy = \dfrac{1}{(x+1)^2}\, dx$.

Let $u_1 = 2 - y$ and $u_2 = x + 1$. Then,

$\dfrac{du_1}{dy} = -1$ and $\dfrac{du_2}{dx} = 1$, or $-du_1 = dy$

and $du_2 = dx$.

Substituting, $-\dfrac{1}{u_1}\, du_1 = \dfrac{1}{(u_2)^2}\, du_2$.

Integrating both sides,

$-\int \dfrac{1}{u_1}\, du_1 = \int \dfrac{1}{(u_2)^2}\, du_2$

$-\int \dfrac{1}{u_1}\, du_1 = \int (u_2)^{-2}\, du_2$

$-\ln|u_1| + C_1 = \dfrac{(u_2)^{-1}}{-1} + C_2$

$-\ln|2-y| = -\dfrac{1}{x+1} + C_2 - C_1$

$-\ln|2-y| = -\dfrac{1}{x+1} + C_3$

$\ln|2-y| = \dfrac{1}{x+1} - C_3$

Solving for y,

$$e^{\ln|2-y|} = e^{\frac{1}{x+1}-C_3}$$

$$2-y = e^{\frac{1}{x+1}} \cdot e^{-C_3}$$

$$2-y = e^{\frac{1}{x+1}} \cdot C_4$$

$$y = 2 - C_4 e^{\frac{1}{x+1}}$$

So, $y = 2 + Ce^{\frac{1}{x+1}}$.

49. $\dfrac{dy}{dx} = \dfrac{2-y^2}{xy}$

Cross-multiplying gives

$xy\,dy = \left(2-y^2\right)dx$.

Multiplying both sides by $\dfrac{1}{x\left(2-y^2\right)}$

gives $\dfrac{y}{2-y^2}\,dy = \dfrac{1}{x}\,dx$.

Let $u_1 = 2-y^2$. Then, $\dfrac{du_1}{dy} = -2y$, or

$-\dfrac{1}{2}\,du_1 = y\,dy$. Substituting,

$-\dfrac{1}{2u_1}\,du_1 = \dfrac{1}{x}\,dx$

Integrating both sides,

$$\int -\dfrac{1}{2u_1}\,du_1 = \int \dfrac{1}{x}\,dx$$

$$-\dfrac{1}{2}\int \dfrac{1}{u_1}\,du_1 = \int \dfrac{1}{x}\,dx$$

$$-\dfrac{1}{2}\ln|u_1| + C_1 = \ln|x| + C_2$$

$$-\dfrac{1}{2}\ln\left|2-y^2\right| = \ln|x| + C_2 - C_1$$

$$-\dfrac{1}{2}\ln\left|2-y^2\right| = \ln|x| + C_3$$

$$\ln\left|2-y^2\right| = -2\ln|x| - 2C_3$$

$$\ln\left|2-y^2\right| = \ln x^{-2} + C_4$$

Solving for y^2,

$$e^{\ln\left|2-y^2\right|} = e^{\ln x^{-2} + C_4}$$

$$2-y^2 = e^{\ln x^{-2}} \cdot e^{C_4}$$

$$2-y^2 = C_5 \cdot x^{-2}$$

$$y^2 = 2 - C_5 x^{-2}$$

So, $y^2 = 2 + Cx^{-2}$.

51. (a) $C(q) = \int C'(q)\,dq = \int 3(q-4)^2\,dq$

Let $u = q-4$. Then $\dfrac{du}{dq} = 1$, or

$du = dq$.

$= 3\int u^2\,du = (q-4)^3 + C$

Let C_0 represent the overhead. Then

$C_0 = C(0) = (0-4)^3 + C$, or

$C = C_0 + 64$.

So, $C(q) = (q-4)^3 + 64 + C_0$.

(b) When $C_0 = 436$,

$C(q) = (q-4)^3 + 500$ and

$C(14) = (14-4)^3 + 500 = \$1,500$.

53. (a) $R(x) = \int R'(x)\,dx$

$$= \int (50 + 3.5xe^{-0.01x^2})\,dx$$

$$= 50\int dx + 3.5\int xe^{-0.01x^2}\,dx$$

Let $u = -0.01x^2$. Then $\dfrac{du}{dx} = -0.02x$,

or $-50\,du = x\,dx$.

$$= 50\int dx + 3.5\int e^{-0.01x^2}x\,dx$$

$$= 50\int dx - 175\int e^u\,du$$

$$= 50x - 175e^{-0.01x^2} + C$$

Since $R(0) = 0$,

$0 = 50(0) - 175e^0 + C$, or

$C = 175$

So, $R(x) = 50x - 175e^{-0.01x^2} + 175$.

(b) $R(1000)$

$$= 50(1000) - 175e^{-0.01(1000)} + 175$$
$$\approx \$50,175$$

55. $R'(x) = x(5-x)^3 \, ; C'(x) = 5 + 2x$

Let P be the profit function. Then,

$$P'(x) = R'(x) - C'(x)$$

$$P(x) = \int \left[R'(x) - C'(x) \right] dx$$

$$= \int \left[x(5-x)^3 - (5+2x) \right] dx$$

$$= \int x(5-x)^3 \, dx - \int (5+2x) \, dx$$

Let $u = 5 - x$. Then $x = 5 - u$, and $\dfrac{du}{dx} = -1$,

or $-du = dx$. Substituting,

$$\int -\left[(5-u)u^3 \right] du - \int (5+2x) \, dx$$

$$= \int \left(u^4 - 5u^3 \right) du - \int (5+2x) \, dx$$

$$= \frac{u^5}{5} - \frac{5u^4}{4} + C_1 - \left(5x + x^2 + C_2 \right)$$

$$= \frac{1}{5}(5-x)^5 - \frac{5}{4}(5-x)^4 - 5x - x^2 + C$$

$P(1)$

$$= \frac{1}{5}(5-1)^5 - \frac{5}{4}(5-1)^4 - 5(1) - (1)^2 + C$$

$$= 204.8 - 320 - 6 + C = -121.2 + C$$

$P(5)$

$$= \frac{1}{5}(5-5)^5 - \frac{5}{4}(5-5)^4 - 5(5) - (5)^2 + C$$

$$= -50 + C$$

$$P(5) - P(1) = (-50 + C) - (-121.2 + C)$$

$$= 71.2$$

Profit will increase by \$7120.

57. (a) $p(x) = \int p'(x) \, dx$

$$= \int \frac{-300x}{(x^2+9)^{3/2}} \, dx$$

Let $u = x^2 + 9$. Then $\dfrac{du}{dx} = 2x$, or

$$\frac{1}{2} du = x \, dx.$$

$$= -300 \int \frac{1}{(x^2+9)^{3/2}} x \, dx$$

$$= -150 \int u^{-3/2} \, du$$

$$= \frac{300}{\sqrt{x^2+9}} + C$$

When the price is \$75, 4 hundred pair are demanded, so

$$75 = \frac{300}{\sqrt{(4)^2+9}} + C \text{ or,}$$

$$C = 15$$

So, $p(x) = \dfrac{300}{\sqrt{x^2+9}} + 15.$

(b) When $x = 5$ hundred,

$$p(5) = \frac{300}{\sqrt{(5)^2+9}} + 15$$

$$= \$66.45 \text{ per pair}$$

$$p(0) = \frac{300}{\sqrt{0+9}} + 15 = \$115 \text{ per pair}$$

(c)

$$90 = \frac{300}{\sqrt{x^2+9}} + 15$$

$$\sqrt{x^2+9} = 4, \text{ or } x \approx 2.65, \text{ or } 265 \text{ pairs.}$$

59. $p'(x) = \dfrac{x}{(x+3)^2}$

(a) $p(x) = \displaystyle\int p'(x)\,dx = \int \dfrac{x}{(x+3)^2}\,dx$

Let $u = x+3$. Then $x = u-3$ and $\dfrac{du}{dx} = 1$, or $du = dx$. Substituting,

$$\int \dfrac{u-3}{u^2}\,du = \int \left(\dfrac{1}{u} - \dfrac{3}{u^2}\right)du$$

$$= \int \left(\dfrac{1}{u} - 3u^{-2}\right)du$$

$$= \ln|u| - \dfrac{3u^{-1}}{-1} + C$$

$$= \ln(x+3) + \dfrac{3}{x+3} + C$$

When 5 thousand units are supplied, the price per unit is \$2.20 so

$$2.2 = \ln(5+3) + \dfrac{3}{5+3} + C$$

$$C = 2.2 - \ln 8 - \dfrac{3}{8} \approx -0.25$$

$$p(x) = \ln(x+3) + \dfrac{3}{x+3} - 0.25$$

(b) $p(10) = \ln 13 + \dfrac{3}{13} - 0.25 \approx 2.546$

So, the unit price should be approximately \$2.55.

61. (a) Since $\dfrac{dV}{dt} = [\text{rate interest is added}]$

$- [\text{rate money is withdrawn}],$

$\dfrac{dV}{dt} = rV - W$. Cross-multiplying gives

$dV = (rV - W)\,dt$. Integrating both

sides, $\displaystyle\int \dfrac{1}{rV - W}\,dV = \int dt$

Let $u = rV - W$. Then, $\dfrac{du}{dV} = r$ or

$\dfrac{1}{r}\,du = dV$. Substituting,

$$\dfrac{1}{r}\int \dfrac{1}{u}\,du = \int dt$$

$$\dfrac{1}{r}\ln|u| + C_1 = t + C_2$$

$$\dfrac{1}{r}\ln|rV - W| = t + C_2 - C_1$$

$$\dfrac{1}{r}\ln|rV - W| = t + C_3$$

$$\ln|rV - W| = rt + rC_3$$

$$\ln|rV - W| = rt + C_4$$

Solving for V,

$$e^{\ln|rV-W|} = e^{rt + C_4}$$

$$rV - W = e^{rt} \cdot e^{C_4}$$

$$rV - W = Ce^{rt}$$

$$rV = W + Ce^{rt}$$

$$V = \dfrac{W}{r} + \dfrac{Ce^{rt}}{r}$$

Since the amount S is deposited initially,

$$S = \dfrac{W}{r} + \dfrac{Ce^0}{r}$$

$$S - \dfrac{W}{r} = \dfrac{C}{r}$$

$$C = Sr - W$$

So,

$$V = \dfrac{W}{r} + \dfrac{(Sr - W)e^{rt}}{r}$$

$$V(t) = \dfrac{W}{r} + \left(S - \dfrac{W}{r}\right)e^{rt}$$

(b) Substituting $r = 0.05$, $W = 50{,}000$, $t = 10$, and $S = 50{,}000$

$V(10)$

$$= \dfrac{50{,}000}{0.05} + \left(500{,}000 - \dfrac{50{,}000}{0.05}\right)e^{0.05(10)}$$

$$= 1{,}000{,}000 - 500{,}000e^{0.5}$$

$$\approx \$175{,}639$$

(c) Need $V(t) = 500,000$ for all t, or

$$500,000$$
$$= \frac{W}{0.05} + \left(500,000 - \frac{W}{0.05}\right)e^{0.05t}$$
$$500,000$$
$$= \frac{W}{0.05} + 500,000e^{0.05t} - \frac{W}{0.05}e^{0.05t}$$
$$500,000 - 500,000e^{0.05t}$$
$$= \frac{W}{0.05} - \frac{W}{0.05}e^{0.05t}$$
$$500,000\left(1 - e^{0.05t}\right)$$
$$= \frac{W}{0.05}\left(1 - e^{0.05t}\right)$$
$$500,000 = \frac{W}{0.05}$$
$$W = \$25,000$$

(d) Need to find t when $V(t) = 0$, if $W = 80,000$. So,
$$0 = \frac{80,000}{0.05}$$
$$+ \left(500,000 - \frac{80,000}{0.05}\right)e^{0.05t}$$
$$0 = 1,600,000 - 1,100,00e^{0.05t}$$
$$1,100,000e^{0.05t} = 1,600,000$$
$$e^{0.05t} = \frac{16}{11}$$
$$\ln e^{0.05t} = \ln\frac{16}{11}$$
$$0.05t = \ln\frac{16}{11}$$
$$t = \frac{\ln\frac{16}{11}}{0.05} \approx 7.5 \text{ years}$$

63. $S(t) = 3 + t; D(t) = 9 - t; k = 0.01$; $p_0 = 1$

(a) $\dfrac{dp}{dt} = k\left[D(t) - S(t)\right]$
$$= 0.01\left[(9 - t) - (3 + t)\right]$$
$$= 0.01(6 - 2t)$$

Since $p(t) = \int \dfrac{dp}{dt}\,dt$,
$$p(t) = \int 0.01(6 - 2t)\,dt$$
$$= 0.01\int(6 - 2t)\,dt$$
$$= 0.01\left(6t - t^2 + C_1\right)$$
$$= 0.06t - 0.01t^2 + 0.01C_1$$
$$= 0.06t - 0.01t^2 + C$$

Since $p_0 = 1$,
$$1 = 0.06(0) - 0.01(0)^2 + C, \text{ or } C = 1.$$
So, $p(t) = 0.06t - 0.01t^2 + 1$.

(b) When $t = 4$,
$$p(4) = 0.06(4) - 0.01(4)^2 + 1$$
$$= 1.08$$

(c) $\lim\limits_{t \to \infty} p(t) = \lim\limits_{t \to \infty}\left(0.06t - 0.01t^2 + 1\right)$

Since this limit does not exist, the unit price increases without bound.

65. $S(t) = 2 + 3p(t);\ D(t) = 10 - p(t);$

$k = 0.02;\ p_0 = 1$

(a) $\dfrac{dp}{dt} = k\big[D(t) - S(t)\big]$

$= 0.02\big[(10 - p) - (2 + 3p)\big]$

$= 0.02(8 - 4p)$

Cross-multiplying gives

$dp = 0.02(8 - 4p)\,dt$.

Multiplying both sides by $\dfrac{1}{8 - 4p}$ gives

$\dfrac{1}{8 - 4p}\,dp = 0.02\,dt$.

Integrating both sides,

$\displaystyle\int \dfrac{1}{8 - 4p}\,dp = \int 0.02\,dt$.

Let $u = 8 - 4p$. Then, $\dfrac{du}{dp} = -4$ or

$-\dfrac{1}{4}\,du = dp$. Substituting,

$-\dfrac{1}{4}\displaystyle\int \dfrac{1}{u}\,du = \int 0.02\,dt$

$-\dfrac{1}{4}\ln|u| + C_1 = 0.02t + C_2$

$-\dfrac{1}{4}\ln|8 - 4p| = 0.02t + C_2 - C_1$

$-\dfrac{1}{4}\ln|8 - 4p| = 0.02t + C_3$

$\ln|8 - 4p| = -0.08t - 4C_3$

$\ln|8 - 4p| = -0.08t + C_4$

Solving for p

$e^{\ln|8 - 4p|} = e^{-0.08t + C_4}$

$8 - 4p = e^{-0.08t} \cdot e^{C_4}$

$8 - 4p = C_5 e^{-0.08t}$

$4p = 8 - C_5 e^{-0.08t}$

$p = \dfrac{1}{4}\left(8 - C_5 e^{-0.08t}\right)$

$p = 2 - Ce^{-0.08t}$

Since $p_0 = 1;\ 1 = 2 - Ce^0$, or $C = 1$.

So, $p(t) = 2 - e^{-0.08t}$.

(b) When $t = 4$,

$p(4) = 2 - e^{-0.08(4)} \approx 1.27$

(c) $\displaystyle\lim_{t \to \infty} p(t) = \lim_{t \to \infty}\left(2 - e^{-0.08t}\right)$

$= 2 - 0 = 2$

So, in the long run, the unit price is approximately 2.

67. Let $G(x)$ represent the height in meters of the tree in x years.

$G(x) = \displaystyle\int G'(x)\,dx$

$= \displaystyle\int \left(1 + \dfrac{1}{(x + 1)^2}\right)dx$

$= \displaystyle\int dx + \int \dfrac{1}{(x + 1)^2}\,dx$

Let $u = x + 1$. Then $\dfrac{du}{dx} = 1$ or, $du = dx$.

$= \displaystyle\int dx + \int \dfrac{1}{u^2}\,du$

$= x - \dfrac{1}{x + 1} + C$

Since the height was 5 meters after 2 years,

$$5 = 2 - \frac{1}{2+1} + C, \text{ or}$$

$$C = \frac{10}{3}$$

So, $G(x) = x - \dfrac{1}{x+1} + \dfrac{10}{3}$ and

$$G(0) = 0 - \frac{1}{0+1} + \frac{10}{3}$$

$$= \frac{7}{3} \text{ meters tall.}$$

69. (a) $C(t) = \displaystyle\int C'(t)\,dt$

$$= \int \frac{-0.01 e^{0.01t}}{\left(e^{0.01t} + 1\right)^2}\,dt$$

Let $u = e^{0.01t} + 1$. Then $\dfrac{du}{dt} = 0.01 e^{0.01t}$,

or $100\,du = e^{0.01t}\,dt$.

$$= -0.01 \int \frac{1}{\left(e^{0.01t} + 1\right)^2}\,e^{0.01t}\,dt$$

$$= -\int \frac{1}{u^2}\,du = \frac{1}{e^{0.01t} + 1} + C$$

When the shot is initially administered, $t = 0$ and

$$0.5 = \frac{1}{e^0 + 1} + C, \text{ or } C = 0$$

So, $C(t) = \dfrac{1}{e^{0.01t} + 1}$.

(b) After one hour, when $t = 60$ minutes, the concentration is

$$C(60) = \frac{1}{e^{0.01(60)} + 1} \approx 0.3543 \text{ mg/cm}^3$$

After three hours, when $t = 180$ minutes,

the concentration is

$$C(180) = \frac{1}{e^{0.01(180)} + 1} \approx 0.1419 \text{ mg/cm}^3$$

(c) To determine how much time passes before next injection is given,

Press $\boxed{y =}$ and input

$C(t) = 1/(e^\wedge(0.01t) + 1)$ for $y_1 =$. Use window dimensions [0, 500]50 by [0, 1]0.02.

Press $\boxed{\text{trace}}$ and move along the curve until $y \approx 0.05$. Use the zoom-in function under the zoom menu to get a more accurate reading. A new injection is given after approximately 294 minutes.

71. (a) $L(t) = \displaystyle\int L'(t)\,dt$

$$= \int \frac{0.24 - 0.03t}{\sqrt{36 + 16t - t^2}}\,dt$$

Let $u = 36 + 16t - t^2$. Then

$$\frac{du}{dt} = 16 - 2t = 2(8 - t), \text{ or}$$

$$\frac{1}{2}\,du = (8 - t)\,dt.$$

$$\int \frac{0.03(8 - t)}{\left(36 + 16t - t^2\right)^{1/2}}\,dt = \frac{0.03}{2} \int u^{-1/2}\,du$$

$$= 0.03(36 + 16t - t^2)^{1/2} + C$$

At 7:00 a.m., $t = 0$ and $L(0) = 0.25$, so

$$0.25 = 0.03\sqrt{36 + 16(0) - (0)} + C, \text{ or}$$

$$C = 0.07$$

So, $L(t) = 0.03\sqrt{36 + 16t - t^2} + 0.07$. To find the peak level,

$L'(t) = 0$ when $0.24 - 0.03t = 0$, or when $t = 8$.

Further, when $0 \le t < 8$, $L'(t) > 0$ so L is increasing; when $t > 8$, $L'(t) < 0$ so L is decreasing so the absolute maximum occurs when $t = 8$, or at 3:00 p.m. The maximum is

$$L(8) = 0.03\sqrt{36 + 16(8) - (8)^2} + 0.07$$

$$= 0.37 \text{ parts per million}$$

(b) To use graphing utility to graph $L(t)$ and answer the questions in part (a), press

$\boxed{y =}$ and input

$$L(t) = 0.03\sqrt{(-t^2 + 16t + 36} + 0.07 \text{ for}$$

$y_1 =$.

Use window dimensions [0, 16]2 by [0.2, 0.4]0.04,

Press $\boxed{\text{graph}}$.

Press $\boxed{\text{trace}}$ and move along curve to the maximum point and use zoom-in if necessary. We find the maximum point occurs when $t = 8$ (at 3:00 p.m.). The ozone level is 0.37 ppm at this time. At 11:00 a.m., $t = 4$. Use the value function under the calc menu to find the ozone level is 0.34 ppm at 11:00 a.m. Trace along the curve to find when the y-value is 0.34 ppm. We find that the ozone level is 0.34 ppm again when $t = 12$, or at 7:00 p.m.

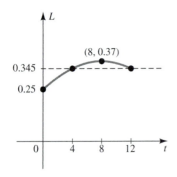

73. **(a)** $x(t) = \int x'(t)\, dt$

$$= \int -2(3t + 1)^{1/2}\, dt$$

Let $u = 3t + 1$. Then

$$\frac{du}{dt} = 3, \text{ or } \frac{1}{3}\, du = dt.$$

$$-\frac{2}{3} \int u^{1/2}\, dt = -\frac{4}{9}(3t + 1)^{3/2} + C$$

When $t = 0$, $x(0) = 4$, so

$$4 = -\frac{4}{9}\left[3(0) + 1\right]^{3/2} + C, \text{ or}$$

$$C = \frac{40}{9}$$

So, $x(t) = -\frac{4}{9}(3t + 1)^{3/2} + \frac{40}{9}$

(b) $x(4) = -\frac{4}{9}\left[3(4) + 1\right]^{3/2} + \frac{4}{9}$

$$\approx -16.4$$

(c) $3 = -\frac{4}{9}(3t + 1)^{3/2} + \frac{40}{9}$

$$\frac{13}{4} = (3t + 1)^{3/2}$$

$$t = \frac{\left(\dfrac{13}{4}\right)^{2/3} - 1}{3} \approx 0.4$$

75. **(a)** $x(t) = \int x'(t)\, dt$

$$= \int \frac{1}{\sqrt{2t + 1}}\, dt$$

Let $u = 2t + 1$. Then, $\frac{1}{2}\, du = dt$.

$$\frac{1}{2} \int u^{-1/2}\, dt$$

$$= \frac{1}{2}\left(2u^{1/2}\right) + C$$

$$= (2t + 1)^{1/2} + C$$

When $t = 0$, $x(0) = 0$ so $C = -1$ and

$$x(t) = (2t+1)^{1/2} - 1.$$

(b) When $t = 4$,

$$x(4) = \left[2(4)+1\right]^{1/2} - 1 = 2$$

(c) $3 = (2t+1)^{1/2} - 1$

$16 = 2t+1$, or

$$t = \frac{15}{2}.$$

77. Let $u = 1 + e^x$. Then

$\dfrac{du}{dx} = e^x$, or $du = e^x\, dx$. Further,

$e^x = u - 1$.

$$\int \frac{e^{2x}}{1+e^x}\, dx = \int \frac{e^x}{1+e^x}\, e^x dx$$

$$= \int \frac{u-1}{u}\, du = \int \left(1 - \frac{1}{u}\right) du$$

$$= 1 + e^x - \ln\left|1 + e^x\right| + C$$

$$= 1 + e^x - \ln\left(1 + e^x\right) + C$$

79. Let $u = x^{2/3} + 1$. Then $\dfrac{du}{dx} = \dfrac{2}{3} x^{-1/3}$, or

$\dfrac{3}{2} du = x^{-1/3} dx$. Further, $x^{2/3} = u - 1$.

$$\int x^{1/3}\left(x^{2/3}+1\right)^{3/2} dx$$

$$= \int \left(x^{2/3}+1\right)^{3/2} x^{2/3} x^{-1/3}\, dx$$

$$= \frac{3}{2}\int u^{3/2}(u-1)\, du = \frac{3}{2}\int u^{5/2} - u^{3/2}\, du$$

$$= \frac{3}{2}\left[\frac{2}{7}\left(x^{2/3}+1\right)^{7/2} - \frac{2}{5}\left(x^{2/3}+1\right)^{5/2}\right] + C$$

$$= \frac{3}{7}\left(x^{2/3}+1\right)^{7/2} - \frac{3}{5}\left(x^{2/3}+1\right)^{5/2} + C$$

81. $\displaystyle\int \frac{dx}{1+e^x}$

$$= \int \frac{dx}{1+\dfrac{1}{e^{-x}}} = \int \frac{dx}{\dfrac{e^{-x}+1}{e^{-x}}}$$

$$= \int \frac{e^{-x}}{e^{-x}+1}\, dx$$

Let $u = e^{-x} + 1$. Then, $\dfrac{du}{dx} = -e^{-x}$ or

$-du = e^{-x}\, dx$. Substituting,

$$= -\int \frac{1}{u}\, du = -\ln|u| + C$$

$$= -\ln\left(e^{-x}+1\right) + C$$

5.3 The Definite Integral and the Fundamental Theorem of Calculus

1.
$$\int_a^b f(x)\,dx \approx \left[f(x_1)+f(x_2)+f(x_3)+\dots+f(x_n)\right]\left(\frac{b-a}{n}\right) \text{ where } \Delta x = \frac{b-a}{n}$$

Here,

$$\approx \left[f(0)+f(0.4)+f(0.8)+f(1.2)+f(1.6)+f(2)\right]\left(\frac{2-0}{6}\right)$$

$$\approx \left[1.1+1.7+2.3+2.5+2.4+2.1\right]\left(\frac{1}{3}\right)$$

$$\approx (12.1)\left(\frac{1}{3}\right) \approx 4$$

3.
$$\int_a^b f(x)\,dx \approx \left[f(x_1)+f(x_2)+f(x_3)+\dots+f(x_n)\right]\left(\frac{b-a}{n}\right)$$

Here, $a = 0$, $b = 4$, $n = 8$, and $f(x) = 4 - x$.

x_j	0	0.5	1.0	1.5	2.0	2.5	3.0	3.5
$4-x_j$	4	3.5	3	2.5	2	1.5	1	0.5

$$\approx (4+3.5+3+2.5+2+1.5+1+0.5)\left(\frac{4-0}{8}\right)$$

$$= (18)\left(\frac{1}{2}\right) = 9$$

$$\int_0^4 (4-x)\,dx = 4x - \frac{x^2}{2}\Big|_0^4 = \left[4(4)-\frac{(4)^2}{2}\right] - [0-0] = 8$$

5.
$$\int_a^b f(x)\,dx \approx \left[f(x_1)+f(x_2)+f(x_3)+\dots+f(x_n)\right]\left(\frac{b-a}{n}\right)$$

Here, $a = 1$, $b = 2$, $n = 8$, and $f(x) = x^2$.

x_j	1	$\dfrac{9}{8}$	$\dfrac{10}{8}$	$\dfrac{11}{8}$	$\dfrac{12}{8}$	$\dfrac{13}{8}$	$\dfrac{14}{8}$	$\dfrac{15}{8}$
$\left(x_j\right)^2$	1	1.27	1.56	1.89	2.25	2.64	3.06	3.52

$$\approx \left(1+1.27+1.56+1.89+2.25+2.64+3.06+3.52\right)\left(\frac{2-1}{8}\right)$$

$$\approx 2.15$$

$$\int_{1}^{2} x^2\,dx = \frac{x^3}{3}\Big|_{1}^{2} = \frac{(2)^3}{3} - \frac{(1)^3}{3} = \frac{7}{3} \approx 2.33$$

7.
$$\int_{a}^{b} f(x)\,dx \approx \left[f(x_1)+f(x_2)+f(x_3)+\rightleftharpoons+f(x_n)\right]\left(\frac{b-a}{n}\right)$$

Here, $a=1$, $b=2$, $n=8$ and $f(x)=\dfrac{1}{x}$.

x_j	1	$\dfrac{9}{8}$	$\dfrac{10}{8}$	$\dfrac{11}{8}$	$\dfrac{12}{8}$	$\dfrac{13}{8}$	$\dfrac{14}{8}$	$\dfrac{15}{8}$
$\dfrac{1}{x_j}$	1	0.89	0.8	0.73	0.67	0.62	0.57	0.53

$$\left(1+0.89+0.8+0.73+0.67+0.62+0.57+0.53\right)\left(\frac{2-1}{8}\right)$$

$$\approx 0.73$$

$$\int_{1}^{2} \frac{1}{x}\,dx = \ln x\Big|_{1}^{2} = \ln 2 - \ln 1$$

$$\approx 0.69$$

9. Here, $a=0$, $b=4$, $n=8$ and $f(x)=x$.

x_j	0	0.5	1.0	1.5	2.0	2.5	3.0	3.5
x_j	0	0.5	1	1.5	2	2.5	3	3.5

$$\approx \left(0 + 0.5 + 1 + 1.5 + 2 + 2.5 + 3 + 3.5\right)\left(\frac{4-0}{8}\right)$$

$$= 7$$

$$\int_0^4 x\,dx = \left.\frac{x^2}{2}\right|_0^4 = \frac{(4)^2}{2} - 0 = 8$$

11. Here, $a = 0$, $b = 4$, $n = 8$ and $f(x) = x^2 + 1$.

x_j	0	0.5	1.0	1.5	2.0	2.5	3.0	3.5
$(x_j)^2 + 1$	1	1.25	2	3.25	5	7.25	10	13.25

$$\left(1 + 1.25 + 2 + 3.25 + 5 + 7.25 + 10 + 13.25\right)\left(\frac{4-0}{8}\right)$$

$$= 21.5$$

$$\int_0^4 \left(x^2 + 1\right)dx = \left.\left(\frac{x^3}{3} + x\right)\right|_0^4$$

$$= \left(\frac{4^3}{3} + 4\right) - 0 \approx 25.33$$

13. Here, $a = 0$, $b = 4$, $n = 8$ and $f(x) = \dfrac{1}{5-x}$.

x_j	0	0.5	1.0	1.5	2.0	2.5	3.0	3.5
$\dfrac{1}{5-x_j}$	0.2	0.22	0.25	0.29	0.33	0.4	0.5	0.67

$$\left(0.2 + 0.22 + 0.25 + 0.29 + 0.33 + 0.4 + 0.5 + 0.67\right)\left(\frac{4-0}{8}\right)$$

$$\approx 1.43$$

$\displaystyle\int_0^4 \frac{1}{5-x}\,dx$ can be solved using substitution. Let $u = 5 - x$. Then, $\dfrac{du}{dx} = -1$ or $-du = dx$. The limits of integration become $5 - 0 = 5$ and $5 - 4 = 1$.

$$-\int_5^1 \frac{1}{u} du = -\left(\ln u \Big|_5^1\right)$$
$$= -(\ln 1 - \ln 5) \approx 1.61$$

15. $\int_{-1}^2 5\,dx = 5x \Big|_{-1}^2 = 5(2) - 5(-1) = 15$

17. $\int_0^5 (3x + 2)dx = \left(\dfrac{3x^2}{2} + 2x\right)\Big|_0^5$
$$= \left[\dfrac{3(5)^2}{2} + 2(5)\right] - 0$$
$$= \dfrac{95}{2}$$

19. $\int_{-1}^1 3t^4 dt = \dfrac{3t^5}{5}\Big|_{-1}^1 = \dfrac{3(1)^5}{5} - \dfrac{3(-1)^5}{5} = \dfrac{6}{5}$

21. $\int_{-1}^1 (2u^{1/3} - u^{2/3})du$
$$= \left(\dfrac{3}{2}u^{4/3} - \dfrac{3}{5}u^{5/3}\right)\Big|_{-1}^1$$
$$= \left[\dfrac{3}{2}(1)^{4/3} - \dfrac{3}{5}(1)^{5/3}\right]$$
$$\qquad - \left[\dfrac{3}{2}(-1)^{4/3} - \dfrac{3}{5}(-1)^{5/3}\right]$$
$$= -\dfrac{6}{5}$$

23. $\int_0^1 e^{-x}(4 - e^x)dx = \int_0^1 (4e^{-x} - e^0)dx$
$$= (-4e^{-x} - x)\Big|_0^1$$
$$= (-4e^{-1} - 1) - (-4e^0 - 0)$$
$$= 3 - \dfrac{4}{e}$$

25. $\int_0^1 (x^4 + 3x^3 + 1)dx = \left(\dfrac{x^5}{5} + \dfrac{3x^4}{4} + x\right)\Big|_0^1$
$$= \left[\dfrac{(1)^5}{5} + \dfrac{3(1)^4}{4} + 1\right] - 0$$
$$= \dfrac{39}{20}$$
$$= 1.95$$

27. $\int_2^5 (2 + 2t + 3t^2)dt$
$$= (2t + t^2 + t^3)\Big|_2^5$$
$$= [(2(5) + (5)^2 + (5)^3]$$
$$\qquad\qquad - [2(2) + (2)^2 + (2)^3]$$
$$= 144$$

29. $\int_1^3 \left(1 + \dfrac{1}{x} + \dfrac{1}{x^2}\right) dx$

$= \left(x + \ln|x| - \dfrac{1}{x}\right)\Big|_1^3$

$= \left(3 + \ln 3 - \dfrac{1}{3}\right) - (1 + \ln 1 - 1)$

$= \dfrac{8}{3} + \ln 3$

31. $\int_{-3}^{-1} \dfrac{t+1}{t^3} dt$

$= \int_{-3}^{-1} \left(\dfrac{1}{t^2} + \dfrac{1}{t^3}\right) dt$

$= \left(-\dfrac{1}{t} - \dfrac{1}{2t^2}\right)\Big|_{-3}^{-1}$

$= \left[\dfrac{-1}{-1} - \dfrac{1}{2(-1)^2}\right] - \left[\dfrac{-1}{-3} - \dfrac{1}{2(-3)^2}\right]$

$= \dfrac{2}{9}$

33. $\int_1^2 (2x-4)^4 dx$

Let $u = 2x - 4$. Then $\dfrac{1}{2} du = dx$, and the limits of integration become $2(1) - 4 = -2$ and $2(2) - 4 = 0$.

$= \dfrac{1}{2}\int_{-2}^0 u^4 du$

$= \dfrac{1}{2}\left(\dfrac{u^5}{5}\right)\Big|_{-2}^0$

$= \dfrac{1}{10}(u^5)\Big|_{-2}^0$

$= \dfrac{1}{10}[0 - (-2)^5]$

$= 3.2$

35. $\int_0^4 \dfrac{1}{\sqrt{6t+1}} dt$

Let $u = 6t + 1$. Then, $\dfrac{1}{6} du = dt$, and the limits of integration become $6(0) + 1 = 1$ and $6(4) + 1 = 25$.

$= \dfrac{1}{6}\int_1^{25} u^{-1/2} du$

$= \dfrac{1}{6}\left(2\sqrt{u}\right)\Big|_1^{25}$

$= \dfrac{1}{3}\left(\sqrt{u}\right)\Big|_1^{25}$

$= \dfrac{1}{3}\left(\sqrt{25} - \sqrt{1}\right)$

$= \dfrac{4}{3}$

37. $\int_0^1 (x^3 + x)\sqrt{x^4 + 2x^2 + 1}\, dx$

Let $u = x^4 + 2x^2 + 1$. Then $\dfrac{1}{4} du = (x^3 + x)dx$, and the limits of integration become $(0) + 2(0) + 1 = 1$ and $(1)^4 + 2(1)^2 + 1 = 4$.

$= \dfrac{1}{4}\int_1^4 u^{1/2} du$

$= \dfrac{1}{4}\left(\dfrac{2}{3} u^{3/2}\right)\Big|_1^4$

$= \dfrac{1}{6}(u^{3/2})\Big|_1^4$

$= \dfrac{1}{6}[(4)^{3/2} - (1)^{3/2}]$

$= \dfrac{7}{6}$

39. $\int_1^{e+1} \dfrac{x}{x-1}\,dx$

Let $u = x - 1$. Then $du = dx$ and $x = u + 1$.
Further, the limits of integration become
$2 - 1 = 1$ and $(e + 1) - 1 = e$.

$= \int_1^e \dfrac{u+1}{u}\,du$

$= \int_1^e \left(1 + \dfrac{1}{u}\right) du$

$= \left(u + \ln|u|\right)\Big|_1^e$

$= (e + \ln e) - (1 + \ln 1)$

$= e$

41. $\int_1^{e^2} \dfrac{(\ln x)^2}{x}\,dx$

Let $u = \ln x$. Then $du = \dfrac{1}{x}\,dx,$ and the
limits of integration become $\ln 1 = 0$ and
$\ln(e)^2 = 2$.

$\int_0^2 u^2\,du = \dfrac{1}{3}(u^3)\Big|_0^2$

$\qquad = \dfrac{1}{3}[(2)^3 - (0)]$

$\qquad = \dfrac{8}{3}$

43. $\int_{1/3}^{1/2} \dfrac{e^{1/x}}{x^2}\,dx$

Let $u = \dfrac{1}{x}$. Then $-du = \dfrac{1}{x^2}\,dx,$ and the

limits of integration become $\dfrac{1}{\frac{1}{3}} = 3$ and

$\dfrac{1}{\frac{1}{2}} = 2$.

$= -\int_3^2 e^u\,du$

$= \int_2^3 e^u\,du$

$= (e^u)\Big|_2^3$

$= e^3 - e^2$

45. $\int_{-3}^2 [-2f(x) + 5g(x)]\,dx$

$= 2\int_{-3}^2 f(x)\,dx + 5\int_{-3}^2 g(x)\,dx$

$= -2(5) + 5(-2)$

$= -20$

47. $\int_4^4 g(x)\,dx = G(4) - G(4) = 0,$ where $G(x)$
is the antiderivative of $g(x)$.

49. $\int_1^2 [3f(x) + 2g(x)]\,dx$

$= 3\int_1^2 f(x)\,dx + 2\int_1^2 g(x)\,dx$

$= 3\left[\int_{-3}^2 f(x)\,dx - \int_{-3}^1 f(x)\,dx\right]$

$\qquad + 2\left[\int_{-3}^2 g(x)\,dx - \int_{-3}^1 g(x)\,dx\right]$

$= 3(5 - 0) + 2(-2 - 4)$

$= 3$

51. $\int_{-1}^2 x^4\,dx = \dfrac{1}{5}(x^5)\Big|_{-1}^2 = \dfrac{1}{5}[(2)^5 - (-1)^5] = \dfrac{33}{5}$

53. $\int_0^4 (3x + 4)^{1/2}\,dx$

Let $u = 3x + 4$. Then $\dfrac{1}{3}\,du = dx,$ and the

limits of integration become $3(0) + 4 = 4$ and $3(4) + 4 = 16$.

$$= \frac{1}{3}\int_4^{16} u^{1/2}\, du$$

$$= \frac{1}{3}\left(\frac{2}{3}u^{3/2}\right)\Bigg|_4^{16}$$

$$= \frac{2}{9}(u^{2/3})\Bigg|_4^{16}$$

$$= \frac{2}{9}\left[(16)^{3/2} - (4)^{3/2}\right]$$

$$= \frac{112}{9}$$

55. $\displaystyle\int_0^{\ln 3} e^{2x}\, dx = \frac{1}{2}(e^{2x})\Bigg|_0^{\ln 3}$

$$= \frac{1}{2}(e^{2\ln 3} - e^0)$$

$$= \frac{1}{2}(e^{\ln 3^2} - 1)$$

$$= 4$$

57. $\displaystyle\int_{-2}^1 \frac{3}{5 - 2x}\, dx$

Let $u = 5 - 2x$. Then, $\dfrac{du}{dx} = -2$ and

$-\dfrac{1}{2}\, du = dx$. When $x = -2$,

$u = 5 - 2(-2) = 9$ and when $x = 1$,

$u = 5 - 2(1) = 3$.

$$-\frac{1}{2}\int_9^3 \frac{3}{u}\, du = -\frac{3}{2}\int_9^3 \frac{1}{u}\, du$$

$$= \frac{3}{2}\int_3^9 \frac{1}{u}\, du$$

$$= \frac{3}{2}\left(\ln|u|\Big\|_3^9\right)$$

$$= \frac{3}{2}(\ln 9 - \ln 3)$$

$$= \frac{3}{2}\left(\ln \frac{9}{3}\right)$$

$$= \frac{3}{2}\ln 3 \approx 1.6479$$

59. $\displaystyle\int_0^5 V'(t)\, dt = V(5) - V(0)$

61. The number of pounds of soybeans stored per week x weeks from now is $12{,}000 - 300x$, a function that decreases linearly from 12,000 to 0 in 40 weeks. The weekly cost rate will be $0.2(12{,}000 - 300x)$ cents per week. The cost over the next 40 weeks

$$= \int_0^{40} 0.2(12{,}000 - 300x)\, dx$$

$$= 0.2(12{,}000x - 150x^2)\Big|_0^{40}$$

$$= 48{,}000 \text{ cents, or } \$480$$

63. Let $V(t)$ be the value of the crop, in dollars, after t days. Then

$$\frac{dV}{dt} = 3(0.3t^2 + 0.6t + 1).$$

The change in value will be

$$V(5) - V(0) = \int_0^5 3(0.3t^2 + 0.6t + 1)\, dt$$

$$= 3(0.1t^3 + 0.3t^2 + t)\Big|_0^5$$

$$= 3[(0.1(5)^3 + 0.3(5)^2 + 5) - 0]$$

$$= \$75$$

65. Let $V(x)$ be the value of the machine, in dollars, after t years. Then,

$$\frac{dV}{dt} = 220(x - 10).$$

$$V(2) - V(1) = \int_1^2 220(x - 10)\, dx$$

$$= \int_1^2 (220x - 2200)\, dx$$

$$= (110x^2 - 2200x)\Big|_1^2$$

$$= [110(2)^2 - 2200(2)]$$
$$\qquad - [110(1) - 2200(1)]$$

$$= -1870,$$

or the machine depreciates by $1,870.

67. $P(3) - P(2)$

$$= \int_2^3 1500\left(2 - \frac{t}{2t + 5}\right)dt$$

$$= 3000\int_2^3 dt - 1500\int_2^3 \frac{t}{2t + 5}\, dt$$

Let $u = 2t + 5$. Then $\dfrac{1}{2}\, du = dt$, and

$t = \dfrac{u-5}{2}$. Further, the limits of integration become $2(2) + 5 = 9$ and $2(3) + 5 = 11$.

$$= 3000\int_2^3 dt - 750\int_9^{11}\dfrac{u-5}{2u}\,du$$

$$= 3000\int_2^3 dt - 375\int_9^{11}\left(1 - \dfrac{5}{u}\right)du$$

$$= 3000(t)\Big|_2^3 - 375\left(u - 5\ln|u|\right)\Big|_9^{11}$$

$$= 3000(3-2)$$
$$\quad - 375[(11 - 5\ln 11) - (9 - 5\ln 9)]$$

$$= 3000 - 375(2 - 5\ln 11 + 5\ln 9)$$

$$\approx 2{,}626 \text{ telephones}$$

69. $V'(t) = 12e^{-0.05t}(e^{0.3t} - 3)$
$$= 12e^{0.25t} - 36e^{-0.05t}$$

(a) In 2006, $t = 0$ and in 2010, $t = 4$.

$$\int_0^4 12e^{0.25t} - 36e^{-0.05t}\,dt$$

$$= 12\int_0^4 e^{0.25t}\,dt - 36\int_0^4 e^{-0.05t}\,dt$$

For the first integral, let $u = 0.25t$.

Then $\dfrac{du}{dt} = 0.25$ and $4\,du = dt$. When $t = 0$, $u = 0$ and when $t = 4$, $u = 1$.
For the second integral, let $u = -0.05t$.

Then, $\dfrac{du}{dt} = -0.05$ and $-20\,du = dt$.
When $t = 0$, $u = 0$ and when $t = 4$, $u = -0.2$.

$$12\left(4\int_0^1 e^u\,du\right) - 36\left(-20\int_0^{-0.2} e^u\,du\right)$$

$$= 48\int_0^1 e^u\,du + 720\int_0^{-0.2} e^u\,du$$

$$= 48\int_0^1 e^u\,du - 720\int_{-0.2}^0 e^u\,du$$

$$= 48\left(e^u\Big|_0^1\right) - 720\left(e^u\Big|_{-0.2}^0\right)$$

$$= 48(e^1 - e^0) - 720(e^0 - e^{-0.2})$$

$$\approx -48.03633 \text{ thousand}$$
So, the value decreases by approximately \$48,036.33.

(b) In 2012, $t = 6$.

$$\int_0^6 12e^{0.25t} - 36e^{-0.05t}\,dt$$

Using the same substitutions as before, for the first integral, when $t = 4$, $u = 1$ and when $t = 6$, $u = 1.5$. For the second integral, when $t = 4$, $u = -0.2$ and when $t = 6$, $u = -0.3$.

$$= 48\int_1^{1.5} e^u\,du - 720\int_{-0.3}^{-0.2} e^u\,du$$

$$= 48\left(e^u\Big|_1^{1.5}\right) - 720\left(e^u\Big|_{-0.3}^{-0.2}\right)$$

$$= 48(e^{1.5} - e^1) - 720(e^{-0.2} - e^{-0.3})$$

$$\approx 28.54652 \text{ thousand}$$
So, the value increases by approximately \$28,546.52.

71. $L(3) - L(0) = \displaystyle\int_0^3 (0.1t + 0.1)\,dt$

$$= (0.05t^2 + 0.1t)\Big|_0^3$$

$$= [0.05(3)^2 + 0.1(3)] - 0$$

$$= 0.75 \text{ ppm}$$

73. $P(8) - P(0) = \displaystyle\int_0^8 (5 + 3t^{2/3})\,dt$

$$= \left(5t - \dfrac{9}{5}t^{5/3}\right)\Big|_0^8$$

$$= \left[5(8) + \dfrac{9}{5}(8)^{5/3}\right] - 0$$

$$= \dfrac{488}{5} \approx 98 \text{ people}$$

75. $P(2) - P(5) = -[P(5) - P(2)]$

$$= -\int_2^5 -\dfrac{2}{t+1}\,dt$$

Let $u = t + 1$. Then $du = dt$, and the limits of integration become $2 + 1 = 3$ and

$5 + 1 = 6.$

$$= 2\int_3^6 \frac{1}{u}\,du$$

$$= 2\left(\ln|u|\right)\Big|_3^6$$

$$= 2(\ln 6 - \ln 3)$$

$$= 2\left(\ln \frac{6}{3}\right)$$

$$= 2\ln 2 \approx 1.386 \text{ grams.}$$

77. $L(10) - L(5) = \int_5^{10} \frac{4}{\sqrt{t+1}}\,dt$

Let $u = t + 1$. Then $du = dt$, and the limits of integration become $5 + 1 = 6$ and $10 + 1 = 11$.

$$= 4\int_6^{11} u^{-1/2}\,du$$

$$= 4\left(2u^{1/2}\right)\Big|_6^{11}$$

$$= 8\left(\sqrt{u}\right)\Big|_6^{11}$$

$$= 8\left(\sqrt{11} - \sqrt{6}\right) \approx 7 \text{ facts.}$$

79. $C(4) - C(0) = \int_0^4 \frac{-0.33t}{\sqrt{0.02t^2 + 10}}\,dt$

$$= -0.33\int_0^4 \frac{t}{\sqrt{0.02t^2 + 10}}\,dt$$

Let $u = 0.02t^2 + 10$. Then $25\,du = t\,dt$, and the limits of integration become $0.02(0) + 10 = 10$ and $0.02(4)^2 + 10 = 10.32.$

$$= -8.25\int_{10}^{10.32} u^{-1/2}\,du$$

$$= -8.25(2u^{1/2})\Big|_{10}^{10.32}$$

$$= -16.5\left(\sqrt{u}\right)\Big|_{10}^{10.32}$$

$$= -16.5\left(\sqrt{10.32} - \sqrt{10}\right)$$

$$\approx -0.8283,$$

or the concentration decreases by approximately 0.8283 mg/cm^3.

81. Let $s(t)$ be the distance traveled, in feet, after t seconds. Since velocity is the

derivative of distance,

$$s(3) - s(0) = \int_0^3 (-32t + 80)\,dt$$

$$= (-16t^2 + 80t)\Big|_0^3$$

$$= [-16(3)^2 + 80(3)] - 0$$

$$= 96 \text{ feet.}$$

83. (a) $\int_0^1 \sqrt{1 - x^2}\,dx$ represents the area under the curve $\sqrt{1 - x^2}$, above the x-axis, from $x = 0$ to $x = 1$. But the graph of $y = \sqrt{1 - x^2}$ is a semi-circle, having radius 1 and center $(0, 0)$ since

$$y = \sqrt{1 - x^2}$$

$$y^2 = 1 - x^2$$

$$x^2 + y^2 = 1$$

The area from $x = 0$ to $x = 1$ corresponds to a quarter of the circle's area.

$$= \frac{1}{4}[\pi(1)^2] = \frac{\pi}{4}$$

(b) Similarly, the graph of $\sqrt{2x - x^2}$ is the same semicircle, shifted one unit to the right since

$$y = \sqrt{2x - x^2}$$

$$y^2 = 2x - x^2$$

$$x^2 - 2x + y^2 = 0$$

$$(x^2 - 2x + 1) + y^2 = 1$$

$$(x - 1)^2 + y^2 = 1$$

So, the area from $x = 1$ to $x = 2$ still corresponds to a quarter of the

$$\text{circle} = \frac{\pi}{4}.$$

5.4   Applying Definite Integration: Distribution of Wealth and Average Value

1. The limits of integration are $x^3 = \sqrt{x}$;

$x^3 - x^{1/2} = 0$; $x^{1/2}(x^{5/2} - 1) = 0$ so $x = 0$

and $x = 1$. The shaded area is

$$\int_0^1 \left(\sqrt{x} - x^3 \right) dx = \left(\frac{2}{3} x^{3/2} - \frac{x^4}{4} \right) \Bigg|_0^1$$

$$= \frac{5}{12}$$

3. The limits of integration are $x = 0$ and

$x = \dfrac{2}{x+1}$, $\ x^2 + x = 2$, $\ x^2 + x - 2 = 0$,

$(x + 2)(x - 1) = 0$, $x = 1$ (rejecting $x = -2$ since shaded area starts at $x = 0$).
The shaded area is

$$\int_0^1 \left(\frac{2}{x+1} - x \right) dx = \left(2\ln|x+1| - \frac{x^2}{2} \right) \Bigg|_0^1$$

$$= 2\ln 2 - \frac{1}{2}$$

5. The shaded area is

$$\int_0^1 [x - (-x)] \, dx = (x^2) \Big|_0^1 = 1.$$

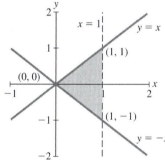

7. The shaded area is

$$\int_1^3 [(-x^2 + 4x - 3) - 0] \, dx$$

$$= \left(-\frac{x^3}{3} + 2x^2 - 3x \right) \Bigg|_1^3$$

$$= \frac{4}{3}$$

9. The shaded area is

$$\int_0^2 [0 - (x^2 - 2x)] \, dx = \left(-\frac{x^3}{3} + x^2 \right) \Bigg|_0^2$$

$$= \frac{4}{3}$$

11. The limits of integration are

$$x^2 - 2x = -x^2 + 4$$

$$2x^2 - 2x + 4 = 0$$

$$2(x - 2)(x + 1) = 0$$

$x = -1$ and $x = 2$.
The shaded area is

$$\int_{-1}^{2}[(-x^2+4)-(x^2-2x)]dx$$

$$=\left(-\frac{2x^3}{3}+x^2+4x\right)\Bigg|_{-1}^{2}$$

$$=9$$

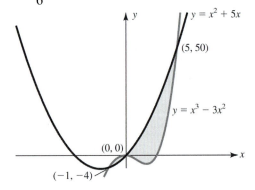

13. The points of intersection are

$$x^3-3x^2=x^2+5x$$

$$x^3-4x^2-5x=0$$

$$x(x-5)(x+1)=0.$$

There are two shaded areas

$$\int_{-1}^{0}[(x^3-3x^2)-(x^2+5x)]dx$$

$$+\int_{0}^{5}[(x^2+5x)-(x^3-3x^2)]dx$$

$$=\left(\frac{x^4}{4}-\frac{4x^3}{3}-\frac{5x^2}{2}\right)\Bigg|_{-1}^{0}$$

$$+\left(-\frac{x^4}{4}+\frac{4x^3}{3}+\frac{5x^2}{2}\right)\Bigg|_{0}^{5}$$

$$=\frac{11}{12}+\frac{825}{12}$$

$$=\frac{443}{6}$$

15. The equation of the top curve is the equation of the line through the points (−4, 0) and (2, 6).

$$m=\frac{6}{6}=1,\ \text{so } y=x+4.$$

The shaded area is

$$\int_{-4}^{2}[(x+4)-0]dx=\left(\frac{x^2}{2}+4x\right)\Bigg|_{-4}^{2}=18.$$

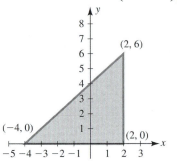

17. The equation of the top curve is the equation of the line through the points $(0, 6)$ and $(2, 8)$.

$$m = \frac{8-6}{2-0} = 1, \text{ so } y = x + 6$$

The shaded area is

$$\int_0^2 [(x+6) - 0]dx = \left(\frac{x^2}{2} + 6x\right)\Bigg|_0^2 = 14.$$

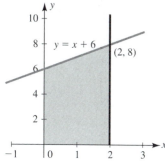

19. $f_{av} = \dfrac{1}{3-(-3)} \displaystyle\int_{-3}^3 (1-x^2)dx$

$$= \frac{1}{6}\left(x - \frac{x^3}{3}\right)\Bigg|_{-3}^3$$

$$= -2$$

21. $f_{av} = \dfrac{1}{1-(-1)} \displaystyle\int_{-1}^1 [e^{-x}(4 - e^{2x})]dx$

$$= \frac{1}{2}\int_{-1}^1 (4e^{-x} - e^x)dx$$

$$= \frac{1}{2}(-4e^{-x} - e^x)\Bigg|_{-1}^1$$

$$= \frac{1}{2}\left(\frac{-3}{e} + 3e\right)$$

$$= \frac{3}{2}\left(e - \frac{1}{e}\right)$$

23. $f_{av} = \dfrac{1}{\ln 3 - 0} \displaystyle\int_0^{\ln 3}\left(\frac{e^x - e^{-x}}{e^x + e^{-x}}\right)dx$

Using substitution with $u = e^x + e^{-x}$,

$$= \frac{1}{\ln 3}\int_2^{10/3} \frac{1}{u}du$$

$$= \frac{1}{\ln 3}(\ln u)\Bigg|_2^{10/3}$$

$$= \frac{1}{\ln 3}\left(\ln \frac{10}{3} - \ln 2\right)$$

$$= \frac{1}{\ln 3}(\ln 10 - \ln 3 - \ln 2)$$

$$= \frac{1}{\ln 3}(\ln 5 - \ln 3)$$

25. $f_{av} = \dfrac{1}{2-0} \displaystyle\int_0^2 (2x - x^2)dx$

$$= \frac{1}{2}\left(x^2 - \frac{x^3}{3}\right)\Bigg|_0^2$$

$$= \frac{2}{3}$$

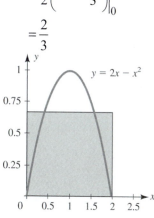

27. $f_{av} = \dfrac{1}{4-2} \displaystyle\int_2^4 \frac{1}{u}du = \frac{1}{2}\Big[\ln|u|\Big]_0^2 = \frac{1}{2}\ln 2$

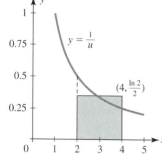

29. $GI = 2\int_0^1 (x - x^3)\,dx = \left(x^2 - \frac{x^4}{2}\right)\Big|_0^1 = \frac{1}{2}$

31. $GI = 2\int_0^1 (x - 0.55x^2 - 0.45x)\,dx$

$\qquad = 2\left(\frac{0.55x^2}{2} - \frac{0.55x^3}{3}\right)\Big|_0^1$

$\qquad = 0.183$

33. $GI = 2\int_0^1 \left(x - \frac{2}{3}x^{3.7} - \frac{1}{3}x\right)dx$

$\qquad = 2\left(\frac{x^2}{3} - \frac{2x^{4.7}}{3(4.7)}\right)\Big|_0^1$

$\qquad = 0.383$

35. Average value of a function is

$\qquad \frac{1}{b-a}\int_a^b f(x)\,dx.$

Here,

$\qquad \frac{1}{5-2}\int_2^5 (0.5p^2 + 3p + 7)\,dp$

$\qquad = \frac{1}{3}\left(\frac{0.5p^3}{3} + \frac{3p^2}{2} + 7p\right)\Big|_2^5$

$\qquad = \frac{1}{3}\left[\left(\frac{0.5(5)^3}{3} + \frac{3(5)^2}{2} + 7(5)\right)\right.$

$\qquad\qquad \left. - \left(\frac{0.5(2)^3}{3} + \frac{3(2)^2}{2} + 7(2)\right)\right]$

$\qquad = 24$ hundred, or 2,400

37. The equation of the function is the equation of the line joining $(0, 60{,}000)$ and $(1, 0)$.

$\qquad m = \frac{60{,}000}{-1},$ so $y = -60{,}000(t-1)$

$\qquad y_{av} = \frac{1}{1-0}\int_0^1 -60{,}000(t-1)\,dt$

$\qquad = -60{,}000\left(\frac{t^2}{2} - t\right)\Big|_0^1$

$\qquad = 30{,}000$ kilograms

39. (a) The average value of a function is

$\qquad \frac{1}{b-a}\int_a^b f(x)\,dx.$

Here, $f(t) = 10{,}000e^{0.05t}$ and

$\qquad \frac{1}{5-0}\int_0^5 10{,}000e^{0.05t}\,dt$

$\qquad = 2{,}000\int_0^5 e^{0.05t}\,dt$

Let $u = 0.05t$. Then, $\frac{du}{dt} = 0.05$ and $20\,du = dt$. When $t = 0$, $u = 0$ and when $t = 5$, $u = 0.25$.

$\qquad = 40{,}000\int_0^{0.25} e^u\,du$

$\qquad = 40{,}000\left(e^u\Big|_0^{0.25}\right)$

$\qquad = 40{,}000(e^{0.25} - e^0)$

$\qquad \approx \$11{,}361.02$

(b) Writing Exercise—Answers will vary.

41. (a) Testing a couple of values shows that P_2' is initially more profitable. It will stay more profitable until $P_2'(t) = P_1'(t)$.

$\qquad 306 + 5t = 130 + t^2$

$\qquad 0 = t^2 - 5t - 176,$

$\qquad 0 = (t - 16)(t + 11)$

or $t = 16$ years (rejecting the negative solution).

(b) Excess $= \int_0^{16} [(306 + 5t) - (130 + t^2)]\,dt$

$\qquad = \left(176t + \frac{5t^2}{2} - \frac{t^3}{3}\right)\Big|_0^{16}$

$\qquad = 2{,}090.67,$ or $\$209{,}067.$

(c)

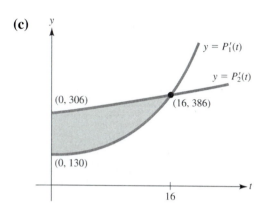

$y = P'_1(t)$

$y = P'_2(t)$

$(0, 306)$

$(16, 386)$

$(0, 130)$

16

43. (a) Testing a couple of values shows that P'_2 is initially more profitable. It will stay more profitable until $P'_2(t) = P'_1(t)$.

$$140e^{0.07t} = 90e^{0.1t}$$

$$\frac{14}{9}e^{0.07t} = e^{0.1t}$$

$$\ln\left(\frac{14}{9}e^{0.07t}\right) = \ln e^{0.1t}$$

$$\ln\frac{14}{9} + \ln e^{0.07t} = 0.1t$$

$$\ln\frac{14}{9} + 0.07t = 0.1t,$$

or $t \approx 14.7$ years.

(b) Excess $= \int_0^{14.7}(140e^{0.07t} - 90e^{0.1t})dt$

$$= (2,000e^{0.07} - 900e^{0.1t})\Big|_0^{14.7}$$

$$\approx 582.22, \text{ or } \$582,220$$

(c)

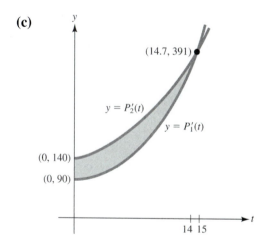

$(14.7, 391)$

$y = P'_2(t)$

$y = P'_1(t)$

$(0, 140)$

$(0, 90)$

14 15

45. $Q(L) = 500L^{2/3}$

(a) The average production is

$$\frac{1}{2,000 - 1,000}\int_{1,000}^{2,000}500L^{2/3}\,dL$$

$$= (0.001)(500)\int_{1,000}^{2,000}L^{2/3}\,dL$$

$$= 0.5\left(\frac{L^{5/3}}{5/3}\Bigg|_{1,000}^{2,000}\right)$$

$$= (0.5)\left(\frac{3}{5}\right)\left(L^{5/3}\Big|_{1,000}^{2,000}\right)$$

$$= 0.3\left[(2,000)^{5/3} - (1,000)^{5/3}\right]$$

$$\approx 65,244 \text{ units}$$

(b) Need to find L so $Q(L) = 65,244$.

$$65,244 = 500L^{2/3}$$

$$L^{2/3} \approx 130.488$$

Raising both sides to the $\frac{3}{2}$ power,

$$L \approx (130.488)^{3/2} \approx 1,491 \text{ worker hours}$$

47. $C(x) = 3x\sqrt{x} + 10 = 3x^{3/2} + 10$

The average cost is

$$\frac{1}{81 - 0}\int_0^{81}\left(3x^{3/2} + 10\right)dx$$

$$= \frac{1}{81}\left[\left(\frac{3x^{5/2}}{5/2} + 10x\right)\Bigg|_0^{81}\right]$$

$$= \frac{1}{81}\left[\left(\frac{6}{5}x^{5/2} + 10x\right)\Bigg|_0^{81}\right]$$

$$= \frac{1}{81}\left[\frac{6}{5}(81)^{5/2} + 10(81) - 0\right]$$

$$\approx 884.8, \text{ or } \$88,480$$

49. Total cost = cost of cabin + cost of land
cost of cabin

= (area of cabin)(price per sq yard)
= (64)(2,000)
= \$128,000
cost of land
= (area of land)(price per sq yard)
area of land
= area under curve − area of cabin

$$= \int_0^{15} 10e^{0.04x}\,dx - 64$$

$$= 250(e^{0.04x})\Big|_0^{15} - 64$$

$$\approx 141.53$$

cost of land = (141.53)(800) = \$113,224
So, the total cost is \$241,224.

51. $GI_1 = 2\int_0^1\left(x - \dfrac{2}{3}x^3 - \dfrac{1}{3}x\right)dx$

$$= 2\left(\dfrac{x^2}{3} - \dfrac{x^4}{6}\right)\Big|_0^1$$

$$= \dfrac{1}{3} \approx 0.33$$

$GI_2 = 2\int_0^1\left(x - \dfrac{5}{6}x^2 - \dfrac{1}{6}x\right)dx$

$$= 2\left(\dfrac{5}{12}x^2 - \dfrac{5}{18}x^3\right)\Big|_0^1$$

$$= \dfrac{5}{18} \approx 0.28$$

$GI_3 = 2\int_0^1\left(x - \dfrac{3}{5}x^4 - \dfrac{2}{5}x\right)dx$

$$= 2\left(\dfrac{3}{10}x^2 - \dfrac{3}{25}x^5\right)\Big|_0^1$$

$$= \dfrac{9}{25} = 0.36$$

So, football is the most equitable, basketball is the least equitable.

53. $Q_{av} = \dfrac{1}{5-0}\int_0^5 2{,}000e^{0.05t}\,dt$

$$= \dfrac{400}{0.05}e^{0.05t}\Big|_0^5 = 2{,}272 \text{ bacteria}$$

55. Excess

$$= \int_0^{10}\left(10e^{0.02t} - \dfrac{20e^{0.02t}}{1+e^{0.02t}}\right)dt$$

$$= 10\int_0^{10} e^{0.02t}\,dt - 20\int_0^{10}\dfrac{e^{0.02t}}{1+e^{0.02t}}\,dt$$

Using substitution with $u = 1 + e^{0.02t}$,

$$= 10\int_0^{10} e^{0.02t}\,dt - 1{,}000\int_2^{1+e^{0.2}}\dfrac{1}{u}\,du$$

$$= 500(e^{0.02t})\Big|_0^{10} - 1{,}000\left(\ln|u|\right)\Big|_2^{1+e^{0.2}}$$

$$\approx 5.710, \text{ or } 5{,}710 \text{ people.}$$

57. (a) M_{av}

$$= \dfrac{1}{12-0}\int_0^{12}(M_0 + 50te^{-0.1t^2})\,dt$$

$$= \dfrac{1}{12}\left[\int_0^{12} M_0\,dt + 50\int_0^{12}(te^{-0.1t^2})\,dt\right]$$

Using substitution with $u = -0.1t^2$,

$$= \dfrac{1}{12}\left[\int_0^{12} M_0\,dt - 250\int_0^{-14.4} e^u\,du\right]$$

$$= \dfrac{1}{12}\left[\int_0^{12} M_0\,dt + 250\int_{-14.4}^0 e^u\,du\right]$$

$$= \dfrac{1}{12}\left[M_0t\Big|_0^{12} + 250(e^u)\Big|_{-14.4}^0\right]$$

$$= M_0 + 20.83 \text{ kilo-Joules per hour.}$$

(b) When $t = 0$, $M(0) = M_0$ so $(0, M_0)$ is an intercept.

$$\lim_{t\to+\infty}(M_0 + 50te^{-0.1t^2}) = M_0, \text{ so}$$

$y = M_0$ is a horizontal asymptote.

$M'(t)$

$$= 50[(t)(e^{-0.1t^2}\cdot -0.2t) + (e^{-0.1t^2})(1)]$$

$$= 50e^{-0.1t^2}(-0.2t^2 + 1)$$

So $M'(t) = 0$ when $-0.2t^2 + 1 = 0$, or $t = \sqrt{5}$. The peak metabolic rate is

$$M\left(\sqrt{5}\right) = M_0 + 50\sqrt{5}e^{-0.5}$$

$$= M_0 + 50\sqrt{\dfrac{5}{e}}$$

$$M''(t) = 50[(e^{-0.1t^2})(-0.4)$$
$$+ (-0.2t^2 + 1)(e^{-0.1t^2} \cdot -0.2t)]$$
$$= -10e^{-0.1t^2}[2 + (-0.2t^2 + 1)]$$

So $M''(t) = 0$ when $3 - 0.2t^2 = 0$, or
$t = \sqrt{15}$.

When $0 < t < \sqrt{5}$,
$M'(t) > 0$ so m is increasing
$M''(t) < 0$ so m is concave down

When $\sqrt{5} < t < \sqrt{15}$,
$M'(t) < 0$ so m is decreasing
$M''(t) < 0$ so m is concave down

When $t > \sqrt{15}$,
$M'(t) < 0$ so m is decreasing
$M''(t) > 0$ so m is concave up

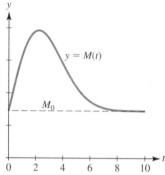

59. $C_{av} = \dfrac{1}{8-0} \displaystyle\int_0^8 \dfrac{3t}{(t^2+36)^{3/2}} \, dt$

Using substitution with $u = t^2 + 36$,

$$= \dfrac{3}{16} \int_{36}^{100} u^{-3/2} \, du$$

$$= \dfrac{3}{16} \left(-\dfrac{2}{\sqrt{u}} \right) \Bigg|_{36}^{100}$$

$$= \dfrac{1}{40} \text{ mg/cm}^3.$$

61. (a) $S = F'(M) = \dfrac{1}{3}(2kM - 3M^2)$

We need to maximize S.

$$F''(M) = \dfrac{1}{3}(2k - 6M)$$

So $F''(M) = 0$ when $2k - 6M = 0$, or
$M = \dfrac{k}{3}$. $F'''(M) = -2$, so $F'''\left(\dfrac{k}{3}\right) < 0$,
so the absolute maximum occurs when
$M = \dfrac{k}{3}$.

(b) $F_{av} = \dfrac{1}{\frac{k}{3}-0} \displaystyle\int_0^{k/3} \dfrac{1}{3}(kM^2 - M^3) \, dM$

$$= \dfrac{1}{k} \left(\dfrac{kM^3}{3} - \dfrac{M^4}{4} \right) \Bigg|_0^{k/3}$$

$$= \dfrac{k^3}{108}$$

63. (a) $T(t) = 3 - \dfrac{1}{3}(t-5)^2$

Since $t = 2$ at 8:00 A.M., $a = 2$. Since
$t = 11$ at 5:00 P.M., $b = 11$. So, the
average temperature is

$$T_{av} = \dfrac{1}{11-2} \int_2^{11} 3 - \dfrac{1}{3}(t-5)^2 \, dt$$

Using substitution for the second term
with $u = t - 5$, $du = dt$, $u_1 = -3$ and
$u_2 = 6$,

$$= \dfrac{1}{9}\left[3t \Big|_2^{11} - \dfrac{1}{9}u^3 \Big|_{-3}^6 \right]$$

$$= \dfrac{1}{9}\left[(33-6) - \dfrac{1}{9}(216+27) \right]$$

$$= 0 ^\circ\text{C}$$

(b) Need to find t when $T(t) = 0$, so

$$0 = -\dfrac{1}{3}(t-5)^2$$

$$\dfrac{1}{3}(t-5)^2 = 3$$

$$(t-5)^2 = 9$$

$$t - 5 = \pm 3$$

$$t = 2.8$$

When $t = 2$, the time is 8:00 A.M. and
when $t = 8$, the time is 2:00 P.M.

65. Press $\boxed{y=}$ and input $\sqrt{\dfrac{2}{5}x^2-2}$ for $y_1 =$,

input $-\sqrt{\dfrac{2}{5}x^2-2}$ for $y_2 =$, and input

$x\wedge 3 - 8.9x^2 + 26.7x - 27$ or $y_3 =$.
Use window dimensions [−5, 5]1 by
[−4, 4]0.5
Press $\boxed{\text{graph}}$.
Use trace and zoom-in to find the points of
intersection are (4.2, 2.25) and
(2.34, −0.44).
An alternative to using trace and zoom is to
use the intersect function under the calc
menu. To find the first point, use $\uparrow$ and $\downarrow$

arrows to verify $y_1 = \sqrt{\dfrac{2}{5}x^2-2}$ is

displayed. Enter and value close to the
point of intersection.
Then, verify $y_3 = x^3 - 8.9x^2 + 26.7x - 27$
is displayed and enter a value close and
finally, enter a guess. This gives the point
(4.2, 2.25).
Repeat this process using

$y_2 = -\sqrt{\dfrac{2}{5}x^2-2}$ and

$y_3 = x^3 - 8.9x^2 + 26.7x - 27$ to find the

second point (2.34, −0.44).
To find the area bounded by the curves, we
also find the positive x-intercept of

$\dfrac{x^2}{5}-\dfrac{y^2}{2}=1$ to be $x = 2.236$.

The area is given by
$$\int_{2.236}^{2.34} y_1 - y_2 + \int_{2.34}^{4.2} y_1 - y_3$$
$$= \int_{2.236}^{2.34} y_1 - \int_{2.236}^{2.34} y_2 + \int_{2.34}^{4.2} y_1 - \int_{2.34}^{4.2} y_3$$
Use the $\int f(x)\,dx$ function under the calc menu

making sure the correct y equation is displayed in
the upper left corner for each integral to find the
area is
0.03008441 − (−0.0300844) + 2.7254917 −
0.68880636 ≈ 2.097
An easier alternative to evaluating each separate

integral is to use the f_nInt function. From the
home screen, select f_nInt from the math menu
and enter
$$f_n\text{Int}(y_1 - y_2, x, 2.236, 2.34)$$
$$+ f_n\text{Int}(y_1 - y_3, x, 2.34, 4.2)$$
to find the area. You input the y equations by
pressing $\boxed{\text{vars}}$ and selecting which y equation you
want from the function window under y-vars.

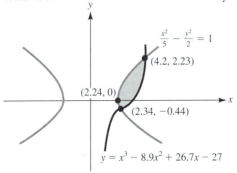

67. Let t_1 represent the starting time of an
arbitrary time interval and let t_2 represent
the ending time. Also, let $S(t)$ represent the
distance function. Then, the average value of

the velocity is $\dfrac{S(t_2)-S(t_1)}{t_2 - t_1}$.

The average velocity is

$$\frac{1}{t_2 - t_1}\int_{t_1}^{t_2} v(t)\,dt$$

Since distance is the integral of velocity,

$$= \frac{1}{t_2 - t_1}\left[S(t)\Big|_{t_1}^{t_2} \right]$$

$$= \frac{1}{t_2 - t_1}[S(t_2)-S(t_1)]$$

$$= \frac{S(t_2)-S(t_1)}{t_2 - t_1}$$

5.5 Additional Applications of Integration to Business and Economics

1. (a) $D(q) = 2(64 - q^2)$

$$A(6) = 2\int_0^6 (64 - q^2)\,dq$$

$$= 2\left(64q - \frac{q^3}{3}\right)\Big|_0^6$$

$$= \$624$$

(b) The consumer's willingness to spend in part (a) is the area under the demand curve from $q = 0$ to $q = 6$.

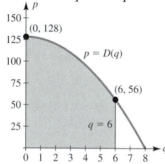

3. (a) $D(q) = \dfrac{400}{0.5q + 2}$

$$A(12) = 400\int_0^{12} \frac{1}{0.5q + 2}\,dq$$

$$= 800\ln|0.5q + 2|\Big|_0^{12}$$

$$= 800\ln 4$$

$$= \$1,109.04$$

(b) The consumer's willingness to spend in part (a) is the area under the demand curve from $q = 0$ to $q = 12$.

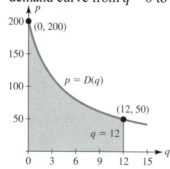

5. (a) $D(q) = 40e^{-0.05q}$

$$A(10) = 40\int_0^{10} e^{-0.05q}\,dq$$

$$= -800e^{-0.05q}\Big|_0^{10}$$

$$= \$314.78$$

(b) The consumer's willingness to spend in part **(a)** is the area under the demand curve from $q = 0$ to $q = 10$.

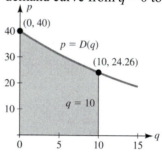

7. $D(q) = p_0$ if $110 = 2(64 - q^2)$ or $q = 3$. The consumer's surplus is

$$CS = \int_0^3 2(64 - q^2)\,dq - 3(110)$$

$$= 2\left(64q - \frac{q^3}{3}\right)\Big|_0^3 - 330$$

$$= \$36$$

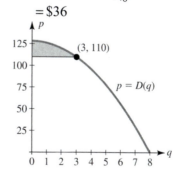

9. $D(q) = p_0$ if $31.15 = 40e^{-0.25}$ or $q = 5$.
The consumer's surplus is

$$CS = \int_0^5 40e^{-0.05q}\,dq - 5(31.15)$$

$$= -800e^{-0.05q}\Big|_0^5 - 93.45$$

$$= \$21.20$$

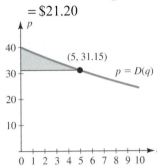

11. $S(q) = 0.3q^2 + 30$, $p_0 = S(4) = \$34.80$.
The producer's surplus is

$$PS = 4(34.80) - \int_0^4 (0.3q^2 + 30)\,dq$$

$$= 139.20 - (0.1q^3 + 30q)\Big|_0^4$$

$$= \$12.80$$

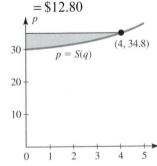

13. $S(q) = 10 + 15e^{0.03q}$, $p_0 = S(3) = \$26.41$.
The producer's surplus is

$$PS = 3(26.41) - \int_0^3 (10 + 15e^{0.03q})\,dq$$

$$= 79.23 - (10q + 500e^{0.03q})\Big|_0^3$$

$$= \$2.14$$

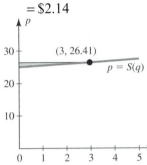

15. (a) The supply equals demand when

$$50 + \frac{2}{3}q^2 = 131 - \frac{1}{3}q^2$$

$$q^2 = 81, \text{ or } q = 9$$

So, the equilibrium price is

$$p_e = D(9) = 131 - \frac{1}{3}(9)^2 = \$104.$$

(b) The corresponding consumer's surplus
is $CS = \int_0^9 \left(131 - \frac{1}{3}q^2\right)dq - 9(104)$

$$= \left(131q - \frac{1}{9}q^3\right)\Big|_0^9 - 936$$

$$= 162, \text{ or } \$162$$

since $q_0 = 9$ means 9,000 units will
be supplied and the corresponding
producer's surplus is

$$PS = (9)(104) - \int_0^9 \left(50 + \frac{2}{3}q^2\right)dq$$

$$= 936 - \left(50q + \frac{2}{9}q^3\right)\Big|_0^9$$

$$= 324, \text{ or } \$324$$

17. (a) The supply equals demand when

$$-0.3q^2 + 70 = 0.1q^2 + q + 20$$
$$0 = 0.4q^2 + q - 50$$
$$q = \frac{-1 \pm \sqrt{1 + 4(0.4)(50)}}{2(0.4)}$$
$$= 10$$

So, the equilibrium price is

$$p_e = D(10) = -0.3(10)^2 + 70 = \$40.$$

(b) The corresponding consumer's surplus

is $CS = \int_0^{10} (-0.3q^2 + 70) \, dq - 10(40)$

$$= (-0.1q^3 + 70q) \Big|_0^{10} - 400$$
$$= 200, \text{ or } \$200$$

since $q_0 = 10$ means 10,000 units will be supplied and the corresponding producer's surplus is

$$PS = 10(40) - \int_0^{10} (0.1q^2 + q + 20) \, dq$$
$$= 400 - \left(\frac{0.1}{3} q^3 + \frac{q^2}{2} + 20q \right) \Big|_0^{10}$$
$$\approx 116.67, \text{ or } \$116.67$$

19. (a) The supply equals demand when

$$\frac{1}{3}(q + 1) = \frac{16}{q + 2} - 3$$
$$\frac{(q + 1)}{3} = \frac{10 - 3q}{q + 2}$$
$$0 = q^2 + 12q - 28$$
$$q = \frac{-12 \pm \sqrt{(12)^2 + 4(1)(28)}}{2(1)}$$

or, $q = 2$.
So, the equilibrium price is

$$p_e = D(2) = \frac{16}{2 + 2} - 3 = \$1.$$

(b) The corresponding consumer's surplus

is $\int_0^2 \left(\frac{16}{q + 2} - 3 \right) dq - 2(1)$

$$= \left(16 \ln|q + 2| - 3q \right) \Big|_0^2 - 2$$
$$= 3.09, \text{ or } \$3.09$$

since $q_0 = 2$ means 2,000 units will be supplied and the corresponding producer's surplus is

$$PS = 2(1) - \int_0^2 \frac{1}{3}(q + 1) \, dq$$
$$= 2 - \frac{1}{3} \left(\frac{q^2}{2} + q \right) \Big|_0^2$$
$$= 0.67, \text{ or } \$0.67$$

21. (a) The use of the machine will be profitable as long as the rate at which revenue is generated is greater than the rate at which costs accumulate. That is, until $R'(t) = C'(t)$

$$7,250 - 18t^2 = 3,620 + 12t^2$$

or $t = 11$ years.

(b) The rate at which net earnings are generated by the machine is $R'(t) - C'(t)$.

So, the net earnings over the next 11 years is

$$\int_0^{11} [R'(t) - C'(t)] dt$$
$$= \int_0^{11} [(7,250 - 18t^2) - (3,620 + 12t^2)] dt$$
$$= \int_0^{11} (3,630 - 30t^2) dt$$
$$= (3,630t - 10t^3) \Big|_0^{11}$$
$$= \$26,620$$

(c)

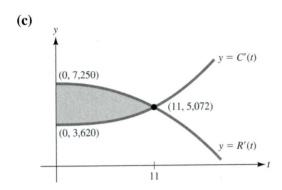

23. (a) The drive is profitable as long as rate of revenue exceeds weekly expenses.

$$e^{-0.3t} = \frac{593}{6,537} = 0.090714,$$
$$-0.3t = \ln 0.090714,$$
or $t = 8$ weeks.

(b) The net earnings during the first 8 weeks are

$$N = \int_0^8 (6,537e^{-0.3t} - 593)dt$$
$$= \left(-\frac{6,537}{0.3}e^{-0.3t} - 593t \right)\Big|_0^8$$
$$= \frac{6,537}{0.3}(1 - 0.09072) - (593)(8)$$
$$= 19,813.26 - 4,744$$
$$= \$15,069.26$$

(c) In geometric terms, the net earnings in part **(b)** is the area of the region between the curves $y = R'(t)$ and $y = E'(t)$.

Rewrite as: $y = R'(t) = 6,537e^{-0.3t}$ and $y = E'(t) = 593$

25.
$$\text{amount} = \int_0^{10} 1000e^{0.1(10-t)}dt$$
$$= 1000e^1 \int_0^{10} e^{-0.1t}dt$$
$$= -10,000e(e^{-0.1t})\Big|_0^{10}$$
$$= -10,000e(e^{-1} - e^0)$$
$$\approx \$17,182.82$$

27. At age 60, Tom would have
$$\int_0^{35} 2500e^{0.05(35-t)}dt$$
$$= 2500e^{1.75} \int_0^{35} e^{-0.05t}dt$$
$$= -50,000e^{1.75}(e^{-0.05t})\Big|_0^{35}$$
$$= -50,000e^{1.75}(e^{-1.75} - e^0)$$
$$\approx \$237,730.13$$
At age 65, Tom would have
$$\int_0^{40} 2500e^{0.05(40-t)}dt$$
$$= 2500e^2 \int_0^{40} e^{-0.05t}dt$$
$$= -50,000e^2(e^{-0.05t})\Big|_0^{40}$$
$$= -50,000e^2(e^{-2} - e^0)$$
$$\approx \$319,452.80$$

29.
$$PV = \int_0^5 1200e^{-0.05t}dt$$
$$= -24,000(e^{-0.05t})\Big|_0^5$$
$$= -24,000(e^{-0.25} - e^0)$$
$$\approx \$5,308.78$$

31. The net income of the first investment is
$$\int_0^5 15,000e^{0.06(5-t)}dt - 50,000$$
$$= 15,000e^{0.3} \int_0^5 e^{-0.06t}dt - 50,000$$
$$= 87,464.70 - 50,000$$
$$= \$37,464.70$$
The net income of the second investment

is $\int_0^5 9000e^{0.06(5-t)}dt - 30,000$

$= 9000e^{0.3}\int_0^5 e^{-0.06t}dt - 30,000$

$= 52,478.82 - 30,000$

$= \$22,478.82$

So, the first investment will generate more income.

33. (a) The profit function is

$P(q)$

$= (110 - q)q - (q^3 - 25q^2 + 2q + 3,000)$

$= 110q - q^2 - q^3 + 25q^2 - 2q - 3,000$

$= -q^3 + 24q^2 + 108q - 3,000$

(b) $P'(q) = -3q^2 + 48q + 108$

$= -3(q^2 - 16q - 36)$

So, $P'(q) = 0$ when

$q = \dfrac{24 \pm \sqrt{24^2 + 3(108)}}{3} = 18$

$P''(q) = -6q + 48$ and $P''(18) < 0$, so $q = 18$ corresponds to the maximum profit.

(c) When $q = 18$, the price is

$p = 110 - 18 = 92$ and the corresponding consumer's surplus is

$CS = \int_0^{18}(110 - q)dq - 18(92)$

$= \left(110q - \dfrac{q^2}{2}\right)\Big|_0^{18} - 1656$

$= \$162$

35. (a) $P(t) = \int P'(t)\,dt$

$= \int 1.3e^{0.04t}\,dt$

$= 1.3\int e^{0.04t}\,dt$

$= 32.5e^{0.04t} + C$

When $t = 0$, $P(0) = 0$ so $C = -32.5$ and

$P(t) = 32.5e^{0.04t} - 32.5$.

When $t = 3$,

$P(3) = 32.5e^{0.04(3)} - 32.5 \approx 4.14$ billion barrels.

Over the following three years, the amount pumped is $P(6) - P(3)$, or

$= (32.5e^{0.04(6)} - 32.5) - 4.14$

≈ 4.68 billion barrels

(b) The field stops operating when it uses up the 20 billion barrels it holds, or when

$20 = 32.5e^{0.04t} - 32.5$

$\dfrac{21}{13} = e^{0.04t}$

$\ln \dfrac{21}{13} = \ln e^{0.04t}$, or

$t = \dfrac{\ln \frac{21}{13}}{0.04}$, or approximately

12 years

(c) $PV = \int V(t)e^{-rt}\,dt$

$= \int 112P'(t)e^{-rt}\,dt$

$= \int_0^{12} 112(1.3e^{0.04t})e^{-0.05t}\,dt$

$= 145.6\int_0^{12} e^{-0.01t}\,dt$

$= -14,560\left(e^{-0.01t}\Big|_0^{12}\right)$

$= -14,560(e^{-0.12} - e^0)$

$\approx 1,646.44$ billion dollars

(d) Writing exercise—Answers will vary.

37. (a) $P(t) = \int P'(t)\,dt$

$= \int 1.2e^{0.02t}\,dt$

$= 1.2\int e^{0.02t}\,dt$

$= 60e^{0.02t} + C$

When $t = 0$, $P(0) = 0$ so $C = -60$ and

$P(t) = 60e^{0.02t} - 60$.

When $t = 3$, $P(3) = 60e^{0.02(3)} - 60$

≈ 3.71 billion barrels

$P(6) - P(3) = (60e^{0.02(6)} - 60) - 3.71$

≈ 3.94 billion barrels

(b) $12 = 60e^{0.02t} - 60$

$$\frac{6}{5} = e^{0.02t}$$

$$\ln\frac{6}{5} = \ln e^{0.02t}$$

$$\ln\frac{6}{5} = 0.02t, \text{ or}$$

$$t = \frac{\ln\frac{6}{5}}{0.02}, \text{ or approximately}$$

9.12 years

(c) $PV = \int V(t)e^{-rt}\,dt$

$$= \int 112e^{0.015t}P'(t)e^{-rt}\,dt$$

$$= \int_0^{9.12} 112e^{0.015t}\left(1.2e^{0.02t}\right)e^{-0.05t}\,dt$$

$$= 134.4\int_0^{9.12} e^{-0.015t}\,dt$$

$$= -8,960\left(e^{-0.015t}\right)\Big|_0^{9.12}$$

$$= -8,960\left(e^{-0.1368} - e^0\right)$$

$$\approx \$1,146 \text{ billion}$$

(d) Writing exercise—Answers will vary.

39. $PV = 10 \text{ million} = \int_0^6 Ae^{-0.05t}\,dt$

$$10 = -20A(e^{-0.05t})\Big|_0^6$$

$$10 = -20A(e^{-0.3} - e^0)$$

$$\frac{1}{-2(e^{-0.3} - 1)} = A \approx 1.929148 \text{ million, or}$$

$$\$1,929,148.$$

41. $A(t) = 10e^{1-0.05t}$

(a) $FV = \int_0^5 10e^{1-0.05t} \cdot e^{0.05(5-t)}\,dt$

$$= 10\int_0^5 e^{(1-0.05t)+(0.25-0.05t)}\,dt$$

$$= 10\int_0^5 e^{1.25-0.1t}\,dt$$

$$= 10e^{1.25}\int_0^5 e^{-0.1t}\,dt$$

$$= 10e^{1.25}\left(-10e^{-0.1t}\Big|_0^5\right)$$

$$= -100e^{1.25}\left(e^{-0.1t}\Big|_0^5\right)$$

$$= -100e^{1.25}(e^{-0.5} - e^0)$$

$$= -100e^{1.25}(e^{0.5} - 1)$$

$$\approx 137.33429$$

or \$137,334.29

(b) $PV = \int_1^3 10e^{1-0.05t} \cdot e^{-0.05t}\,dt$

$$= 10\int_1^3 e^{(1-0.05t)-0.05t}\,dt$$

$$= 10\int_1^3 e^{1-0.1t}\,dt$$

$$= 10e^1\int_1^3 e^{-0.1t}\,dt$$

$$= 10e\left(-10e^{-0.1t}\Big|_1^3\right)$$

$$= -100e\left(e^{-0.1t}\Big|_1^3\right)$$

$$= -100e(e^{-0.3} - e^{0.1})$$

$$\approx 44,585.04$$

or \$44,585.04

43. (a) $R'(t) = 300(18 + 0.3t^{1/2})$

$$FV = \int_0^{36} 300(118 + 0.3t^{1/2})\,dt$$

$$= 300\int_0^{36} (118 + 0.3t^{1/2})\,dt$$

$$= 300(118t + 0.2t^{3/2})\Big|_0^{36}$$

$$= 300(4248 + 43.2)$$

$$= \$1,287,360$$

(b) Writing exercise—Answers will vary.

45. (a) $FV = \int_0^T f(t)e^{r(T-t)}\,dt$

$$= \int_0^T Me^{r(T-t)}\,dt$$

$$= \int_0^T Me^{rT} \cdot e^{-rt}\,dt$$

$$= Me^{rT}\int_0^T e^{-rt}\,dt$$

$$= \frac{Me^{rT}}{-r}\left(e^{-rt}\Big|_0^T\right)$$

$$= -\frac{Me^{rT}}{r}(e^{-rT} - e^0)$$

$$= \frac{M}{r}\left(-e^0 + e^{rT}\right)$$

$$= \frac{M}{r}(e^{rT} - 1)$$

(b) $PV = \int_0^T f(t)e^{-rt}\,dt$

$$= \int_0^T Me^{-rt}\,dt$$

$$= M\int_0^T e^{-rt}\,dt$$

$$= -\frac{M}{r}\left(e^{-rt}\Big|_0^T\right)$$

$$= -\frac{M}{r}\left(e^{-rT} - e^0\right)$$

$$= \frac{M}{r}\left(-e^{-rT} + e^0\right)$$

$$= \frac{M}{r}\left(1 - e^{-rT}\right)$$

5.6 Additional Applications of Integration to the Life and Social Sciences

1. After 5 months, the number of the original population surviving is $50,000e^{-0.1(5)}$. The number of new members surviving after 5 months is $\int_0^5 40e^{-0.1(5-t)}\,dt$.

So, the total will be

$$= 50,000e^{-0.5} + 40e^{-0.5}\int_0^5 e^{0.1t}\,dt$$

$$= e^{-0.5}\left[50,000 + 400(e^{0.1t})\Big|_0^5\right]$$

$$\approx 30,484 \text{ members.}$$

3. After 3 years, the number of the original population surviving is $500,000e^{-0.011(3)}$. The number of new members surviving after 3 years is $\int_0^3 800e^{-0.011(3-t)}\,dt$.

So, the total will be

$$500,000e^{-0.033} + 800e^{-0.033}\int_0^3 e^{0.011t}\,dt$$

$$= 800e^{-0.033}\left[625 + \frac{1}{0.011}(e^{0.011t})\Big|_0^3\right]$$

$$\approx 486,130 \text{ members}$$

5. After 8 years, the number of the original population surviving is $500,000e^{-0.013(8)}$. The number of new members surviving after 8 years is

$$\int_0^8 100e^{0.01t} - e^{-0.013(8-t)}\,dt$$

$$= 100\int_0^8 e^{0.01t-0.104+0.013t}\,dt$$

$$= 100e^{-0.104}\int_0^8 e^{0.023t}\,dt$$

So, the total will be

$$500,000e^{-0.104} + 100e^{-0.104}\int_0^8 e^{0.023t}\,dt$$

$$= 100e^{-0.104}\left[5000 + \frac{1}{0.023}e^{0.023t}\Big|_0^8\right]$$

$$= 100e^{-0.104}\left[5000 + \frac{1000}{23}\left(e^{0.023t}\Big|_0^8\right)\right]$$

$$= 100,000e^{-0.104}\left[5 + \frac{1}{23}(e^{0.184} - e^0)\right]$$

$$= 100,000e^{-0.104}\left[5 + \frac{1}{23}(e^{0.184} - 1)\right]$$

$$\approx 451,404 \text{ members}$$

7. Volume of $S = \pi \int_0^1 (3x+1)^2 \, dx$

$$= \pi \int_0^1 (9x^2 + 6x + 1) \, dx$$

$$= \pi \left(3x^3 + 3x^2 + x \Big|_0^1 \right)$$

$$= \pi [(3+3+1) - (0)]$$

$$= 7\pi$$

9. Volume of S

$$= \pi \int_{-1}^3 (x^2 + 2)^2 \, dx$$

$$= \pi \int_{-1}^3 (x^4 + 4x^2 + 4) \, dx$$

$$= \pi \left(\frac{x^5}{5} + \frac{4x^3}{3} + 4x \Big|_{-1}^3 \right)$$

$$= \pi \left[\left(\frac{243}{5} + \frac{108}{3} + 12 \right) - \left(-\frac{1}{5} - \frac{4}{3} - 4 \right) \right]$$

$$= \pi \left(\frac{729}{15} + \frac{540}{15} + \frac{180}{15} + \frac{3}{15} + \frac{20}{15} + \frac{60}{15} \right)$$

$$= \frac{1532}{15} \pi$$

11. Volume of $S = \pi \int_{-2}^2 \left(\sqrt{4 - x^2} \right)^2 dx$

$$= \pi \int_{-2}^2 (4 - x^2) \, dx$$

$$= \pi \left(4x - \frac{x^3}{3} \Big|_{-2}^2 \right)$$

$$= \left[\left(8 - \frac{8}{3} \right) - \left(-8 + \frac{8}{3} \right) \right]$$

$$= \pi \left(\frac{24}{3} - \frac{8}{3} + \frac{24}{3} - \frac{8}{3} \right)$$

$$= \frac{32}{3} \pi$$

13. Volume of $S = \pi \int_1^{e^2} \left(\frac{1}{\sqrt{x}} \right)^2 dx$

$$= \pi \int_1^{e^2} \frac{1}{x} \, dx$$

$$= \pi \left(\ln x \Big|_1^{e^2} \right)$$

$$= \pi (\ln e^2 - \ln 1)$$

$$= \pi (2 - 0)$$

$$= 2\pi$$

15. $P(t) = \int P'(t) \, dt = \int e^{0.02t} \, dt = 50e^{0.02t} + C$

When $t = 0$ $P(0) = 50$

$50 = 50e^0 + C$, or $C = 0$.

So, $P(t) = 50e^{0.02t}$ and

$P(10) = 50e^{0.02(10)} \approx 61.07$ million, or

61,070,138 people.

17. After 8 months, the number of the original members remaining is $200e^{-0.2(8)}$.

The number of new members remaining is

$\int_0^8 10e^{-0.2(8-t)} \, dt$.

So, the total will be

$$200e^{-1.6} + 10e^{-1.6} \int_0^8 e^{0.2t} \, dt$$

$$= 10e^{-1.6} \left[20 + 5(e^{0.2t}) \Big|_0^8 \right]$$

$$\approx 80 \text{ members}$$

19. After 30 days, the number of those originally infected who still have the disease is $5000e^{-0.02(30)}$.

The number of those since infected who still have the disease is

$\int_0^{30} 60e^{-0.02(30-t)} \, dt$.

So, the total still infected will be

$$20e^{-0.6}\left(250+3\int_0^{30}e^{0.02t}\,dt\right)$$

$$=20e^{-0.6}\left[250+150(e^{0.02t})\Big|_0^{30}\right]$$

$$=1{,}000e^{-0.6}\left[5+3(e^{0.02t})\Big|_0^{30}\right]$$

$$\approx 4{,}098\text{ people}$$

21. $\displaystyle\int_0^{10}30e^{0.1t}\,dt=300(e^{0.1t})\Big|_0^{10}$
$$\approx 515.48\text{ billion barrels.}$$

23. After 10 months, the number of the original members remaining is $8{,}000e^{-10/10}$.
The number of new members remaining is $\displaystyle\int_0^{10}200e^{-(10-t)/10}\,dt$.
So, the total will be

$$200e^{-1}\left(40+\int_0^{10}e^{t/10}\,dt\right)$$

$$=200e^{-1}\left[40+10(e^{t/10})\Big|_0^{10}\right]$$

$$\approx 4{,}207\text{ members}$$

25. From text Example 5.6.4, the total quantity of blood flowing through an artery is $\dfrac{\pi kR^4}{2}$.

The average velocity is $\dfrac{\text{total quantity}}{\text{area}}$

$$=\frac{\pi kR^4/2}{\pi R^2}=\frac{\pi kR^4}{2}\cdot\frac{1}{\pi R^2}$$

$$=\frac{kR^2}{2}$$

Need to show that the maximum velocity is kR^2. Maximum flow occurs at the central axis, where $r=0$. So, the maximum velocity is
$$S(0)=k\left(R^2-0^2\right)=kR^2.$$ The average velocity is half of the maximum velocity.

27. # people $=\displaystyle\int_1^2 2\pi r(25{,}000e^{-0.05r^2})\,dr$
$$=50{,}000\pi\int_1^2 re^{-0.05r^2}\,dr$$

Let $u=-0.05r^2$; then $-10\,du=r\,dr$ and the limits of integration become $-0.05(1)^2=-0.05$ and $-0.05(2)^2=-0.2$.
So,

$$=50{,}000\pi\int_{-0.05}^{-0.2}e^u\cdot-10\,du$$

$$=500{,}000\pi\int_{-0.2}^{-0.05}e^u\,du$$

$$=500{,}000\pi\left(e^u\Big|_{-0.2}^{-0.05}\right)$$

$$=500{,}000\pi(e^{-0.05}-e^{-0.2})$$

$$\approx 208{,}128\text{ people}$$

29. (a) $\displaystyle\int_0^3 0.3t(49-t^2)^{0.4}\,dt$

Using substitution with $u=49-t^2$,

$$=\frac{-0.3}{2}\int_{49}^{40}u^{0.4}\,du$$

$$=\frac{0.3}{2}\int_{40}^{49}u^{0.4}\,du$$

$$=\frac{0.3}{2.8}(u^{1.4})\Big|_{40}^{49}$$

$$\approx 6.16,$$

so LDL decreases by approximately 6.16 units.

(b) $L(t)=\displaystyle\int L'(t)\,dt$
$$=\int 0.3t(49-t^2)^{0.4}\,dt$$
$$=\frac{3}{28}(49-t^2)^{1.4}+C$$
When $t=0$, $L(t)=120$ so
$$120=\frac{3}{28}(49)^{1.4}+C,\text{ or}$$
$$C=120-\frac{3}{28}(49)^{1.4}$$

So,

$L(t)$

$= \dfrac{3}{28}(49 - t^2)^{1.4} + 120 - \dfrac{3}{28}(49)^{1.4}$

$= \dfrac{3}{28}(49 - t^2)^{1.4} + 120 - \dfrac{21}{4}(49)^{0.4}$

(c) To find how many days it takes for patient's LDL level to be safe,

Press $\boxed{y =}$

Input $(3/28)(49 - x^2)\wedge(1.4) + 120$
$- (21/4)(49)\wedge(0.4)$ for $y_1 =$.

Use window dimensions [0, 10]1 by [0, 200]20.

Press $\boxed{\text{graph}}$.

Use $\boxed{\text{trace}}$ and zoom-in to find that $y = 100$ when $x \approx 5.8$.

Therefore, it takes approximately 5.8 days for the LDL level to be safe.

31. Using the result of problem #24,

$P(10) = 3{,}000e^{-0.07(10)}$

$\qquad + \displaystyle\int_0^{10} 10e^{0.01t} e^{-0.07(10-t)} \, dt$

$= 3{,}000e^{-0.7} + 10e^{-0.7} \displaystyle\int_0^{10} e^{0.08t} \, dt$

$= 10e^{-0.7} \left[300 + \dfrac{1}{0.08}(e^{0.08t}) \Big|_0^{10} \right]$

$\approx 1{,}566$ members of the species.

33. Using the result of problem #24,

$P(10) = 85{,}000 \left(\dfrac{1}{10 + 1} \right)$

$\qquad + \displaystyle\int_0^{10} 1{,}000 \dfrac{1}{(10 - t) + 1} \, dt$

$= \dfrac{85{,}000}{11} + 1{,}000 \displaystyle\int_0^{10} \dfrac{1}{11 - t} \, dt$

$= \dfrac{85{,}000}{11} - 1{,}000 \displaystyle\int_{11}^1 \dfrac{1}{u} \, du$

$= \dfrac{85{,}000}{11} + 1{,}000 \displaystyle\int_1^{11} \dfrac{1}{u} \, du$

$= \dfrac{85{,}000}{11} + 1{,}000 \ln|u| \Big|_1^{11}$

$\approx 10{,}125$ people.

35. $D(t) = \displaystyle\int D'(t) \, dt$

$\qquad = \displaystyle\int 0.12 + \dfrac{0.08}{t + 1} \, dt$

$\qquad = 0.12t + 0.08 \ln|t + 1| + C$

When $t = 0$, $D(0) = 0$ so $C = 0$ and
$D(t) = 0.12t + 0.08 \ln|t + 1|$.

When $t = 12$ months (1 year),

$D(12) = 0.12(12) + 0.08 \ln|12 + 1|$
$\qquad \approx 1.65$, or 165 infected people

of those inoculated.

Of those not inoculated,

$W(t) = \displaystyle\int W'(t) \, dt$

$\qquad = \displaystyle\int \dfrac{0.8e^{0.13t}}{(1 + e^{0.13t})^2} \, dt$

Using substitution, with $u = 1 + e^{0.13t}$,

$= 0.8 \displaystyle\int \dfrac{1}{(1 + e^{0.13t})^2} e^{0.13t} \, dt$

$= \dfrac{0.8}{0.13} \displaystyle\int u^{-2} \, du$

$= \dfrac{80}{13} \left[\dfrac{-1}{(1 + e^{0.13t})} \right] + C$

When $t = 0$, $W(0) = 0$, so

$0 = \dfrac{80}{13} \left(\dfrac{-1}{2} \right) + C$, or $C = \dfrac{40}{13}$ and

$W(t) = \dfrac{-80}{13(1 + e^{0.13t})} + \dfrac{40}{13}$.

So, after 12 months,

$W(12) = \dfrac{-80}{13(1 + e^{0.13(12)})} + \dfrac{40}{13}$

$\qquad \approx 2.01$, or approximately

201 people infected.

So, approximately $201 - 165 = 36$ people protected by the drug, or

$\dfrac{W(12) - D(12)}{W(12)} \approx 18.1\%$.

37. (a) At birth,

$$L(0) = \frac{110e^0}{1+e^0} = 55 \text{ years of age.}$$

$$L(50) = \frac{110e^{0.015(50)}}{1+e^{0.015(50)}} \approx 74.7 \text{ years of age.}$$

(b) $L_{av} = \dfrac{1}{70-10} \displaystyle\int_{10}^{70} \dfrac{110e^{0.015t}}{1+e^{0.015t}} dt$

Using substitution, with $u = 1 + e^{0.015t}$,

$$= \frac{110}{60} \int_{10}^{70} \frac{1}{1+e^{0.015t}} e^{0.015t} dt$$

$$= \frac{11}{6(0.015)} \int_{1+e^{0.15}}^{1+e^{1.05}} \frac{1}{u} du$$

$$= \frac{11}{0.09} \left(\ln|u| \right) \Big|_{1+e^{0.15}}^{1+e^{1.05}}$$

$$\approx 70.78 \text{ years of age}$$

(c) To find the age T such that $L(T) = T$, we must find T such that

$$\frac{110e^{0.015T}}{1+e^{0.015T}} = T$$

$$110e^{0.015T} - T(1+e^{0.015T}) = 0$$

Press $\boxed{y=}$ and input $110e \wedge (0.015x) - (x * (1 + e \wedge (0.015x)))$ for $y_1 =$.
Use window dimensions [0, 100]10 by [−10, 120]20.
Press $\boxed{\text{graph}}$.
Use the zero function under the calc menu to find that $T \approx 86.4$ years.
On the average, this is how long people in this country live.

(d) $L_e = \dfrac{1}{86.4-0} \displaystyle\int_0^{86.4} \dfrac{110e^{0.015t}}{1+e^{0.015t}} dt$

Using substitution as before,

$$= \frac{110}{(86.4)(0.015)} \int_2^{1+e^{1.296}} \frac{1}{u} du$$

$$= \frac{110}{1.296} [\ln(1+e^{1.296}) - \ln 2]$$

$$\approx 71.7 \text{ years of age}$$

39. (a) $0 = -0.41t^2 + 0.97t = t(0.97 - 0.41t)$
so $R(t) = 0$ when $t = 0$ and when $t \approx 2.37$ sec.

(b) Volume $= \displaystyle\int_0^{2.37} (-0.41t^2 + 0.97t) dt$

$$= \left(\frac{-0.41}{3} t^3 + \frac{0.97}{2} t^2 \right) \Big|_0^{2.37}$$

$$\approx 0.905 \text{ liters}$$

(c) R_{av}

$$= \frac{1}{2.37-0} \int_0^{2.37} (-0.41t^2 + 0.97t) dt$$

$$\approx \frac{0.905}{2.37} \approx 0.382 \text{ liters/sec.}$$

41. $T(r) = \dfrac{3}{2+r} = 3(2+r)^{-1}$

(a) domain: $[0, \infty)$
intercepts:
when $r = 0$, $T(0) = \dfrac{3}{2}$; point $\left(0, \dfrac{3}{2} \right)$

when $T(r) = 0$, no solution
vertical asymptote outside of domain
$(r = -2)$
horizontal asymptote

$$\lim_{r \to \infty} \frac{\frac{3}{r}}{\frac{2}{r}+1} = 0, \text{ or } y = 0$$

$$T'(r) = -\frac{3}{(2+r)^2} = -3(2+r)^{-2}$$

$$T''(r) = \frac{6}{(2+r)^3}$$

When $r \geq 0$,

$T'(r) < 0$ so T is decreasing
$T''(r) > 0$ so T is concave up

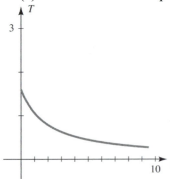

(b) $T(r) = \dfrac{3}{2+r}$

$2 + r = \dfrac{3}{T}$

$r(T) = \dfrac{3}{t} - 2$

Graph is reflection of the graph in part
(a) over the line $y = x$.

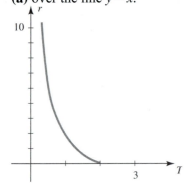

(c) When $r = 0$, $T = \dfrac{3}{2}$ and when $r = 7$,

$T = \dfrac{1}{3}$.

Volume

$= \pi \int_{1/3}^{3/2} \left(\dfrac{3}{T} - 2 \right)^2 dT$

$= \pi \int_{1/3}^{3/2} \left(\dfrac{9}{T^2} - \dfrac{12}{T} + 4 \right) dT$

$= \pi \left[-\dfrac{9}{T} - 12 \ln T + 4T \right]_{1/3}^{3/2}$

$= \pi \left[\left(-6 - 12 \ln \dfrac{3}{2} + 6 \right) \right.$

$\left. \qquad - \left(-27 - 12 \ln \dfrac{1}{3} + \dfrac{4}{3} \right) \right]$

$= \pi \left[-12 \ln \dfrac{3}{2} + \dfrac{81}{3} + 12 \ln \dfrac{1}{3} - \dfrac{4}{3} \right]$

$= \pi \left[12 \ln \dfrac{1}{3} - 12 \ln \dfrac{3}{2} + \dfrac{77}{3} \right]$

$\approx 23.93 \text{ ft}^3$

43. $p(r) = \dfrac{200}{5 + 2r^2}$

(a) Since the pollution is distributed in a circular fashion about the smoke stack,

$\text{pollution} = 2\pi \int_0^3 r \left(\dfrac{200}{5 + 2r^2} \right) dr$

$\qquad = 400\pi \int_0^3 \dfrac{r}{5 + 2r^2} dr$

Using substitution with $u = 5 + 2r^2$,

$\dfrac{1}{4} du = r \, dr$ and limits of integration

$u_1 = 5$ and $u_2 = 23$,

$$= 400\pi \int_5^{23} \frac{1}{u} \cdot \frac{1}{4} \, du$$

$$= 100\pi \int_5^{23} \frac{1}{u} \, du$$

$$= 100\pi \left(\ln u \Big|_5^{23} \right)$$

$$= 100\pi (\ln 23 - \ln 5)$$

$$= 100\pi \ln \frac{23}{5} \approx 479.42 \text{ units}$$

(b) $4 = \dfrac{200}{5 + 2r^2}$

$$L = r = \sqrt{\frac{45}{2}} = \frac{3\sqrt{10}}{2} \approx 4.74 \text{ miles}$$

amount of pollution

$$= 2\pi \int_0^{\frac{3\sqrt{10}}{2}} r \left(\frac{200}{5 + 2r^2} \right) dr$$

$$= 100\pi \left(\ln u \Big|_5^{50} \right)$$

$$= 100\pi (\ln 50 - \ln 5)$$

$$= 100\pi \ln 10 \approx 723.38 \text{ units}$$

45. $y = \sqrt{x}\,(3 - x) = 3x^{1/2} - x^{3/2}$

The graph of y crosses the x-axis when $0 = \sqrt{x}$ or $0 = 3 - x$. That is, $x = 0$ and $x = 3$. The volume of the tumor is approximately

$$= \pi \int_0^3 \left(3x^{1/2} - x^{3/2} \right)^2 dx$$

$$= \pi \int_0^3 \left(9x - 6x^2 + x^3 \right) dx$$

$$= \pi \left[\left(\frac{9x^2}{2} - \frac{6x^3}{3} + \frac{x^4}{4} \right) \Big|_0^3 \right]$$

$$= \pi \left[\frac{9(3)^2}{2} - 2(3)^3 + \frac{(3)^4}{4} - 0 \right]$$

$$= \pi (6.75) \approx 21.21 \text{ cm}^3$$

47. For the first colony, the number of bacteria after 50 days will be

$$100{,}000e^{-0.011(50)} + \int_0^{50} 50e^{-0.011(50-t)} \, dt$$

$$= 100{,}000e^{-0.55} + 50e^{-0.55} \int_0^{50} e^{0.011t} \, dt$$

$$= 50e^{-0.55} \left[2{,}000 + \frac{1}{0.011} (e^{0.011t}) \Big|_0^{50} \right]$$

$$\approx 59{,}618$$

The number in the second colony will be

$$P(50) = \frac{5{,}000}{1 + 49e^{0.009(50)}}$$

$$\approx 64.228, \text{ or } 64{,}228$$

So, after 50 days, the population is larger in the second colony.

Similarly, after 100 days, the first colony's population will be

$$100{,}000e^{-0.011(100)} + \int_0^{100} 50e^{-0.011(100-t)} \, dt$$

$$\approx 36{,}320$$

and the second colony will be

$$P(100) = \frac{5{,}000}{1 + 49e^{0.009(100)}}$$

$$\approx 41.145, \text{ or } 41{,}145$$

So, the second colony is still larger after 100 days. Similarly, after 300 days, the first will be

$$100{,}000e^{-0.011(300)}$$

$$+ \int_0^{300} 50e^{-0.011(300-t)} \, dt$$

$$\approx 8{,}066$$

and the second will be

$$P(300) = \frac{5{,}000}{1 + 49e^{0.009(300)}} \approx 6{,}848.$$

So, after 300 days, the first colony is now larger.

49. Volume $= \int_0^h \pi y^2 \, dx$

Since the hypotenuse of the triangle is

along the line $y = \dfrac{r}{h}x$,

$$= \pi \int_0^h \left(\frac{r}{h}x\right)^2 dx$$

$$= \frac{\pi r^2}{h^2} \int_0^h x^2 dx$$

$$= \frac{\pi r^2}{h^2} \left(\frac{x^3}{3}\bigg|_0^h\right)$$

$$= \frac{\pi r^2}{h^2} \left(\frac{h^3}{3} - 0\right)$$

$$= \frac{1}{3}\pi r^2 h$$

Checkup for Chapter 5

1. **(a)** $\displaystyle\int \left(x^3 - \sqrt{3x} + 5e^{-2x}\right) dx$

$$= \int x^3 dx - \sqrt{3} \int x^{1/2} dx + 5 \int e^{-2x} dx$$

$$= \frac{x^4}{4} - \frac{2\sqrt{3}}{3} x^{3/2} - \frac{5}{2} e^{-2x} + C$$

(b) $\displaystyle\int \frac{x^2 - 2x + 4}{x} dx$

$$= \int \left(x - 2 + \frac{4}{x}\right) dx$$

$$= \int x\, dx - 2\int dx + 4\int \frac{1}{x} dx$$

$$= \frac{x^2}{2} - 2x + 4\ln|x| + C$$

(c) $\displaystyle\int \sqrt{x}\left(x^2 - \frac{1}{x}\right) dx = \int (x^{5/2} - x^{-1/2}) dx$

$$= \frac{2}{7}x^{7/2} - 2x^{1/2} + C$$

(d) $\displaystyle\int \frac{x\, dx}{(3 + 2x^2)^{3/2}}$

Let $u = 3 + 2x^2$; then $\dfrac{1}{4} du = x\, dx$

$$= \frac{1}{4}\int u^{-3/2} du$$

$$= \frac{1}{4}(-2u^{-1/2}) + C$$

$$= \frac{-1}{2\sqrt{3 + 2x^2}} + C$$

(e) $\displaystyle\int \frac{\ln \sqrt{x}}{x} dx = \int \frac{\frac{1}{2}\ln x}{x} dx$

Let $u = \ln x$; then $du = \dfrac{1}{x} dx$

$$\frac{1}{2}\int (\ln x)\frac{1}{x} dx = \frac{1}{2}\int u\, du$$

$$= \frac{1}{4}(\ln x)^2 + C$$

(f) $\displaystyle\int x e^{1+x^2} dx$

Let $u = 1 + x^2$; then $\dfrac{1}{2} du = x\, dx$

$$= \int (e^{1+x^2})x\, dx$$

$$= \frac{1}{2}\int e^u du$$

$$= \frac{1}{2}e^{1+x^2} + C$$

2. (a) $\int_1^4 \left(x^{3/2} + \dfrac{2}{x} \right) dx$

$= \int_1^4 x^{3/2} dx + 2\int_1^4 \dfrac{1}{x} dx$

$= \dfrac{2}{5} x^{5/2} \Big|_1^4 + 2\big(\ln|x| \big)\Big|_1^4$

$= \dfrac{2}{5}[(4)^{5/2} - (1)^{5/2}] + 2[\ln 4 - \ln 1]$

$= \dfrac{62}{5} + 2\ln 4$

$= \dfrac{62}{5} + 2\ln 2^2$

$= \dfrac{62}{5} + 4\ln 2$

(b) $\int_0^3 e^{3-x} dx$

Let $u = 3 - x$; then $-du = dx$ and the limits of integration become $3 - 3 = 0$ and $3 - 0 = 3$

$= -\int_3^0 e^u du$

$= \int_0^3 e^u du$

$= e^3 - e^0$

$= e^3 - 1$

(c) $\int_0^1 \dfrac{x}{x+1} dx$

Let $u = x + 1$; then $du = dx$ and $x = u - 1$. Further, the limits of integration become $0 + 1 = 1$ and $1 + 2 = 2$

$= \int_1^2 \dfrac{u-1}{u} du$

$= \int_1^2 \left(1 - \dfrac{1}{u} \right) du$

$= \big(u - \ln|u| \big)\Big|_1^2$

$= (2 - \ln 2) - (1 - \ln 1)$

$= 1 - \ln 2$

(d) $\int_0^3 \dfrac{(x+3)dx}{\sqrt{x^2 + 6x + 4}}$

Let $u = x^2 + 6x + 4$; then

$du = (2x + 6)dx$ or, $\dfrac{du}{2} = (x + 3)dx.$

Further, the limits of integration become $0 + 6(0) + 4 = 4$ and $(3)^2 + 6(3) + 4 = 31$

$= \dfrac{1}{2} \int_4^{31} u^{-1/2} du$

$= \dfrac{1}{2} (2u^{1/2}) \Big|_4^{31}$

$= u^{1/2} \Big|_4^{31}$

$= \sqrt{31} - 2$

3. (a) Area $= \int_1^4 \left[\left(x + \sqrt{x} \right) - 0 \right] dx$

$= \int_1^4 x + x^{1/2} dx$

$= \left(\dfrac{x^2}{2} + \dfrac{2}{3} x^{3/2} \right)\Big|_1^4$

$= \left[\dfrac{(4)^2}{2} + \dfrac{2}{3}(4)^{3/2} \right]$

$\qquad - \left[\dfrac{1}{2} + \dfrac{2}{3}(1)^{3/2} \right]$

$= \dfrac{73}{6}$ sq units

(b) The limits of integration are

$x^2 - 3x = x + 5$

$x^2 - 4x - 5 = 0$

$(x - 5)(x + 1) = 0$

$\qquad x = -1, 5$

Further, from a sketch of the graphs, or by comparing function values between $-1 < x < 5$, $y = x + 5$ is the top curve.

$$\text{Area} = \int_{-1}^{5} [(x+5) - (x^2 - 3x)]dx$$

$$= \int_{-1}^{5} (4x + 5 - x^2)dx$$

$$= \left(2x^2 + 5x - \frac{x^3}{3}\right)\Bigg|_{-1}^{5}$$

$$= \left[2(5)^2 + 5(5) - \frac{(5)^3}{3}\right]$$

$$- \left[2(-1)^2 + 5(-1) - \frac{(-1)^3}{3}\right]$$

$$= 36 \text{ sq units}$$

4. $f_{av} = \dfrac{1}{2-1}\displaystyle\int_{1}^{2}\dfrac{x-2}{x}dx$

$$= \int_{1}^{2}\left(1 - \frac{2}{x}\right)dx$$

$$= \left(x - 2\ln|x|\right)\Big|_{1}^{2}$$

$$= (2 - 2\ln 2) - (1 - 2\ln 1)$$

$$= 1 - 2\ln 2$$

5. Net change $= \displaystyle\int_{a}^{b} R'(q)dq$

$$= \int_{4}^{9} q(10 - q)dq$$

$$= \int_{4}^{9} (10q - q^2)dq$$

$$= \left(5q^2 - \frac{q^3}{3}\right)\Bigg|_{4}^{9}$$

$$= \left[5(9)^2 - \frac{(9)^3}{3}\right]$$

$$- \left[5(4)^2 - \frac{(4)^3}{3}\right]$$

$$= \frac{310}{3} \text{ hundred,}$$

or approximately $10,333.33.

6. The rate the trade deficit is changing
= rate of change of imports
    − rate of change of exports.
$$D'(t) = I'(t) - E'(t)$$
So, the change over the next five years is
$$\int_{0}^{5} [E'(t) - I'(t)]dt$$

$$= \int_{0}^{5} [12.5e^{0.2t} - (1.7t + 3)]dt$$

$$= \left[12.5\left(\frac{1}{0.2}e^{0.2t}\right) - \frac{1.7}{2}t^2 - 3t\right]_{0}^{5}$$

$$= \left[62.5e^{0.2(5)} - \frac{1.7}{2}(5)^2 - 3(5)\right]$$

$$- [62.5e^0 - 0 - 0]$$

$$\approx 71.14, \text{ or the trade deficit will}$$
increase by approximately
71.14 billion dollars.

7. When $q_0 = 4$, $p_0 = 25 - (4)^2 = 9$.
$$CS = \int_{0}^{4} (25 - q^2)dq - (4)(9)$$

$$= \left(25q - \frac{q^3}{3}\right)\Bigg|_{0}^{4} - 36$$

$$= \left[25(4) - \frac{(4)^3}{3}\right] - 36$$

$$\approx 42.6667, \text{ or approximately}$$
$4,266.67

8. $FV = \displaystyle\int_{0}^{3} 5,000e^{0.05(3-t)}dt$

$$= 5,000e^{0.15}\int_{0}^{3} e^{-0.05t}dt$$

$$= \frac{5,000e^{0.15}}{-0.05}(e^{-0.05t})\Bigg|_{0}^{3}$$

$$= \frac{5,000e^{0.15}}{-0.05}(e^{-0.05(3)} - e^0)$$

$$\approx \$16,183.42$$

9. The number of the original 50,000 people
remaining after 20 years is
$$50,000e^{-0.02(20)}.$$
The number of new arrivals remaining

after 20 years is $\int_0^{20} 700e^{-0.02(20-t)} dt$.

So, the total will be

$50,000e^{-0.4} + 700e^{-0.4}\int_0^{20} e^{0.02t} dt$

$= 100e^{-0.4}\left[500 + 7\left(\frac{1}{0.02}e^{0.02t}\right)\Big|_0^{20}\right]$

$= 100e^{-0.4}\left[500 + 350(e^{0.02t})\Big|_0^{20}\right]$

$\approx 45{,}055$ people

10. $C_{av} = \frac{1}{3-0}\int_0^3 \frac{0.3t}{(t^2+16)^{1/2}} dt$

Let $u = t^2 + 16$; then $\frac{1}{2} du = t\, dt$, and the limits of integration become $0 + 16 = 16$ and $(3)^2 + 16 = 25$.

$= \frac{0.3}{3}\int_0^3 \frac{1}{(t^2+16)^{1/2}} t\, dt$

$= \frac{0.1}{2}\int_{16}^{25} u^{-1/2} du$

$= 0.05(2u^{1/2})\Big|_{16}^{25}$

$= 0.1(u^{1/2})\Big|_{16}^{25}$

$= 0.1\left(\sqrt{25} - \sqrt{16}\right)$

$= 0.1 \text{ mg/cm}^3$

Review Exercises

1. $\int\left(x^3 + \sqrt{x} - 9\right) dx$

$= \int x^3 dx + \int x^{1/2} dx - 9\int dx$

$= \frac{x^4}{4} + \frac{2}{3}x^{3/2} - 9x + C$

3. $\int(x^4 - 5e^{-2x})dx = \int x^4 dx - 5\int e^{-2x} dx$

$= \frac{x^5}{5} + \frac{5}{2}e^{-2x} + C$

5. $\int\left(\frac{5x^3 - 3}{x}\right)dx = \int\left(5x^2 - \frac{3}{x}\right)dx$

$= 5\int x^2 dx - 3\int\frac{1}{x} dx$

$= \frac{5x^3}{3} - 3\ln|x| + C$

7. $\int\left(t^5 - 3t^2 + \frac{1}{t^2}\right)dt$

$= \int t^5 dt - 3\int t^2 dt + \int t^{-2} dt$

$= \frac{t^6}{6} - t^3 - \frac{1}{t} + C$

9. $\int\sqrt{3x+1}\, dx = \int(3x+1)^{1/2} dx$

Let $u = 3x + 1$; then $\frac{1}{3} du = dx$

$= \frac{1}{3}\int u^{1/2} du$

$= \frac{2}{9}(3x+1)^{3/2} + C$

11. $\int(x+2)(x^2+4x+2)^5 dx$

Let $u = x^2 + 4x + 2$; then $du = (2x + 4)dx$, or $\frac{1}{2} du = (x+2)dx$

$= \int(x^2+4x+2)^5(x+2)dx$

$= \frac{1}{2}\int u^5 du$

$= \frac{1}{12}(x^2+4x+2)^6 + C$

13. $\int\frac{3x+6}{(2x^2+8x+3)^2} dx$

Let $u = 2x^2 + 8x + 3$; then,

$du = (4x+8)dx$, or $\dfrac{1}{4}du = (x+2)dx$

$$= \int \frac{3(x+2)}{(2x^2+8x+3)^2}dx$$

$$= \frac{3}{4}\int u^{-2}du$$

$$= \frac{-3}{4(2x^2+8x+3)} + C$$

15. $\int v(v-5)^{12}\,dv$

Let $u = v-5$; then, $du = dv$ and $v = u+5$

$$= \int (u+5)u^{12}du$$

$$= \int (u^{13}+5u^{12})du$$

$$= \frac{(v-5)^{14}}{14} + \frac{5(v-5)^{13}}{13} + C$$

17. $\int 5xe^{-x^2}\,dx$

Let $u = -x^2$; then $-\dfrac{1}{2}du = x\,dx$

$$= 5\int (e^{-x^2})x\,dx$$

$$= -\frac{5}{2}\int e^u du$$

$$= -\frac{5}{2}e^{-x^2} + C$$

19. $\int \left(\dfrac{\sqrt{\ln x}}{x}\right)dx$

Let $u = \ln x$; then $du = \dfrac{1}{x}dx$

$$= \int (\ln x)^{1/2}\cdot\frac{1}{x}dx$$

$$= \int u^{1/2}du$$

$$= \frac{2}{3}(\ln x)^{3/2} + C$$

21. $\int_0^1 (5x^4-8x^3+1)dx = (x^5-2x^4+x)\Big|_0^1$

$$= (1-2+1)-0$$

$$= 0$$

23. $\int_0^1 \left(e^{2x}+4\sqrt[3]{x}\right)dx$

$$= \int_0^1 (e^{2x}+4x^{1/3})dx$$

$$= \left(\frac{1}{2}e^{2x}+3x^{4/3}\right)\Big|_0^1$$

$$= \left[\frac{1}{2}e^2+3(1)\right]-\left[\frac{1}{2}e^0+3(0)\right]$$

$$= \frac{1}{2}e^2+\frac{5}{2}$$

25. $\int_{-1}^2 30(5x-2)^2\,dx$

Let $u = 5x-2$; then $\dfrac{1}{5}du = dx$, and the

limits of integration become
$5(-1)-2 = -7$ and $5(2)-2 = 8$

$$= \frac{30}{5}\int_{-7}^8 u^2du$$

$$= 6\left(\frac{u^3}{3}\right)\Big|_{-7}^8$$

$$= 6\left[\frac{(8)^3}{3}-\frac{(-7)^3}{3}\right]$$

$$= 1,710$$

27. $\int_0^1 2te^{t^2-1}\,dt$

Let $u = t^2-1$; then $du = 2t\,dt$, and the
limits of integration become $(0)-1 = -1$
and $(1)^2-1 = 0$

$$= \int_{-1}^0 e^u du$$

$$= (e^u)\Big|_{-1}^0$$

$$= e^0-e^{-1}$$

$$= 1-\frac{1}{e}$$

29. $\int_0^{e-1}\left(\dfrac{x}{x+1}\right)dx$

Let $u = x + 1$; then $du = dx$, $x = u - 1$, and the limits of integration become $0 + 1 = 1$ and $(e - 1) + 1 = e$

$= \int_1^e\left(\dfrac{u-1}{u}\right)du$

$= \int_1^e\left(1 - \dfrac{1}{u}\right)du$

$= \left(u - \ln|u|\right)\Big|_1^e$

$= (e - \ln e) - (1 - \ln 1)$

$= e - 2$

31. Area $= \int_1^4\Big[\left(x + 2\sqrt{x}\right) - 0\Big]dx$

$= \int_1^4(x + 2x^{1/2})dx$

$= \left(\dfrac{x^2}{2} + \dfrac{4}{3}x^{3/2}\right)\Big|_1^4$

$= \left[\dfrac{(4)^2}{2} + \dfrac{4}{3}(4)^{3/2}\right] - \left[\dfrac{1}{2} + \dfrac{4}{3}(1)\right]$

$= \dfrac{101}{6}$

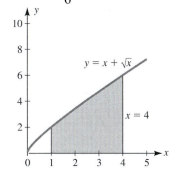

33. Area $= \int_1^2\left(\dfrac{1}{x} + x^2\right)dx$

$= \left(\ln|x| + \dfrac{x^3}{3}\right)\Big|_1^2$

$= \left(\ln 2 + \dfrac{(2)^3}{3}\right) - \left(\ln 1 + \dfrac{1}{3}\right)$

$= \ln 2 + \dfrac{7}{3}$

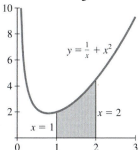

35. The limits of integration are

$$\dfrac{4}{x} = 5 - x$$

$$4 = 5x - x^2$$

$$x^2 - 5x + 4 = 0$$

$$(x - 4)(x - 1) = 0$$

$$x = 1, 4$$

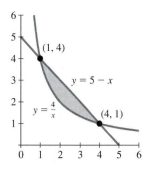

Noting that $y = 5 - x$ is the top curve,

$$\text{Area} = \int_1^4 \left[(5-x) - \left(\frac{4}{x} \right) \right] dx$$

$$= \int_1^4 \left(5 - x - \frac{4}{x} \right) dx$$

$$= \left(5x - \frac{x^2}{2} - 4\ln|x| \right) \Bigg|_1^4$$

$$= \left[5(4) - \frac{(4)^2}{2} - 4\ln 4 \right]$$

$$\qquad - \left[5(1) - \frac{1}{2} - 4\ln 1 \right]$$

$$= \frac{15}{2} - 4\ln 4$$

$$= \frac{15}{2} - 4\ln(2)^2$$

$$= \frac{15}{2} - 8\ln 2$$

37. The graph of $y = 2 + x - x^2$ intersects

$y = 0$ when $0 = 2 + x - x^2$

$$x^2 - x - 2 = 0$$

$$(x-2)(x+1) = 0 \text{ or,}$$

$$x = -1, 2$$

So, the limits of integration are $x = -1$ and $x = 2$

$$\text{Area} = \int_{-1}^2 [(2+x-x^2) - 0]\, dx$$

$$= \int_{-1}^2 (2+x-x^2)\, dx$$

$$= \left(2x + \frac{x}{2} - \frac{x^3}{3} \right) \Bigg|_{-1}^2$$

$$= \left[2(2) + \frac{2}{2} - \frac{(2)^3}{3} \right]$$

$$\qquad - \left[2(-1) + \frac{-1}{2} - \frac{(-1)^3}{3} \right]$$

$$= \frac{9}{2}$$

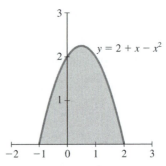

39. $f_{av} = \frac{1}{8-1} \int_1^8 \left(x^3 - 3x + \sqrt{2x} \right) dx$

$$= \frac{1}{7} \int_1^8 \left(x^3 - 3x + \sqrt{2}x^{1/2} \right) dx$$

$$= \frac{1}{7} \left(\frac{x^4}{4} - \frac{3x^2}{2} + \frac{2\sqrt{2}}{3} x^{3/2} \right) \Bigg|_1^8$$

$$= \frac{1}{7} \left[\left(\frac{(8)^4}{4} - \frac{3(8)^2}{2} + \frac{2\sqrt{2}}{3}(8)^{3/2} \right) \right.$$

$$\left. - \left(\frac{1}{4} - \frac{3(1)}{2} + \frac{2\sqrt{2}}{3}(1) \right) \right]$$

$$= \frac{11,407}{84} - \frac{2\sqrt{2}}{21}$$

41. $g_{av} = \frac{1}{2-0} \int_0^2 v e^{-v^2} dv$

Let $u = -v^2$; then $du = -2v\, dv$, or

$-\frac{1}{2} du = v\, dv$. Further, the limits of

integration become 0 and $-(2)^2 = -4$

$$= -\frac{1}{4} \int_0^{-4} e^u \, du$$

$$= \frac{1}{4} \int_{-4}^0 e^u \, du$$

$$= \frac{1}{4} (e^u) \Bigg|_{-4}^0$$

$$= \frac{1}{4}(e^0 - e^{-4})$$

$$= \frac{1}{4} \left(1 - \frac{1}{e^4} \right)$$

43. When $q_0 = 2$, $p_0 = 4[36-(2)^2] = \$128$

$$CS = \int_0^2 4(36-q^2)dq - 2(128)$$

$$= 4\left(36q - \frac{q^3}{3}\right)\Bigg|_0^2 - 256$$

$$= 4\left[\left(36(2) - \frac{(2)^3}{3}\right) - 0\right] - 256$$

$$= \frac{64}{3}, \text{ or approximately } \$21.33$$

45. When $q_0 = 4$, $p_0 = 10e^{-0.1(4)} \approx \$6.70$

$$CS = \int_0^4 10e^{-0.1q}dq - 4(6.70)$$

$$= 10\int_0^4 e^{-0.1q}dq - 26.80$$

$$= -100(e^{-0.1q})\Big|_0^4 - 26.80$$

$$= -100(e^{-0.1(4)} - e^0) - 26.80$$

$$\approx \$6.17$$

47. $GI = 2\int_0^1 (x - x^{3/2})dx$

$$= 2\left(\frac{x^2}{2} - \frac{2}{5}x^{5/2}\right)\Bigg|_0^1$$

$$= 2\left[\left(\frac{1}{2} - \frac{2}{5}(1)\right) - 0\right]$$

$$= \frac{1}{5}$$

49. $GI = 2\int_0^1 [x - (0.3x^2 + 0.7x)]dx$

$$= 2\int_0^1 (0.3x - 0.3x^2)dx$$

$$= 0.6\int_0^1 (x - x^2)dx$$

$$= 0.6\left(\frac{x^2}{2} - \frac{x^3}{3}\right)\Bigg|_0^1$$

$$= 0.6\left[\left(\frac{1}{2} - \frac{1}{3}\right) - 0\right]$$

$$= 0.1$$

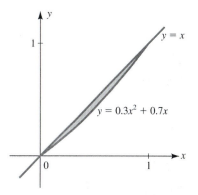

51. After 6 months, the number of the original population surviving is $75,000e^{-0.09(6)}$. The number of new members surviving is

$$\int_0^6 60e^{-0.09(6-t)}dt = 60e^{-0.54}\int_0^6 e^{0.09t}dt$$

So, the total will be

$$75,000e^{-0.54} + 60e^{-0.54}\int_0^6 e^{0.09t}dt$$

$$= 60e^{-0.54}\left[1250 + \frac{1}{0.09}e^{0.09t}\Big|_0^6\right]$$

$$= 60e^{-0.54}\left[1250 + \frac{100}{9}\left(e^{0.09t}\Big|_0^6\right)\right]$$

$$= 60e^{-0.54}\left[1250 + \frac{100}{9}(e^{0.54} - 1)\right]$$

$$\approx 43,984 \text{ members}$$

53. After 10 years, the number of the original population surviving is $100,000e^{-0.2(10)}$ The number of new members surviving is

$\int_0^{10} [90e^{0.1t}][e^{-0.2(10-t)}]dt$

$= 90\int_0^{10} e^{0.1t-2+0.2t}\,dt$

$= 90e^{-2}\int_0^{10} e^{0.3t}\,dt$

So, the total will be

$100,000e^{-2} + 90e^{-2}\int_0^{10} e^{0.3t}\,dt$

$= 10e^{-2}\left[10,000 + 9\left(\frac{1}{0.3}e^{0.3t}\Big|_0^{10}\right)\right]$

$= 10e^{-2}\left[10,000 + 30\left(e^{0.3t}\Big|_0^{10}\right)\right]$

$= 100e^{-2}[1,000 + 3(e^3 - 1)]$

$\approx 14,308$ members

55. Volume of S

$= \pi\int_{-1}^{2} (x^2 + 1)^2\,dx$

$= \pi\int_{-1}^{2} (x^4 + 2x^2 + 1)\,dx$

$= \pi\left[\frac{x^5}{5} + \frac{2x^3}{3} + x\right]\Big|_{-1}^{2}$

$= \pi\left[\left(\frac{32}{5} + \frac{16}{3} + 2\right) - \left(-\frac{1}{5} - \frac{2}{3} - 1\right)\right]$

$= \pi\left[\frac{33}{5} + \frac{18}{3} + 3\right]$

$= \frac{78}{5}\pi$

≈ 49.01

57. Volume of $S = \pi\int_1^3 \left(\frac{1}{\sqrt{x}}\right)^2 dx$

$= \pi\int_1^3 \frac{1}{x}\,dx$

$= \pi\left(\ln x\Big|_1^3\right)$

$= \pi(\ln 3 - \ln 1)$

$= \pi\ln 3 \approx 3.45$

59. $y = \int \frac{dy}{dx}\,dx = \int 2\,dx = 2x + C$

$4 = 2(-3) + C$, or $C = 10$

so, $y = 2x + 10$

61. $x = \int \frac{dx}{dt}\,dt = \int e^{-2t}\,dt = -\frac{1}{2}e^{-2t} + C$

$4 = -\frac{1}{2}e^0 + C$, or $C = \frac{9}{2}$ so,

$x = \frac{1}{2}(9 - e^{-2t})$

63. Since slope $= \frac{dy}{dx}$,

$y = \int x(x^2 + 1)^{-1}\,dx$.

Let $u = x^2 + 1$; then $du = 2x\,dx$, or

$\frac{1}{2}du = x\,dx$

$= \frac{1}{2}\int \frac{1}{u}\,du$

$= \frac{1}{2}\ln|x^2 + 1| + C$

Since the graph of y passes through the point $(1, 5)$

$5 = \frac{1}{2}\ln 2 + C$, or $C = 5 - \frac{1}{2}\ln 2$ so,

$y = \frac{1}{2}\ln(x^2 + 1) + 5 - \frac{1}{2}\ln 2$

65. $V'(t) = 2[0.5t^2 + 4(t + 1)^{-1}]$

increase $= \int_0^6 \left[t^2 + \frac{8}{(t + 1)}\right]dt$

Let $u = t + 1$; then $du = dt$, and the limits of integration become $0 + 1 = 1$ and $6 + 1 = 7$

$= \int_0^6 t^2\,dt + 8\int_1^7 \frac{1}{u}\,du$

$= \left(\frac{t^3}{3}\right)\Big|_0^6 + 8(\ln|u|)\Big|_1^7$

$= (72 - 0) + 8(\ln 7 - \ln 1)$

$= 72 + 8\ln 7 \approx \$87.57$

67. Since $t = 1$ at 10:00 A.M., and $t = 3$ at noon, the number of people will be

$\int_1^3 [-4(t + 2)^3 + 54(t + 2)^2]\,dt$.

Let $u = t + 2$; then $du = dt$, and the limits of integration become $1 + 2 = 3$ and

$$3 + 2 = 5$$

$$= \int_3^5 (-4u^3 + 54u^2)\,du$$

$$= (-u^4 + 18u^3)\Big|_3^5$$

$$= [-(5)^4 + 18(5)^3] - [-(3)^4 + 18(3)^3]$$

$$= 1,220 \text{ people}$$

69. $C(x) = \int C'(x)\,dx$

$$= \int (18x^2 + 500)\,dx$$

$$= 6x^3 + 500x + C$$

When $x = 0$, $C(0) = 8,000$ so $C = 8,000$,

and $C(x) = 6x^3 + 500x + 8,000$

So, $C(5) = 6(5)^3 + 500(5) + 8,000$

$$= 11,250 \text{ commuters}$$

71. $D'(t) = \dfrac{1}{1 + 2t}$

The amount of oil demanded during the

year 2013 will be $D(t) = \int_1^2 \dfrac{1}{1 + 2t}\,dt$.

Using substitution with $u = 1 + 2t$,

$\dfrac{1}{2}\,du = dt$ and the limits of integration

become $1 + 2(1) = 3$ and $1 + 2(2) = 5$.

$$\frac{1}{2}\int_3^5 \frac{1}{u}\,du$$

$$= \frac{1}{2}\left(\ln|u|\Big|_3^5\right)$$

$$= \frac{1}{2}(\ln 5 - \ln 3) \approx 0.2554 \text{ billion barrels}$$

Similarly, the amount of oil demanded
during the year 2014 will be

$$\int_2^3 \frac{1}{1 + 2t}\,dt$$

$$= \frac{1}{2}\int_5^7 \frac{1}{u}\,du = \frac{1}{2}\left(\ln|u|\Big|_5^7\right)$$

$$= \frac{1}{2}(\ln 7 - \ln 5)$$

$$\approx 0.1682 \text{ billion barrels}$$

So, approximately $0.2554 - 0.1682 =$

0.0872 billion more barrels will be
demanded in 2013 than in 2014.

73. $FV = \int_0^5 1,200e^{0.08(5-t)}\,dt$

$$= 1,200e^{0.4}\int_0^5 e^{-0.08t}\,dt$$

$$= \frac{1,200e^{0.4}}{-0.08}(e^{-0.08t})\Big|_0^5$$

$$= -15,000e^{0.4}(e^{-0.4} - e^0)$$

$$\approx \$7,377.37$$

75. The number of the original houses still on
the market after 10 weeks is $200e^{-0.2(10)}$.
The number of new listings which will still
be on the market after 10 weeks is

$$\int_0^{10} 8e^{-0.2(10-t)}\,dt.$$

So, the total will be

$$200e^{-2} + 8e^{-2}\int_0^{10} e^{0.2t}\,dt$$

$$= 8e^{-2}\left(25 + \int_0^{10} e^{0.02t}\,dt\right)$$

$$= 8e^{-2}\left[25 + \frac{1}{0.2}(e^{0.2t})\Big|_0^{10}\right]$$

$$= 8e^{-2}[25 + 5(e^2 - e^0)]$$

$$\approx 62 \text{ houses.}$$

77. The decay function is of the form

$$Q(t) = Q_0 e^{-kt}.$$

Since the half-life is 35 years,

$$\frac{Q_0}{2} = Q_0 e^{-k(35)}$$

$$\ln \frac{1}{2} = \ln e^{-35k}$$

$$k = \frac{\ln \frac{1}{2}}{-35} = \frac{-\ln \frac{1}{2}}{35} = \frac{\ln 2}{35}$$

$$\approx 0.0198$$

The amount remaining

$$= \int_0^{200} 300 e^{-0.0198(200-t)}\, dt$$

$$= 300 e^{-3.96} \int_0^{200} e^{0.0198t}\, dt$$

$$= 300^{-3.96} \left(\frac{1}{0.0198} e^{0.0198t} \right) \Bigg|_0^{200}$$

$$= \frac{300 e^{-3.96}}{0.0198} (e^{3.96} - e^0)$$

$$\approx 14{,}863 \text{ pounds}$$

79. Rate revenue changes
= (#barrels)(rate selling price changes)
$R'(t) = 900(92 + 0.8t)$
Since time is measured in months,

$$\text{revenue} = \int_0^{36} 900(92 + 0.8t)\, dt$$

$$= 900(92t + 0.4t^2) \Big|_0^{36}$$

$$= 900[(92(36) + 0.4(36)^2) - 0]$$

$$\approx \$3{,}447{,}360$$

81. P_{av}

$$= \frac{1}{6-0} \int_0^6 (0.06t^2 - 0.2t + 6.2)\, dt$$

$$= \frac{1}{6}(0.02t^3 - 0.1t^2 + 6.2t) \Big|_0^6$$

$$= \frac{1}{6}[(0.02(6)^3 - 0.1(6)^2 + 6.2(6)) - 0]$$

$$= \$6.32 \text{ per pound}$$

83. At 8:00 A.M., $t = 8$ and at 8:00 P.M., $t = 20$ so the change in temperature will be

$$\int_8^{20} -0.02(t - 7)(t - 14)\, dt$$

$$= -0.02 \int_8^{20} (t^2 - 21t + 98)\, dt$$

$$= -0.02 \left(\frac{t^3}{3} - \frac{21t^2}{2} + 98t \right) \Bigg|_8^{20}$$

$$= -0.02 \left[\left(\frac{(20)^3}{3} - \frac{21(20)^2}{2} + 98(20) \right) \right.$$

$$\left. - \left(\frac{(8)^3}{3} - \frac{21(8)^2}{2} + 98(8) \right) \right]$$

≈ -2.88, or a decrease of approximately $2.88°C$

85. (a) $p(x) = \int p'(x)\, dx$

$$= \int (0.2 + 0.003x^2)\, dx$$

$$= 0.2x + 0.001x^3 + C$$

When $x = 0$, $p(0) = 250$ cents, so
$C = 250$ and

$$p(x) = 0.2x + 0.001x^3 + 250.$$

Press $\boxed{y=}$ and input $p(x)$ for $y_1 =$.
Use window dimensions [0, 50]10 by [240, 340]20.
Press $\boxed{\text{graph}}$.
Use the value function under the calc menu and input $x = 10$ to find the price of eggs 10 weeks from now is 253 cents or $2.53.

(b) $p(x) = \int (0.3 + 0.003x^2)\, dx$

$$= 0.3x + 0.001x^3 + C$$

$$= 0.3x + 0.001x^3 + 250$$

Press $\boxed{y=}$ and input $p_2(x)$ for $y_2 =$.
Use window dimensions [0, 50]10 by [240, 340]20.
Press $\boxed{\text{graph}}$.
Use the value function under the calc menu and input $x = 10$. Verify that

$p_2(x) = 0.3x + 0.001x^3 + 250$ is

displayed on the upper left corner.

$P_2(10) = 254$ cents or \$2.54.

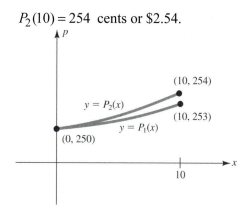

87. Let $s(t)$ be the distance traveled, in meters, after time t, in minutes. Then,

$$s(t) = \int_2^3 v(t)dt.$$

$$\int_2^3 (1 + 4t + 3t^2)dt$$

$$= (t + 2t^2 + t^3)\Big|_2^3$$

$$= [3 + 2(3)^2 + (3)^3] - [2 + 2(2)^2 + (2)^3]$$

$$= 30 \text{ meters}$$

89. $GI_{sw} = 2\int_0^1 (x - x^{1.6})dx$

$$= 2\left(\frac{x^2}{2} - \frac{x^{2.6}}{2.6}\right)\Big|_0^1$$

$$= 2\left[\left(\frac{1}{2} - \frac{1}{2.6}\right) - 0\right] \approx 0.2308$$

$GI_{PT} = 2\int_0^1 [x - (0.65x^2 + 0.35x)]dx$

$$= 2\int_0^1 (0.65x - 0.65x^2)dx$$

$$= 1.3\int_0^1 (x - x^2)dx$$

$$= 1.3\left(\frac{x^2}{2} - \frac{x^3}{3}\right)\Big|_0^1$$

$$= 1.3\left[\left(\frac{1}{2} - \frac{1}{3}\right) - 0\right] \approx 0.2167$$

So, income is more equitably distributed for physical therapists.

91. $2x^2 + 3y^2 = 6$

The equation for the bottom half of the curve is $y = -\sqrt{\dfrac{6 - 2x^2}{3}}$.

The volume, in cubic miles, of the lake is *half* the volume of the solid generated by this curve. Since when $y = 0$, $x = \pm\sqrt{3}$, want

$$\frac{\pi}{2}\int_{-\sqrt{3}}^{\sqrt{3}} \left(-\sqrt{\frac{6 - 2x^2}{3}}\right)^2 dx$$

$$= \frac{\pi}{2}\int_{-\sqrt{3}}^{\sqrt{3}} \left(2 - \frac{2}{3}x^2\right) dx$$

$$= \frac{\pi}{2}\left[2x - \frac{2x^3}{9}\right]_{-\sqrt{3}}^{\sqrt{3}}$$

$$= \frac{\pi}{2}\left[\left(2\sqrt{3} - \frac{2(\sqrt{3})^3}{9}\right)\right.$$

$$\left. - \left(-2\sqrt{3} + \frac{2(\sqrt{3})^3}{9}\right)\right]$$

$$= \frac{\pi}{2}\left[4\sqrt{3} - \frac{4(3^{3/2})}{9}\right] \approx 7.255$$

To have 1,000 trout per cubic mile, need $1,000(7.255) = 7,255$ trout. So, need an additional 2,255 trout.

93. (a) $S_{av} = \dfrac{1}{N - 0}\int_0^N S(t)dt$

(b) Since velocity is the derivative of distance, $D(t) = \int_0^N S(t)\,dt$.

(c) Average speed $= \dfrac{\text{distance traveled}}{\text{time elapsed}}$

95. Press $\boxed{y=}$ and input $(x - 2)/(x + 1)$ for $y_1 =$ and input $\sqrt{\ }(25 - x^2)$ for $y_2 =$. Use window dimensions $[-5, 5]1$ by $[-1, 6]1$.
Press $\boxed{\text{graph}}$.
Use trace and zoom-in to find the points of

intersection are (−4.66, 1.82),
(−1.82, 4.66), and (4.98, 0.498).
An alternative to using trace and zoom is to
use the intersect function under the calc
menu. Enter a value close to the point of
intersection on $y_1 =$ and enter a value close
to the same point of intersection on $y_2 =$
and finally, enter a guess for the point of
intersection. Repeat this process for the
other two points of intersections.
The curves are bounded by the points of
intersection given by $x = -4.66$ and
$x = -1.82$. To find the area bounded by the
curves, we need to find

$$\int_{-4.66}^{-1.82} (y_2 - y_1)dx$$

$$= \int_{-4.66}^{-1.82} y_2 \, dx - \int_{-4.66}^{-1.82} y_1 \, dx$$

For each separate integral, use the
$\int f(x) \, dx$ function under the calc menu
making sure that the correct y equation is

displayed in the upper left corner. We find
the area to be
$10.326439 - 7.32277423 \approx 3$.
An alternative to finding each separate
integral is to use the fnInt function from
the home screen. Select fnInt function
from the math menu and enter
$\text{fnInt}(y_2 - y_1, x, -4.66, -1.82)$. You
input the y equations by pressing vars and
selecting which y equation you want from
the function window under y-vars.

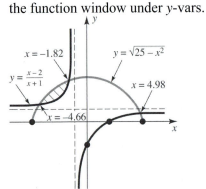

Chapter 6

Additional Topics in Integration

6.1 Integration by Parts; Integral Tables

1. Both terms are easy to integrate; however, the derivative of x becomes simpler while the derivative of e^{-x} does not. So,

$$u = x \quad \text{and} \quad dV = e^{-x} dx$$
$$du = dx \qquad V = -e^{-x}$$

and

$$\int xe^{-x} dx = -xe^{-x} - \int -e^{-x} dx$$
$$= -xe^{-x} + \int e^{-x} dx$$
$$= -xe^{-x} - e^{-x} + C$$
$$= -(x+1)e^{-x} + C$$

3. Both terms are easy to integrate; however, the derivative of $1 - x$ becomes simpler while the derivative of e^x does not. So,

$$u = 1 - x \qquad dV = e^x dx$$
$$du = -dx \qquad V = e^x$$

and

$$\int (1-x)e^x dx = (1-x)e^x - \int e^x - dx$$
$$= (1-x)e^x + \int e^x dx$$
$$= (1-x)e^x + e^x + C$$
$$= [(1-x)+1]e^x + C$$
$$= (2-x)e^x + C$$

5. $\ln 2t$ cannot be easily integrated. So,
$$u = \ln 2t \qquad \text{and} \quad dV = t\, dt$$
$$du = \frac{1}{2t} \cdot 2\, dt = \frac{1}{t} dt \qquad V = \frac{t^2}{2}$$

and

$$\int t \ln t\, dt = \frac{t^2}{2} \ln 2t - \int \frac{t^2}{2} \cdot \frac{1}{t} dt$$
$$= \frac{t^2}{2} \ln 2t - \frac{1}{2} \int t\, dt$$
$$= \frac{t^2}{2} \ln 2t - \frac{1}{4} t^2 + C$$
$$= \frac{t^2}{2}\left(\ln 2t - \frac{1}{2} \right) + C$$

7. Both terms are easy to integrate; however, the derivative of v becomes simpler while the derivative of $e^{-v/5}$ does not. So,

$$u = v \quad \text{and} \quad dV = e^{-v/5} dv$$
$$du = dv \qquad V = -5e^{-v/5}$$

and

$$\int ve^{-v/5} dv = -5ve^{-v/5} - \int -5e^{-v/5} dv$$
$$= -5ve^{-v/5} + 5\int e^{-v/5} dv$$
$$= -5ve^{-v/5} - 25e^{-v/5} + C$$
$$= -5(v+5)e^{-v/5} + C$$

9. Both terms are easy to integrate; however, the derivative of x becomes simpler while the derivative of $\sqrt{x-6}$ does not. So,

$$u = x \quad \text{and} \quad dV = (x-6)^{1/2} dx$$
$$du = dx \qquad V = \frac{2}{3}(x-6)^{3/2}$$

and

$$\int x\sqrt{x-6}\, dx$$
$$= \frac{2}{3} x(x-6)^{3/2} - \int \frac{2}{3}(x-6)^{3/2} dx$$
$$= \frac{2}{3} x(x-6)^{3/2} - \frac{2}{3}\int (x-6)^{3/2} dx$$
$$= \frac{2}{3} x(x-6)^{3/2} - \frac{4}{15}(x-6)^{5/2} + C$$

11. Both terms are easy to integrate; however, the derivative of x becomes simpler while the derivative of $(x+1)^8$ does not. So,

$$u = x \quad \text{and} \quad dV = (x+1)^8\, dx$$
$$du = dx \qquad V = \frac{1}{9}(x+1)^9$$

and

$$\int x(x+1)^8\, dx$$
$$= \frac{1}{9}x(x+1)^9 - \int \frac{1}{9}(x+1)^9\, dx$$
$$= \frac{1}{9}x(x+1)^9 - \frac{1}{9}\int (x+1)^9\, dx$$
$$= \frac{1}{9}x(x+1)^9 - \frac{1}{90}(x+1)^{10} + C$$

13. Rewriting, $\displaystyle\int \frac{x}{\sqrt{x+2}}\, dx = \int x(x+2)^{-1/2}\, dx$,

both terms are easy to integrate; however, the derivative of x becomes simpler while the derivative of $(x+2)^{-1/2}$ does not. So,

$$u = x \quad \text{and} \quad dV = (x+2)^{-1/2}\, dx$$
$$du = dx \qquad V = 2(x+2)^{1/2}$$

and

$$\int \frac{x}{\sqrt{x+2}}\, dx$$
$$= 2x(x+2)^{1/2} - \int 2(x+2)^{1/2}\, dx$$
$$= 2x(x+2)^{1/2} - 2\int (x+2)^{1/2}\, dx$$
$$= 2x\sqrt{x+2} - \frac{4}{3}(x+2)^{3/2} + C$$

15. Rewriting,

$$\int_{-1}^{4} \frac{x}{\sqrt{x+5}}\, dx = \int_{-1}^{4} x(x+5)^{-1/2}\, dx, \text{ both}$$

terms are easy to integrate; however, the derivative of x becomes simpler while the derivative of $(x+5)^{-1/2}$ does not. So,

$$u = x \quad \text{and} \quad dV = (x+5)^{-1/2}\, dx$$
$$du = dx \qquad V = 2(x+5)^{1/2}$$

and

$$\int_{-1}^{4} \frac{x}{\sqrt{x+5}}\, dx$$
$$= 2x(x+5)^{1/2}\Big|_{-1}^{4} - \int_{-1}^{4} 2(x+5)^{1/2}\, dx$$
$$= 2x(x+5)^{1/2}\Big|_{-1}^{4} - 2\int_{-1}^{4} (x+5)^{1/2}\, dx$$
$$= \left[2x\sqrt{x+5} - \frac{4}{3}(x+5)^{3/2} \right]\Bigg|_{-1}^{4}$$
$$= \left[2(4)\sqrt{4+5} - \frac{4}{3}(4+5)^{3/2} \right]$$
$$\quad - \left[2(-1)\sqrt{-1+5} - \frac{4}{3}(-1+5)^{3/2} \right]$$
$$= \frac{8}{3}$$

17. Rewriting, $\displaystyle\int_0^1 \frac{x}{e^{2x}}\, dx = \int_0^1 xe^{-2x}\, dx$, both

terms are easy to integrate; however, the derivative of x becomes simpler while the derivative of e^{-2x} does not. So,

$$u = x \quad \text{and} \quad dV = e^{-2x}\, dx$$
$$du = dx \qquad V = -\frac{1}{2}e^{-2x}$$

and

$$\int_0^1 \frac{x}{e^{2x}}\, dx = -\frac{x}{2}e^{-2x}\Big|_0^1 - \int_0^1 -\frac{1}{2}e^{-2x}\, dx$$
$$= -\frac{x}{2}e^{-2x}\Big|_0^1 + \frac{1}{2}\int_0^1 e^{-2x}\, dx$$
$$= \left[-\frac{x}{2}e^{-2x} - \frac{1}{4}e^{-2x} \right]\Bigg|_0^1$$
$$= \left[-\frac{1}{2}e^{-2} - \frac{1}{4}e^{-2} \right] - \left[0 - \frac{1}{4}e^0 \right]$$
$$= -\frac{3}{4}e^{-2} + \frac{1}{4}$$
$$= \frac{1}{4}(1 - 3e^{-2})$$

19. $\ln \sqrt[3]{x}$ cannot be easily integrated. So,

$$u = \ln \sqrt[3]{x} \qquad \text{and} \quad dV = x\,dx$$
$$= \ln(x)^{1/3}$$
$$V = \frac{x^2}{2}$$
$$= \frac{1}{3}\ln x$$
$$du = \frac{1}{3x}dx$$

and

$$\int_0^{e^2} x \ln \sqrt[3]{x}\,dx$$

$$= \frac{x^2}{6}\ln x \Big|_1^{e^2} - \int_0^{e^2} \frac{x^2}{2}\cdot\frac{1}{3x}dx$$

$$= \frac{x^2}{6}\ln x \Big|_1^{e^2} - \frac{1}{6}\int_1^{e^2} x\,dx$$

$$= \left(\frac{x^2}{6}\ln x - \frac{x^2}{12}\right)\Big|_1^{e^2}$$

$$= \left[\frac{(e^2)^2}{6}\ln(e^2) - \frac{(e^2)^2}{12}\right] - \left[\frac{1}{6}\ln 1 - \frac{1}{12}\right]$$

$$= \frac{1}{12}(3e^4 + 1)$$

21. $\ln 2t$ cannot be easily integrated. So,

$$u = \ln 2t \qquad\qquad \text{and} \quad dV = t\,dt$$
$$du = \frac{1}{2t}\cdot 2\,dt = \frac{1}{t}dt \qquad\qquad V = \frac{t^2}{2}$$

and

$$\int_{1/2}^{e/2} t \ln 2t\,dt = \frac{t^2}{2}\ln 2t \Big|_{1/2}^{e/2} - \int_{1/2}^{e/2} \frac{t^2}{2}\cdot\frac{1}{t}dt$$

$$= \frac{t^2}{2}\ln 2t \Big|_{1/2}^{e/2} - \frac{1}{2}\int_{1/2}^{e/2} t\,dt$$

$$= \left(\frac{t^2}{2}\ln 2t - \frac{t^2}{4}\right)\Big|_{1/2}^{e/2}$$

$$= \left[\frac{\left(\frac{e}{2}\right)^2}{2}\ln 2\left(\frac{e}{2}\right) - \frac{\left(\frac{e}{2}\right)^2}{4}\right]$$

$$\quad - \left[\frac{\left(\frac{1}{2}\right)^2}{2}\ln 2\left(\frac{1}{2}\right) - \frac{\left(\frac{1}{2}\right)^2}{4}\right]$$

$$= \frac{1}{16}(e^2 + 1)$$

23. Rewriting, $\displaystyle\int \frac{\ln x}{x^2}dx = \int x^{-2}\ln x\,dx$, $\ln x$

cannot be easily integrated. So,

$$u = \ln x \quad \text{and} \quad dV = x^{-2}dV$$
$$du = \frac{1}{x}dx \qquad V = -\frac{1}{x}$$

and

$$\int \frac{\ln x}{x^2}dx = -\frac{1}{x}\ln x - \int -\frac{1}{x}\cdot\frac{1}{x}dx$$

$$= -\frac{1}{x}\ln x + \int x^{-2}dx$$

$$= -\frac{1}{x}\ln x - \frac{1}{x} + C$$

$$= -\frac{1}{x}(\ln x + 1) + C$$

25. Using the hint,

$$u = x^2 \quad \text{and} \quad dV = xe^{x^2}\, dx$$
$$du = 2x\, dx$$
$$\text{let } u = x^2; \frac{1}{2} du = x\, dx$$
$$V = \frac{1}{2} e^{x^2}$$

and

$$\int x^3 e^{x^2}\, dx = \frac{x^2}{2} e^{x^2} - \int \frac{1}{2} e^{x^2} \cdot 2x\, dx$$
$$= \frac{x^2}{2} e^{x^2} - \frac{1}{2} e^{x^2} + C$$
$$= \frac{1}{2} e^{x^2} (x^2 - 1) + C$$

27. Rewriting, $\displaystyle\int \frac{x\, dx}{3 - 5x} = \int \frac{x\, dx}{3 + -5x}$ which is of

the form $\displaystyle\int \frac{u\, du}{a + bu}$ (formula #1). Using

$u = x$, $du = dx$, $a = 3$, and $b = -5$, the
formula yields

$$\int \frac{x\, dx}{3 - 5x}$$
$$= \frac{1}{(-5)^2}\left[3 + -5x - 3\ln|3 + -5x|\right] + C$$
$$= \frac{1}{25}\left(3 - 5x - 3\ln|3 - 5x|\right) + C$$

29. Rewriting,

$$\int \frac{\sqrt{4x^2 - 9}}{x^2}\, dx = \int \frac{\sqrt{(2x)^2 - (3)^2}}{x^2}\, dx \text{ most}$$

closely resembles $\displaystyle\int \frac{\sqrt{u^2 - a^2}}{u^2}\, du$ (formula

#19). Now,

$$\int \frac{\sqrt{(2x)^2 - (3)^2}}{x^2}\, dx = \int \frac{4\sqrt{(2x)^2 - (3)^2}}{4x^2}\, dx$$
$$= 2\int \frac{\sqrt{(2x)^2 - (3)^2}}{(2x)^2}\, 2\, dx$$

and formula #19 can be used with $u = 2x$,

$du = 2\, dx$, and $a = 3$. So,

$$\int \frac{\sqrt{4x^2 - 9}}{x^2}\, dx$$
$$= 2\left[\frac{-\sqrt{4x^2 - 9}}{2x} + \ln\left|2x + \sqrt{4x^2 - 9}\right|\right] + C$$
$$= \frac{-\sqrt{4x^2 - 9}}{x} + 2\ln\left|2x + \sqrt{4x^2 - 9}\right| + C$$

31. As written, $\displaystyle\int \frac{dx}{x(2 + 3x)}$ is of the form

$\displaystyle\int \frac{du}{u(a + bu)}$ (formula #6). Using $u = x$,

$du = dx$, $a = 2$, and $b = 3$, the formula yields

$$\int \frac{dx}{x(2 + 3x)} = \frac{1}{2}\ln\left|\frac{x}{2 + 3x}\right| + C.$$

33. Rewriting, $\displaystyle\int \frac{du}{16 - 3u^2} = \int \frac{du}{3\left(\frac{16}{3} - u^2\right)}$

$$= \frac{1}{3}\int \frac{du}{\frac{16}{3} - u^2}$$
$$= \frac{1}{3}\int \frac{du}{\left(\frac{4}{\sqrt{3}}\right)^2 - u^2}$$

which is of the form $\displaystyle\int \frac{du}{a^2 - u^2}$ (formula

#16). Using $a = \dfrac{4}{\sqrt{3}}$, the formula yields

$$\int \frac{du}{16 - 3u^2} = \frac{1}{3}\left[\frac{1}{2\left(\frac{4}{\sqrt{3}}\right)}\ln\left|\frac{\frac{4}{\sqrt{3}} + u}{\frac{4}{\sqrt{3}} - u}\right|\right] + C$$
$$= \frac{\sqrt{3}}{24}\ln\left|\frac{\frac{4 + \sqrt{3}u}{\sqrt{3}}}{\frac{4 - \sqrt{3}u}{\sqrt{3}}}\right| + C$$
$$= \frac{\sqrt{3}}{24}\ln\left|\frac{4 + \sqrt{3}u}{4 - \sqrt{3}u}\right| + C$$

35. $\int (\ln x)^3 \, dx$ is of the form $\int (\ln u)^n \, du$

(formula #27). Using $u = x$, the formula

yields $\int (\ln x)^3 \, dx = x(\ln x)^3 - 3\int (\ln x)^2 \, dx$

Applying the formula again to the last term

$= x(\ln x)^3 - 3\left[x(\ln x)^2 - 2\int \ln x \, dx \right]$

$= x(\ln x)^3 - 3x(\ln x)^2 + 6\int \ln x \, dx$

Applying the formula one more time (or using formula #23),

$= x(\ln x)^3 - 3x(\ln x)^2 + 6[x \ln x - x] + C$

$= x(\ln x)^3 - 3x(\ln x)^2 + 6x \ln x - 6x + C$

37. $\int \dfrac{dx}{x^2 (5 + 2x)^2}$ is of the form $\int \dfrac{du}{u^2 (a + bu)^2}$

(formula #8). Using $u = x$, $du = dx$, $a = 5$, and $b = 2$, the formula yields

$\int \dfrac{dx}{x^2 (5 + 2x)^2}$

$= -\dfrac{1}{25}\left[\dfrac{5 + 4x}{x(5 + 2x)} + \dfrac{4}{5}\ln\left| \dfrac{x}{5 + 2x} \right| \right] + C$

39. $\dfrac{dy}{dx} = xe^{-2x}$; $y = 0$ when $x = 0$

$y = \int \dfrac{dy}{dx} dx$

$= \int xe^{-2x} \, dx$

Both terms are easy to integrate. However the derivative of x becomes simpler while the derivative of e^{-2x} does not. So,

$u = x$ and $dV = e^{-2x} \, dx$

$du = dx$ $V = -\dfrac{1}{2}e^{-2x}$

and

$\int xe^{-2x} \, dx$

$= -\dfrac{x}{2}e^{-2x} - \int -\dfrac{1}{2}e^{-2x} \, dx$

$= -\dfrac{x}{2}e^{-2x} + \dfrac{1}{2}\int e^{-2x} \, dx$

$= -\dfrac{x}{2}e^{-2x} + \dfrac{1}{2}\left(-\dfrac{1}{2}e^{-2x} \right) + C$

$= -\dfrac{x}{2}e^{-2x} - \dfrac{1}{4}e^{-2x} + C$

So, $y = -\dfrac{x}{2}e^{-2x} - \dfrac{1}{4}e^{-2x} + C$

Since, $y = 0$ when $x = 0$,

$0 = 0 - \dfrac{1}{4}e^0 + C$, or $C = \dfrac{1}{4}$ and

$y = -\dfrac{x}{2}e^{-2x} - \dfrac{1}{4}e^{-2x} + \dfrac{1}{4}$

41. $\dfrac{dy}{dx} = \dfrac{xy}{\sqrt{x+1}}$; $y = 1$ when $x = 0$

Separating the variables,

$\dfrac{dy}{y} = \dfrac{x}{(x+1)^{1/2}} \, dx$

Integrating both sides,

$\int \dfrac{dy}{y} = \int \dfrac{x}{(x+1)^{1/2}} \, dx$

Now,

$\int \dfrac{dy}{y} = \ln|y| + C_1$

To integrate the right hand side, use

$u = x$ and $dV = (x+1)^{-1/2} \, dx$

$du = dx$ $V = 2(x+1)^{1/2}$

Then,

$$\int \frac{x}{(x+1)^{1/2}} dx$$

$$= 2x(x+1)^{1/2} - \int 2(x+1)^{1/2} dx$$

$$= 2x(x+1)^{1/2} - 2\int (x+1)^{1/2} dx$$

$$= 2x(x+1)^{1/2} - 2\left[\frac{2}{3}(x+1)^{3/2}\right] + C_2$$

$$= 2x(x+1)^{1/2} - \frac{4}{3}(x+1)^{3/2} + C_2$$

and $\ln|y| + C_1$

$$= 2x(x+1)^{1/2} - \frac{4}{3}(x+1)^{3/2} + C_2$$

$\ln|y|$

$$= 2x(x+1)^{1/2} - \frac{4}{3}(x+1)^{3/2} + C$$

Since $y = 1$ when $x = 0$, $0 = 0 - \frac{4}{3}(1) + C$

or $C = \frac{4}{3}$ and

$\ln|y|$

$$= 2x(x+1)^{1/2} - \frac{4}{3}(x+1)^{3/2} + \frac{4}{3}$$

43. Slope $= y' = (x+1)e^{-x}$

$$y = \int y' dx = \int (x+1)e^{-x} dx$$

$u = x+1$ and $dV = e^{-x} dx$
$du = dx$ $\qquad V = -e^{-x}$

$$y = -(x+1)e^{-x} - \int -e^{-x} dx$$

$$= -(x+1)e^{-x} + \int e^{-x} dx$$

$$= -(x+1)e^{-x} - e^{-x} + C$$

Since the graph of y passes through the
point $(1, 5)$, $5 = -(1+1)e^{-1} - e^{-1} + C$ or,

$C = 5 + \frac{3}{e}$. So,

$$y = -(x+1)e^{-x} - e^{-x} + 5 + \frac{3}{e}$$

$$= 5 + \frac{3}{e} - \frac{x+2}{e^x}$$

45. $Q(t) = \int_0^3 100te^{-0.5t} dt = 100\int_0^3 te^{-0.5t} dt$

$u = t$ and $dV = e^{-0.5t} dt$
$du = dt$ $\qquad V = -\frac{1}{-0.5}e^{-0.5t} = -2e^{-0.5t}$

$$Q(t) = 100\left[-2te^{-0.5t}\Big|_0^3 - \int_0^3 -2e^{-0.5t} dt\right]$$

$$= 100\left[-2te^{-0.5t}\Big|_0^3 + 2\int_0^3 e^{-0.5t} dt\right]$$

$$= 100\left[-2te^{-0.5t} - 4e^{-0.5t}\right]\Big|_0^3$$

$$= 100[(-6e^{-1.5} - 4e^{-1.5}) - (0 - 4e^0)]$$

$$= 100\left(\frac{-10}{e^{1.5}} + 4\right) \approx \$176.87,$$

or approximately 176 units.

47. $FV = \int_0^{10} (3,000 + 5t) e^{0.05(10-t)} dt$

$= e^{0.5} \int_0^{10} (3,000 + 5t) e^{-0.05t} dt$

$u = 3,000 + 5t \quad \text{and} \quad dV = e^{-0.05t} dt$

$du = 5 dt \qquad\qquad\qquad V = -20 e^{-0.05t}$

So,

$FV = e^{0.5} \left[-20(3,000 + 5t) e^{-0.05t} \Big|_0^{10} - \int_0^{10} -20 e^{-0.05t} \cdot 5 dt \right]$

$= e^{0.5} \left[-20(3,000 + 5t) e^{-0.05t} \Big|_0^{10} + 100 \int_0^{10} e^{-0.05t} dt \right]$

$= e^{0.5} \left[-20(3,000 + 5t) e^{-0.05t} - 2,000 e^{-0.05t} \right]_0^{10}$

$= e^{0.5} \left(\left[-20(3,000 + 5(10)) e^{-0.5t} - 2,000 e^{-0.5} \right] - \left[-20(3,000 + 0) e^0 - 2,000 e^0 \right] \right)$

$= e^{0.5} (-63,000 e^{-0.5} + 62,000)$

$\approx \$39,220.72$

49. $PV = \int_0^5 (20 + 3t) e^{-0.07t} dt$

$u = 20 + 3t \quad \text{and} \quad dV = e^{-0.07t} dt$

$du = 3 dt \qquad\qquad\qquad V = -\frac{100}{7} e^{-0.07t}$

So,

$PV = -\frac{100}{7} (20 + 3t) e^{-0.07t} \Big|_0^5 - \int_0^5 -\frac{100}{7} e^{-0.07t} \cdot 3 dt$

$= -\frac{100}{7} (20 + 3t) e^{-0.07t} \Big|_0^5 + \frac{300}{7} \int_0^5 e^{-0.07t} dt$

$= \left[-\frac{100}{7} (20 + 3t) e^{-0.07t} - \frac{30,000}{49} e^{-0.07t} \right]_0^5$

$= \left[-\frac{100}{7} (20 + 3(5)) e^{-0.35} - \frac{30,000}{49} e^{-0.35} \right] - \left[-\frac{100}{7} (20 + 0) e^0 - \frac{30,000}{49} e^0 \right]$

$= \left(-500 e^{-0.35} - \frac{30,000}{49} e^{-0.35} \right) - \left(-\frac{2,000}{7} - \frac{30,000}{49} \right)$

$\approx 114.17345 \text{ hundred, or } \$11,417.35$

51. (a) $p = D(q)$

$D(q) = 10 - q e^{0.02q}$

$D(5) = 10 - (5) e^{0.02(5)} = \$4.47 \text{ each}$

(b) $CS = \int_0^5 (10 - qe^{0.02q})dq - 5(4.47)$

$= \int_0^5 10\,dq - \int_0^5 qe^{0.02q}\,dq - 22.35$

$u = q$ and $dV = e^{0.02q}\,dq$
$du = dq$ $V = 50e^{0.02q}$

$= 10q\big|_0^5 - \left[50qe^{0.02q}\big|_0^5 - \int_0^5 50e^{0.02q}\,dq\right] - 22.35$

$= 10q\big|_0^5 - 50qe^{0.02q}\big|_0^5 + 50\int_0^5 e^{0.02q}\,dq - 22.35$

$= (10q - 50qe^{0.02q} + 2{,}500e^{0.02q})\big|_0^5 - 22.35$

$= \left[10(5) - 50(5)e^{0.02(5)} + 2{,}500e^{0.02(5)}\right] - [0 - 0 + 2{,}500e^0] - 22.35$

≈ 14.28456 thousand, or \$14,284.56

53. $GI = 2\int_0^1 (x - xe^{x-1})dx$

$= 2\left[\int_0^1 x\,dx - \int_0^1 xe^{x-1}dx\right]$

$u = x$ and $dV = e^{x-1}dx$
$du = dx$ $V = e^{x-1}$

$= 2\left[\frac{x^2}{2}\bigg|_0^1 - \left(xe^{x-1}\big|_0^1 - \int_0^1 e^{x-1}dx\right)\right]$

$= 2\left(\frac{x^2}{2} - xe^{x-1} + e^{x-1}\right)\bigg|_0^1$

$= 2\left[\left(\frac{1}{2} - 1e^0 + e^0\right) - (0 - 0 + e^{-1})\right]$

$= 1 - \frac{2}{e} \approx 0.2642$

55. $P(x) = \frac{500\ln(x+1)}{(x+1)^2}; \ 0 \le x \le 10$

$P_{ave}(x) = \frac{1}{10-0}\int_0^{10} \frac{500\ln(x+1)}{(x+1)^2}\,dx$

$= 50\int_0^{10} \frac{\ln(x+1)}{(x+1)^2}\,dx$

Since $\ln(x+1)$ cannot be integrated,

$u = \ln(x+1)$ and $dV = (x+1)^{-2}\,dx$

$du = \frac{1}{x+1}\,dx$ $V = -\frac{1}{x+1}$

and

$50\int_0^{10} \frac{\ln(x+1)}{(x+1)^2}\,dx$

$= 50\left[-\frac{1}{x+1}\ln(x+1)\bigg|_0^{10} - \int_0^{10} -\frac{1}{(x+1)^2}\,dx\right]$

$= 50\left[\left(-\frac{1}{11}\ln 11 - 0\right) + \int_0^{10}(x+1)^{-2}\,dx\right]$

$= 50\left[\left(-\frac{1}{11}\ln 11\right) - \left(\frac{1}{x+1}\bigg|_0^{10}\right)\right]$

$= 50\left[-\frac{1}{11}\ln 11 - \left(\frac{1}{11} - 1\right)\right]$

≈ 34.555, or \$34,555

57. $P(t) = \int P'(t)\,dt = \int t \ln\sqrt{t+1}\,dt$

$u = \ln\sqrt{t+1}$ and $dV = t\,dt$

$\quad = \ln(t+1)^{1/2}$ $dV = \dfrac{t^2}{2}$

$\quad = \dfrac{1}{2}\ln(t+1)$

$du = \dfrac{1}{2(t+1)}\,dt$

$P(t) = \dfrac{t^2}{4}\ln(t+1) - \int \dfrac{t^2}{2}\cdot\dfrac{1}{2(t+1)}\,dt$

$\quad = \dfrac{t^2}{4}\ln(t+1) - \dfrac{1}{4}\int \dfrac{t^2}{t+1}\,dt$

Rewriting,

$\int \dfrac{t^2}{t+1}\,dt = \int \dfrac{1+t^2-1}{t+1}\,dt$

$\quad = \int \dfrac{1+(t+1)(t-1)}{t+1}\,dt$

$\quad = \int \dfrac{1}{t+1}\,dt + \int (t-1)\,dt$

So,

$P(t) = \dfrac{t^2}{4}\ln(t+1)$

$\qquad - \dfrac{1}{4}\left[\ln|t+1| + \dfrac{(t-1)^2}{2}\right] + C$

$\quad = \dfrac{t^2}{4}\ln(t+1) - \dfrac{1}{4}\ln|t+1|$

$\qquad\qquad\qquad\qquad - \dfrac{(t-1)^2}{8} + C$

When $t = 0$, $P(0) = 2000$ thousand, so

$2000 = 0 - \dfrac{1}{4}\ln 1 - \dfrac{1}{8} + C$, or $C = 2000.125$.

So,

$P(t) = \dfrac{t^2}{4}\ln(t+1) - \dfrac{1}{4}\ln|t+1| - \dfrac{(t-1)^2}{8}$

$\qquad\qquad\qquad\qquad\qquad + 2000.125$

and when $t = 5$,

$P(5) = \dfrac{25}{4}\ln 6 - \dfrac{1}{4}\ln 6 - 2 + 2000.152$

$\quad = 6\ln 6 + 1998.125$

$\quad \approx 2,008.8756$ thousand.

The population will be approximately 2,008,876 people.

59. $C_{av} = \dfrac{1}{6-0}\displaystyle\int_0^6 4t e^{(2-0.3t)}\,dt$

$\quad = \dfrac{2}{3}e^2 \displaystyle\int_0^6 t e^{-0.3t}\,dt$

$u = t$ and $dV = e^{-0.3t}\,dt$

$du = dt$ $V = -\dfrac{10}{3}e^{-0.3t}$

So,

$C_{av} = \dfrac{2}{3}e^2\left[-\dfrac{10}{3}t e^{-0.3t}\Big|_0^6 - \int_0^6 -\dfrac{10}{3}e^{-0.3t}\,dt \right]$

$\quad = \dfrac{2}{3}e^2\left[-\dfrac{10}{3}t e^{-0.3t}\Big|_0^6 + \dfrac{10}{3}\int_0^6 e^{-0.3t}\,dt \right]$

$\quad = \dfrac{2}{3}e^2\left[-\dfrac{10}{3}t e^{-0.3t} - \dfrac{100}{9}e^{-0.3t} \right]_0^6$

$\quad = \dfrac{2}{3}e^2\left[\left(-\dfrac{10}{3}(6)e^{-1.8} - \dfrac{100}{9}e^{-1.8} \right) \right.$

$\qquad\qquad\qquad\qquad \left. - \left(0 - \dfrac{100}{9}e^0 \right) \right]$

$\quad = \dfrac{2}{3}e^2\left(-\dfrac{280}{9}e^{-1.8} + \dfrac{100}{9} \right) \approx 29.4$ mg/ml

61. With N as the number of months, in general the population

$= P_0 S(N) + \displaystyle\int_0^N R(t)S(N-t)\,dt$

Here,

members

$= 5,000 e^{-0.02(9)} + \displaystyle\int_0^9 5t e^{-0.02(9-t)}\,dt$

$= 5,000 e^{-0.18} + 5e^{-0.18}\displaystyle\int_0^9 t e^{0.02t}\,dt$

$= 5e^{-0.18}\left[1,000 + \displaystyle\int_0^9 t e^{0.02t}\,dt \right]$

$u = t$ and $dV = e^{0.02t}\,dt$

$du = dt$ $V = 50 e^{0.02t}$

$$= 5e^{-0.18}\left(1{,}000 + 50te^{0.02t}\Big|_0^9 - \int_0^9 50e^{0.02t}\,dt\right)$$

$$= 5e^{-0.18}\left(1{,}000 + 50te^{0.02t}\Big|_0^9 - 50\int_0^9 e^{0.02t}\,dt\right)$$

$$= 5e^{-0.18}\left[1{,}000 + \left(50te^{0.02t} - 2{,}500e^{0.02t}\right)\Big|_0^9\right]$$

$$= 5e^{-0.18}[1{,}000 + (50(9)e^{0.02(9)} - 2{,}500e^{0.02(9)})$$
$$-(0 - 2{,}500e^0)]$$

$$\approx 4{,}367 \text{ members}$$

63. $\dfrac{dN}{dt} = kN(60 - N)$

Need to find t when $N(t) = 20$. To find N, separate the variables.

$$\frac{dN}{N(60 - N)} = k\,dt$$

Integrating both sides,

$$\int \frac{dN}{N(60 - N)} = \int k\,dt$$

To integrate the left hand side, note that this is of the form $\displaystyle\int \frac{du}{u(a + bu)}$ (formula #6) where $u = N$, $a = 60$, and $b = -1$.

$$\int \frac{dN}{N(60 - N)} = \frac{1}{60}\ln\left|\frac{N}{60 - N}\right| + C_1$$

So,

$$\frac{1}{60}\ln\left|\frac{N}{60 - N}\right| + C_1 = kt + C_2$$

$$\ln\left|\frac{N}{60 - N}\right| = 60kt + C_3$$

To solve for N,

$$e^{\ln\left|\frac{N}{60-N}\right|} = e^{60kt + C_3}$$

$$\frac{N}{60 - N} = e^{60kt} \cdot e^{C_3}$$

$$\frac{N}{60 - N} = Ce^{60kt}$$

When $t = 0$, $N = 2$ (the twins), so

$$\frac{2}{60 - 2} = Ce^0 \text{ or, } C = \frac{1}{29} \text{ and}$$

$$\frac{N}{60 - N} = \frac{1}{29}e^{60kt}$$

$$N = (60 - N)\frac{1}{29}e^{60kt}$$

Multiplying both sides by e^{-60kt} and distributing the $\dfrac{1}{29}$,

$$Ne^{-60kt} = \frac{60}{29} - \frac{N}{29}$$

$$N\left(\frac{1}{29} + e^{-60kt}\right) = \frac{60}{29}$$

$$N(t) = \frac{60/29}{1/29 + e^{-60kt}}$$

$$= \frac{60}{1 + 29e^{-60kt}}$$

Since $N = 3$ when $t = 1$,

$$3 = \frac{60}{1 + 29e^{-60k}}$$

$$1 + 29e^{-60k} = 20$$

$$29e^{-60k} = 19$$

$$e^{-60k} = \frac{19}{29}$$

$$\ln e^{-60k} = \ln\frac{19}{29}$$

$$-60k = \ln\frac{19}{29} \approx -0.42$$

and $N(t) = \dfrac{60}{1 + 29e^{-0.42t}}$

When $N = 20$,

$$20 = \frac{60}{1+29e^{-0.42t}}$$

$$1+29e^{-0.42t} = 3$$

$$29e^{-0.42t} = 2$$

$$e^{-0.42t} = \frac{2}{29}$$

$$\ln e^{-0.42t} = \ln\frac{2}{29}$$

$$-0.42t = \ln\frac{2}{29}$$

$$t = \frac{\ln(2/29)}{-0.42} \approx 6.37 \text{ days}$$

So, Scelerat will know our spy's identity before a week passes, and our spy will be a dead duck plucker.

65. $\text{area} = \int_0^{\ln 2}(2-e^x)dx$

$$= (2x - e^x)\Big|_0^{\ln 2}$$

$$= (2\ln 2 - e^{\ln 2}) - (0 - e^0)$$

$$= 2\ln 2 - 1 \approx 0.38629$$

$$\bar{x} = \frac{1}{0.38629}\int_0^{\ln 2} x(2-e^x)dx$$

$$= 2.5887\left[\int_0^{\ln 2}2x\,dx - \int_0^{\ln 2}xe^x\,dx\right]$$

Let $u = x$ and $dV = e^x dx$

65 (continued).

$$= 2.5887\left[x^2\Big|_0^{\ln 2} - \left(xe^x\Big|_0^{\ln 2} - \int_0^{\ln 2}e^x\,dx\right)\right]$$

$$= 2.5887(x^2 - xe^x + e^x)\Big|_0^{\ln 2}$$

$$= 2.5887\Big[((\ln 2)^2 - (\ln 2)(e^{\ln 2}) + e^{\ln 2}) - (0 - 0 + e^0)\Big]$$

$$\approx 0.244$$

$$\bar{y} = \frac{1}{2(0.38629)}\int_0^{\ln 2}(2-e^x)^2\,dx$$

$$= 1.2944\int_0^{\ln 2}(4 - 4e^x + e^{2x})\,dx$$

$$= 1.2944\left(4x - 4e^x + \frac{1}{2}e^{2x}\right)\Big|_0^{\ln 2}$$

$$= 1.2944\left[\left(4\ln 2 - 4e^{\ln 2} + \frac{1}{2}e^{2(\ln 2)}\right) - \left(0 - 4e^0 + \frac{1}{2}e^0\right)\right]$$

$$\approx 0.353$$

So, the centroid is (0.244, 0.353).

67. (a) The kiosk should be located at the centroid. Using $y = \sqrt{2x^2-1}$,

$$\text{Area} = \int_1^5 \sqrt{2x^2-1}\,dx = \int_1^5 \sqrt{\left(\sqrt{2}x\right)^2 - (1)^2}\,dx$$

which most closely resembles $\int\sqrt{u^2-a^2}\,du$ (formula #18). Rewriting,

$$\int_1^5\sqrt{\left(\sqrt{2}x\right)^2-(1)^2}\,dx = \frac{1}{\sqrt{2}}\int_1^5\sqrt{\left(\sqrt{2}x\right)^2-(1)^2}\cdot\sqrt{2}\,dx$$

The formula can be used with $u = \sqrt{2}x$, $du = \sqrt{2}\,dx$, and $a = 1$.

$$= \frac{1}{\sqrt{2}} \left[\frac{\sqrt{2}x}{2} \sqrt{2x^2-1} - \frac{1}{2} \ln\left|\sqrt{2}x + \sqrt{2x^2-1}\right| \right]_1^5$$

$$= \frac{1}{\sqrt{2}} \left[\left(\frac{\sqrt{2}(5)}{2} \sqrt{2(5)^2-1} - \frac{1}{2} \ln\left|\sqrt{2}(5) + \sqrt{2(5)^2-1}\right| \right) - \left(\frac{\sqrt{2}(1)}{2} \sqrt{2(1)^2-1} - \frac{1}{2} \ln\left|\sqrt{2}(1) + \sqrt{2(1)^2-1}\right| \right) \right]$$

$$= \frac{1}{\sqrt{2}} \left[\frac{35\sqrt{2}}{2} - \frac{1}{2} \ln\left|5\sqrt{2}+7\right| - \frac{\sqrt{2}}{2} + \frac{1}{2} \ln\left(\sqrt{2}+1\right) \right]$$

$$\approx 16.3768$$

$$\bar{x} = \frac{1}{16.3768} \int_1^5 x\sqrt{2x^2-1}\,dx$$

Using substitution with $u = 2x^2 - 1$, $\frac{1}{4}du = x\,dx$, and limits of integration of $2(1)^2 - 1 = 1$ and

$2(5)^2 - 1 = 49$,

$$\bar{x} = 0.06106 \left(\frac{1}{4} \int_1^{49} u^{1/2}\,du \right)$$

$$= 0.01527 \left(\frac{2}{3} u^{3/2} \right) \Big|_1^{49}$$

$$= 0.010177[(49)^{3/2} - (1)^{3/2}] \approx 3.48$$

$$\bar{y} = \frac{1}{2(16.3768)} \int_1^5 \left(\sqrt{2x^2-1} \right)^2 dx$$

$$= 0.030531 \int_1^5 (2x^2 - 1)\,dx$$

$$= 0.030531 \left[\frac{2x^3}{3} - x \right]_1^5$$

$$= 0.030531 \left[\left(\frac{2(5)^3}{3} - 5 \right) - \left(\frac{2(1)^3}{3} - 1 \right) \right]$$

$$\approx 2.40$$

So, the kiosk should be located at the coordinates (3.48, 2.40).

(b) Writing Exercise—Answers will vary.

69. $\int u^n e^{au}\,du$

Let $f = u^n$ and $dV = e^{au}\,du$

$\qquad df = nu^{n-1}\,du \qquad\qquad V = \dfrac{1}{a}e^{au}$

$= \dfrac{1}{a}u^n e^{au} - \int \dfrac{1}{a}e^{au} \cdot nu^{n-1}\,du$

$= \dfrac{1}{a}u^n e^{au} - \dfrac{n}{a}\int u^{n-1}e^{au}\,du$

71. $\int \dfrac{e^{kx}}{x^n}\,dx$

Both terms are easy to integrate, however the derivative of e^{kx} is a little easier to work with than the derivative of x^{-n}, so use

$\qquad u = e^{kx} \qquad$ and $\quad dV = x^{-n}\,dx$

$\qquad du = ke^{kx}\,dx \qquad V = \dfrac{x^{(-n+1)}}{(-n+1)}$

$\int x^{-n}e^{kx}\,dx$

$= \dfrac{1}{1-n}e^{kx}x^{-n+1} - \int ke^{kx}\cdot\dfrac{1}{1-n}x^{-n+1}\,dx$

$= \dfrac{1}{1-n}e^{kx}x^{-n+1} - \dfrac{k}{1-n}\int e^{kx}x^{-n+1}\,dx$

$= \dfrac{1}{n-1}\left[-e^{kx}x^{-n+1} + k\int e^{kx}x^{-n+1}\,dx\right]$

73. To use graphing utility to find where curves intersect and compute the area of region bounded by the curves,

Press $\boxed{y=}$ and input $\sqrt{\left(\left(\dfrac{2}{5}\right)x^2 - 2\right)}$ for

$y_1 =$.

Input $-\sqrt{\left(\left(\dfrac{2}{5}\right)x^2 - 2\right)}$ for $y_2 =$, and

input $x^{\wedge}3 - 3.5x^2 + 2x$ for $y_3 =$.

Use window dimensions $[-1, 4]1$ by $[-3, 5]1$ for a good view of where the graphs intersect.

Use trace and zoom to find the points of intersection or use the intersect function

under the calc menu to find $(2.966, 1.232)$ and $(2.608, -0.850)$ are the two points of intersection.

To find the area bounded by the curves, we must find that the x-intercept of the hyperbola is $x \approx 2.236$. Then we need

$\displaystyle\int_{2.236}^{2.608} y_1 - y_2 + \int_{2.608}^{2.966} y_1 - y_3$

$= \displaystyle\int_{2.236}^{2.608} y_1 - \int_{2.236}^{2.608} y_2 + \int_{2.608}^{2.966} y_1 - \int_{2.608}^{2.966} y_3$

Use the $\int f(x)\,dx$ function under the calc

menu making sure the current equation is activated for each integral. The area is approximately 0.75834. Alternatively, you can use the *fnInt* function from the home screen under the math menu and enter

$\quad fnInt(y_1 - y_2, x, 2.236, 2.608)$

$\qquad + fnInt(y_1 - y_3, x, 2.608, 2.966)$

You can insert y_1, y_2, y_3 by pressing $\boxed{\text{vars}}$ and select Function under y-vars and then select which y function to insert.

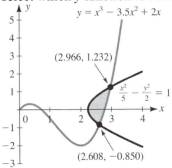

75. Press $\boxed{y=}$. Input $e^{\wedge}(2x) + 4$ for $y_1 =$ and $5e^{\wedge}(x)$ for y_2.

Use window dimensions $[-1, 3]1$ by $[-5, 25]5$.

Press $\boxed{\text{graph}}$.

Use the intersect function under the calc menu to find the two points of intersection. Enter a value close to the first point of intersection on $y_1 =$ and also on

$y_2 =$. Then enter a guess. The first point of intersection is $(0, 5)$. Repeat this process for the second point of intersection to find $(1.386, 20)$. To find the area bounded by these two curves, we must find

$$\int_0^{1.386} y_2 - y_1 = \int_0^{1.386} y_2 - \int_0^{1.386} y_1.$$

Use the $\int f(x)\,dx$ function under the calc menu making sure the current equation is activated for each integral. The area is approximately 1.9548. Alternatively, you can use the *fnInt* function from the home screen under the math menu and enter

fnInt$(y_2 - y_1, x, 0, 1.386)$

You can insert y_1 and y_2 by pressing vars and selecting function under the y-vars.

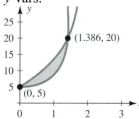

77. To use the numeric integration feature to evaluate the integral,

Press $\boxed{y=}$ and input $\sqrt{(4x^2 - 7)}$ for $y_1 =$.
Use window dimensions $[-1, 4]1$ by $[-3, 5]1$.
Press graph.
Use the $\int f(x)\,dx$ function under the calc menu. Enter $x = 2$ for the lower limit and $x = 3$ for the upper limit. We see that

$$\int_2^3 \sqrt{4x^2 - 7}\,dx \approx 4.227.$$

To verify, we use formula #18 on the table of integrals with

$u = 2x$

$du = 2\,dx$

$dx = \dfrac{1}{2}\,du$

When $x = 2$, $u = 4$; when $x = 3$, $u = 6$.
So,

$$\int_2^3 \sqrt{4x^2 - 7}\,dx$$

$$= \frac{1}{2}\int_4^6 \sqrt{u^2 - 7}\,du$$

$$= \frac{1}{2}\left[\frac{u}{2}\sqrt{u^2 - 7} - \frac{7}{2}\ln\left|u + \sqrt{u^2 - 7}\right|\right]\Bigg|_4^6$$

$$= \frac{1}{2}\left[3\sqrt{29} - \frac{7}{2}\ln\left(6 + \sqrt{29}\right) - 2(3) + \frac{7}{2}\ln 7\right]$$

$$= \frac{1}{2}(8.45309083)$$

$$\approx 4.227$$

79. To use the numeric integration feature to evaluate the integral,
Press $\boxed{y=}$ and input
$\sqrt{((x^2 + 2x))/((x+1)^2)}$ for $y_1 =$.
Use window dimensions $[-1, 3]1$ by $[-1, 2]1$. Press graph.
Use the $\int f(x)\,dx$ function under the calc menu with $x = 0$ as the lower limit and $x = 1$ the upper limit. We see that

$$\int_0^1 \frac{\sqrt{x^2 + 2x}}{(x+1)^2}\,dx \approx 0.4509$$

To verify, we use formula #19 on the table of integrals:

$$\int_0^1 \frac{\sqrt{x^2 + 2x}}{(x+1)^2}\,dx = \int_0^1 \frac{\sqrt{(x+1)^2 - 1}}{(x+1)^2}\,dx$$

Let $u = x + 1$
     $du = dx$
When $x = 0$, $u = 1$; when $x = 1$, $u = 2$.
So,

$$\int_1^2 \frac{\sqrt{u^2 - 1}}{u^2}\,du$$

$$= \left[-\frac{\sqrt{u^2 - 1}}{u} + \ln\left|u + \sqrt{u^2 - 1}\right|\right]\Bigg|_1^2$$

$$= -\frac{\sqrt{3}}{2} + \ln\left|2 + \sqrt{3}\right| - \left(-\frac{\sqrt{0}}{1} + \ln\left|1 + \sqrt{0}\right|\right)$$

$$= -\frac{\sqrt{3}}{2} + \ln\left(2 + \sqrt{3}\right) - \ln 1$$

$$\approx 0.4509$$

6.2   Numerical Integration

1. For $\int_1^2 x^2\,dx$ with $n = 4$, $\Delta x = \dfrac{2-1}{4} = 0.25$, and $x_1 = 1$, $x_2 = 1.25$, $x_3 = 1.50$, $x_4 = 1.75$, $x_5 = 2$.

   (a) By the trapezoidal rule,

$$\int_1^2 x^2\,dx = \frac{\Delta x}{2}[f(x_1) + 2f(x_2) + 2f(x_3) + 2f(x_4) + f(x_5)]$$
$$= \frac{0.25}{2}[1^2 + 2(1.25)^2 + 2(1.5)^2 + 2(1.75)^2 + 2^2]$$
$$\approx 2.3438.$$

   (b) By Simpson's rule,

$$\int_1^2 x^2\,dx = \frac{\Delta x}{3}[f(x_1) + 4f(x_2) + 2f(x_3) + 4f(x_4) + f(x_5)]$$
$$= \frac{0.25}{3}[1^2 + 4(1.25)^2 + 2(1.5)^2 + 4(1.75)^2 + 2^2]$$
$$\approx 2.3333.$$

3. For $\int_0^1 \dfrac{1}{1+x^2}\,dx$ with $n = 4$, $\Delta x = \dfrac{1-0}{4} = 0.25$, and $x_1 = 0$, $x_2 = 0.25$, $x_3 = 0.50$, $x_4 = 0.75$, $x_5 = 1$.

   (a) By the trapezoidal rule,

$$\int_0^1 \frac{1}{1+x^2}\,dx = \frac{\Delta x}{2}[f(x_1) + 2f(x_2) + 2f(x_3) + 2f(x_4) + f(x_5)]$$
$$= \frac{0.25}{2}\left[1 + \frac{2}{1+(0.25)^2} + \frac{2}{1+(0.5)^2} + \frac{2}{1+(0.75)^2} + \frac{1}{2}\right]$$
$$\approx 0.7828$$

   (b) By Simpson's rule,

$$\int_0^1 \frac{1}{1+x^2}\,dx = \frac{\Delta x}{3}[f(x_1) + 4f(x_2) + 2f(x_3) + 4f(x_4) + f(x_5)]$$
$$= \frac{0.25}{3}\left[1 + \frac{4}{1+(0.25)^2} + \frac{2}{1+(0.5)^2} + \frac{4}{1+(0.75)^2} + \frac{1}{2}\right]$$
$$\approx 0.7854.$$

5. For $\int_{-1}^0 \sqrt{1+x^2}\,dx$ with $n = 4$, $\Delta x = \dfrac{0-(-1)}{4} = 0.25$, and $x_1 = -1$, $x_2 = -0.75$, $x_3 = -0.5$, $x_4 = -0.25$,

   $x_5 = 0$.

(a) By the trapezoidal rule,

$$\int_{-1}^{0} \sqrt{1+x^2}\, dx = \frac{\Delta x}{2}[f(x_1)+2f(x_2)+2f(x_3)+2f(x_4)+f(x_5)]$$

$$= \frac{0.25}{2}\left[\sqrt{1+(-1)^2}+2\sqrt{1+(-0.75)^2}+2\sqrt{1+(-0.5)^2}+2\sqrt{1+(-0.25)^2}+\sqrt{1+(0)^2}\right]$$

$$\approx 1.1515.$$

(b) By Simpson's rule,

$$\int_{-1}^{0} \sqrt{1+x^2}\, dx$$

$$= \frac{\Delta x}{3}[f(x_1)+4f(x_2)+2f(x_3)+4f(x_4)+f(x_5)]$$

$$= \frac{0.25}{3}\left[\sqrt{1+(-1)^2}+4\sqrt{1+(-0.75)^2}+2\sqrt{1+(-0.5)^2}+4\sqrt{1+(-0.25)^2}+\sqrt{1+(0)^2}\right]$$

$$\approx 1.1478.$$

7. For $\int_{0}^{1} e^{-x^2}\, dx$ with $n=4$, $\Delta x = \frac{1-0}{4}=0.25$, and $x_1=0$, $x_2=0.25$, $x_3=0.50$, $x_4=0.75$, $x_5=1$.

(a) By the trapezoidal rule,

$$\int_{0}^{1} e^{-x^2}\, dx = \frac{\Delta x}{2}[f(x_1)+2f(x_2)+2f(x_3)+2f(x_4)+f(x_5)]$$

$$= \frac{0.25}{2}[1+2e^{-(0.25)^2}+2e^{-(0.5)^2}+2e^{-(0.75)^2}+e^{-1}]$$

$$\approx 0.7430.$$

(b) By Simpson's rule,

$$\int_{0}^{1} e^{-x^2}\, dx = \frac{\Delta x}{3}[f(x_1)+4f(x_2)+2f(x_3)+4f(x_4)+f(x_5)]$$

$$= \frac{0.25}{3}[1+4e^{-(0.25)^2}+2e^{-(0.5)^2}+4e^{-(0.75)^2}+e^{-1}]$$

$$\approx 0.7469.$$

9. For $\int_{2}^{4} \frac{dx}{\ln x}$ with $n=6$, $\Delta x = \frac{4-2}{6}=\frac{1}{3}$ and $x_1=2$, $x_2=\frac{7}{3}$, $x_3=\frac{8}{3}$,

$x_4=3, x_5=\frac{10}{3}, x_6=\frac{11}{3}, x_7=4.$

(a) By the trapezoidal rule,

$$\int_{2}^{4} \frac{dx}{\ln x} \approx \frac{\Delta x}{2}[f(x_1)+2f(x_2)+2f(x_3)+2f(x_4)+2f(x_5)+2f(x_6)+f(x_7)]$$

$$= \frac{\frac{1}{3}}{2}\left[\frac{1}{\ln 2}+\frac{2}{\ln \frac{7}{3}}+\frac{2}{\ln \frac{8}{3}}+\frac{2}{\ln 3}+\frac{2}{\ln \frac{10}{3}}+\frac{2}{\ln \frac{11}{3}}+\frac{1}{\ln 4}\right]$$

$$\approx 1.9308$$

(b) By Simpson's rule,

$$\int_2^4 \frac{dx}{\ln x} \approx \frac{\Delta x}{3}[f(x_1)+4f(x_2)+2f(x_3)+4f(x_4)+2f(x_5)+4f(x_6)+f(x_7)]$$

$$= \frac{\frac{1}{3}}{3}\left[\frac{1}{\ln 2}+\frac{4}{\ln \frac{7}{3}}+\frac{2}{\ln \frac{8}{3}}+\frac{4}{\ln 3}+\frac{2}{\ln \frac{10}{3}}+\frac{4}{\ln \frac{11}{3}}+\frac{1}{\ln 4}\right]$$

$$\approx 1.9228$$

11. For $\int_0^1 \sqrt[3]{1+x^2}\,dx$ with $n=4$, $\Delta x = \frac{1-0}{4}=0.25$ and $x_1=0$, $x_2=0.25$, $x_3=0.05$, $x_4=0.75$, $x_5=1$.

(a) By the trapezoidal rule,

$$\int_0^1 \sqrt[3]{1+x^2}\,dx \approx \frac{\Delta x}{2}[f(x_1)+4f(x_2)+2f(x_3)+4f(x_4)+2f(x_5)]$$

$$= \frac{0.25}{2}\left[1+\sqrt[3]{1.0625}+2\sqrt[3]{1.25}+2\sqrt[3]{1.5625}+\sqrt[3]{2}\right]$$

$$\approx 1.0970.$$

(b) By Simpson's rule,

$$\int_0^1 \sqrt[3]{1+x^2}\,dx \approx \frac{\Delta x}{3}[f(x_1)+4f(x_2)+2f(x_3)+4f(x_4)+f(x_5)]$$

$$= \frac{0.25}{3}\left[1+4\sqrt[3]{1.0625}+2\sqrt[3]{1.25}+4\sqrt[3]{1.5625}+\sqrt[3]{2}\right]$$

$$\approx 1.0948$$

13. For $\int_0^2 e^{-\sqrt{x}}\,dx$ with $n=8$, $\Delta x = \frac{2-0}{8}=0.25$ and $x_1=0$, $x_2=0.25$, $x_3=0.5$, $x_4=0.75$, $x_5=1$, $x_6=1.25$, $x_7=1.5$, $x_8=1.75$, $x_9=2$.

(a) By the trapezoidal rule,

$$\int_0^2 e^{-\sqrt{x}}\,dx$$

$$\approx \frac{\Delta x}{2}[f(x_1)+2f(x_2)+2f(x_3)+2f(x_4)+2f(x_5)+2f(x_6)+2f(x_7)+2f(x_8)+f(x_9)]$$

$$= \frac{0.25}{2}\left[1+2e^{-\sqrt{0.25}}+2e^{-\sqrt{0.5}}+2e^{-\sqrt{0.75}}+2e^{-1}+2e^{-\sqrt{1.25}}+2e^{-\sqrt{1.5}}+2e^{-\sqrt{1.75}}+e^{-\sqrt{2}}\right]$$

$$\approx 0.8492.$$

(b) By Simpson's rule,

$$\int_0^2 e^{-\sqrt{x}}\,dx$$

$$\approx \frac{\Delta x}{3}[f(x_1)+4f(x_2)+2f(x_3)+4f(x_4)+2f(x_5)+4f(x_6)+2f(x_7)+4f(x_8)+f(x_9)]$$

$$= \frac{0.25}{3}\left[1+4e^{-\sqrt{0.25}}+2e^{-\sqrt{0.5}}+4e^{-\sqrt{0.75}}+2e^{-1}+4e^{-\sqrt{1.25}}+2e^{-\sqrt{1.5}}+4e^{-\sqrt{1.75}}+e^{-\sqrt{2}}\right]$$

$$\approx 0.8362.$$

15. For $\int \dfrac{1}{x^2}\,dx$ with $n = 4$, $\Delta x = \dfrac{2-1}{4} = 0.25$, and $x_1 = 1$, $x_2 = 1.25$, $x_3 = 1.50$, $x_4 = 1.75$, $x_5 = 2$.

(a) By the trapezoidal rule,

$$\int_1^2 \frac{1}{x^2}\,dx = \frac{\Delta x}{2}[f(x_1) + 2f(x_2) + 2f(x_3) + 2f(x_4) + f(x_5)]$$

$$= \frac{0.25}{2}\left[1 + \frac{2}{(1.25)^2} + \frac{2}{(1.5)^2} + \frac{2}{(1.75)^2} + \frac{1}{2^2}\right]$$

$$\approx 0.5090.$$

The error estimate is $|E_n| \le \dfrac{M(b-a)^3}{12n^2}$. For $n = 1$, $a = 1$, and $b = 2$, $|E_4| \le \dfrac{M(2-1)^2}{12(4^2)} = \dfrac{M}{192}$,

where M is the maximum value of $|f''(x)|$ on $1 \le x \le 2$. Now $f(x) = x^{-2}$, $f'(x) = -2x^{-3}$, and

$f''(x) = 6x^{-4}$. For $1 \le x \le 2$, $|f''(x)| = \dfrac{6}{x^4} \le \dfrac{6}{1^4} = 6$. So, $|E_4| = \dfrac{6}{192} \approx 0.03125$.

(b) By Simpson's rule,

$$\int_1^2 \frac{1}{x^2}\,dx = \frac{\Delta x}{3}[f(x_1) + 4f(x_2) + 2f(x_3) + 4f(x_4) + f(x_5)]$$

$$= \frac{0.25}{3}\left[1 + \frac{4}{(1.25)^2} + \frac{2}{(1.5)^2} + \frac{4}{(1.75)^2} + \frac{1}{2^2}\right]$$

$$\approx 0.5004.$$

The error estimate is $|E_n| \le \dfrac{M(b-a)^5}{180n^4}$. For $n = 4$, $a = 1$, and $b = 2$, $|E_4| \le \dfrac{M(2-1)^5}{180(4^4)} = \dfrac{M}{46{,}080}$

where M is the maximum value of $|f^{(4)}(x)|$ on $1 \le x \le 2$. Now $f''(x) = 6x^{-4}$,

$f^{(3)}(x) = -24x^{-5}$, and $f^{(4)}(x) = 120x^{-6}$. For $1 \le x \le 2$, $|f^{(4)}(x)| = \dfrac{120}{x^6} \le \dfrac{120}{1^6} = 120$. So,

$|E_4| \le \dfrac{120}{46{,}080} \approx 0.0026$.

17. For $\int_1^3 \sqrt{x}\,dx$ with $n = 10$, $\Delta x = \dfrac{3-1}{10} = 0.2$, and $x_1 = 1$, $x_2 = 1.2$, $x_3 = 1.4$, ..., $x_{10} = 2.8$, $x_{11} = 3$.

(a) By the trapezoidal rule,

$$\int_1^3 \sqrt{x}\,dx$$

$$= \frac{\Delta x}{2}[f(x_1) + 2f(x_2) + 2f(x_3) + \cdots + 2f(x_{10}) + f(x_{11})]$$

$$= \frac{0.2}{2}\left[1 + 2\sqrt{1.2} + 2\sqrt{1.4} + 2\sqrt{1.6} + 2\sqrt{1.8} + 2\sqrt{2} + \sqrt{2.2} + 2\sqrt{2.4} + 2\sqrt{2.6} + 2\sqrt{2.8} + \sqrt{3}\right]$$

$$\approx 2.7967.$$

The error estimate is $|E_n| \le \dfrac{M(b-a)^3}{12n^2}$. For $n = 10$, $a = 1$, and $b = 3$,

$$|E_{10}| \le \frac{M(3-1)^3}{12(10^2)} = \frac{8M}{1,200} = \frac{M}{150},$$ where M is the maximum value of $|f''(x)|$ on $1 \le x \le 3$. Now,

$$f(x) = x^{1/2}, \quad f'(x) = \frac{1}{2}x^{-1/2}, \text{ and } f''(x) = -\frac{1}{4}x^{-3/2}. \text{ For } 1 \le x \le 3,$$

$$\left|f''(x)\right| = \left|-\frac{1}{4}x^{-3/2}\right| \le \frac{1}{4}(1^{-3/2}) = \frac{1}{4}. \text{ So, } |E_{10}| = \frac{1}{150}\left(\frac{1}{4}\right) \approx 0.0017.$$

(b) By Simpson's rule,

$$\int_1^3 \sqrt{x}\, dx$$
$$= \frac{\Delta x}{3}[f(x_1) + 4f(x_2) + 2f(x_3) + \cdots + 4f(x_{10}) + f(x_{11})]$$
$$= \frac{0.2}{3}\left[1 + 4\sqrt{1.2} + 2\sqrt{1.4} + 4\sqrt{1.6} + 2\sqrt{1.8} + 4\sqrt{2} + 2\sqrt{2.2} + 4\sqrt{2.4} + 2\sqrt{2.6} + 4\sqrt{2.8} + \sqrt{3}\right]$$
$$\approx 2.7974.$$

The error estimate is $|E_n| \le \dfrac{M(b-a)^5}{180n^4}$. For $n = 10$, $a = 1$, and $b = 3$,

$$|E_{10}| \le \frac{M(3-1)^5}{180(10^4)} = \frac{32M}{180(10^4)},$$ where M is the maximum value of $|f^{(4)}(x)|$ on $1 \le x \le 3$. Now,

$$f''(x) = -\frac{1}{4}x^{-3/2}, \quad f^{(3)}(x) = \frac{3}{8}x^{-5/2}, \text{ and } f^{(4)}(x) = -\frac{15}{16}x^{-7/2}. \text{ For } 1 \le x \le 3,$$

$$\left|f^{(4)}(x)\right| = \left|-\frac{15}{16}x^{-7/2}\right| \le \frac{15}{16}(1^{-7/2}) = \frac{15}{16}. \text{ So, } |E_{10}| = \frac{32}{180(10,000)}\left(\frac{15}{16}\right) \approx 0.0000167.$$

19. For $\int_0^1 e^{x^2}\, dx$ with $n = 4$, $\Delta x = \dfrac{1-0}{4} = 0.25$, and $x_1 = 0$, $x_2 = 0.25$, $x_3 = 0.50$, $x_4 = 0.75$, $x_5 = 1$.

(a) By the trapezoidal rule,
$$\int_1^2 e^{x^2}\, dx = \frac{\Delta x}{2}[f(x_1) + 2f(x_2) + 2f(x_3) + 2f(x_4) + f(x_5)]$$
$$= \frac{0.25}{2}[1 + 2e^{(0.25)^2} + 2e^{(0.5)^2} + 2e^{(0.75)^2} + e^1]$$
$$\approx 1.4907.$$

The error estimate is $|E_n| \le \dfrac{M(b-a)^3}{12n^2}$. For $n = 4$, $a = 10$, and $b = 1$, $|E_4| \le \dfrac{M(1-0)^3}{12(4^2)} = \dfrac{M}{192}$,

where M is the maximum value of $|f''(x)|$ on $0 \le x \le 1$. Now, $f(x) = e^{x^2}$, $f'(x) = -2xe^{-x^2}$,

and $f''(x) = (4x^2 + 2)e^{x^2}$. For $0 \le x \le 1$, $|f''(x)| = [4(1^2) + 2]e^{1^2} = 6e$. So, $|E_4| = \dfrac{6e}{192} \approx 0.0849$.

(b) By Simpson's rule,

$$\int_1^2 e^{x^2}\,dx = \frac{\Delta x}{3}[f(x_1)+4f(x_2)+2f(x_3)+4f(x_4)+f(x_5)]$$

$$= \frac{0.25}{3}[1+4e^{(0.25)^2}+2e^{(0.5)^2}+4e^{(0.75)^2}+e^1]$$

$$\approx 1.4637.$$

The error estimate is $\left|E_n\right| \le \dfrac{M(b-a)^5}{180n^4}$. For $n=4$, $a=0$, and $b=1$, $\left|E_4\right| \le \dfrac{M(1-0)^5}{180(4^4)} = \dfrac{M}{46{,}080}$,

where M is the maximum value of $\left|f^{(4)}(x)\right|$ on $0 \le x \le 1$. Now, $f''(x)=(4x^2+2)e^{x^2}$,

$f^{(3)}(x)=(8x^3+12x)e^{x^2}$, and $f^{(4)}(x)=(16x^4+48x^2+12)e^{x^2}$. For $0 \le x \le 1$,

$\left|f^{(4)}(x)\right| = [16(1^4)+48(1^2)+12]e^{1^2} = 76e$. So, $\left|E_4\right| \le \dfrac{76e}{46{,}080} \approx 0.0045$.

21. The integral to be approximated is

$\int_1^3 \dfrac{1}{x}\,dx$. The derivatives of

$f(x)=\dfrac{1}{x}=x^{-1}$ are $f'(x)=-x^{-2}$,

$f''(x)=2x^{-3}$, $f^{(3)}(x)=-6x^{-4}$, and

$f^{(4)}(x)=24x^{-5}$.

(a) For the trapezoidal rule,

$\left|E_n\right| \le \dfrac{M(b-a)^3}{12n^2}$, where M is the

maximum value of $\left|f''(x)\right|$ on

$1 \le x \le 3$. Now $\left|f''(x)\right| = \dfrac{2}{x^3} \le \dfrac{2}{1^3} = 2$

on $1 \le x \le 3$. $\left|E_n\right| \le \dfrac{2(3-1)^3}{12n^2} = \dfrac{4}{3n^2}$,

which is less than 0.00005 if

$4 < 3(0.00005)n^2$ or

$n > \sqrt{\dfrac{4}{3(0.00005)}} \approx 163.3$. So,

164 intervals should be used.

(b) For Simpson's rule,

$\left|E_n\right| \le \dfrac{M(b-a)^5}{180n^4}$, where M is the

maximum value of $\left|f^{(4)}(x)\right|$ on

$1 \le x \le 3$. Now,

$\left|f^{(4)}(x)\right| = \left|\dfrac{24}{x^5}\right| \le \dfrac{24}{1^5} = 24$ on

$1 \le x \le 3$. $\left|E_n\right| \le \dfrac{24(3-1)^5}{180n^4} = \dfrac{768}{180n^4}$

which is less than 0.00005 if

$768 < 180(0.00005)n^4$ or

$n > \sqrt[4]{\dfrac{768}{180(0.00005)}} \approx 17.1$. So,

18 subintervals should be used.

23. The integral to be approximated is

$\int_1^2 \dfrac{1}{\sqrt{x}}\,dx$. The derivatives of

$f(x)=\dfrac{1}{\sqrt{x}}=x^{-1/2}$ are $f'(x)=-\dfrac{1}{2}x^{-3/2}$,

$f''(x)=\dfrac{3}{4}x^{-5/2}$, $f^{(3)}(x)=-\dfrac{15}{8}x^{-7/2}$,

and $f^{(4)}(x)=\dfrac{105}{16}x^{-9/2}$.

(a) For the trapezoidal rule,

$$|E_n| \le \frac{M(b-a)^3}{12n^2},$$ where M is the

maximum value of $|f''(x)|$ on

$1 \le x \le 2$. Now $|f''(x)| = \frac{3}{4}x^{-5/2} \le \frac{3}{4}$

on $1 \le x \le 2$. $|E_n| \le \frac{3}{4}\frac{(2-1)^3}{12n^2} = \frac{1}{16n^2}$,

which is less than 0.00005 if

$1 < 16(0.00005)n^2$ or

$$n > \sqrt{\frac{1}{16(0.00005)}} \approx 35.4.$$ So,

36 intervals should be used.

(b) For Simpson's rule,

$$|E_n| \le \frac{M(b-a)^5}{180n^4},$$ where M is the

maximum value of $|f^{(4)}(x)|$ on

$1 \le x \le 2$. Now,

$$\left|f^{(4)}(x)\right| = \left|\frac{105}{16}x^{-9/2}\right| \le \frac{105}{16}$$ on

$1 \le x \le 2$. $|E_n| \le \frac{105(2-1)^5}{16(180)n^4} = \frac{7}{162n^4}$,

which is less than 0.00005 if

$7 < 192(0.00005)n^4$ or

$$n > \sqrt[4]{\frac{7}{192(0.00005)}} \approx 5.2.$$ So,

6 subintervals should be used.

25. The integral to be approximated is $\int_{1.2}^{2.4} e^x\,dx$.

(a) For the trapezoidal rule,

$$|E_n| \le \frac{M(b-a)^3}{12n^2},$$ where M is the

maximum value of $|f''(x)|$ on

$1.2 \le x \le 2.4$. Now $|f''(x)| = |e^x| \le e^{2.4}$

on $1.2 \le x \le 2.4$.

$$|E_n| \le \frac{e^{2.4}(2.4-1.2)^3}{12n^2} = \frac{1.728e^{2.4}}{12n^2}$$

which is less than 0.00005 if

$1.728e^{2.4} < 12(0.00005)n^2$ or

$$n > \sqrt{\frac{1.728e^{2.4}}{12(0.00005)}} \approx 178.2.$$ So,

179 intervals should be used.

(b) For Simpson's rule,

$$|E_n| \le \frac{M(b-a)^5}{180n^4},$$ where M is the

maximum value of $|f^{(4)}(x)|$ on

$1.2 \le x \le 2.4$. Now $\left|f^{(4)}(x)\right| = |e^x| \le e^{2.4}$

on $1.2 \le x \le 2.4$.

$$|E_n| \le \frac{e^{2.4}(2.4-1.2)^5}{180n^4}$$ which is less

than 0.00005 is

$e^{2.4}(1.2)^5 < 180(0.00005)n^4$ or

$$n > \sqrt[4]{\frac{e^{2.4}(1.2)^5}{180(0.00005)}} \approx 7.4.$$ So,

8 subintervals should be used.

27. For $\int_0^1 \sqrt{1-x^2}\,dx$ with $n = 8$, $\Delta x = \dfrac{1-0}{8} = 0.125$, and $x_1 = 0$, $x_2 = 0.125$, $x_3 = 0.25$, ..., $x_8 = 0.875$, $x_9 = 1$.

(a) By the trapezoidal rule,

$$\int_0^1 \sqrt{1-x^2}\,dx$$

$$= \frac{\Delta x}{2}[f(x_1) + 2f(x_2) + 2f(x_3) + \cdots + 2f(x_8) + f(x_9)]$$

$$= 0.0625\Big[\sqrt{1-(0)^2} + 2\sqrt{1-(0.125)^2} + 2\sqrt{1-(0.25)^2} + 2\sqrt{1-(0.375)^2} + 2\sqrt{1-(0.5)^2}$$

$$+ 2\sqrt{1-(0.625)^2} + 2\sqrt{1-(0.75)^2} + 2\sqrt{1-(0.875)^2} + \sqrt{1-(1)^2}\,\Big]$$

$$\approx 0.7725$$

$(0.7725)(4) = 3.090$ as an approximation of π.

(b) By Simpson's rule,

$$\int_0^1 \sqrt{1-x^2}\,dx$$

$$= \frac{\Delta x}{3}[f(x_1) + 4f(x_2) + 2f(x_3) + 4f(x_4) + \cdots + 4f(x_8) + f(x_9)]$$

$$= \frac{1}{24}\Big[\sqrt{1-(0)^2} + 4\sqrt{1-(0.125)^2} + 2\sqrt{1-(0.25)^2} + 4\sqrt{1-(0.375)^2} + 2\sqrt{1-(0.5)^2}$$

$$+ 4\sqrt{1-(0.625)^2} + 2\sqrt{1-(0.75)^2} + 4\sqrt{1-(0.875)^2} + \sqrt{1-(1)^2}\,\Big]$$

$$\approx 0.7803$$

$(0.7803)(4) = 3.121$ as an approximation of π.

29. For $\int_1^6 \dfrac{e^{-0.4x}}{x}\,dx$ with $n = 10$, $\Delta x = \dfrac{6-1}{10} = 0.5$, and $x_1 = 1$, $x_2 = 1.5$, $x_3 = 2.0$, $x_4 = 2.5$, $x_5 = 3.0$, $x_6 = 3.5$, $x_7 = 4.0$, $x_8 = 4.5$, $x_9 = 5.0$, $x_{10} = 5.5$, $x_{11} = 6.0$. By the Trapezoidal rule,

$$\int_1^6 \frac{e^{-0.4x}}{x}\,dx$$

$$= \frac{\Delta x}{2}[f(x_1) + 2f(x_2) + 2f(x_3) + 2f(x_4) + 2f(x_5) + 2f(x_6) + 2f(x_7) + 2f(x_8) + 2f(x_9)$$

$$+ 2f(x_{10}) + f(x_{11})]$$

$$= 0.25\left[\left(\frac{e^{-0.4(1)}}{1}\right) + 2\left(\frac{e^{-0.4(1.5)}}{1.5}\right) + 2\left(\frac{e^{-0.4(2)}}{2}\right) + 2\left(\frac{e^{-0.4(2.5)}}{2.5}\right) + 2\left(\frac{e^{-0.4(3)}}{3}\right) + 2\left(\frac{e^{-0.4(3.5)}}{3.5}\right)\right.$$

$$\left. + 2\left(\frac{e^{-0.4(4)}}{4}\right) + 2\left(\frac{e^{-0.4(4.5)}}{4.5}\right) + 2\left(\frac{e^{-0.4(5)}}{5}\right) + 2\left(\frac{e^{-0.4(5.5)}}{5.5}\right) + \left(\frac{e^{-0.4(6)}}{6}\right)\right]$$

$$\approx 0.6929$$

So, the estimate of the average value is $\dfrac{1}{6-1}(0.6929) = 0.1386$.

31. Volume of $S = \pi \int_0^1 \left(\dfrac{x}{1+x} \right)^2 dx$

Using the trapezoidal rule with $n = 7$, $\Delta x = \dfrac{1-0}{7}$ and $x_1 = 0$, $x_2 = \dfrac{1}{7}$, $x_3 = \dfrac{2}{7}$, $x_4 = \dfrac{3}{7}$, $x_5 = \dfrac{4}{7}$,

$x_6 = \dfrac{5}{7}$, $x_7 = \dfrac{6}{7}$, $x_8 = 1$.

$$\int_0^1 \left(\dfrac{x}{1+x} \right)^2 dx \approx \dfrac{\Delta x}{2} [f(x_1) + 2f(x_2) + 2f(x_3) + 2f(x_4) + 2f(x_5) + 2f(x_6) + 2f(x_7) + f(x_8)]$$

$$= \dfrac{\frac{1}{7}}{2} \left[0 + 2\left(\dfrac{1}{8}\right)^2 + 2\left(\dfrac{2}{9}\right)^2 + 2\left(\dfrac{3}{10}\right)^2 + 2\left(\dfrac{4}{11}\right)^2 + 2\left(\dfrac{5}{12}\right)^2 + 2\left(\dfrac{6}{13}\right)^2 + \left(\dfrac{1}{2}\right)^2 \right]$$

$$\approx 0.114124$$

So, the volume is $\approx \pi(0.114124) \approx 0.3585$.

33. $FV = e^{rT} \displaystyle\int_0^T f(t) e^{-rt} dt$

$$= e^{0.06(10)} \int_0^{10} \sqrt{t}\, e^{-0.06t} dt$$

$$= e^{0.6} \int_0^{10} \sqrt{t}\, e^{-0.06t} dt$$

Using the trapezoidal rule with $n = 5$, $\Delta t = \dfrac{10-0}{5}$ and $t_1 = 0$, $t_2 = 2$, $t_3 = 4$, $t_4 = 6$, $t_5 = 8$, $t_6 = 10$

$$\int_0^{10} \sqrt{t}\, e^{-0.06t} dt \approx \dfrac{\Delta t}{2} [f(t_1) + 2f(t_2) + 2f(t_3) + 2f(t_4) + 2f(t_5) + f(t_6)]$$

$$= \dfrac{2}{2} \left[0 + 2\sqrt{2}e^{-0.12} + 2\sqrt{4}e^{-0.24} + 2\sqrt{6}e^{-0.36} + 2\sqrt{8}e^{-0.48} + \sqrt{10}e^{-0.6} \right]$$

$$\approx 14.308884$$

So, $FV \approx e^{0.6}(14.308884) \approx 26.07249$ or \$26,072.

35. $\displaystyle\int_0^{24} p(q)\, dq$

$$\approx \dfrac{\Delta q}{3} [f(0) + 4f(4) + 2f(8) + 4f(12) + 2f(16) + 4f(20) + f(24)]$$

$$= \dfrac{4}{3} \left[49.12 + 4(42.9) + 2(31.32) + 4(19.83) + 2(13.87) + 4(10.58) + 7.25 \right]$$

$$\approx 586.653 \text{ thousand, or } \$586,653$$

37. We need to approximate $FV = \int_a^b (\text{rate income enters})e^{r(b-t)}\,dt$.

Since the readings are every 2 months, $\Delta t = 2$, $r = \dfrac{0.04}{12}$, $a = 0$, and $b = 12$.

Future value
$$\approx \frac{2}{3}[(437)e^{(0.04/12)(12-0)} + 4(357)e^{(0.04/12)(12-2)} + 2(615)e^{(0.04/12)(12-4)} + 4(510)e^{(0.04/12)(12-6)}$$
$$+ 2(415)e^{(0.04/12)(12-8)} + 4(550)e^{(0.04/12)(12-10)} + (593)e^{(0.04/12)(12-12)}]$$
$$\approx \$5949.70$$

39. We need to approximate $PS = p_0 q_0 - \int_0^{q_0} S(q)\,dq$ using the trapezoidal rule. Since data was collected in increments of 1 thousand units, $\Delta q = 1$;

$$\int_0^7 S(q)\,dq \approx \frac{1}{2}[1.21 + 2(3.19) + 2(3.97) + 2(5.31) + 2(6.72) + 2(8.16) + 2(9.54) + 11.03]$$
$$= 43.01$$

So, $PS \approx (11.03)(7) - 43.01 = 34.2$ or $\$34,200$.

41. We need to approximate $\int_0^1 [x - L(x)]\,dx$ using the trapezoidal rule, with $\Delta x = 0.125$.

$$\approx \frac{0.125}{2}[0 + 2(0.125 - 0.0063) + 2(0.25 - 0.0631) + 2(0.375 - 0.1418) + 2(0.5 - 0.2305)$$
$$+ 2(0.625 - 0.3342) + 2(0.75 - 0.4713) + 2(0.875 - 0.6758) + (1 - 1)]$$
$$\approx 0.197125$$
$$GI = 2\int_0^1 [x - L(x)]\,dx \approx 2(0.197125) \approx 0.394$$

43.
$$P(T) = P_0 S(T) + \int_0^T RS(T - t)\,dt$$
$$= 3000e^{-0.01(8)} + \int_0^8 50\sqrt{t} \cdot e^{-0.01(8-t)}\,dt$$
$$= 3000e^{-0.08} + 50e^{-0.08}\int_0^8 \sqrt{t}\,e^{0.01t}\,dt$$
$$= 50e^{-0.08}\left[60 + \int_0^8 \sqrt{t}\,e^{0.01t}\,dt\right]$$

Using Simpson's rule with $n = 8$, $\Delta t = \dfrac{8 - 0}{8}$ and $t_1 = 0$, $t_2 = 1$, $t_3 = 2$, ..., $t_9 = 8$.

$$\int_0^8 \sqrt{t}\,e^{0.01t}\,dt$$
$$\approx \frac{\Delta t}{3}[f(t_1) + 4f(t_2) + 2f(t_3) + 4f(t_4) + 2f(t_5) + 4f(t_6) + 2f(t_7) + 4f(t_8) + f(t_9)]$$
$$= \frac{1}{3}\left[0 + 4e^{0.01} + 2\sqrt{2}e^{0.02} + 4\sqrt{3}e^{0.03} + 2\sqrt{4}e^{0.04} + 4\sqrt{5}e^{0.05} + 2\sqrt{6}e^{0.06} + 4\sqrt{7}e^{0.07} + \sqrt{8}e^{0.08}\right]$$
$$\approx 15.749112$$

So, the number of people with the flu is $\approx 50e^{-0.08}[60 + 15.749112] \approx 3,496$ people.

45. We need to approximate $\int_a^b f(x) - g(x)\,dx$ using the trapezoidal rule. Since readings are made every 5 feet, $\Delta t = 5$.

Area $\approx \dfrac{5}{2}[0 + 2 + 2(5) + 2(7) + 2(8) + 2(8) + 2(5) + 2(6) + 2(4) + 2(3) + 0]$

≈ 230 square feet

47. We need to approximate $2\pi \int_0^{10} r \cdot D(r)\,dr$ using the trapezoidal rule. Since measurements were made every 2 miles, $\Delta r = 2$;

$\int_0^{10} rD(r)\,dr = \dfrac{2}{2}[0 + 2(2)(2844) + 2(4)(2087) + 2(6)(1752) + 2(8)(1109) + (10)(879)]$

$= 75{,}630$

So, the total population is $\approx 2\pi(75{,}630) \approx 475{,}197$ people.

49. Using Simpson's rule with $\Delta x = 6$,

$A \approx \dfrac{6}{3}\Big[0 + 4(10) + 2(8) + 4(8) + 2(12) + 4(12) + 0\Big]$

$= 2(160) = 320$ sq ft

6.3 Improper Integrals

1. $\displaystyle\int_1^\infty \frac{1}{x^3}\,dx = \lim_{N\to\infty} \int_1^N x^{-3}\,dx$

$\displaystyle = \lim_{N\to\infty} -\frac{1}{2}\left(\frac{1}{x^2}\right)\Big|_1^N$

$\displaystyle = -\frac{1}{2}\lim_{N\to\infty}\left(\frac{1}{N^2} - \frac{1}{1}\right)$

$\displaystyle = -\frac{1}{2}(0 - 1)$

$= \dfrac{1}{2}$

3. $\displaystyle\int_1^\infty \frac{1}{\sqrt{x}}\,dx = \lim_{N\to\infty}\int_1^N x^{-1/2}\,dx$

$\displaystyle = \lim_{N\to\infty} 2(x^{1/2})\Big|_1^N$

$\displaystyle = 2\lim_{N\to\infty}(x^{1/2})\Big|_1^N$

$\displaystyle = \left(\sqrt{N} - 1\right)$

$= \infty$

So, the integral diverges.

5. $\displaystyle\int_3^\infty \frac{1}{2x-1}\,dx = \lim_{N\to\infty}\int_3^N \frac{1}{2x-1}\,dx$

$\displaystyle = \lim_{N\to\infty} \frac{1}{2}\ln|2x-1|\Big|_3^N$

$\displaystyle = \frac{1}{2}\lim_{N\to\infty}\ln|2x-1|\Big|_3^N$

$\displaystyle = \frac{1}{2}\lim_{N\to\infty}[\ln(2N-1) - \ln 7]$

$= \infty$

So, the integral diverges.

7. $\int_3^\infty \frac{1}{(2x-1)^2}\,dx = \lim_{N\to\infty}\int_3^N (2x-1)^{-2}\,dx$

$$= \lim_{N\to\infty}\frac{1}{2}\left(-\frac{1}{2x-1}\right)\Big|_3^N$$

$$= \frac{1}{2}\lim_{N\to\infty}\left(-\frac{1}{2N-1}+\frac{1}{5}\right)$$

$$= \frac{1}{2}\cdot\frac{1}{5}$$

$$= \frac{1}{10}$$

9. $\int_0^\infty 5e^{-2x}\,dx = \lim_{N\to\infty}5\int_0^N e^{-2x}\,dx$

$$= 5\lim_{N\to\infty}-\frac{1}{2}(e^{-2x})\Big|_0^N$$

$$= -\frac{5}{2}\lim_{N\to\infty}(e^{-2N}-e^0)$$

$$= -\frac{5}{2}\cdot -1$$

$$= \frac{5}{2}$$

11. $\int_1^\infty \frac{x^2}{(x^3+2)^2}\,dx = \lim_{N\to\infty}\int_1^N \frac{x^2}{(x^3+2)^2}\,dx$

Using substitution with $u = x^3 + 2$,

$$= \lim_{N\to\infty}\frac{1}{3}\int_3^{N^3+2}u^{-2}\,du$$

$$= \frac{1}{3}\lim_{N\to\infty}\left(-\frac{1}{u}\right)\Big|_3^{N^3+2}$$

$$= \frac{1}{3}\lim_{N\to\infty}\left(-\frac{1}{N^3+2}+\frac{1}{3}\right)$$

$$= \frac{1}{3}\cdot\frac{1}{3}$$

$$= \frac{1}{9}$$

13. $\int_1^\infty \frac{x^2}{\sqrt{x^3+2}}\,dx = \lim_{N\to\infty}\int_1^N \frac{x^2}{(x^3+2)^{1/2}}\,dx$

Using substitution with $u = x^3 + 2$,

$$= \lim_{N\to\infty}\frac{1}{3}\int_3^{N^3+2}u^{-1/2}\,du$$

$$= \frac{1}{3}\lim_{N\to\infty}2(u^{1/2})\Big|_3^{N^3+2}$$

$$= \frac{2}{3}\lim_{N\to\infty}\left(\sqrt{N^3+2}-\sqrt{3}\right)$$

$$= \infty$$

So, the integral diverges.

15. $\int_1^\infty \frac{e^{-\sqrt{x}}}{\sqrt{x}}\,dx = \lim_{N\to\infty}\int_1^N \frac{e^{-\sqrt{x}}}{\sqrt{x}}\,dx$

Using substitution with $u = -\sqrt{x}$,

$$= \lim_{N\to\infty}-2\int_{-1}^{-\sqrt{N}}e^u\,du$$

$$= 2\int_{-\sqrt{N}}^{-1}e^u\,du$$

$$= 2\lim_{N\to\infty}(e^u)\Big|_{-\sqrt{N}}^{-1}$$

$$= 2\lim_{N\to\infty}\left(\frac{1}{e}-\frac{1}{e^{\sqrt{N}}}\right)$$

$$= \frac{2}{e}$$

17. $\int_0^\infty 2xe^{-3x}\,dx = \lim_{N\to\infty}2\int_0^N xe^{-3x}\,dx$

Use integration by parts, with $u = x$ and $dV = e^{-3x}\,dx$,

$$= 2 \lim_{N \to \infty} \left[-\frac{x}{3} e^{-3x} \Big|_0^N - \int_0^N -\frac{1}{3} e^{-3x} dx \right]$$

$$= 2 \lim_{N \to \infty} \left[-\frac{x}{3} e^{-3x} \Big|_0^N + \frac{1}{3} \int_0^N e^{-3x} dx \right]$$

$$= 2 \lim_{N \to \infty} \left(-\frac{x}{3} e^{-3x} - \frac{1}{9} e^{-3x} \right) \Big|_0^N$$

$$= 2 \lim_{N \to \infty} \left[\left(-\frac{N}{3} e^{-3N} - \frac{1}{9} e^{-3N} \right) \right.$$
$$\left. - \left(0 - \frac{1}{9} e^0 \right) \right]$$

$$= 2 \lim_{N \to \infty} \left(-\frac{N}{3} e^{-3N} - \frac{1}{9} e^{-3N} + \frac{1}{9} \right)$$

$$= 2 \cdot \frac{1}{9}$$

$$= \frac{2}{9}$$

19. $\int_1^\infty \frac{\ln x}{x} dx = \lim_{N \to \infty} \int_1^N \frac{\ln x}{x} dx$

Using substitution with $u = \ln x$,

$$= \lim_{N \to \infty} \int_0^{\ln N} u \, du$$

$$= \lim_{N \to \infty} \left(\frac{u^2}{2} \right) \Big|_0^{\ln N}$$

$$= \frac{1}{2} \lim_{N \to \infty} (u^2) \Big|_0^{\ln N}$$

$$= \frac{1}{2} \lim_{N \to \infty} [(\ln N)^2 - 0]$$

$$= \infty$$

So, the integral diverges.

21. $\int_2^\infty \frac{1}{x \ln x} dx = \lim_{N \to \infty} \int_2^N \left(\frac{1}{\ln x} \right) \frac{1}{x} dx$

Using substitution with $u = \ln x$,

$$= \lim_{N \to \infty} \int_{\ln 2}^{\ln N} \frac{1}{u} du$$

$$= \lim_{N \to \infty} \left(\ln|u| \right) \Big|_{\ln 2}^{\ln N}$$

$$= \lim_{N \to \infty} [\ln(\ln N) - \ln(\ln 2)]$$

$$= \infty$$

So, the integral diverges.

23. $\int_0^\infty x^2 e^{-x} dx = \lim_{N \to \infty} \int_0^N x^2 e^{-x} dx$

Using integration by parts with $u = x^2$ and $dV = e^{-x} dx$,

$$= \lim_{N \to \infty} \left[-x^2 e^{-x} \Big|_0^N - \int_0^N -2x e^{-x} dx \right]$$

$$= \lim_{N \to \infty} \left[-x^2 e^{-x} \Big|_0^N + 2 \int_0^N x e^{-x} dx \right]$$

Using integration by parts with $u = x$ and $dV = e^{-x} dx$,

$$= \lim_{N \to \infty} \left[-x^2 e^{-x} \Big|_0^N \right.$$
$$\left. + 2 \left(-x e^{-x} \Big|_0^N - \int_0^N -e^{-x} dx \right) \right]$$

$$= \lim_{N \to \infty} \left[-x^2 e^{-x} \Big|_0^N \right.$$
$$\left. + 2 \left(-x e^{-x} \Big|_0^N + \int_0^N e^{-x} dx \right) \right]$$

$$= \lim_{N \to \infty} \left[-x^2 e^{-x} - 2x e^{-x} - 2e^{-x} \right] \Big|_0^N$$

$$= \lim_{N \to \infty} [(-N^2 e^{-N} - 2N e^{-N} - 2e^{-N})$$
$$- (0 - 0 - 2e^0)]$$

$$= 2$$

$$\int_{-\infty}^{0} 3e^{4x}\, dx = \lim_{N\to-\infty} \int_{N}^{0} 3e^{4x}\, dx$$

25.

$$= 3 \lim_{N\to-\infty} \int_{N}^{0} e^{4x}\, dx$$

$$= 3 \lim_{N\to-\infty} \left(\frac{1}{4} e^{4x} \Big|_{N}^{0} \right)$$

$$= \frac{3}{4} \lim_{N\to-\infty} \left(e^{0} - e^{4N} \right)$$

$$= \frac{3}{4}(1-0) = \frac{3}{4}$$

27. $\displaystyle \int_{-\infty}^{-1} \frac{1}{x^2}\, dx = \lim_{N\to-\infty} \int_{N}^{-1} x^{-2}\, dx$

$$= \lim_{N\to-\infty} \left(-\frac{1}{x} \Big|_{N}^{-1} \right)$$

$$= \lim_{N\to-\infty} \left(1 + \frac{1}{N} \right) = 1 + 0 = 1$$

29. $\displaystyle \int_{-\infty}^{\infty} xe^{-x}\, dx$

First, find the indefinite integral $\int xe^{-x}\, dx$ using integration by parts.

$$u = x \qquad \text{and} \qquad dV = e^{-x} dx$$
$$du = dx \qquad\qquad\qquad V = -e^{-x}$$

$$\int xe^{-x}\, dx = -xe^{-x} - \int -e^{-x}\, dx$$

$$= -xe^{-x} - e^{-x} + C$$

Letting

$$\int_{-\infty}^{\infty} xe^{-x}\, dx = \int_{C}^{\infty} xe^{-x}\, dx + \int_{-\infty}^{C} xe^{-x}\, dx$$

choose $C = 0$.

$$\int_{0}^{\infty} xe^{-x}\, dx = \lim_{N\to\infty} \int_{0}^{N} xe^{-x}\, dx$$

$$= \lim_{N\to\infty} \left[\left(-xe^{-x} - e^{-x} \right) \Big|_{0}^{N} \right]$$

$$= \lim_{N\to\infty} \left[\left(-Ne^{-N} - e^{-N} \right) \right.$$
$$\left. -\left(0 - e^{0} \right) \right]$$

$$= 0 + 1 = 1$$

Similarly,

$$\int_{-\infty}^{0} xe^{-x}\, dx$$

$$= \lim_{N\to-\infty} \int_{N}^{0} xe^{-x}\, dx$$

$$= \lim_{N\to-\infty} \left(-xe^{-x} - e^{-x} \Big|_{N}^{0} \right)$$

$$= \lim_{N\to-\infty} \left[\left(0 - e^{0} \right) - \left(-Ne^{-N} - e^{-N} \right) \right]$$

$$= \lim_{N\to-\infty} \left(-1 + Ne^{-N} - e^{-N} \right)$$

Since $\displaystyle \lim_{N\to-\infty} e^{-N}$ does not exist, the integral diverges.

31. $PV = \int_0^\infty 2,400e^{-0.04t}\,dt$

$\qquad = \lim_{N \to \infty} 2,400 \int_0^N e^{-0.04t}\,dt$

$\qquad = 2,400 \lim_{N \to \infty} (-25e^{-0.04t})\Big|_0^N$

$\qquad = -60,000 \lim_{N \to \infty} (e^{-0.04t})\Big|_0^N$

$\qquad = -60,000 \lim_{N \to \infty} (e^{-0.04N} - e^0)$

$\qquad = -60,000 \cdot -1$

$\qquad = \$60,000$

33. $PV = \int_0^\infty (100,000 + 900t)e^{-0.05t}\,dt$

$\qquad = \lim_{N \to \infty} \int_0^N (100,000 + 900t)e^{-0.05t}\,dt$

Using integration by parts with $u = 100,000 + 900t$ and $dV = e^{-0.05t}\,dt$.

$\qquad = \lim_{N \to \infty} \left[-20(100,000 + 900t)e^{-0.05t}\Big|_0^N - \int_0^N -18,000e^{-0.05t}\,dt \right]$

$\qquad = \lim_{N \to \infty} \left[-20(100,000 + 900t)e^{-0.05t}\Big|_0^N + 18,000 \int_0^N e^{-0.05t}\,dt \right]$

$\qquad = -20 \lim_{N \to \infty} [(100,000 + 900t)e^{-0.05t} + 18,000e^{-0.05t}]\Big|_0^N$

$\qquad = -20 \lim_{N \to \infty} [(100,000 + 900N)e^{-0.05N} + 18,000e^{-0.05N}) - (100,000e^0 + 18,000e^0)]$

$\qquad = -20 \lim_{N \to \infty} [118,000e^{-0.05N} + 900Ne^{-0.05N} - 118,000]$

$\qquad = -20(-118,000)$

$\qquad = \$2,360,000$

35. (a) The capitalized cost of the first machine is

$\qquad M_1 = 10,000 + \int_0^\infty 1,000(1 + 0.06t)e^{-0.09t}\,dt$

$\qquad\qquad = 10,000 + 1,000 \lim_{N \to \infty} \int_0^N (1 + 0.06t)e^{-0.09t}\,dt$

Using integration by parts,

$u = 1 + 0.06t$	$dV = e^{-0.09t}\,dt$
$du = 0.06\,dt$	$V = -\dfrac{1}{0.09}e^{-0.09t}$

$$\int_0^N (1+0.06t)e^{-0.09t}\, dt$$

$$= -\frac{100}{9}(1+0.06t)e^{-0.09t}\Big|_0^N - \int_0^N -\frac{0.06}{0.09}e^{-0.09t}\, dt$$

$$= -\frac{100}{9}(1+0.06t)e^{-0.09t}\Big|_0^N + \frac{2}{3}\int_0^N e^{-0.09t}\, dt$$

$$= -\frac{100}{9}(1+0.06t)e^{-0.09t}\Big|_0^N + \frac{2}{3}\left(-\frac{1}{0.09}e^{-0.09t}\Big|_0^N\right)$$

$$= -\frac{100}{9}(1+0.06t)e^{-0.09t} - \frac{200}{27}e^{-0.09t}\Big|_0^N$$

$$= \left[-\frac{100}{9}(1+0.06t) - \frac{200}{27}\right]e^{-0.09t}\Big|_0^N$$

$$= \left(-\frac{500}{27} - \frac{2}{3}t\right)e^{-0.09t}\Big|_0^N$$

$$= \left(-\frac{500}{27} - \frac{2}{3}N\right)e^{-0.09N} - \left(-\frac{500}{27} - 0\right)e^0$$

$$= \left(-\frac{500}{27} - \frac{2}{3}N\right)e^{-0.09N} + \frac{500}{27}$$

So,

$$M_1 = 10{,}000 + 1{,}000 \lim_{N\to\infty}\left(\left[-\frac{500}{27} - \frac{2}{3}N\right]e^{-0.09N} + \frac{500}{27}\right)$$

$$= \$10{,}000 + 1{,}000\left(\frac{500}{27}\right) \approx \$28{,}518.50$$

The capitalized cost of the second machine is

$$M_2 = 8,000 + \int_0^\infty 1,100e^{-0.09t}\,dt$$

$$= 8,000 + 1,100 \lim_{N\to\infty} \int_0^N e^{-0.09t}\,dt$$

$$= 8,000 + 1,100 \lim_{N\to\infty}\left(-\frac{1}{0.09}e^{-0.09t}\Big|_0^N\right)$$

$$= 8,000 - \frac{110,000}{9}\lim_{N\to\infty}\left(e^{-0.09t}\Big|_0^N\right)$$

$$= 8,000 - \frac{110,000}{9}\lim_{N\to\infty}\left(e^{-0.09N}-e^0\right)$$

$$= 8,000 + \frac{110,000}{9} \approx \$20,222.22$$

So, the second machine should be purchased.

(b) Writing exercise; answers will vary.

37. $$PV = \int_0^\infty f(t)e^{-rt}\,dt$$

Here, $f(t) = Q$, which is constant. So,

$$PV = Q\lim_{N\to\infty}\int_0^N e^{-rt}\,dt$$

$$= Q\lim_{N\to\infty}\left(\frac{-1}{r}e^{-rt}\Big|_0^N\right)$$

$$= -\frac{Q}{r}\lim_{N\to\infty}\left(e^{-rt}\Big|_0^N\right)$$

$$= -\frac{Q}{r}\lim_{N\to\infty}\left(e^{-rN}-e^0\right)$$

$$= -\frac{Q}{r}(0-1) = \frac{Q}{r}$$

39. Number of patients

$$= \lim_{N\to\infty}\int_0^N 10e^{-(N-t)/20}\,dt$$

$$= \lim_{N\to\infty} 10e^{-N/20}\int_0^N e^{t/20}\,dt$$

$$= 10\lim_{N\to\infty} e^{-N/20}(20e^{t/20})\Big|_0^N$$

$$= 200\lim_{N\to\infty} e^{-N/20}(e^{N/20}-e^0)$$

$$= 200\lim_{N\to\infty}(e^0 - e^{-N/20})$$

$$= 200\cdot 1$$

$$= 200 \text{ patients}$$

41. Amount of drug

$$= \lim_{N \to \infty} \int_0^N 5e^{-(N-t)/10} dt$$

$$= \lim_{N \to \infty} 5e^{-N/10} \int_0^N e^{t/10} dt$$

$$= 5 \lim_{N \to \infty} e^{-N/10} (10e^{t/10}) \Big|_0^N$$

$$= 50 \lim_{N \to \infty} e^{-N/10} (e^{N/10} - e^0)$$

$$= 50 \lim_{N \to \infty} (e^0 - e^{-N/10})$$

$$= 50 \cdot 1$$

$$= 50 \text{ units}$$

43. $\int_0^t C(e^{-ax} - e^{-bx}) dx$

Since all people are eventually infected,

$$\int_0^\infty C(e^{-ax} - e^{-bx}) dx = 1$$

Now,

$$\int_0^\infty C(e^{-ax} - e^{-bx}) dx$$

$$= C \lim_{N \to \infty} \int_0^N \left(e^{-ax} - e^{-bx} \right) dx$$

$$= C \lim_{N \to \infty} \left[\left(-\frac{1}{a} e^{-ax} + \frac{1}{b} e^{-bx} \right) \Big|_0^N \right]$$

$$= C \lim_{N \to \infty} \left[-\frac{1}{a} e^{-aN} + \frac{1}{b} e^{-bN} \right.$$

$$\left. - \left(-\frac{1}{a} e^0 + \frac{1}{b} e^0 \right) \right]$$

$$= C \lim_{N \to \infty} \left[-\frac{1}{a} e^{-aN} + \frac{1}{b} e^{-bN} + \frac{1}{a} - \frac{1}{b} \right]$$

$$= C \left(0 + 0 + \frac{1}{a} - \frac{1}{b} \right)$$

Since the integral equals one,

$$1 = C \left(\frac{1}{a} - \frac{1}{b} \right)$$

$$1 = C \left(\frac{b-a}{ab} \right), \text{ or } C = \frac{ab}{b-a}$$

45. $p(r) = 100e^{-0.02r} + 2,000e^{-0.001r^2}$

The population within an R mile radius is

$$P = \int_0^R 2\pi r p(r) dr$$

Letting $R \to \infty$,

$$P = \lim_{R \to \infty} \int_0^R 2\pi r \left(100e^{-0.02r} + 2,000e^{-0.001r^2} \right) dr$$

$$= 2\pi \lim_{R \to \infty} \int_0^R \left(100re^{-0.02r} + 2,000re^{-0.001r^2} \right) dr$$

Separating the integral into two,

$$= 2\pi \lim_{R \to \infty} \left[\int_0^R 100re^{-0.02r} dr + \int_0^R 2,000re^{-0.001r^2} dr \right]$$

Now,

$$\int_0^R 100re^{-0.02r} dr = 100 \int_0^R re^{-0.02r} dr$$

Using integration by parts,

$$dV = e^{-0.02r} dr$$

$$u = r \quad \text{and} \quad V = -\frac{1}{0.02} e^{-0.02r}$$

$$du = dr \qquad = -50e^{-0.02r}$$

$$= 100 \left[-50re^{-0.02r} \Big|_0^R - \int_0^R -50e^{-0.02r} dr \right]$$

$$= 100 \left[-50re^{-0.02r} \Big|_0^R + 50 \left(-50e^{-0.02r} \Big|_0^R \right) \right]$$

$$= -5,000 \left[\left(re^{-0.02r} + 50e^{-0.02R} \right) \Big|_0^R \right]$$

$$= -5,000 \left[\left(Re^{-0.02R} + 50e^{-0.02R} \right) - \left(0 + 50e^0 \right) \right]$$

$$= -5,000 \left(Re^{-0.02R} + 50e^{-0.02R} - 50 \right)$$

Also,

$$\int_0^R 2,000re^{-0.001r^2} dr = 2,000 \int_0^R re^{-0.001r^2} dr$$

can be solved using the substitution
$u = -0.001r^2$. Then, $du = -0.002r\, dr$,

$-\dfrac{1}{0.002}\,du = r\,dr$, or $-500\,du = r\,dr$.

The limits of integration become

$-0.001(0)^2 = 0$ and $-0.001R^2$.

$= 2{,}000\left(-500\displaystyle\int_0^{-0.001R^2} e^u\,du\right)$

$= -1{,}000{,}000\left(e^u\Big|_0^{-0.001R^2}\right)$

$= -1{,}000{,}000\left(e^{-0.001R^2} - e^0\right)$

$= -1{,}000{,}000\left(e^{-0.001R^2} - 1\right)$

Putting the pieces together,

$\begin{aligned}
P &= 2\pi\lim_{R\to\infty}\Big[-5{,}000\left(Re^{-0.02R} + 50e^{-0.02R} - 50\right)\\
&\qquad -1{,}000{,}000\left(e^{-0.001R^2} - 1\right)\Big]\\
&= 2\pi\Big[-5{,}000(0 + 0 - 50) - 1{,}000{,}000(0 - 1)\Big]\\
&= 2\pi(1{,}250{,}000)\\
&\approx 7{,}853{,}982 \text{ people}
\end{aligned}$

6.4 Introduction to Continuous Probability

1. Since $f(x) \ge 0$ for all x, the first condition is met.

$\begin{aligned}
\int_{-\infty}^{\infty} f(x)\,dx &= \int_{30}^{40}\frac{1}{10}\,dx\\
&= \frac{1}{10}x\Big|_{30}^{40}\\
&= \frac{1}{10}(40 - 30)\\
&= \frac{1}{10}(10)\\
&= 1
\end{aligned}$

So, the second condition is also met and this is a probability density function.

3. The first condition, $f(x) \ge 0$ for all x, is not met since $f(4) = \dfrac{1}{4}[7 - 2(4)] = -\dfrac{1}{4}$.

So, this is not a probability function.

5. Since $f(x) \ge 0$ for all x, the first condition is met. Checking the second condition,

$\begin{aligned}
\int_{-\infty}^{\infty} f(x)\,dx &= \int_0^{\infty}\frac{10}{(x+10)^2}\,dx\\
&= \lim_{N\to\infty}\int_0^N \frac{10}{(x+10)^2}\,dx\\
&= 10\lim_{N\to\infty}\int_0^N \frac{1}{(x+10)^2}\,dx
\end{aligned}$

Using substitution with $u = x + 10$ and $du = dx$,

$\begin{aligned}
&= 10\lim_{N\to\infty}\int_{10}^{N+10}\frac{1}{u^2}\,du\\
&= 10\lim_{N\to\infty}\left(-\frac{1}{u}\Big|_{10}^{N+10}\right)\\
&= 10\lim_{N\to\infty}\left(-\frac{1}{N+10} + \frac{1}{10}\right)\\
&= 10\left(0 + \frac{1}{10}\right)\\
&= 1
\end{aligned}$

The third condition is also met, so f is a probability density function.

7. Since $f(x) \ge 0$ for all x, the first condition is met. Checking the second condition,

$\begin{aligned}
\int_{-\infty}^{\infty} f(x)\,dx &= \int_0^{\infty} xe^{-x}\,dx\\
&= \lim_{N\to\infty}\int_0^N xe^{-x}\,dx
\end{aligned}$

Using integration by parts with $u = x$ and $dV = e^{-x}\,dx$,

$$= \lim_{N \to \infty} \left[-xe^{-x} \Big|_0^N - \int_0^N 1 \cdot -e^{-x} dx \right]$$

$$= \lim_{N \to \infty} \left[-xe^{-x} \Big|_0^N + \int_0^N e^{-x} dx \right]$$

$$= \lim_{N \to \infty} \left[(-xe^{-x} - e^{-x}) \Big|_0^N \right]$$

$$= \lim_{N \to \infty} \left[(-Ne^{-N} - e^{-N}) - (0 - 1) \right]$$

$$= (0 - 0) - (0 - 1)$$

$$= 1$$

The third condition is also met, so f is a probability density function.

9. The first condition is not met. For example

$f(-1) = \dfrac{3}{2}(-1)^2 + 2(-1) = -\dfrac{1}{2}$. Since it is

not the case that $f(x) \geq 0$ for all x, f is *not* a probability density function.

11. For $f(x) \geq 0$ for all x, it must be the case that $k - 3(1) \geq 0$, or $k \geq 3$.

$$\int_{-\infty}^{\infty} f(x) dx = \int_0^1 (k - 3x) dx$$

$$= \left(kx - \frac{3x^2}{2} \right) \Big|_0^1$$

$$= k - \frac{3}{2}$$

For this integral to equal 1, $k - \dfrac{3}{2} = 1$, or

$k = \dfrac{5}{2}$

Since it cannot be the case that $k \geq 3$ and

$k = \dfrac{5}{2}$, no such number exists.

13. Need $\displaystyle\int_{-\infty}^{\infty} f(x) dx = 1$, or

$$\int_0^2 (kx - 1)(x - 2) dx = 1$$

$$\int_0^2 (kx^2 - 2kx - x + 2) dx = 1$$

$$\left(\frac{kx^3}{3} - kx^2 - \frac{x^2}{2} + 2x \right) \Big|_0^2 = 1$$

$$\left(\frac{8}{3}k - 4k - 2 + 4 \right) - 0 = 1$$

$$-\frac{4}{3}k + 2 = 1, \text{ or } k = \frac{3}{4}$$

But, using this k, can find a value for x

such that $f(x) < 0$ since when $x = \dfrac{3}{2}$,

$$f\left(\frac{3}{2} \right) = \left[\frac{3}{4}\left(\frac{3}{2} \right) - 1 \right]\left[\frac{3}{2} - 2 \right] = -\frac{1}{16}$$

So, no such number exists.

15. Need $\displaystyle\int_{-\infty}^{\infty} f(x) dx = 1$, or

$$\int_0^1 (x - kx^2) dx = 1$$

$$\left(\frac{x^2}{2} - \frac{kx^3}{3} \right) \Big|_0^1 = 1$$

$$\left(\frac{1}{2} - \frac{k}{3} \right) - 0 = 1, \text{ or } k = -\frac{3}{2}$$

Then, for $0 \leq x \leq 1$, $f(x) = x + \dfrac{3}{2}x^2$ which

is always greater than or equal to zero.

Since both conditions are met, $k = -\dfrac{3}{2}$.

17. Need $\displaystyle\int_0^{\infty} xe^{-kx} dx = 1$, or

$$\lim_{n \to \infty} \int_0^n xe^{-kx} dx = 1$$

Using integration by parts with

$$u = x \qquad dV = e^{-kx} dx$$

$$du = dx \qquad V = -\frac{1}{k}e^{-kx}$$

$$= \lim_{n \to \infty} \left(-\frac{x}{k} e^{-kx} \Big|_0^n - \int_0^n -\frac{1}{k} e^{-kx} \, dx \right)$$

$$= \lim_{n \to \infty} \left(-\frac{x}{k} e^{-kx} \Big|_0^n - \frac{1}{k^2} e^{-kx} \Big|_0^n \right)$$

$$= \lim_{n \to \infty} \left[\left(-\frac{n}{k} e^{-kn} - 0 \right) \right.$$
$$\left. - \left(\frac{1}{k^2} e^{-kn} - \frac{1}{k^2} e^0 \right) \right]$$

$$= \lim_{n \to \infty} \left(-\frac{n}{k} e^{-kn} - \frac{1}{k^2} e^{-kn} + \frac{1}{k^2} \right)$$

$$= \frac{1}{k^2}$$

Setting this equal to one, $k = 1$. Then, for $x \geq 0$, $f(x) = xe^{-x}$ which is always greater than or equal to zero. Since both conditions are met, $k = 1$.

19. (a) $P(2 \leq x \leq 5) = \int_2^5 \frac{1}{3} \, dx = \frac{x}{3} \Big|_2^5 = 1$

Note: $\int_{-\infty}^{\infty} f(x) \, dx = \int_2^5 f(x) \, dx$ in this problem, so needn't even integrate to conclude that the probability is 1.

(b) $P(3 \leq x \leq 4) = \int_3^4 \frac{1}{3} \, dx = \frac{x}{3} \Big|_3^4 = \frac{1}{3}$

(c) $P(X \geq 4) = \int_4^5 \frac{1}{3} \, dx = \frac{x}{3} \Big|_4^5 = \frac{1}{3}$

Note: can also calculate as $1 - \int_0^4 \frac{1}{3} \, dx$.

21. (a) $P(0 \leq x \leq 4) = \int_0^4 \frac{1}{8} (4 - x) \, dx$

$$= \int_0^4 \frac{1}{2} - \frac{1}{8} x \, dx$$

$$= \left(\frac{x}{2} - \frac{x^2}{16} \right) \Big|_0^4$$

$$= (2 - 1) - 0$$

$$= 1$$

Note: $\int_{-\infty}^{\infty} f(x) \, dx = \int_0^4 f(x) \, dx$ in this problem, so needn't even integrate to conclude that the probability is 1.

(b) $P(2 \leq x \leq 3) = \int_2^3 \frac{1}{8} (4 - x) \, dx$

$$= \left(\frac{x}{2} - \frac{x^2}{16} \right) \Big|_2^3$$

$$= \left(\frac{3}{2} - \frac{9}{16} \right) - \left(1 - \frac{1}{4} \right)$$

$$= \frac{3}{16}$$

(c) $P(X \geq 1) = \int_1^4 \frac{1}{8} (4 - x) \, dx$

$$= \left(\frac{x}{2} - \frac{x^2}{16} \right) \Big|_1^4$$

$$= (2 - 1) - \left(\frac{1}{2} - \frac{1}{16} \right)$$

$$= \frac{9}{16}$$

Note: can also calculate as
$$1 - \int_0^1 \frac{1}{8} (4 - x) \, dx.$$

23. (a) $P(1 \le x < \infty) = \int_1^\infty \frac{3}{x^4} dx$

$$= 3 \lim_{N \to \infty} \int_1^N \frac{1}{x^4} dx$$

$$= 3 \lim_{N \to \infty} \left(-\frac{1}{3x^3} \Big|_1^N \right)$$

$$= 3 \lim_{N \to \infty} \left(-\frac{1}{3N^3} + \frac{1}{3} \right)$$

$$= 3 \left(0 + \frac{1}{3} \right)$$

$$= 1$$

Note: $\int_{-\infty}^\infty f(x) dx = \int_1^\infty \frac{3}{x^4} dx$ in this problem, so needn't even integrate to conclude that the probability is 1.

(b) $P(1 \le x \le 2) = \int_1^2 \frac{3}{x^4} dx$

$$= 3 \left(-\frac{1}{3x^3} \Big|_1^2 \right)$$

$$= 3 \left(-\frac{1}{24} + \frac{1}{3} \right)$$

$$= \frac{7}{8}$$

(c) $P(X \ge 2) = 1 - \int_1^2 \frac{3}{x^4} dx = 1 - \frac{7}{8} = \frac{1}{8}$

25. (a) $P(X \ge 0) = \int_0^\infty 2xe^{-x^2} dx$

$$= \lim_{N \to \infty} \int_0^N 2xe^{-x^2} dx$$

Using substitution with $u = -x^2$ and $-du = 2x\, dx$,

$$= \lim_{N \to \infty} -\int_0^{-N^2} e^u du$$

$$= \lim_{N \to \infty} \left(-e^u \Big|_0^{-N^2} \right)$$

$$= \lim_{N \to \infty} (-e^{-N^2} + e^0)$$

$$= 0 + 1$$

$$= 1$$

Note: $\int_{-\infty}^\infty f(x) dx = \int_0^\infty f(x) dx$ in this problem, so needn't even integrate to conclude that the probability is 1.

(b) $P(1 \le x \le 2) = \int_1^2 2xe^{-x^2} dx$

$$= -e^u \Big|_{-1}^{-4}$$

$$= -e^{-4} + e^{-1}$$

$$= \frac{1}{e} - \frac{1}{e^4}$$

(c) $P(X \le 2) = \int_0^2 2xe^{-x^2} dx$

$$= -e^u \Big|_0^{-4}$$

$$= -e^{-4} + 1$$

$$= 1 - \frac{1}{e^4}$$

27. $E(X) = \int_{-\infty}^\infty x f(x) dx$

$$= \int_2^5 x \cdot \frac{1}{3} dx$$

$$= \frac{1}{3} \int_2^5 x\, dx$$

$$= \frac{1}{3} \left(\frac{x^2}{2} \right) \Big|_2^5$$

$$= \frac{1}{6} (25 - 4)$$

$$= \frac{21}{6}$$

$$= \frac{7}{2}$$

29. $E(X) = \int_{-\infty}^{\infty} x f(x) dx$

$= \int_0^4 x \cdot \frac{1}{8}(4-x) dx$

$= \frac{1}{8} \int_0^4 (4x - x^2) dx$

$= \frac{1}{8} \left(2x^2 - \frac{x^3}{3} \right) \Big|_0^4$

$= \frac{1}{8} \left[\left(2(4)^2 - \frac{(4)^3}{3} \right) - 0 \right]$

$= \frac{1}{8} \cdot \frac{32}{3}$

$= \frac{4}{3}$

31. $E(X) = \int_{-\infty}^{\infty} x f(x) dx$

$= \int_1^{\infty} x \cdot \frac{3}{x^4} dx$

$= \lim_{N \to \infty} 3 \int_1^N x^{-3} dx$

$= 3 \lim_{N \to \infty} \left(-\frac{1}{2x^2} \right) \Big|_1^N$

$= -\frac{3}{2} \lim_{N \to \infty} \left(\frac{1}{x^2} \right) \Big|_1^N$

$= -\frac{3}{2} \lim_{N \to \infty} \left(\frac{1}{N^2} - \frac{1}{1} \right)$

$= -\frac{3}{2} \cdot -1$

$= \frac{3}{2}$

33. (a) $P(10 \le X \le 15)$

$= \int_{10}^{15} 0.02 e^{-0.02x} dx$

$= 0.02 \left(\frac{1}{-0.02} e^{-0.02x} \right) \Big|_{10}^{15}$

$= (-e^{-0.02x}) \Big|_{10}^{15}$

$= -e^{-0.3} + e^{-0.2}$

$= \frac{1}{e^{0.2}} - \frac{1}{e^{0.3}} \approx 0.078$

(b) $P(X < 8) = \int_0^8 0.02 e^{-0.02x} dx$

$= (-e^{-0.02x}) \Big|_0^8$

$= -e^{-0.16} + e^0$

$= 1 - \frac{1}{e^{0.16}} \approx 0.148$

(c) $P(X) > 12 = \int_{12}^{\infty} 0.02 e^{-0.02x} dx$

$= \lim_{N \to \infty} 0.02 \int_{12}^N e^{-0.02x} dx$

$= \lim_{N \to \infty} (-e^{-0.02x}) \Big|_{12}^N$

$= \lim_{N \to \infty} (-e^{-0.02N} + e^{-0.24})$

$= \frac{1}{e^{0.24}} \approx 0.787$

(d) $E(X) = \int_{-\infty}^{\infty} x f(x) dx$

$= \lim_{N \to \infty} \int_0^N x \cdot 0.02 e^{-0.02x} dx$

$= \lim_{N \to \infty} (0.02) \int_0^N x e^{-0.02x} dx$

Using integration by parts with $u = x$ and $dV = e^{-0.02x} dx$,

$$= 0.02 \lim_{N \to \infty} \left[-50xe^{-0.02x} \Big|_0^N \right.$$

$$\left. - \int_0^N -50e^{-0.02x}\,dx \right]$$

$$= \lim_{N \to \infty} \left[-xe^{-0.02x} \Big|_0^N - \int_0^N -e^{-0.02x}\,dx \right]$$

$$= \lim_{N \to \infty} \left[-xe^{-0.02x} - 50e^{-0.02x} \right]_0^N$$

$$= \lim_{N \to \infty} \left[(-Ne^{-0.02N} - 50e^{-0.02N}) \right.$$

$$\left. - (0 - 50e^0) \right]$$

$$= 50 \text{ months.}$$

35. (a) $P(x > 4) = \int_4^\infty f(x)\,dx$

Since, $f(x) = 0$ for $x > 7$,

$$= \int_4^7 \left(\frac{3}{28} + \frac{3}{x^2} \right) dx$$

$$= \left(\frac{3}{28}x - \frac{3}{x} \right) \Big|_4^7$$

$$= \left[\frac{3}{28}(7) - \frac{3}{7} \right] - \left[\frac{3}{28}(4) - \frac{3}{4} \right]$$

$$= \frac{9}{14} \approx 0.643$$

(b) $P(x < 5) = \int_0^5 f(x)\,dx$

Since $f(x) = 0$ for $x < 3$,

$$= \int_3^5 \left(\frac{3}{28} + \frac{3}{x^2} \right) dx$$

$$= \left(\frac{3}{28}x - \frac{3}{x} \right) \Big|_3^5$$

$$= \left[\frac{3}{28}(5) - \frac{3}{5} \right] - \left[\frac{3}{28}(3) - \frac{3}{3} \right]$$

$$= \frac{43}{70} \approx 0.614$$

(c) $P(4 \le X \le 6) = \int_4^6 f(x)\,dx$

$$= \int_4^6 \left(\frac{3}{28} + \frac{3}{x^2} \right) dx$$

$$= \left(\frac{3}{28}x - \frac{3}{x} \right) \Big|_4^6$$

$$= \left[\frac{3}{28}(6) - \frac{3}{6} \right] - \left[\frac{3}{28}(4) - \frac{3}{4} \right]$$

$$= \frac{13}{28} \approx 0.464$$

(d) $E(X) = \int_{-\infty}^\infty x f(x)\,dx$

Here,

$$= \int_3^7 x \left(\frac{3}{28} + \frac{3}{x^2} \right) dx$$

$$= \int_3^7 \left(\frac{3}{28}x + \frac{3}{x} \right) dx$$

$$= \left(\frac{3}{56}x^2 + 3\ln x \right) \Big|_3^7$$

$$= \left[\frac{3}{56}(7)^2 + 3\ln 7 \right]$$

$$\quad - \left[\frac{3}{56}(3)^2 + 3\ln 3 \right]$$

$$\approx 8.4627 - 3.7780$$

$$\approx 4.685 \text{ years}$$

37. (a) $P(X > x) = \int_x^\infty f(x)\,dx$

$$= 1 - \int_0^x f(x)\,dx$$

(b) Since $f(x)$ is an exponential function

with parameter $\lambda = \dfrac{1}{20}$,

$$P(X > 10) = 1 - \int_0^{10} \frac{1}{20} e^{-1/20x}\, dx$$

$$= 1 - \frac{1}{20} \int_0^{10} e^{-1/20x}\, dx$$

$$= 1 - \frac{1}{20} \left(-20 e^{-1/20x} \Big|_0^{10} \right)$$

$$= 1 + \left(e^{-1/20x} \Big|_0^{10} \right)$$

$$= 1 + (e^{-1/2} - e^0) = 1 + e^{-1/2} - 1$$

$$= e^{-1/2} \approx 0.61$$

39. The uniform density function for x, in

seconds is $f(x) = \begin{cases} \dfrac{1}{45} & \text{if } 0 \le x \le 45 \\ 0 & \text{otherwise} \end{cases}$

(a) $P(0 \le X \le 15) = \int_0^{15} \frac{1}{45}\, dx$

$$= \frac{1}{45}(x) \Big|_0^{15}$$

$$= \frac{1}{45}(15 - 0)$$

$$= \frac{1}{3}$$

(b) $P(5 \le X \le 10) = \int_5^{10} \frac{1}{45}\, dx$

$$= \frac{1}{45}(x) \Big|_5^{10}$$

$$= \frac{1}{45}(10 - 5)$$

$$= \frac{1}{9}$$

(c) $E(X) = \int_{-\infty}^{\infty} x f(x)\, dx$

$$= \int_0^{45} x \cdot \frac{1}{45}\, dx$$

$$= \frac{1}{45} \left(\frac{x^2}{2} \right) \Big|_0^{45}$$

$$= \frac{1}{90} (x^2) \Big|_0^{45}$$

$$= \frac{1}{90} [(45)^2 - 0]$$

$$= \frac{45}{2} \text{ seconds}$$

41. (a) $E(X) = \int_{-\infty}^{\infty} x f(x)\, dx$

$$= \int_0^{\infty} x \cdot k e^{-kx}\, dx$$

$$= \lim_{N \to \infty} k \int_0^N x e^{-kx}\, dx$$

Using integration by parts with $u = x$
and $dV = e^{-kx} dx$.

$$= k \lim_{N \to \infty} \left[\frac{-x}{k} e^{-kx} \Big|_0^N - \int_0^N -\frac{1}{k} e^{-kx}\, dx \right]$$

$$= \lim_{N \to \infty} \left[-x e^{-kx} \Big|_0^N + \int_0^N e^{-kx}\, dx \right]$$

$$= \lim_{N \to \infty} \left[-x e^{-kx} - \frac{1}{k} e^{-kx} \right] \Big|_0^N$$

$$= \lim_{N \to \infty} \left[\left(-N e^{-kN} - \frac{1}{k} e^{-kN} \right) - \left(0 - \frac{1}{k} e^0 \right) \right]$$

$$= \frac{1}{k}$$

So, $\frac{1}{k} = 5$, or $k = \frac{1}{5}$.

(b) $P(X < 2) = \int_0^2 \frac{1}{5} e^{-x/5} dx$

$$= (-e^{-x/5}) \Big|_0^2$$

$$= -e^{-2/5} + e^0$$

$$= 1 - \frac{1}{e^{2/5}} \approx 0.330$$

(c) $P(X > 7) = \lim_{N \to \infty} \int_7^N \frac{1}{5} e^{-x/5} dx$

$$= \lim_{N \to \infty} (-e^{-x/5}) \Big|_7^N$$

$$= \lim_{N \to \infty} (-e^{-N/5} + e^{-7/5})$$

$$= \frac{1}{e^{7/5}} \approx 0.247$$

43. (a) $P(0 \le X \le 5) = \int_0^5 0.2 e^{-0.2x} dx$

$$= 0.2 \left(\frac{1}{-0.2} e^{-0.2x} \right) \Big|_0^5$$

$$= (-e^{-0.2x}) \Big|_0^5$$

$$= -e^{-1} + e^0$$

$$= 1 - \frac{1}{e} \approx 0.632$$

(b) $P(X > 6) = \lim_{N \to \infty} \int_6^N 0.2 e^{-0.2x} dx$

$$= \lim_{N \to \infty} (-e^{-0.2x}) \Big|_6^N$$

$$= \lim_{N \to \infty} (-e^{-0.2N} + e^{-1.2})$$

$$= e^{-1.2} \approx 0.301$$

(c) $E(X) = \int_{-\infty}^{\infty} x f(x) dx$

$$= \lim_{N \to \infty} \int_0^N x(0.2 e^{-0.2x}) dx$$

$$= \lim_{N \to \infty} 0.2 \int_0^N x e^{-0.2x} dx$$

Using integration by parts with $u = x$ and $dV = e^{-0.2x} dx$,

$$= \lim_{N \to \infty} \frac{1}{5} \left[-5x e^{-0.2x} \Big|_0^N - \int_0^N -5 e^{-0.2x} dx \right]$$

$$= \lim_{N \to \infty} \left[-x e^{-0.2x} \Big|_0^N + \int_0^N e^{-0.2x} dx \right]$$

$$= \lim_{N \to \infty} [-x e^{-0.2x} - 5 e^{-0.2x}] \Big|_0^N$$

$$= \lim_{N \to \infty} [(-N^{-0.2N} - 5 e^{-0.2N}) - (0 - 5 e^0)]$$

$$= 5 \text{ minutes.}$$

45. (a) $P(X > 3) = \int_3^{\infty} \frac{1}{3} e^{-x/3} dx$

$$= \lim_{N \to \infty} \frac{1}{3} \int_3^N e^{-x/3} dx$$

$$= \frac{1}{3} \lim_{N \to \infty} (-3 e^{-x/3}) \Big|_3^N$$

$$= -1 \lim_{N \to \infty} (e^{-N/3} - e^{-1})$$

$$= -1 \left(\frac{-1}{3} \right)$$

$$= \frac{1}{e} \approx 0.368$$

(b) $P(2 \le X \le 5) = \int_2^5 \frac{1}{3} e^{-x/3} dx$

$$= -1(e^{-x/3}) \Big|_1^5$$

$$= -1(e^{-5/3} - e^{-2/3})$$

$$= \frac{1}{e^{2/3}} - \frac{1}{e^{5/3}} \approx 0.325$$

(c) $E(X) = \int_{-\infty}^{\infty} x f(x) dx$

$$= \lim_{N \to \infty} \int_0^N x \cdot \frac{1}{3} e^{-x/3} dx$$

$$= \lim_{N \to \infty} \frac{1}{3} \int_0^N x e^{-x/3} dx$$

Using integration by parts with $u = x$ and $dV = e^{-x/3} dx$.

$$= \frac{1}{3} \lim_{N \to \infty} \left[-3xe^{-x/3} \Big|_0^N - \int_0^N -3e^{-x/3} dx \right]$$

$$= \lim_{N \to \infty} \left[-xe^{-x/3} \Big|_0^N + \int_0^N e^{-x/3} dx \right]$$

$$= \lim_{N \to \infty} \left[-xe^{-x/3} - 3e^{-x/3} \right]_0^N$$

$$= \lim_{N \to \infty} \left[(-Ne^{N/3} - 3e^{-N/3}) - (0 - 3e^0) \right]$$

$$= 3 \text{ minutes.}$$

47. The probability density function for this

situation is $f(x) = \begin{cases} 0.5e^{-0.5} & x \geq 0 \\ 0 & x < 0 \end{cases}$

$$P(X > 2) = 1 - \int_0^2 0.5e^{-0.5x} dx$$

$$= 1 - \left(-e^{-0.5x} \Big|_0^2 \right)$$

$$= 1 + \left(e^{-0.5x} \Big|_0^2 \right)$$

$$= 1 + (e^{-1} - e^0)$$

$$= e^{-1}$$

$$= \frac{1}{e} \approx 0.3679$$

49. (a) $P(X \geq 6)$

$$= \lim_{N \to \infty} \int_6^N 0.0866e^{-0.0866t} dt$$

$$= \lim_{N \to \infty} 0.0866 \left(\frac{1}{-0.0866t} e^{-0.0866t} \right) \Big|_6^N$$

$$= \lim_{N \to \infty} (-e^{-0.0866t}) \Big|_6^N$$

$$= \lim_{N \to \infty} (e^{-0.0866N} + e^{0.5196})$$

$$= \frac{1}{e^{0.5196}}$$

(b) $P(X \geq 6)$

$$= \lim_{N \to \infty} \int_6^N 0.135e^{-0.135t} dt$$

$$= \lim_{N \to \infty} 0.135 \left(\frac{1}{-0.135t} e^{-0.135t} \right) \Big|_6^N$$

$$= \lim_{N \to \infty} (-e^{-0.135t}) \Big|_6^N$$

$$= \lim_{N \to \infty} (-e^{-0.135N} + e^{-0.81})$$

$$= \frac{1}{e^{0.81}} \approx 0.445$$

(c) Writing Exercise—Answers will vary.

51. $E(x) = \int_{-\infty}^{\infty} x f(x) dx$

$$= \int_0^{\infty} kxe^{kx} dx$$

$$= \lim_{N \to \infty} \int_0^N kxe^{-kx} dx$$

$$= \lim_{N \to \infty} \left(-xe^{-kx} \Big|_0^N + \int_0^N e^{-kx} dx \right)$$

$$= \lim_{N \to \infty} \left(-xe^{-kx} - \frac{1}{k} e^{-kx} \right) \Big|_0^N$$

$$= \lim_{N \to \infty} \left(-Ne^{-kN} - \frac{1}{k} e^{-kN} + \frac{1}{k} \right)$$

$$= \frac{1}{k}$$

53. Need to find m such that

$$\int_0^m \frac{1}{40} dt = \frac{1}{2}$$

$$\frac{t}{40} \Big|_0^m = \frac{1}{2}$$

$$\frac{m}{40} - 0 = \frac{1}{2}$$

$$m = 20 \text{ seconds}$$

55. Need to find m such that

$$\int_a^m \frac{1}{b-a}\,dt = \frac{1}{2}$$

$$\frac{t}{b-a}\Big|_a^m = \frac{1}{2}$$

$$\frac{m}{b-a} - \frac{a}{b-a} = \frac{1}{2}$$

$$m-a = \frac{1}{2}(b-a)$$

$$m = \frac{1}{2}b - \frac{1}{2}a + a$$

$$m = \frac{1}{2}b + \frac{1}{2}a$$

$$m = \frac{b+a}{2}$$

57. (a)

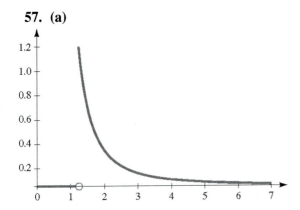

(b) $\int_{10}^{20} \frac{1.2}{x^2}\,dx = 1.2\int_{10}^{20}\frac{1}{x^2}\,dx$

$$= 1.2\left(-\frac{1}{x}\Big|_{10}^{20}\right)$$

$$= 1.2\left(-\frac{1}{20} + \frac{1}{10}\right)$$

$$= 0.06$$

(c) Writing Exercise—Answers will vary.

Checkup for Chapter 6

1. (a) $\int \sqrt{2x}\,\ln x^2\,dx$

Let $u = \ln x^2$ and $dV = \sqrt{2}x^{1/2}\,dx$
$\quad = 2\ln x$
$\quad du = \frac{2}{x}\,dx \qquad V = \frac{2\sqrt{2}}{3}x^{3/2}$

$$= \frac{4\sqrt{2}}{3}x^{3/2}\ln x - \int \frac{2\sqrt{2}}{3}x^{3/2}\cdot\frac{2}{x}\,dx$$

$$= \frac{4\sqrt{2}}{3}x^{3/2}\ln x - \frac{4\sqrt{2}}{3}\int x^{1/2}\,dx$$

$$= \frac{4\sqrt{2}}{3}x^{3/2}\ln x - \frac{8\sqrt{2}}{9}x^{3/2} + C$$

$$= \frac{4\sqrt{2}}{9}x^{3/2}\left[3\ln|x| - 2\right] + C$$

(b) $\int_0^1 xe^{0.2x}\,dx$

Let $u = x$ and $dV = e^{0.2x}\,dx$
$\quad du = dx \qquad\quad = 5e^{0.2x}$

$$= 5xe^{0.2x}\Big|_0^1 - \int_0^1 5e^{0.2x}\,dx$$

$$= (5xe^{0.2x} - 25e^{0.2x})\Big|_0^1$$

$$= [5(1)e^{0.2(1)} - 25e^{0.2(1)}] - [0 - 25e^0]$$

$$= 25 - 20e^{0.2}$$

(c) $\int_{-4}^0 x\sqrt{1-2x}\,dx$

Let $u = x$ and
$\quad du = dx$
$dV = (1-2x)^{1/2}\,dx$
$$= -\frac{1}{2}\cdot\frac{2}{3}(1-2x)^{3/2}$$

$$= -\frac{x}{3}(1-2x)^{3/2}\Big|_{-4}^{0} - \int_{-4}^{0}-\frac{1}{3}(1-2x)^{3/2}dx$$

$$= -\frac{x}{3}(1-2x)^{3/2}\Big|_{-4}^{0} + \frac{1}{3}\int_{-4}^{0}(1-2x)^{3/2}dx$$

$$= \left[-\frac{x}{3}(1-2x)^{3/2} - \frac{1}{15}(1-2x)^{5/2}\right]\Big|_{-4}^{0}$$

$$= \left[0 - \frac{1}{15}(1)\right] - \left[\frac{4}{3}(9)^{3/2} - \frac{1}{15}(9)^{5/2}\right]$$

$$= -\frac{298}{15}$$

(d) $\displaystyle\int\frac{x-1}{e^{x}}dx = \int(x-1)e^{-x}dx$

Let $u = x-1$ and $dV = e^{-x}dx$
$du = dx$ $\qquad V = -e^{-x}$

$$= -(x-1)e^{-x} - \int -e^{-x}dx$$

$$= -(x-1)e^{-x} + \int e^{-x}dx$$

$$= -(x-1)e^{-x} - e^{-x} + C$$

$$= [(-x+1)-1]e^{-x} + C$$

$$= -xe^{-x} + C$$

2. (a) $\displaystyle\int_{1}^{\infty}\frac{1}{x^{1.1}}dx$

$$= \lim_{N\to\infty}\int_{1}^{N}x^{-1.1}dx$$

$$= \lim_{N\to\infty}(-10x^{-0.1})\Big|_{1}^{N}$$

$$= \lim_{N\to\infty}[-10N^{-0.1} + 10(1)^{-0.1}]$$

$$= 0 + 10$$

$$= 10$$

(b) $\displaystyle\int_{1}^{\infty}xe^{-2x}dx = \lim_{N\to\infty}\int_{1}^{N}xe^{-2x}dx$

Let $u = x$ and $dV = e^{-2x}$
$du = dx$ $\qquad V = -\frac{1}{2}e^{-2x}$

$$= \lim_{N\to\infty}\left[-\frac{x}{2}e^{-2x}\Big|_{1}^{N} - \int_{1}^{N}-\frac{1}{2}e^{-2x}dx\right]$$

$$= \lim_{N\to\infty}\left[-\frac{x}{2}e^{-2x}\Big|_{1}^{N} + \frac{1}{2}\int_{1}^{N}e^{-2x}dx\right]$$

$$= \lim_{N\to\infty}\left[-\frac{x}{2}e^{-2x} - \frac{1}{4}e^{-2x}\right]\Big|_{1}^{N}$$

$$= \lim_{N\to\infty}\left[\left(-\frac{N}{2}e^{-2N} - \frac{1}{4}e^{-2N}\right)\right.$$
$$\left. - \left(-\frac{1}{2}e^{-2(1)} - \frac{1}{4}e^{-2(1)}\right)\right]$$

$$= 0 + \frac{1}{2}e^{-2} + \frac{1}{4}e^{-2}$$

$$= \frac{3}{4}e^{-2}$$

(c) $\displaystyle\int_{1}^{\infty}\frac{x}{(x+1)^{2}}dx = \lim_{N\to\infty}\int_{1}^{N}x(x+1)^{-2}dx$

Let $u = x$ and $dV = (x+1)^{-2}dx$
$du = dx$ $\qquad V = -\frac{1}{(x+1)}$

$$= \lim_{N\to\infty}\left[-\frac{x}{x+1}\Big|_{1}^{N} - \int_{1}^{N}-\frac{1}{x+1}dx\right]$$

$$= \lim_{N\to\infty}\left[-\frac{x}{x+1}\Big|_{1}^{N} + \int_{1}^{N}\frac{1}{x+1}dx\right]$$

$$= \lim_{N\to\infty}\left[-\frac{x}{x+1} + \ln|x+1|\right]\Big|_{1}^{N}$$

$$= \lim_{N\to\infty}\left[\left(-\frac{N}{N+1} + \ln(N+1)\right)\right.$$
$$\left. - \left(-\frac{1}{2} + \ln 2\right)\right]$$

Since $\displaystyle\lim_{N\to\infty}-\frac{N}{N+1} = \lim_{N\to\infty}-\frac{1}{1} = -1,$

and $\lim_{N\to\infty} \ln(N+1) = \infty$,

$$= \lim_{N\to\infty}\left[-\frac{N}{N+1} + \ln(N+1) + \frac{1}{2} - \ln 2\right]$$

$$= \infty$$

so, the integral diverges.

(d) $\int_{-\infty}^{\infty} xe^{-x^2}\, dx$

$$= \lim_{M\to\infty}\int_M^0 xe^{-x^2}\, dx$$

$$\qquad\qquad + \lim_{N\to\infty}\int_0^N xe^{-x^2}\, dx$$

Using substitution with $u = -x^2$ and

$$-\frac{1}{2}\, du = x\, dx,$$

$$= \lim_{M\to-\infty} -\frac{1}{2}\int_{-M^2}^0 e^u\, du + \lim_{N\to\infty} -\frac{1}{2}\int_0^{-N^2} e^u\, du$$

$$= -\frac{1}{2}\lim_{M\to-\infty}\int_{-M^2}^0 e^u\, du + \frac{1}{2}\lim_{N\to\infty}\int_{-N^2}^0 e^u\, du$$

$$= -\frac{1}{2}\lim_{M\to-\infty}\left(e^u\Big|_{-M^2}^0\right) + \frac{1}{2}\lim_{N\to\infty}\left(e^u\Big|_{-N^2}^0\right)$$

$$= -\frac{1}{2}\lim_{M\to-\infty}\left(e^0 - e^{-M^2}\right) + \frac{1}{2}\lim_{N\to\infty}\left(e^0 - e^{-N^2}\right)$$

$$= \frac{1}{2}(1-0) + \frac{1}{2}(1-0)$$

$$= 0$$

3. (a) $\int\left(\ln\sqrt{3x}\right)^2 dx$

$$= \int \ln(3x)^{1/2}\cdot\ln(3x)^{1/2}\, dx$$

$$= \int \frac{1}{2}\ln(3x)\cdot\frac{1}{2}\ln(3x)\, dx$$

$$= \frac{1}{4}\int(\ln 3x)^2\, dx$$

which most resembles $\int(\ln u)^n\, du$

(formula #27). Let $u = 3x$; then

$$du = 3\, dx \text{ or } \frac{1}{3}\, du = dx,$$

$$= \frac{1}{4}\int(\ln u)^2\cdot\frac{1}{3}\, du$$

$$= \frac{1}{12}\int(\ln u)^2\, du$$

$$= \frac{1}{12}\left[u(\ln u)^2 - 2\int \ln u\, du\right]$$

Using formula #23,

$$= \frac{1}{12}\left[u(\ln u)^2 - 2\left(u\ln|u| - u\right)\right] + C$$

$$= \frac{1}{12}\left[3x(\ln|3x|)^2 - 2(3x)\ln|3x| + 6x\right] + C$$

$$= \frac{x}{4}(\ln|3x|)^2 - \frac{x}{2}\ln|3x| + \frac{x}{2} + C$$

$$= \frac{x}{4}\left[\left(\ln|3x|\right)^2 - 2\ln|3x| + 2\right] + C$$

(b) $\int\dfrac{dx}{x\sqrt{4+x^2}}$ is of the form

$$\int\frac{du}{u\sqrt{a^2+u^2}} \text{ (formula #11).}$$

Let $x = u$, $dx = du$, and $a = 2$,

$$= -\frac{1}{2}\ln\left|\frac{\sqrt{4+x^2}+2}{x}\right| + C$$

(c) $\int\dfrac{dx}{x^2\sqrt{x^2-9}}$ is of the form

$$\int\frac{du}{u^2\sqrt{u^2-a^2}} \text{ (formula #21).}$$

Let $x = u$, $dx = du$, and $a = 3$,

$$= \frac{\sqrt{x^2-9}}{9x} + C$$

(d) $\int\dfrac{dx}{3x^2-4x}$ can be written as

$$\int\frac{dx}{x(-4+3x)} \text{ so it is of the form}$$

$$\int\frac{du}{u(a+bu)} \text{ (formula #6).}$$

Let $x = u$, $dx = du$, $a = -4$, and $b = 3$,

4. At 10:00a.m., $t = 2$ and at noon, $t = 4$. Need

$$\int_2^4 100te^{-0.5t}\,dt$$

$$= 100\int_2^4 te^{-0.5t}\,dt$$

Using integration by parts,

$$u = t \qquad dV = e^{-0.5t}\,dt$$

$$du = dt \qquad V = -\frac{1}{0.5}e^{-0.5t}$$

$$= -2e^{-0.5t}$$

$$100\int_2^4 te^{-0.5t}\,dt$$

$$= 100\left[-2te^{-0.5t}\Big|_2^4 + 2\int_2^4 e^{-0.5t}\,dt\right]$$

$$= 100\left[-2te^{-0.5t}\Big|_2^4 - 4e^{-0.5t}\Big|_2^4\right]$$

$$= -200\left[\left(te^{-0.5t} + 2e^{-0.5t}\right)\Big|_2^4\right]$$

$$= -200\left[\left(4e^{-2} + 2e^{-2}\right) - \left(2e^{-1} + 2e^{-1}\right)\right]$$

$$= -200\left(6e^{-2} - 4e^{-1}\right) \approx 132 \text{ units}$$

5.

$$\frac{dA}{dt} = 0.05A$$

$$\int \frac{1}{A}\,dA = \int 0.05\,dt$$

$$\ln|A| = 0.05t + C_1$$

$$e^{\ln|A|} = e^{0.05t + C_1}$$

$$|A| = e^{C_1}\cdot e^{0.05t}$$

$$A = \pm e^{C_1}\cdot e^{0.05t}$$

$$A(t) = Ce^{0.05t}$$

Since $A = 10,000$ when $t = 0$, $C = 10,000$.

$$= -\frac{1}{4}\ln\left|\frac{x}{3x-4}\right| + C$$

So $A(t) = 10,000e^{0.05t}$. When $t = 10$,

$$A(10) = 10,000e^{0.05(10)} \approx \$16,487.21.$$

6. $PV = \int_0^\infty (50 + 3t)e^{-0.06t}\,dt$

$$\lim_{N\to\infty}\int_0^N (50 + 3t)e^{-0.06t}\,dt$$

Using integration by parts with
$u = 50 + 3t$ and $dV = e^{-0.06t}\,dt$,

$$= \lim_{N\to\infty}\left[(50 + 3t) - \frac{50}{3}e^{-0.06t}\Big|_0^N\right.$$

$$\left. - \int_0^N -\frac{50}{3}e^{-0.06t}3\,dt\right]$$

$$= \lim_{N\to\infty}\left[-\frac{50}{3}(50 + 3t)e^{-0.06t}\Big|_0^N\right.$$

$$\left. + 50\int_0^N e^{-0.06t}\,dt\right]$$

$$= \lim_{N\to\infty} -\frac{50}{3}[(50 + 3t)e^{-0.06t}$$

$$+ 50e^{-0.06t}]\Big|_0^N$$

$$= -\frac{50}{3}\lim_{N\to\infty}[50e^{-0.06t} + 3te^{-0.06t}$$

$$+ 50e^{-0.06t}]\Big|_0^N$$

$$= -\frac{50}{3}\lim_{N\to\infty}(100e^{-0.06t} + 3te^{-0.06t})\Big|_0^N$$

$$= -\frac{50}{3}\lim_{N\to\infty}[(100e^{-0.06N} + 3Ne^{-0.06N})$$

$$- (100e^0 + 0)]$$

$$= -\frac{50}{3}\cdot -100$$

$$= \frac{5,000}{3} \approx 1,666.6667 \text{ thousand,}$$

or approximately \$1,666,666.67.

7. Since x is measured in months, we need

  (a) $P(X > 12)$

$$= \int_{12}^{\infty} 0.03 e^{-0.03x}\,dx$$

$$= \lim_{N \to \infty} 0.03 \int_{12}^{N} e^{-0.03x}\,dx$$

$$= \lim_{N \to \infty} 0.03 \cdot \frac{1}{-0.03} e^{-0.03x} \Big|_{12}^{N}$$

$$= \lim_{N \to \infty} \left(-e^{-0.03x}\right) \Big|_{12}^{N}$$

$$= \lim_{N \to \infty} \left[-e^{-0.03N} + e^{-0.03(12)}\right]$$

$$= e^{-0.36} \approx 0.6977$$

  (b) $P(3 \le x \le 6) = \int_{3}^{6} 0.03 e^{-0.03x}\,dx$

$$= \left(-e^{-0.03x}\right) \Big|_{3}^{6}$$

$$= -e^{-0.18} + e^{-0.09}$$

$$\approx 0.07866$$

  (c) $E(X) = \int_{-\infty}^{\infty} x f(x)\,dx$

$$= \lim_{N \to \infty} 0.03 \int_{0}^{N} x e^{-0.03x}\,dx$$

Using integration by parts with $u = x$ and $dV = e^{-0.03x}\,dx$.

$$= \lim_{N \to \infty} 0.03 \left[\frac{x}{-0.03} e^{-0.03x} \Big|_{0}^{N} \right.$$

$$\left. - \int_{0}^{N} \frac{1}{-0.03} e^{-0.03x}\,dx \right]$$

$$= \lim_{N \to \infty} \left[-x e^{-0.03x} \Big|_{0}^{N} + \int_{0}^{N} e^{-0.03x}\,dx \right]$$

$$= \lim_{N \to \infty} \left[-x e^{-0.03x} - \frac{1}{0.03} e^{-0.03x} \right]\Big|_{0}^{N}$$

$$= \lim_{N \to \infty} \left[\left(-N e^{-0.03N} - \frac{1}{0.03} e^{-0.03N} \right) \right.$$

$$\left. - \left(0 - \frac{1}{0.03} e^{0} \right) \right]$$

$$= \frac{100}{3} \approx 33.3 \text{ months}$$

8. (a) The probability density function for this situation is $f(t) = \begin{cases} \dfrac{1}{60} & 0 \le t \le 60 \\ 0 & \text{otherwise} \end{cases}$

  (b) $P(X \ge 45) = \int_{45}^{60} \frac{1}{60}\,dt$

$$= \frac{t}{60} \Big|_{45}^{60}$$

$$= 1 - \frac{3}{4}$$

$$= \frac{1}{4}$$

  (c) $P(5 \le x \le 15) = \int_{5}^{15} \frac{1}{60}\,dt$

$$= \frac{t}{60} \Big|_{5}^{15}$$

$$= \frac{1}{4} - \frac{1}{12}$$

$$= \frac{1}{6}$$

9. Amount of drug $= \lim_{N \to \infty} \int_{0}^{N} 0.7 e^{-0.2(N-t)}\,dt$

$$= 0.7 \lim_{N \to \infty} e^{-0.2N} \int_{0}^{N} e^{0.2t}\,dt$$

$$= 0.7 \lim_{N \to \infty} e^{-0.2N} \left(5 e^{0.2t} \Big|_{0}^{N} \right)$$

$$= 3.5 \lim_{N \to \infty} e^{-0.2N} \left(e^{0.2N} - e^{0} \right)$$

$$= 3.5 \lim_{N \to \infty} \left(e^{0} - e^{-0.2N} \right)$$

$$= 3.5(1 - 0) = 3.5 \text{ mg}$$

10. To approximate $\int_{3}^{4} \frac{\sqrt{25 - x^2}}{x}\,dx$ using the trapezoidal rule with $n = 8$,

$$\Delta x = \frac{4 - 3}{8} = 0.125,$$

$$\approx \frac{0.125}{2}\left[\left(\frac{\sqrt{25-(3)^2}}{3}\right)+2\left(\frac{\sqrt{25-(3.125)^2}}{3.125}\right)\right.$$

$$+2\left(\frac{\sqrt{25-(3.25)^2}}{3.25}\right)+2\left(\frac{\sqrt{25-(3.375)^2}}{3.375}\right)$$

$$+2\left(\frac{\sqrt{25-(3.5)^2}}{3.5}\right)+2\left(\frac{\sqrt{25-(3.625)^2}}{3.625}\right)$$

$$+2\left(\frac{\sqrt{25-(3.75)^2}}{3.75}\right)+2\left(\frac{\sqrt{25-(3.875)^2}}{3.875}\right)$$

$$\left.+\left(\frac{\sqrt{25-(4)^2}}{4}\right)\right]\approx 1.027552$$

Using formula #17 with $x = u$, $dx = du$, and $a = 5$,

$$=\left[\sqrt{25-x^2}-5\ln\left|\frac{5+\sqrt{25-x^2}}{x}\right|\right]\Bigg|_3^4$$

$$=\left[\sqrt{25-4^2}-5\ln\left|\frac{5+\sqrt{25-4^2}}{4}\right|\right]$$

$$=-\left[\sqrt{25-3^2}-5\ln\left|\frac{5+\sqrt{25-3^2}}{3}\right|\right]$$

$$=(3-5\ln 2)-(4-5\ln 3)$$

$$=-1-\ln 2^5+\ln 3^5$$

$$=-1+\ln\left(\frac{3}{2}\right)^5$$

$$=-1+5\ln\left(\frac{3}{2}\right)\approx 1.02736$$

Review Exercises

1. $\int te^{1-t}\,dt$

Let $\quad u=t \quad$ and $\quad dV=e^{1-t}\,dt$
$\qquad du=dt \qquad\qquad V=-e^{1-t}$

$$=-te^{1-t}-\int -e^{1-t}\,dt$$

$$=-te^{1-t}+\int e^{1-t}\,dt$$

$$=-te^{1-t}-e^{1-t}+C$$

$$=-e^{1-t}(t+1)+C$$

3. $\int x(2x+3)^{1/2}\,dx$

Let $\quad u=x \quad$ and $\quad dV=(2x+3)^{1/2}$
$\qquad du=dx$

Using substitution with $u = 2x + 3$,

$$V=\frac{1}{2}\left[\frac{2}{3}(2x+3)^{3/2}\right]=\frac{1}{3}(2x+3)^{3/2}$$

So, $\int x(2x+3)^{1/2}\,dx$

$$=\frac{x}{3}(2x+3)^{3/2}-\int \frac{1}{3}(2x+3)^{3/2}\,dx$$

Using substitution with $u = 2x + 3$,

$$=\frac{x}{3}(2x+3)^{3/2}-\frac{1}{3}\left(\frac{1}{2}\right)\left(\frac{2}{5}\right)(2x+3)^{5/2}+C$$

$$=\frac{x}{3}(2x+3)^{3/2}-\frac{1}{15}(2x+3)^{5/2}+C$$

5. $\int_1^4 \frac{\ln\sqrt{S}}{\sqrt{S}}\,dS = \int_1^4 S^{-1/2}\ln S^{1/2}\,dS$

Let $\quad u = \ln S^{1/2} \quad$ and $\quad dV = S^{-1/2}\,dS$

$\qquad\qquad = \frac{1}{2}\ln S \qquad\qquad V = 2S^{1/2}$

$\qquad du = \frac{1}{2S}\,dS$

$= S^{1/2}\ln S\Big|_1^4 - \int_1^4 2S^{1/2}\cdot\frac{1}{2S}\,dS$

$= S^{1/2}\ln S\Big|_1^4 - \int_1^4 S^{-1/2}\,dS$

$= (S^{1/2}\ln S - 2S^{1/2})\Big|_1^4$

$= \left[\sqrt{4}\ln 4 - 2\sqrt{4}\right] - [1\ln 1 - 2(1)]$

$= 2\ln 4 - 2$

$= 2\ln(2)^2 - 2$

$= 4\ln 2 - 2$

7. $\int_{-2}^1 (2x+1)(x+3)^{3/2}\,dx$

Let $\quad u = 2x+1 \quad$ and $\quad dV = (x+3)^{3/2}\,dx$

$\qquad du = 2\,dx \qquad\qquad\quad V = \frac{2}{5}(x+3)^{5/2}$

$= \frac{2}{5}(2x+1)(x+3)^{5/2}\Big|_{-2}^1 - \int_{-2}^1 \frac{2}{5}(x+3)^{5/2}\cdot 2\,dx$

$= \left[\frac{2}{5}(2x+1)(x+3)^{5/2} - \left(\frac{4}{5}\right)\left(\frac{2}{7}\right)(x+3)^{7/2}\right]\Big|_{-2}^1$

$= \left[\frac{2}{5}(2(1)+1)(1+3)^{5/2} - \frac{8}{35}(1+3)^{7/2}\right] - \left[\frac{2}{5}(2(-2)+1)(-2+3)^{5/2} - \frac{8}{35}(-2+3)^{7/2}\right]$

$= \frac{74}{7}$

9. $\int x^3(3x^2+2)^{1/2}\,dx = \int x^2\cdot x(3x^2+2)^{1/2}\,dx$

Let $\quad u = x^2 \quad$ and $\quad dV = x(3x^2+2)^{1/2}\,dx$

$\qquad du = 2x\,dx$

Using substitution with $u = 3x^2+2$,

$V = \left(\frac{1}{6}\right)\left(\frac{2}{3}\right)(3x^2+2)^{3/2}$

$V = \frac{1}{9}(3x^2+2)^{3/2}$

So, $\int x^2 \cdot x(3x^2+2)^{1/2}\,dx = \dfrac{x^2}{9}(3x^2+2)^{3/2} - \int \dfrac{1}{9}(3x^2+2)^{3/2}2x\,dx$

$\qquad\qquad = \dfrac{x^2}{9}(3x^2+2)^{3/2} - \dfrac{2}{9}\int(3x^2+2)^{3/2}x\,dx$

$\qquad\qquad = \dfrac{x^2}{9}(3x^2+2)^{3/2} - \left(\dfrac{2}{9}\right)\left(\dfrac{1}{6}\right)\left(\dfrac{2}{5}\right)(3x^2+2)^{5/2} + C$

$\qquad\qquad = \dfrac{x^2}{9}(3x^2+2)^{3/2} - \dfrac{2}{135}(3x^2+2)^{5/2} + C$

11. $\displaystyle\int \frac{5\,dx}{8-2x^2} = \int \frac{5\,dx}{2(4-x^2)} = \frac{5}{2}\int \frac{dx}{4-x^2}$

which is of the form $\displaystyle\int \frac{du}{a^2-u^2}$ (formula
#16). Let $x = u$, $dx = du$, and $a = 2$,

$\quad = \dfrac{5}{2}\left[\dfrac{1}{2(2)}\ln\left|\dfrac{2+x}{2-x}\right| + C\right]$

$\quad = \dfrac{5}{8}\ln\left|\dfrac{2+x}{2-x}\right| + C$

13. $\displaystyle\int w^2 e^{-w/3}\,dw = \int w^2 e^{-\frac{1}{3}w}\,dw$

which is of the form $\displaystyle\int u^n e^{au}\,du$ (formula

#26). Let $w = u$, $dw = du$, and $a = -\dfrac{1}{3}$,

$\quad = \dfrac{1}{-\frac{1}{3}}w^2 e^{-w/3} - \dfrac{2}{-\frac{1}{3}}\int we^{-w/3}\,dw$

$\quad = -3w^2 e^{-w/3} + 6\int we^{-w/3}\,dw$

Using formula #22,

$\quad = -3w^2 e^{-w/3}$

$\qquad + 6\left[\dfrac{1}{\left(-\frac{1}{3}\right)^2}\left(-\dfrac{1}{3}w-1\right)e^{-w/3}\right] + C$

$\quad = -3w^2 e^{-w/3} + 54\left(-\dfrac{1}{3}w-1\right)e^{-w/3} + C$

$\quad = -3w^2 e^{-w/3} - 18we^{-w/3} - 54e^{-w/3} + C$

15. $\displaystyle\int (\ln 2x)^3\,dx = \frac{1}{2}\int(\ln 2x)^3 \cdot 2\,dx$

which is of the form $\displaystyle\int(\ln u)^n\,du$ (formula
#27). Let $u = 2x$, $du = 2\,dx$, and $n = 3$,

$\quad = \dfrac{1}{2}\left[2x(\ln 2x)^3 - 3\int(\ln 2x)^2 2\,dx\right]$

$\quad = x(\ln 2x)^3$

$\qquad - \dfrac{3}{2}\left[2x(\ln 2x)^2 - 2\int(\ln 2x)2\,dx\right]$

$\quad = x(\ln 2x)^3 - 3x(\ln 2x)^2$

$\qquad\qquad + 3\left[2x\ln|2x| - 2x\right] + C$

$\quad = x(\ln 2x)^3 - 3x(\ln 2x)^2 + 6x\ln 2x - 6x$

$\qquad\qquad\qquad + C$

$\quad = x[(\ln 2x)^3 - 3(\ln 2x)^2 + 6(\ln 2x) - 6]$

$\qquad\qquad\qquad + C$

17. $\dfrac{dy}{dx} = x\ln\sqrt{x}$; $y = 3$ when $x = 1$

$\quad y = \displaystyle\int \frac{dy}{dx}\,dx$

$\qquad = \displaystyle\int x\ln\sqrt{x}\,dx$

Since $\ln\sqrt{x} = \ln x^{1/2} = \dfrac{1}{2}\ln x$

$\qquad = \dfrac{1}{2}\displaystyle\int x\ln x\,dx$

Need to use integration by parts, and since
cannot integrate $\ln x$,

$\quad u = \ln x \qquad dV = x\,dx$

$\quad du = \dfrac{1}{x}\,dx \qquad V = \dfrac{1}{2}x^2$

$$\frac{1}{2}\int x \ln x\, dx$$

$$= \frac{1}{2}\left[\frac{x^2}{2}\ln x - \int \frac{1}{2}x\, dx\right]$$

$$= \frac{x^2}{4}\ln x - \frac{1}{8}x^2 + C$$

Since $y = 3$ when $x = 1$,

$$3 = \frac{1}{4}\ln 1 - \frac{1}{8} + C$$

$$3 = -\frac{1}{8} + C, \text{ or } C = \frac{25}{8}$$

and

$$y = \frac{x^2}{4}\ln x - \frac{x^2}{8} + \frac{25}{8} \text{ or,}$$

$$y = \frac{1}{2}x^2 \ln x^{1/2} - \frac{x^2}{8} + \frac{25}{8}$$

$$= \frac{1}{2}x^2 \ln \sqrt{x} - \frac{1}{8}x^2 + \frac{25}{8}$$

19. $\dfrac{dy}{dx} = \dfrac{e^y}{xy}$; $y = 0$ when $x = 1$

Separating the variables gives

$$\frac{y}{e^y}\, dy = \frac{1}{x}\, dx$$

$$ye^{-y}\, dy = \frac{1}{x}\, dx$$

Integrating both sides,

$$\int ye^{-y}\, dy = \int \frac{1}{x}\, dx$$

For the left hand side, use integration by parts with

$$u = y \qquad dV = e^{-y}\, dy$$

$$du = dy \qquad V = -e^{-y}$$

$$-ye^{-y} - \int -e^{-y}\, dy = \int \frac{1}{x}\, dx$$

$$-ye^{-y} + \int e^{-y}\, dy = \int \frac{1}{x}\, dx$$

$$-ye^{-y} - e^{-y} + C_1 = \ln|x| + C_2$$

$$-ye^{-y} - e^{-y} = \ln|x| + C$$

$$e^{-y}(-y-1) = \ln|x| + C$$

Since $y = 0$ when $x = 1$,

$$e^0(-1) = \ln 1 + C, \text{ or } C = -1$$

$$e^{-y}(-y-1) = \ln|x| - 1$$

Multiplying both sides by -1 gives

$$e^{-y}(y+1) = 1 - \ln|x|.$$

21. $\displaystyle\int_0^\infty \frac{1}{\sqrt[3]{1+2x}}\, dx$

Using substitution with $u = 1 + 2x$,

$$= \lim_{N\to\infty} \int_0^N (1+2x)^{-1/3}\, dx$$

$$= \lim_{N\to\infty} \frac{3}{4}(1+2x)^{2/3}\Big|_0^N$$

$$= \frac{3}{4}\lim_{N\to\infty}\left[(1+2N)^{2/3} - 1\right]$$

$$= \infty.$$

So, the interval diverges.

23. $\displaystyle\int_0^\infty \frac{3t}{t^2+1}\, dx$

Using substitution with $u = t^2 + 1$,

$$= 3\lim_{N\to\infty} \int_0^N \frac{t}{t^2+1}\, dx$$

$$= 3\lim_{N\to\infty} \frac{1}{2}\ln(t^2+1)\Big|_0^N$$

$$= \frac{3}{2}\lim_{N\to\infty}\left[\ln(N^2+1) - \ln 1\right]$$

$$= \infty.$$

So, the interval diverges.

25. $\displaystyle\int_0^\infty xe^{-2x}\, dx$

Using integration by parts with $u = x$ and

$dV = e^{-2x}dx,$

$= \lim_{N \to \infty} \int_0^N xe^{-2x}dx$

$= \lim_{N \to \infty} \left[\left(-\frac{1}{2}xe^{-2x} \right) \Big|_0^N + \frac{1}{2}\int_0^N e^{-2x}dx \right]$

$= \lim_{N \to \infty} \left(-\frac{1}{2}xe^{-2x} - \frac{1}{4}e^{-2x} \right) \Big|_0^N$

$= \frac{1}{4}$

27. $\int_0^\infty x^2 e^{-2x}dx = \lim_{N \to \infty} \int_0^N x^2 e^{-2x}dx$

Using integration by parts twice, first with $u = x^2$ and $dV = e^{-2x}dx$ and second with $u = x$ and $dV = e^{-2x}dx$,

$= \lim_{N \to \infty} \left[-\frac{1}{2}x^2 e^{-2x} \Big|_0^N + \int_0^N xe^{-2x}dx \right]$

$= \lim_{N \to \infty} \left[-\frac{1}{2}x^2 e^{-2x} \Big|_0^N - \frac{1}{2}xe^{-2x} \Big|_0^N - \int_0^N -\frac{1}{2}e^{-2x}dx \right]$

$= \lim_{N \to \infty} \left[-\frac{1}{2}x^2 e^{-2x} \Big|_0^N - \frac{1}{2}xe^{-2x} \Big|_0^N - \frac{1}{4}e^{-2x} \Big|_0^N \right]$

$= \lim_{N \to \infty} \left[\left(-\frac{1}{2}N^2 e^{-2N} + 0 \right) - \left(\frac{1}{2}Ne^{-2x} + 0 \right) - \left(\frac{1}{4}e^{-2N} - \frac{1}{4}e^0 \right) \right]$

$= 0 - 0 + \frac{1}{4} = \frac{1}{4}$

29. $\int_1^\infty \frac{\ln x}{\sqrt{x}}dx = \lim_{N \to \infty} \int_1^N x^{-1/2} \ln x\, dx$

Using integration by parts with $u = \ln x$ and $dV = x^{-1/2}dx$,

$= \lim_{N \to \infty} \left[2x^{1/2} \ln x \Big|_1^N - \int_1^N 2x^{1/2} \cdot \frac{1}{x}dx \right]$

$= \lim_{N \to \infty} \left[2x^{1/2} \ln x \Big|_1^N - 2\int_1^N x^{-1/2}dx \right]$

$= \lim_{N \to \infty} 2[x^{1/2} \ln x - 2x^{1/2}] \Big|_1^N$

$= 2 \lim_{N \to \infty} [(N^{1/2} \ln N - 2N^{1/2}) - (\ln 1 - 2)]$

$= \infty$, so the integral diverges.

31. $\int_{-\infty}^{-1} xe^{x+1}dx = \lim_{N \to -\infty} \int_N^{-1} xe^{x+1}dx$

Using integration by parts with

$u = x \qquad dV = e^{x+1}dx$

$du = dx \qquad V = e^{x+1}$

$= \lim_{N \to -\infty} \left[xe^{x+1} \Big|_N^{-1} - \int_N^{-1} e^{x+1}dx \right]$

$= \lim_{N \to -\infty} \left[\left(xe^{x+1} - e^{x+1} \right) \Big|_N^{-1} \right]$

$= \lim_{N \to -\infty} \left[\left(-e^0 - e^0 \right) - \left(Ne^{N+1} - e^{N+1} \right) \right]$

$= -1 - 1 - 0 + 0 = -2$

33. $\int_{-\infty}^{\infty} x^3 e^{-x^2} dx$

$= \int_{-\infty}^{C} x^3 e^{-x^2} dx + \int_C^{\infty} x^3 e^{-x^2} dx$

Choose $C = 0$. Then,

$= \int_{-\infty}^0 x^3 e^{-x^2} dx + \int_0^\infty x^3 e^{-x^2} dx$

$= \lim_{N \to -\infty} \int_N^0 x^3 e^{-x^2} dx + \lim_{N \to -\infty} \int_0^N x^3 e^{-x^2} dx$

Let $u = x^2$. Then, $\frac{1}{2}du = x\, dx$ and $x^3 = x \cdot u$. So,

$\int x^3 e^{-x^2} dx = \int u \cdot e^{-u} \cdot x\, dx$

$= \frac{1}{2}\int ue^{-u} du$

Now use integration by parts with

$u = u \qquad dV = e^{-u}du$

$du = du \qquad V = -e^{-u}$

$= \frac{1}{2}\left[-ue^{-u} - \int -e^{-u}du \right]$

$= \frac{1}{2}\left(-ue^{-u} + \int e^{-u}du \right)$

$$= \frac{1}{2}\left(-ue^{-u} - e^{-u} + C\right)$$

$$= \frac{1}{2}\left(-x^2 e^{-x^2} - e^{-x^2} + C\right)$$

So,

$$\lim_{N \to -\infty} \int_N^0 x^3 e^{-x^2}\,dx$$

$$= \lim_{N \to -\infty}\left[\frac{1}{2}\left(-x^2 e^{-x^2} - e^{-x^2}\right)\Big|_N^0\right]$$

$$= \frac{1}{2}\lim_{N \to -\infty}\left[\left(0 - e^0\right) - \left(-N^2 e^{-N^2} - e^{-N^2}\right)\right]$$

$$= \frac{1}{2}(0 - 1 + 0 + 0) = -\frac{1}{2}$$

Similarly,

$$\lim_{N \to \infty} \int_0^N x^3 e^{-x^2}\,dx$$

$$= \lim_{N \to \infty}\left[\frac{1}{2}\left(-x^2 e^{-x^2} - e^{-x^2}\right)\Big|_0^N\right]$$

$$= \frac{1}{2}\lim_{N \to \infty}\left[\left(-N^2 e^{-N^2} - e^{-N^2}\right) - \left(0 - e^0\right)\right]$$

$$= \frac{1}{2}(0 - 0 - 0 + 1) = \frac{1}{2}$$

So $\displaystyle\int_{-\infty}^{\infty} x^3 e^{-x^2}\,dx = -\frac{1}{2} + \frac{1}{2} = 0$.

35. (a) $\displaystyle P(1 \le X \le 4) = \int_1^4 f(x)\,dx$

$$= \int_1^4 \frac{1}{3}\,dx$$

$$= \frac{x}{3}\Big|_1^4$$

$$= 1$$

(b) $\displaystyle P(2 \le X \le 3) = \int_2^3 f(x)\,dx$

$$= \int_2^3 \frac{1}{3}\,dx$$

$$= \frac{x}{3}\Big|_2^3$$

$$= \frac{1}{3}$$

(c) $\displaystyle P(X \le 2) = \int_{-\infty}^2 f(x)\,dx$

$$= \int_1^2 \frac{1}{3}\,dx$$

$$= \frac{x}{3}\Big|_1^2$$

$$= \frac{1}{3}$$

37. (a) $\displaystyle P(X \ge 0) = \int_0^\infty f(x)\,dx$

$$= \lim_{N \to \infty} \int_0^N 0.2 e^{-0.2x}\,dx$$

$$= \lim_{N \to \infty} (-e^{-0.2x})\Big|_0^N$$

$$= \lim_{N \to \infty} (-e^{-0.2N} + 1)$$

$$= 1$$

(b) $\displaystyle P(1 \le X \le 4) = \int_1^4 f(x)\,dx$

$$= \int_1^4 0.2 e^{-0.2x}\,dx$$

$$= -e^{-0.2x}\Big|_1^4$$

$$= -e^{-0.8} + e^{-0.2} \approx 0.3694$$

(c) $\displaystyle P(X \ge 5) = \int_5^\infty f(x)\,dx$

$$= \lim_{N \to \infty} \int_5^N 0.2 e^{-0.2x}\,dx$$

$$= \lim_{N \to \infty} -e^{-0.2x}\Big|_5^N$$

$$= \lim_{N \to \infty} \left[-e^{-0.2N} + e^{-1}\right]$$

$$\approx 0.3679$$

39. In N years, the population of the city will be

$$P_0 f(N) + \int_0^N r(t) f(N-t)\, dt \quad \text{where}$$

$P_0 = 100{,}000$ is the current population,

$f(t) = e^{-t/20}$ is the fraction of the residents remaining for at least t years, and $r(t) = 100t$ is the rate of new arrivals. In the long run, the number of residents will be

$$\lim_{N \to \infty} \Big[100{,}000 e^{-N/20}$$
$$+ \int_0^N 100 t e^{-(N-t)/20}\, dt \Big]$$

$$= 0 + \lim_{N \to \infty} \Big[100 e^{-N/20} \int_0^N t e^{t/20}\, dt \Big]$$

$$= \lim_{N \to \infty} 100 e^{-N/20} [20 t e^{t/20} - 400 e^{t/20}]\Big|_0^N$$

$$= \lim_{N \to \infty} 100 e^{-N/20} \Big[\left(20 N e^{N/20} - 400 e^{N/20} \right) - 0 - 400 e^0 \Big]$$

$$= \lim_{N \to \infty} 100 \left(20 N e^0 - 400 e^0 - 0 + 400 e^{-N/20} \right)$$

$$= \lim_{N \to \infty} 100 (20N - 400 + 400 e^{-N/20})$$

$$= \infty$$

So, the population will increase without bound.

41. Let x denote the number of minutes since the start of the movie at the time of your arrival. The uniform density function for x is

$$f(x) = \begin{cases} \dfrac{1}{120} & \text{if } 0 \le x \le 120 \\ 0 & \text{otherwise} \end{cases}$$

So, the probability that you arrive within 10 minutes (before or after) of the start of a movie is

$$P(0 \le X \le 10) + P(110 \le X \le 120)$$
$$= 2P(0 \le X \le 10)$$
$$= 2\int_0^{10} \frac{1}{120}\, dx$$
$$= \frac{x}{60}\Big|_0^{10}$$
$$= \frac{1}{6}$$

43. $PV = \displaystyle\lim_{N \to \infty} \int_0^N (8{,}000 + 400t) e^{-0.05t}\, dt$

Using integration by parts with

$u = 8{,}000 + 400t$ and $dV = e^{-0.05t}\, dt$

$$= \lim_{N \to \infty} \Big[-20(8{,}000 + 400t) e^{0.05t} \Big|_0^N$$
$$- \int_0^N -8{,}000 e^{-0.05t}\, dt \Big]$$

$$= \lim_{N \to \infty} \Big[-20(8{,}000 + 400t) e^{-0.05t} \Big|_0^N$$
$$+ 8{,}000 \int_0^N e^{0.05t} \Big]$$

$$= -20 \lim_{N \to \infty} [(8{,}000 + 400t) e^{-0.05t}$$
$$+ 8{,}000 e^{-0.05t}] \Big|_0^N$$

$$= -20 \lim_{N \to \infty} [(8{,}000 + 400N) e^{-0.05N}$$
$$+ 8{,}000 e^{-0.05N} - (8{,}000 e^0 + 8{,}000 e^0)]$$

$$= -20 \lim_{N \to \infty} [16{,}000 e^{-0.05N}$$
$$+ 400 N e^{-0.05N} - 16{,}000]$$

$$= -20(-16{,}000)$$
$$= \$320{,}000$$

45. Let x denote the time (in minutes) between your arrival and the next batch of cookies. Then x is uniformly distributed with probability density function

$$f(x) = \begin{cases} \dfrac{1}{45} & \text{if } 0 \le x \le 45 \\ 0 & \text{otherwise} \end{cases}$$

So, the probability that you arrive within 5 minutes (before or after) the cookies were baked is

$$P(0 \le X \le 5) + P(40 \le X \le 45)$$
$$= 2P(0 \le X \le 5)$$
$$= 2\int_0^5 \frac{1}{45}\, dx$$
$$= \frac{2x}{45}\Big|_0^5$$
$$= \frac{2}{9}$$

47. (a) $\displaystyle\int_5^\infty 0.07e^{-0.07u}\,du = -\lim_{N\to\infty} e^{-0.07u}\Big|_5^N$

$$= 0.7047$$

47. (b) $\displaystyle\int_{10}^{15} 0.07e^{-0.07u}\,du = -e^{-0.07u}\Big|_{10}^5$

$$= -e^{-0.35} + e^{-0.7} = 0.1466$$

49. $\displaystyle\lim_{N\to\infty}\int_0^N (300 - 200e^{-0.03t})e^{-0.02(N-t)}\,dt$

$$= \lim_{N\to\infty}\int_0^N 300e^{-0.02(N-t)} - 200e^{-0.02N-0.01t}\,dt$$

$$= \lim_{N\to\infty} 300e^{-0.02N}\int_0^N e^{0.02t}\,dt - \lim_{N\to\infty} 200e^{-0.02N}\int_0^N e^{-0.01t}\,dt$$

$$= 300\lim_{N\to\infty} e^{-0.02N}\int_0^N e^{0.02t}\,dt - 200\lim_{N\to\infty} e^{-0.02N}\int_0^N e^{-0.01t}\,dt$$

$$= 300\lim_{N\to\infty} e^{-0.02N}\left(50e^{0.02t}\Big|_0^N\right) - 200\lim_{N\to\infty} e^{-0.02N}\left(-100e^{-0.01t}\Big|_0^N\right)$$

$$= 15{,}000\lim_{N\to\infty} e^{-0.02N}(e^{0.02N} - e^0) + 20{,}000\lim_{N\to\infty} e^{-0.02N}(e^{-0.01N} - e^0)$$

$$= 15{,}000\lim_{N\to\infty} (e^0 - e^{-0.02N}) + 20{,}000\lim_{N\to\infty} (e^{-0.03N} - e^{-0.02N})$$

$$= 15{,}000(1 - 0) + 20{,}000(0 - 0)$$

$$= 15{,}000 \text{ pounds}$$

51. For $\displaystyle\int_0^2 e^{x^2}\,dx$ with $n = 8$, $\Delta x = \dfrac{2-0}{8} = 0.25$, and $x_1 = 0$, $x_2 = 0.25$, $x_3 = 0.50$, ... $x_8 = 1.75$, $x_9 = 2$.

(a) By the trapezoidal rule,

$$\int_0^2 e^{x^2}\,dx = \frac{\Delta x}{2}[f(x_1) + 2f(x_2) + 2f(x_3) + \cdots + f(x_9)]$$

$$= \frac{0.25}{2}[1 + 2e^{(0.25)^2} + 2e^{(0.5)^2} + 2e^{(0.75)^2} + 2e^1 + 2e^{(1.25)^2} + 2e^{(1.5)^2} + 2e^{(1.75)^2} + e^2]$$

$$= 17.5651$$

The error estimate is $|E_n| \geq \dfrac{M(b-a)^3}{12n^2}$. For $n = 8$, $a = 0$ and $b = 2$, $|E_8| \leq \dfrac{M(2-0)^3}{12(8)^2} = \dfrac{M}{96}$

where M is the maximum value of $|f''(x)|$ on $0 \leq x \leq 2$. Now $f(x) = e^{x^2}$, $f'(x) = 2xe^{x^2}$, and

$f''(x) = (2 + 4x^2)e^{x^2}$. For $0 \leq x \leq 2$, $|f''(x)| \leq [2 + 4(2)^2]e^{(2)^2} = 18e^4$. So,

$$|E_8| \leq \frac{18e^4}{96} \approx 10.2372.$$

(b) By Simpson's rule,

$$\int_0^2 e^{x^2}\,dx = \frac{\Delta x}{3}[f(x_1)+4f(x_2)+2f(x_3)+4f(x_4)+\cdots+f(x_9)]$$

$$= \frac{0.25}{3}[1+4e^{(0.25)^2}+2e^{(0.5)^2}+4e^{(0.75)^2}+2e^1+4e^{(1.25)^2}+2e^{(1.5)^2}+4e^{(1.75)^2}+e^{(2)^2}]$$

$$=16{,}536.$$

The error estimate is $\left|E_n\right| \le \dfrac{M(b-a)^5}{180n^4}$. For $n=8$, $a=0$, and $b=2$, $\left|E_8\right| \le \dfrac{M(2-0)^5}{180(8)^4} = \dfrac{M}{23{,}040}$

where M is the maximum value of $\left|f^{(4)}(x)\right|$ and $0 \le x \le 2$. Now $f^{(3)}(x)=(8x^3+12x)e^{x^2}$, and

$f^{(4)}(x)=(16x^4+48x^2+12)e^{x^2}$. For $0 \le x \le 2$, $\left|f^{(4)}(x)\right| \le [16(2)^4+48(2)^2+12]e^{2^2} \le 460e^4$.

So, $\left|E_8\right| \le \dfrac{460e^4}{23{,}040} \approx 1.0901$.

53. For $\int_1^2 xe^{1/x}\,dx$ with $n=8$, $\Delta x = \dfrac{2-1}{8} = \dfrac{1}{8} = 0.125$, and $x_1=1$, $x_2=1.125$, $x_3=1.25$, ..., $x_8=1.875$, $x_9=2$.

(a) By the trapezoidal rule,

$$\int_1^2 xe^{1/x}\,dx \approx \frac{\Delta x}{2}[f(x_1)+2f(x_2)+2f(x_3)+\cdots+2f(x_8)+f(x_9)]$$

$$= 0.0625[(1)e^1+2(1.125)e^{1/1.125}+2(1.25e^{1/1.25})+2(1.375)e^{1/1.375}+2(1.5)e^{1/1.5}$$
$$+2(1.625)e^{1/1.625}+2(1.75)^{1/1.75}+2(1.875)e^{1/1.875}+(2)e^{1/2}]$$

$$\approx 2.9495$$

For the error estimate $\left|E_8\right| \le \dfrac{M(2-1)^3}{12(8)^2} = \dfrac{M}{768}$ where M is the maximum value of $\left|f''(x)\right|$ on

$1 \le x \le 2$.

$$f'(x)=(x)\left(e^{1/x}\cdot\frac{-1}{x^2}\right)+(e^{1/x})(1)=e^{1/x}\left(-\frac{1}{x}+1\right)$$

$$f''(x)=(e^{1/x})\left(\frac{1}{x^2}\right)+\left(-\frac{1}{x}+1\right)\left(e^{1/x}\cdot\frac{1}{x^2}\right)$$

$$=\frac{1}{x^2}d^{1/x}\left[1+-\left(-\frac{1}{x}+1\right)\right]$$

$$=\frac{1}{x^3}e^{1/x}$$

Since $f''(x)$ is always positive and decreasing on $1 \le x \le 2$, $M=\left|f''(1)\right|=e$.

$\left|E_8\right| \le \dfrac{e}{768} \approx 0.003539$

(b) By Simpson's rule,

$$\int_1^2 xe^{1/x}\,dx = \frac{\Delta x}{3}[f(x_1)+4f(x_2)+2f(x_3)+4f(x_4)+\cdots+4f(x_8)+f(x_9)]$$

$$= \frac{1}{24}[(1)e^1 + 4(1.125)e^{1/1.125} + 2(1.25)e^{1/125} + 4(1.375)e^{1/1.375} + 2(1.5)e^{1/1.5}$$

$$+ 4(1.625)e^{1/1.625} + 2(1.75)e^{1/1.75} + 4(1.875)e^{1/1.875} + (2)e^{1/2}]$$

$$\approx 2.94834$$

For the error estimate, $|E_8| \le \dfrac{M(2-1)^5}{180(8)^4} = \dfrac{M}{737,280}$ where M is the maximum value of $\left|f^{(4)}(x)\right|$

on $1 \le x \le 2$.

$$f^{(3)}(x) = \left(\frac{1}{x^3}\right)\left(e^{1/x}\cdot -\frac{1}{x^2}\right)+(e^{1/x})\left(-\frac{3}{x^4}\right) = e^{1/x}\left(-\frac{1}{x^5}-\frac{3}{x^4}\right)$$

$$f^{(4)} = (e^{1/x})\left(\frac{5}{x^6}+\frac{12}{x^5}\right)+\left(-\frac{1}{x^5}-\frac{3}{x^4}\right)\left(e^{1/x}\cdot -\frac{1}{x^2}\right)$$

$$= e^{1/x}\left(\frac{5+12x}{x^6}\right)+e^{1/x}\left(\frac{1+3x}{x^7}\right)$$

$$= \frac{1}{x^7}e^{1/x}[x(5+12x)+(1+3x)]$$

$$= \frac{1}{x^7}e^{1/x}(12x^2+8x+1)$$

Since $\left|f^{(4)}(x)\right|$ is always positive and decreasing on $1 \le x \le 2$, $M = \left|f^{(4)}(1)\right| = 21e$.

$$|E_8| \le \frac{21e}{737,280} \approx 0.000077$$

55. (a) $|E_n| \le \dfrac{M(b-a)^3}{12n^2} < 0.00005$

$$\frac{M(1-0.5)^3}{12n^2} < 0.00005$$

$$n^2 > \frac{0.125M}{12(0.00005)}$$

$$n^2 > 208.33333M$$

$$f(x) = e^{-1.1x}$$

$$f'(x) = -1.1e^{-1.1x}$$

$$f''(x) = 1.21e^{-1.1x}$$

Since $f''(x)$ is always decreasing for $0.5 \le x \le 1$ but greater than zero. The maximum value of

$$|f''(x)| = 1.21e^{1.1(0.5)}$$

$$= 1.21e^{-0.55} \approx 0.69811.$$

So, $n^2 > 145.439$

$$n > 12.0598, \text{ or } n = 13.$$

(b) $|E_n| \le \dfrac{M(b-a)^5}{180n^4} < 0.00005$

$$\dfrac{M(1-0.5)^5}{180n^4} < 0.00005$$

$$n^4 > \dfrac{0.03125M}{180(0.00005)}$$

$$n^4 > 3.4722M$$

$f^{(3)}(x) = -1.331e^{-1.1x}$

$f^{(4)}(x) = -1.4641e^{-1.1x}$

which again is always decreasing for $0.5 \le x \le 1$ and greater than zero. So the maximum value of

$$\left| f^{(4)}(x) \right| = 1.4641e^{-1.1(0.5)} \approx 0.8447.$$

So, $n^4 > (3.4722)(0.8447) \approx 2.9330$

$$n > 1.30867, \text{ or } n = 2$$

57. To use the graphing utility to find where the curves intersect, and then find the area region bounded by the curves,

Press $\boxed{y=}$ and input $-x\wedge 3 - 2x^2 + 5x - 2$ for $y_1 =$ and input $x \ln(x)$ for $y_2 =$.

Use window dimensions $[-4, 3]0.5$ by $[-0.8, 0.4]0.1$
Press $\boxed{\text{graph}}$.

Use trace and zoom to find the points of intersection or use the intersect function under the calc menu to find that $(0.406, -0.37)$ and $(1, 0)$ are the two points of intersection.

To find the area bounded by the curves, we must find

$$\int_{0.406}^{1} (-x^3 - 2x^2 + 5x - 2 - x \ln x)dx$$

Use the $\int f(x)\,dx$ function under the calc menu (making sure that y_1 is shown in the upper left corner) with $x = 0.406$ as the lower limit and $x = 1$ as the upper limit to

find that $\int_{0.406}^{1} (-x^3 - 2x^2 + 5x - 2)dx$
$$= 0.03465167$$

Repeat this process with y_2 activated to

find that $\int_{0.406}^{1} x \ln x\, dx \approx -.1344992.$

The area is
$0.03465167 - (-0.1344992) \approx 0.1692.$
Alternatively, you can use fnInt function under the math menu:
fnInt$(y_1 - y_2, x, 0.406, 1)$

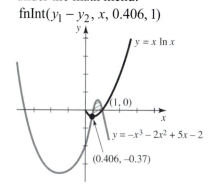

59. To use numeric integration feature to evaluate the integral,

Press $\boxed{y=}$ and input $\dfrac{2+3x}{(9-x^2)}$ for $y_1 =$.

Use window dimensions $[-5, 5]1$ by $[-5, 5]1$.
Press $\boxed{\text{graph}}$.
Use the $\int f(x)\,dx$ function under the calc menu with $x = -1$ as the lower limit and $x = 1$ as the upper limit to find

$$\int_{-1}^{1} \dfrac{2+3x}{9-x^2}\,dx \approx 0.4621.$$

61. To use numeric integration feature to compute the integral,

Press $\boxed{y=}$ and enter $\left(\dfrac{1}{\sqrt{\pi}}\right) * e\wedge(-x^2)$ for

$y_1 =$.
Use window dimensions $[-50, 50]20$ by $[-3, 3]1$.
Press $\boxed{\text{graph}}$.
Use the $\int f(x)\,dx$ function under the calc menu with $x = 0$ as the lower limit and $x = 1$ as the upper limit to find

$$\int_{0}^{1} \dfrac{1}{\sqrt{\pi}} e^{-x^2}\,dx = 0.4214$$

Repeat this process with $x = 10$ as the

upper limit to find $\int_{0}^{10} \dfrac{1}{\sqrt{\pi}} e^{-x^2}\,dx = 0.5$

Repeat this process with $x = 50$ as the upper limit to find $\int_0^{50} \frac{1}{\sqrt{\pi}} e^{-x^2} \, dx = 0.5$

The improper integral $\int_0^{\infty} \frac{1}{\sqrt{\pi}} e^{-x^2}$

appears to 0.5.

63.

$$\frac{dS}{dt} = \frac{aS}{b + cS + S^2}$$

$$\int \frac{S^2 + cS + b}{aS} \, dS = \int dt$$

$$\int \left(\frac{1}{a} S + \frac{c}{a} + \frac{b}{a} \frac{1}{S} \right) dS = \int dt$$

$$\frac{1}{2a} S^2 + \frac{c}{a} S + \frac{b}{a} \ln S = t + c$$

Chapter 7

Calculus of Several Variables

7.1 Functions of Several Variables

1. $f(x, y) = 5x + 3y$
$f(-1, 2) = 5(-1) + 3(2) = 1$
$f(3, 0) = 5(3) + 3(0) = 15$

3. $g(x, y) = x(y - x^3)$
$g(1, 1) = 1(1 - (1)^3) = 0$
$g(-1, 4) = -1(4 - (-1)^3) = -5$

5. $f(x, y) = (x - 1)^2 + 2xy^3$
$f(2, -1) = (2 - 1)^2 + 2(2)(-1)^3 = -3$
$f(1, 2) = (1 - 1)^2 + 2(1)(2)^3 = 16$

7. $g(x, y) = \sqrt{y^2 - x^2}$
$g(4, 5) = \sqrt{5^2 - 4^2} = \sqrt{9} = 3$
$g(-1, 2) = \sqrt{2^2 - (-1)^2} = \sqrt{3} \approx 1.732$

9. $f(r, s) = \dfrac{s}{\ln r}$
$f(e^2, 3) = \dfrac{3}{\ln e^2} = \dfrac{3}{2}$
$f(\ln 9, e^3) = \dfrac{e^3}{\ln(\ln 9)} \approx 25.515$

11. $g(x, y) = \dfrac{y}{x} + \dfrac{x}{y}$
$g(1, 2) = \dfrac{2}{1} + \dfrac{1}{2} = \dfrac{5}{2}$
$g(2, -3) = -\dfrac{3}{2} + -\dfrac{2}{3} = -\dfrac{13}{16} \approx -2.167$

13. $f(x, y, z) = xyz$
$f(1, 2, 3) = (1)(2)(3) = 6$
$f(3, 2, 1) = (3)(2)(1) = 6$

15. $F(r, s, t) = \dfrac{\ln(r + t)}{r + s + t}$
$f(1, 1, 1) = \dfrac{\ln(2)}{3} \approx 0.2310$

$f(0, e^2, 3e^2) = \dfrac{\ln(3e^2)}{4e^2}$
$= \dfrac{2 + \ln 3}{4e^2} \approx 0.1048$

17. $f(x, y) = \dfrac{5x + 2y}{4x + 3y}$
The domain of f is the set of all real pairs (x, y) such that $4x + 3y \neq 0$, or $y \neq -\dfrac{4}{3}x$.

19. $f(x, y) = \sqrt{x^2 - y}$
The domain of f is the set of all real pairs (x, y) such that $x^2 - y \geq 0$, or $y \leq x^2$.

21. $f(x, y) = \ln(x + y - 4)$
The domain of f is the set of all real pairs (x, y) such that $x + y - 4 > 0$, or $y > 4 - x$.

23. $f(x, y) = x + 2y$
With $C = 1$, $C = 2$, and $C = -3$, the three sketched level curves have equations $x + 2y = 1$, $x + 2y = 2$, and $x + 2y = -3$.

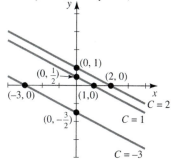

25. $f(x, y) = x^2 - 4x - y$
With $C = -4$, and $C = 5$, the two sketched

level curves have equations

$$x^2 - 4x - y = -4 \quad \text{and} \quad x^2 - 4x - y = 5.$$

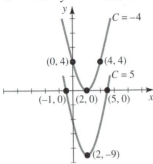

27. $f(x, y) = xy$

With $C = 1$, $C = -1$, $C = 2$, and $C = -2$, the four sketched level curves have equations $xy = 1$, $xy = -1$, $xy = 2$, and $xy = -2$.

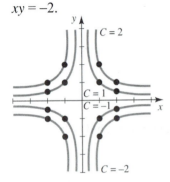

29. $f(x, y) = xe^y$

With $C = 1$, and $C = e$, the two sketched level curves have equations $xe^y = 1$ and $xe^y = e$.

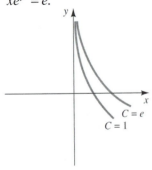

31. (a) $Q(x, y) = 10x^2 y$ and $x = 20$, $y = 40$.

$$Q(20, 40) = 10(20)^2 (40)$$
$$= 160,000 \text{ units.}$$

(b) With one more skilled worker, $x = 21$ and the additional output is
$$Q(21, 40) - Q(20, 40) = 16,400 \text{ units.}$$

(c) With one more unskilled worker, $y = 41$ and the additional output is
$$Q(20, 41) - Q(20, 40) = 4,000 \text{ units.}$$

(d) With one more skilled worker and one more unskilled worker, $x = 21$ and $y = 41$, so the additional output is
$$Q(21, 41) - Q(20, 40) = 20,810 \text{ units.}$$

33. (a) Let R denote the total monthly revenue. Then
$$R = \text{(revenue from the first brand)}$$
$$+ \text{(revenue from the second brand)}$$
$$= x_1 D_1(x_1, x_2) + x_2 D_2(x_1, x_2).$$
So,
$$R(x_1, x_2) = x_1(200 - 10x_1 + 20x_2)$$
$$+ x_2(100 + 5x_1 - 10x_2)$$
$$= 200x_1 - 10x_1^2 + 25x_1 x_2$$
$$+ 100x_2 - 10x_2^2.$$

(b) If $x_1 = 21$ and $x_2 = 16$, then
$$R(21, 16)$$
$$= 200(21) - 10(21)^2 + 25(21)(16)$$
$$+ 100(16) - 10(16)^2$$
$$= \$7,230$$

35. $f(x, y) = Ax^a y^b$
$$f(2x, 2y) = A(2x)^a (2y)^b$$
$$= A(2)^a x^a (2)^b y^b$$
$$= (2^{a+b}) Ax^a y^b$$
$x \geq 0$, $y \geq 0$, and $A > 0$

(a) If $a + b > 1$, $2^{a+b} > 2$ and f more than doubles.

(b) If $a + b < 1$, $2^{a+b} < 2$ and f increases but does not double.

(c) If $a + b = 1$, $2^{a+b} = 2$ and f doubles (exactly).

37. Let R denote the manufacturer's revenue. Then

$R = $ (revenue from domestic sales)
$\quad\quad + $ (revenue from sales abroad)

$$R(x, y) = x\left(60 - \frac{x}{5} + \frac{y}{20}\right)$$

$$+ y\left(50 - \frac{y}{10} + \frac{x}{20}\right)$$

$$= 60x + 50y - \frac{x^2}{5} - \frac{y^2}{10} + \frac{xy}{10}.$$

39. (a) $Q(10, 20) = 30 + 40 = 70$ units

(b) $3x + 2y = 70$ or $y = -\dfrac{3}{2}x + 35$

(c)

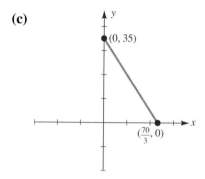

(d)
$$70 = 3 \cdot (12) + 2(20 + \Delta y)$$
$$2\Delta y = 70 - 36 - 40$$
$$\Delta y = -\frac{6}{2} = -3,$$

or decrease unskilled labor by 3 workers.

41. $U(25, 8) = (25 + 1)(8 + 2) = 260$

The indifference curve shows all combinations of close friends and interesting projects that lead to the same satisfaction for Beverly. So, you could say she is "indifferent" to change whenever the combination stays the same.

43. $M(A, n, i) = \dfrac{Ai}{1 - (1 + i)^{-12n}}$

(a) $M\left(250000, 15, \dfrac{0.052}{12}\right)$

$$= \frac{250,000\left(\frac{0.052}{12}\right)}{1 - \left(1 + \frac{0.052}{12}\right)^{-12(15)}} \approx \$2,003.13$$

The total amount paid is
$(2003.13)(12)(15) = \$360,563.40$
Since the original loan is for
$\$250,000$, the interest paid is
$360,563.4 - 250,000 = \$110,563.40.$

(b) $M\left(250000, 30, \dfrac{0.056}{12}\right)$

$$= \frac{250,000\left(\frac{0.056}{12}\right)}{1 - \left(1 + \frac{0.056}{12}\right)^{-12(30)}} \approx \$1,435.20$$

The total amount paid is
$(1435.20)(12)(30) = \$516,672$
Since the original loan is for
$\$250,000$, the interest paid is
$516,672 - 250,000 = \$266,672.$

45. $Q(K, L) = A[\alpha K^{-\beta} + (1 - \alpha)L^{-\beta}]^{-1/\beta}$
$Q(sK, sL)$
$= A[\alpha(sK)^{-\beta} + (1 - \alpha)(sL)^{-\beta}]^{-1/\beta}$
$= A[\alpha s^{-\beta}K^{-\beta} + (1 - \alpha)s^{-\beta}L^{-\beta}]^{-1/\beta}$
$= A(s^{-\beta})^{-1/\beta}[\alpha K^{-\beta} + (1 - \alpha)L^{-\beta}]^{-1/\beta}$
$= sA[\alpha K^{-\beta} + (1 - \alpha)L^{-\beta}]^{-1/\beta}$
$= sQ(K, L)$

47. (a) $S(15.83, 87.11)$
$= 0.0072(15.83 \,\hat{}\, 0.425)$
$\quad\quad\quad\quad (87.11 \,\hat{}\, 0.725)$
Input into home screen to find
$S(15.83, 87.11) \approx 0.5938.$
To sketch several additional level curves of $S(W, H)$, we will use the list feature of the calculator.
In general, $0.0072W^{0.425}H^{0.725} = S$

$$H^{0.725} = \frac{S}{0.0072} W^{-0.425}$$

$$H = \left(\frac{S}{0.0072} W^{-0.425} \right)^{1/0.725}$$

We will use $S = 0.3$, 0.5938, and 1.5.
Press $\boxed{y =}$.
Input $((L_1 / 0.0072) * x \wedge (-0.425))$
$\wedge (1 / 0.725)$ for $y_1 =$.
From the home screen, input
$\{0.3, 0.5938, 1.5\}$
$\boxed{STO \rightarrow}$ $\boxed{2nd}$ L_1.
Use window dimensions [0, 400]50 by
[0, 150]25.
Press $\boxed{graph}$.
Different combinations of height and
weight that result in the same surface
area.

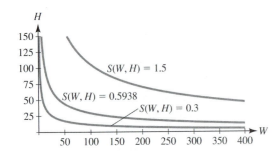

(b) $0.648 = 0.0072(18.37)^{0.425} H^{0.725}$,

$H^{0.725} = 26.121$, $H = 90.05$ cm.

(c) Let W_0, H_0 be Jenny's weight and
height at birth. Then,
$$S(W_0, H_0) = 0.0072 W_0^{0.425} H_0^{0.725}$$
When $W = 6W_0$ and $H = 2H_0$,
$$S(6W_0, 2H_0)$$
$$= 0.0072(6W_0)^{0.425}(2H_0)^{0.725}$$
$$= 0.0072(6)^{0.425} W_0^{0.425} (2)^{0.725} H_0^{0.725}$$
$$\approx 3.53966 S(W_0, H_0)$$

The % change in surface area is:
$$100 \frac{3.53966 S(W_0, H_0) - S(W_0, H_0)}{S(W_0, H_0)}$$
$$= 100 \frac{2.53966 S(W_0, H_0)}{S(W_0, H_0)}$$
$$\approx 253.97\% \text{ increase.}$$

(d) Writing Exercise—Answers will vary.

49. (a) $B_m(90, 190, 22)$
$$= 66.47 + 13.75(90)$$
$$+ 5.00(190) - 6.77(22)$$
$$= 2,105.03 \text{ kilocalories}$$

(b) $B_f(61, 170, 27)$
$$= 655.10 + 9.60(61) + 1.85(170)$$
$$- 4.68(27)$$
$$= 1,428.84 \text{ kilocalories.}$$

(c) $B_m(85, 193, A) = 66.47 + 13.75(85)$
$$+ 5.00(193) - 6.77A$$
$$2,018 = 2,200.22 - 6.77A$$
$$A \approx 26.9 \text{ years old}$$

(d) $B_f(67, 173, A) = 655.10 + 9.60(67)$
$$+ 1.85(173) - 4.68A$$
$$1,504 = 1,618.35 - 4.68A$$
$$A \approx 24.4 \text{ years old}$$

51. (a) $V(3,875, 1.675, 0.004)$
$$= \frac{9.3(3,875)}{1.675}[(0.0075)^2 - (0.004)^2]$$
$$\approx 0.866 \text{ cm/sec}$$

(b) For the fixed values of L and R,
$$V(P, r) = \frac{9.3P}{1.675}[(0.0075)^2 - r^2]$$
$$= 5.55P(0.0000563 - r^2)$$
To sketch several level curves of V,
set $V(P, r) = C$ for several values of C
and solve for P. We will use the list
feature of the calculator with $C = 100$,
200 and 300. Setting $V(P, r) = C$
$$5.55(0.0000563 - r^2) = C$$

In general, $P = \dfrac{0.1802C}{0.0000563 - r^2}$

Press $\boxed{y =}$.

Input $(0.1802L_1)/(0.0000563 - x^2)$

for $y_1 =$.

From the home screen, enter

$\{100, 200, 300\} \boxed{STO \rightarrow} \boxed{2nd} L_1$.

Use z-standard function under the zoom menu for the standard window dimensions.

Press $\boxed{graph}$.

Note that there are vertical asymptotes when $r = \pm\sqrt{0.0000563}$ but that the graph is defined in between these asymptotes as well.

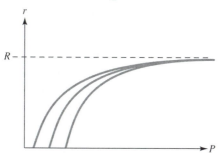

The curves represent different combinations of pressure and distance from the axis that result in the same speed.

53. $P(2, 0.53, 23)$

$= 0.075(2)(0.53)(273.15 + 23)$

≈ 23.54 atmospheres

55. (a) To sketch graphs of several level curves, for simplicity's sake, we will choose $a = b = 1$. We use the list feature of the calculator to sketch level curves for $T(P, V) = C$ for

$C = -100, 0, 100$.

In general,

$$0.0122\left(P + \frac{1}{V^2}\right)(V - 1) - 273.15 = C$$

and $P = \dfrac{C + 273.15}{0.0122(V - 1)} - \dfrac{1}{V^2}$.

Press $\boxed{y =}$.

Input

$(L_1 + 273.15)/(0.0122(x - 1)) - \left(\dfrac{1}{x^2}\right)$

for $y_1 =$.

From the home screen, enter

$\{-100, 0, 100\} \boxed{Sto \rightarrow} \boxed{2nd} L_1$.

Use window dimensions

$[0, 35000]5{,}000$ by $[0, 2.9]0.3$.

Press $\boxed{graph}$.

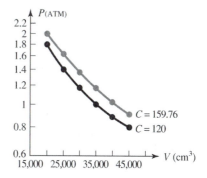

(b) To find $T(1.13, 31.275 \times 10^3)$,

From the home screen, enter

$0.0122(1.13 + (6.49 \times 10^6)/31{,}275^2)$

$(31{,}275 - 56.2) - 273.15 \approx 159.76$.

Thus, the temperature is $159.76°C$.

7.2 Partial Derivatives

1. $f(x, y) = 7x - 3y + 4$

$f_x = 7 \qquad f_y = -3$

3. $f(x, y) = 4x^3 - 3x^2 y + 5x$

$f_x = 12x^2 - 6xy + 5$

$f_y = -3x^2$

5. $f(x, y) = 2xy^5 + 3x^2 y + x^2$

$f_x = 2y^5 + 6xy + 2x$

$f_y = 2x(5y^4) + 3x^2 = 10xy^4 + 3x^2$

7. $z = (3x + 2y)^5$

$$\frac{\partial z}{\partial x} = 5(3x + 2y)^4 \frac{\partial}{\partial x}(3x + 2y)$$

$$= 15(3x + 2y)^4$$

$$\frac{\partial z}{\partial y} = 5(3x + 2y)^4 \frac{\partial}{\partial y}(3x + 2y)$$

$$= 10(3x + 2y)^4$$

9. $f(s, t) = \dfrac{3t}{2s} = \dfrac{3}{2}s^{-1}t$

$$f_s = \frac{3}{2}(-1)s^{-2}t = -\frac{3t}{2s^2}$$

$$f_t = \frac{3}{2}s^{-1} = \frac{3}{2s}$$

11. $z = xe^{xy}$

$$\frac{\partial z}{\partial x} = x(ye^{xy}) + e^{xy}(1) = (xy + 1)e^{xy}$$

$$\frac{\partial z}{\partial y} = x(e^{xy})(x) = x^2 e^{xy}$$

13. $f(x, y) = \dfrac{e^{2-x}}{y^2} = e^{2-x}y^{-2}$

$$f_x = -e^{2-x}y^{-2} = -\frac{e^{2-x}}{y^2}$$

$$f_y = e^{2-x}(-2y^{-3}) = -\frac{2e^{2-x}}{y^3}$$

15. $f(x, y) = \dfrac{2x + 3y}{y - x}$

$$f_x = \frac{(y - x)(2) - (2x + 3y)(-1)}{(y - x)^2}$$

$$= \frac{5y}{(y - x)^2}$$

$$f_y = \frac{(y - x)(3) - (2x + 3y)(1)}{(y - x)^2} = -\frac{5x}{(y - x)^2}$$

17. $z = u \ln v$

$$\frac{\partial z}{\partial u} = (1)\ln v = \ln v$$

$$\frac{\partial z}{\partial v} = u\left(\frac{1}{v}\right) = \frac{u}{v}$$

19. $f(x, y) = \dfrac{\ln(x + 2y)}{y^2}$

$$f_x = \frac{(y^2)\left[\frac{1}{(x+2y)}\right] - \ln(x + 2y)(0)}{y^4}$$

$$= \frac{1}{y^2(x + 2y)}$$

$$f_y = \frac{(y^2)\left[\frac{2}{(x+2y)}\right] - \ln(x + 2y)(2y)}{y^4}$$

$$= \frac{(y)(2) - (x + 2y)\ln(x + 2y)(2)}{(x + 2y)y^3}$$

$$= \frac{2[y - (x + 2y)\ln(x + 2y)]}{y^3(x + 2y)}$$

21. $f(x, y) = x^2 + 3y$

$$f_x(x, y) = 2x \qquad f_x(1, -1) = 2(1) = 2$$

$$f_y(x, y) = 3 \qquad f_y(1, -1) = 3$$

23. $f(x, y) = \dfrac{y}{2x + y} = y(2x + y)^{-1}$

$$f_x(x, y) = -y(2x + y)^{-2}(2) = -\frac{2y}{(2x + y)^2}$$

$$f_x(0, -1) = -\frac{2(-1)}{(2(0) + -1)^2} = 2$$

$$f_y(x, y) = \frac{(2x + y)(1) - (y)(1)}{(2x + y)^2} = \frac{2x}{(2x + y)^2}$$

$$f_y(0, -1) = \frac{2(0)}{(2(0) + (-1))^2} = 0$$

25. $f(x, y) = 3x^2 - 7xy + 5y^3 - 3(x + y) - 1$

$$f_x = 6x - 7y - 3$$

$$f_y = -7x + 15y^2 - 3$$

$f_x(-2, 1) = -12 - 7 - 3 = -22$

$f_y(-2, 1) = 14 + 15 - 3 = 26$

27. $f(x, y) = xe^{-2y} + ye^{-x} + xy^2$

$f_x = e^{-2y} - ye^{-x} + y^2$

$f_y = -2xe^{-2y} + e^{-x} + 2xy$

$f_x(0, 0) = 1 - 0 + 0 = 1$

$f_y(0, 0) = 0 + 1 + 0 = 1$

29. $f(x, y) = 5x^4y^3 + 2xy$

$f_x = 5(4x^3)y^3 + 2y = 20x^3y^3 + 2y$

$f_y = 5x^4(3y^2) + 2x = 15x^4y^2 + 2x$

$f_{xx} = \dfrac{\partial}{\partial x}(f_x) = 20(3x^2)y^3 + 0 = 60x^2y^3$

$f_{yy} = \dfrac{\partial}{\partial y}(f_y) = 15x^4(2y) + 0 = 30x^4y$

$f_{xy} = \dfrac{\partial}{\partial y}(f_x)$

$\quad = 20x^3(3y^2) + 2(1)$

$\quad = 60x^3y^2 + 2$

$f_{yx} = \dfrac{\partial}{\partial x}(f_y)$

$\quad = 15(4x^3)y^2 + 2(1)$

$\quad = 60x^3y^2 + 2$

$\quad = f_{xy}$

31. $f(x, y) = e^{x^2y}$

$f_x = 2xye^{x^2y}$ and $f_y = x^2e^{x^2y}$

$f_{xx} = \dfrac{\partial}{\partial x}(f_x)$

$\quad = 2xy(e^{x^2y})(2xy) + e^{x^2y}(2y)$

$\quad = 2y(2x^2y + 1)e^{x^2y}$

$f_{yy} = \dfrac{\partial}{\partial y}(f_y) = x^2(e^{x^2y})(x^2) = x^4e^{x^2y}$

$f_{xy} = \dfrac{\partial}{\partial y}(f_x)$

$\quad = 2xy(e^{x^2y})(x^2) + e^{x^2y}(2x)$

$\quad = 2x(x^2y + 1)e^{x^2y}$

$f_{yx} = \dfrac{\partial}{\partial x}(f_y)$

$\quad = x^2(e^{x^2y})(2xy) + e^{x^2y}(2x)$

$\quad = 2x(x^2y + 1)e^{x^2y}$

$\quad = f_{xy}$

33. $f(s, t) = \sqrt{s^2 + t^2} = (s^2 + t^2)^{1/2}$

$f_s = \dfrac{1}{2}(s^2 + t^2)^{-1/2}(2s) = s(s^2 + t^2)^{-1/2}$

$f_t = \dfrac{1}{2}(s^2 + t^2)^{-1/2}(2t) = t(s^2 + t^2)^{-1/2}$

$f_{ss} = s\left[-\dfrac{1}{2}(s^2 + t^2)^{-3/2}(2s)\right]$

$\qquad\quad + (s^2 + t^2)^{-1/2}(1)$

$\quad = \dfrac{-s^2}{(s^2 + t^2)^{3/2}} + \dfrac{1}{(s^2 + t^2)^{1/2}}\dfrac{(s^2 + t^2)}{(s^2 + t^2)}$

$\quad = \dfrac{t^2}{(s^2 + t^2)^{3/2}}$

$f_{tt} = t\left[-\dfrac{1}{2}(s^2 + t^2)^{-3/2}(2t)\right]$

$\qquad\quad + (s^2 + t^2)^{-1/2}(1)$

$\quad = \dfrac{s^2}{(s^2 + t^2)^{3/2}}$

$$f_{st} = \frac{\partial}{\partial t}(f_s)$$

$$= s\left[-\frac{1}{2}(s^2 + t^2)^{-3/2}(2t)\right]$$

$$= \frac{-st}{(s^2 + t^2)^{3/2}}$$

$$f_{ts} = \frac{\partial}{\partial s}(f_t)$$

$$= t\left[-\frac{1}{2}(s^2 + t^2)^{-3/2}(2s)\right]$$

$$= \frac{-st}{(s^2 + t^2)^{3/2}}$$

$$= f_{st}$$

35. $z = 2x + 3y;\ x = t^2;\ y = 5t$

$$\frac{dz}{dt} = \frac{\partial z}{\partial x} \cdot \frac{dx}{dt} + \frac{\partial z}{\partial y} \cdot \frac{dy}{dt}$$

$$= (2)(2t) + (3)(5)$$

$$= 4t + 15$$

37. $z = \dfrac{3x}{y};\ x = t;\ y = t^2$

$$\frac{dz}{dt} = \frac{\partial z}{\partial x} \cdot \frac{dx}{dt} + \frac{\partial z}{\partial y} \cdot \frac{dy}{dt}$$

$$= \left(\frac{3}{y}\right)(1) + \left(-\frac{3x}{y^2}\right)(2t)$$

$$= \frac{3}{y} - \frac{6xt}{y^2}$$

39. $z = xy;\ x = e^{2t};\ y = e^{-3t}$

$$\frac{dz}{dt} = \frac{\partial z}{\partial x} \cdot \frac{dx}{dt} + \frac{\partial z}{\partial y} \cdot \frac{dy}{dt}$$

$$= (y)(2e^{2t}) + (x)(-3e^{-3t})$$

$$= 2ye^{2t} - 3xe^{-3t}$$

41. $D_1(p_1, p_2) = 500 - 6p_1 + 5p_2$
$D_2(p_1, p_2) = 200 + 2p_1 - 5p_2$

$$\frac{\partial D_1}{\partial p_2} = 5 \text{ and } \frac{\partial D}{\partial p_1} = 2$$

Since both partial derivatives are positive for all p_1 and p_2, the commodities are substitute commodities.

43. $D_1(p_1, p_2) = 3,000 + \dfrac{400}{p_1 + 3} + 50p_2$

$$D_2(p_1, p_2) = 2,000 - 100p_1 + \frac{500}{p_2 + 4}$$

$$\frac{\partial D_1}{\partial p_2} = 50 \text{ and } \frac{\partial D_2}{\partial p_1} = -100$$

Since the partial derivatives are opposite in sign for all p_1 and p_2, the commodities are neither substitute nor complementary.

45. $D_1(p_1, p_2) = \dfrac{7p_2}{1 + p_1^2}$

$$D_2(p_1, p_2) = \frac{p_1}{1 + p_2^2}$$

$$\frac{\partial D_1}{\partial p_2} = \frac{7}{1 + p_1^2} > 0 \text{ and } \frac{\partial D_2}{\partial p_1} = \frac{1}{1 + p_2^2}$$

Since both partial derivatives are positive for all p_1 and p_2, the commodities are substitute commodities.

47. The partial derivative

$$Q_K = \frac{\partial Q}{\partial K} = 30K^{-1/2}L^{1/3} = \frac{30L^{1/3}}{K^{1/2}}$$

is the rate of change of the output with respect to the capital investment. This is an approximation to the additional number of units that will be produced each week if the capital investment is increased from K to $K + 1$ while the size of the labor force is not changed. In particular, if the capital investment K is increased from 900 (thousand) to 901 (thousand) and the size of the labor force is $L = 1,000$, the resulting change in output is

$$\Delta Q = Q_K(900, 1000)$$
$$= \frac{30(1,000)^{1/3}}{(900)^{1/2}}$$
$$= \frac{30(10)}{30}$$
$$= 10,$$

or daily output will increase by 10 units.

49. $Q(K, L) = 150[0.4K^{-1/2} + 0.6L^{-1/2}]^{-2}$

(a) $Q_K = -300[0.4K^{-1/2} + 0.6L^{-1/2}]^{-3}$
$$(-0.2K^{-3/2})$$
$$= 60k^{-3/2}[0.4K^{-1/2} + 0.6L^{-1/2}]^{-3}$$

$Q_L = -300[0.4K^{-1/2} + 0.6L^{-1/2}]^{-3}$
$$(-0.3K^{-3/2})$$
$$= 90L^{-3/2}[0.4K^{-1/2} + 0.6L^{-1/2}]^{-3}$$

(b) $Q_K(5041, 4900)$
$$= 60(5041)^{-3/2}[0.4(5041)^{-1/2}$$
$$+ 0.6(4900)^{-1/2}]^{-3}$$
$$= \frac{60}{\left(\sqrt{5041}\right)^3}\left[\frac{0.4}{\sqrt{5041}} + \frac{0.6}{\sqrt{4900}}\right]^{-3}$$
$$\approx 58.48$$
$Q_L(5041, 4900)$
$$= 90(4900)^{-3/2}[0.4(5041)^{-1/2}$$
$$+ 0.6(4900)^{-1/2}]^{-3}$$
$$= \frac{90}{\left(\sqrt{4900}\right)^3}\left[\frac{0.4}{\sqrt{5041}} + \frac{0.6}{\sqrt{4900}}\right]^{-3}$$
$$\approx 91.54$$

(c) additional labor employment

51. $F(x, y) = 200 - 24\sqrt{x} + 4(0.1y + 3)^{3/2}$
$$\frac{\partial F}{\partial y} = 6(0.1y + 3)^{1/2}(0.1) = 0.6(0.1y + 3)^{1/2}$$

is the rate of change of demand with

respect to the price of gasoline. When the selling price is kept constant, $\Delta f \approx \frac{\partial F}{\partial y}\Delta y$.

Since $y = 380$ cents and $\Delta y = -1$,
$\Delta F \approx 0.6[0.1(380) + 3]^{1/2}(-1) \approx -3.84$ or demand will decrease by approximately 4 bicycles.

53. (a) If the price x of the first lawnmower increases, the demand for that same lawnmower should fall. If the price y of the second (competing) lawnmower increases, the demand for the first lawnmower should increase.

(b) $D_x < 0, D_y > 0$

(c) With $D = a + bx + cy$, $D_x = b < 0$ and $D_y = c > 0$.

55. $Q(K, L) = 120K^{1/2}L^{1/3}$

(a) $Q_L = 120K^{1/2}\left(\frac{1}{3}L^{-2/3}\right)$
$$= 40K^{1/2}L^{-2/3}$$
$$Q_{LL} = -\frac{80}{3}K^{1/2}L^{-5/3}$$

$Q_{LL} < 0$; for a fixed level of capital investment, the effect on output of the addition of one worker hour is greater when the work force is small, than when it is large.

(b) $Q_K = 60K^{-1/2}L^{1/3}$
$$Q_{KK} = -30K^{-3/2}L^{1/3}$$
$Q_{KK} < 0$; for a fixed work force, the effect on output of the addition of $1,000 in capital investment is greater when the capital investment is small, than when it is large.

57. $D_1(p_1, p_2) = 700 - 4p_1^2 + 7p_1p_2$
$$D_2(p_1, p_2) = 300 - 2\sqrt{p_2} + 5p_1p_2$$

$$\frac{\partial D_1}{\partial p_2} = 7p_1 \text{ and } \frac{\partial D_2}{\partial p_1} = 5p_2$$

Since $p_1 > 0$ and $p_2 > 0$,

$$\frac{\partial D_1}{\partial p_2} > 0 \text{ and } \frac{\partial D_2}{\partial p_1} > 0$$

So, these are substitute commodities and the second commodity is pencils.

59. (a) To store the output function,
Press $\boxed{y=}$ and input
$1,175x + 483L_1 + 3.1(x \wedge 2) * L_1$
$-1.2(x \wedge 3) - 2.7(L_1 \wedge 2)$ for $y_1 =$.
From the home screen, input $\{71\}$
$\boxed{sto\rightarrow}\,\boxed{2nd}\,L_1$.
Use window dimensions $[0, 400]25$ by $[0, 250]25$.
Use the value function under the calc menu and enter $x = 37$ to find $Q(37, 71) \approx 304,691$ units.
Use the value function again and enter $x = 38$ to find $Q(38, 71) \approx 317,310$ units.
From the home screen, input $\{72\}$
$\boxed{sto\rightarrow}\,\boxed{2nd}\,L_1$. Use the value function under the calc menu and enter $x = 37$ to find $Q(37, 72) \approx 309,031$ units.

(b) $Q_x(x, y) = 1,175 + 6.2xy - 3.6x^2$
To estimate the change in output when x is increased from 37 to 38 while y remains at 71, we find $Q_x(37, 71)$.
Press $\boxed{y=}$.
Input $1,175 + 6.2xL_1 - 3.6x^2$ in $y_2 =$.
Deactivate $y_1 =$ so that only y_2 is activated.
From the home screen, input $\{71\}$
$\boxed{sto\rightarrow}\,\boxed{2nd}\,L_1$. Use the value function under the calc menu and enter $x = 37$ and $Q_x(37, 71) = 12,534$ units. Thus, if the skilled workforce is increased from 37 to 38 and the unskilled remains constant at 71, the output is approximately increased by 12,534 units.

The actual change is
$Q(38, 71) - Q(37, 71)$
$= 317,310 - 304,691$
$= 12,619$ units.

(c) $Q_y(x, y) = 483 + 3.1x^2 - 5.4y$
To estimate the change in output when y is increased from 71 to 72 while x remains at 37, we find $Q_y(37, 71)$.
Press $\boxed{y=}$.
Input $483 + 3.1x^2 - 5.4L_1$ for $y_3 =$.
Deactivate $y_1 =$ and $y_2 =$ so only $y_3 =$ is activated.
From the home screen, input $\{71\}$
$\boxed{sto\rightarrow}\,\boxed{2nd}\,L_1$. Use the value function under the calc menu and enter $x = 37$ to find $Q_y(37, 71) \approx 4,344$ units.

Thus, if the unskilled workforce is increased from 71 to 72 and the skilled remains at 37, the output is approximately increased by 4,344 units. The actual change is
$Q(37, 72) - Q(37, 71)$
$= 309,031 - 304,691$
$= 4,340$ units.

61. $Q(x, y) = 10xy^{1/2}$

$$\Delta Q \approx \frac{\partial Q}{\partial x}\Delta x + \frac{\partial Q}{\partial y}\Delta y$$

$$= (10y^{1/2})\Delta x + \left(\frac{5x}{y^{1/2}}\right)\Delta y$$

$$= \left(10\sqrt{36}\right)(-3) + \left(\frac{5 \cdot 30}{\sqrt{36}}\right)(5)$$

$$= -55$$

or the number of units produced will decrease by 55.

63. $Q(x, y) = 200 - 10x^2 + 20xy$
$x(t) = 10 + 0.5t$
$y(t) = 12.8 + 0.2t^2$

(a) $\dfrac{dQ}{dt} = \dfrac{\partial Q}{\partial x} \cdot \dfrac{dx}{dt} + \dfrac{\partial Q}{\partial y} \cdot \dfrac{dy}{dt}$

$= (-20x + 20y)(0.5) + (20x)(0.4t)$

When $t = 4$, $x(4) = 12$ and $y(4) = 16$. So,

$\dfrac{dQ}{dt} = [-20(12) + 20(16)](0.5)$

$+ [20(12)][0.4(4)]$

$= 424$ units per month/month

(b) When $t = 4$, $Q(12, 16) = 2{,}600$ so

$100 \dfrac{Q'(t)}{Q(t)} = 100 \dfrac{424}{2{,}600} \approx 16.31\%.$

65. $Q(x, y) = 0.08x^2 + 0.12xy + 0.03y^2$

$\Delta Q \approx \dfrac{\partial Q}{\partial x} \Delta x + \dfrac{\partial Q}{\partial y} \Delta y$

$= (0.16x + 0.12y)\Delta x$

$\qquad + (0.12x + 0.06y)\Delta y$

$= [0.16(80) + 0.12(200)](0.5)$

$\qquad + [0.12(80) + 0.06(200)](2)$

$= 61.6$

or an increase of 61.6 units produced per day.

67. $P(x, y) = (x - 40)(55 - 4x + 5y)$

$\qquad + (y - 45)(70 + 5x - 7y)$

(a) $P_x = (x - 40)(-4) + (55 - 4x + 5y)(1)$

$\qquad + (y - 45)(5)$

$= -4x + 160 + 55 - 4x + 5y$

$\qquad + 5y - 225$

$= -8x + 10y - 10$

$P_y = (x - 40)(5) + (y - 45)(-7)$

$\qquad + (70 + 5x - 7y)(1)$

$= 5x - 200 - 7y + 315 + 70$

$\qquad + 5x - 7y$

$= 10x - 14y + 185$

(b) $P_x(70, 73) = -8(70) + 10(73) - 10$

$= 160$

$P_y(70, 73) = 10(70) - 14(73) + 185$

$= -137$

(c) $\Delta P \approx (P_x)(\Delta x) + (P_y)(\Delta y)$

$\approx (160)(1) + (-137)(2)$

$= -114,$

or the daily profit will decrease by 114 cents.

(d) $\Delta P \approx (160)(2) + (-137)(-1) = 457$ or the daily profit will increase by 457 cents.

69. $F(L, r) = \dfrac{kL}{r^4}$

(a) $F(3.17, 0.085) = 60{,}727.24k$

$\dfrac{\partial F}{\partial L} = \dfrac{k}{r^4} = 19{,}156.86k$

$\dfrac{\partial F}{\partial r} = -\dfrac{4kL}{r^5} = -2{,}857{,}752.58k$

(b) $F(1.2L, 0.8r) = \dfrac{k(1.2L)}{(0.8r)^4}$

$= 2.93F(L, r)$

$\dfrac{\partial F}{\partial L}(1.2L, 0.8r) = 2.44\dfrac{\partial F}{\partial L}(L, r)$

$\dfrac{\partial F}{\partial r}(1.2L, 0.8r) = 3.66\dfrac{\partial F}{\partial r}(L, r)$

71. $\dfrac{\partial F}{\partial z} = \dfrac{c\pi x^2}{4}\left[\dfrac{1}{2}(y - z)^{-1/2}(-1)\right]$

$= \dfrac{-c\pi x^2}{8\sqrt{y - z}}$

is the rate of change of blood flow with respect to the pressure in the capillary. Since this rate is negative, the blood flow is decreasing.

73. $z = x^2 - y^2$

$\dfrac{\partial z}{\partial x} = 2x$ and $\dfrac{\partial^2 z}{\partial x^2} = 2$

$\dfrac{\partial z}{\partial y} = -2y$ and $\dfrac{\partial^2 z}{\partial y^2} = -2$

Since $\dfrac{\partial^2 z}{\partial x^2} + \dfrac{\partial^2 z}{\partial y^2} = 0$ the function satisfies Laplace's equation.

75. $z = xe^y - ye^x$

$\dfrac{\partial z}{\partial x} = e^y - ye^x$ and $\dfrac{\partial^2 z}{\partial x^2} = -ye^x$

$\dfrac{\partial z}{\partial y} = xe^y - e^x$ and $\dfrac{\partial^2 z}{\partial y^2} = xe^y$

Since $\dfrac{\partial^2 z}{\partial x^2} + \dfrac{\partial^2 z}{\partial y^2} = -ye^x + xe^y \neq 0$ the

function does not satisfy Laplace's equation.

77. $\dfrac{\partial V}{\partial R} = 2\pi RH$ is the rate of change of the volume with respect to the radius. When the height is kept constant, $\Delta V \approx \dfrac{\partial V}{\partial R}\Delta R$

Since $R = 3$, $H = 12$, and $\Delta R = 1$,
$\Delta V \approx [2\pi(3)(12)](1) = 72\pi$, or an increase in volume of approximately 226 cubic cm.

79. $PV = nRT$
Solving for V,

$V = \dfrac{nRT}{P}$ and $\dfrac{\partial V}{\partial T} = \dfrac{nR}{P}$, or $= \dfrac{V}{T}$

Solving for T,

$T = \dfrac{PV}{nR}$ and $\dfrac{\partial T}{\partial P} = \dfrac{V}{nR}$, or $= \dfrac{T}{P}$

Solving for P,

$P = \dfrac{nRT}{V}$ and $\dfrac{\partial P}{\partial V} = -\dfrac{nRT}{V^2}$, or $-\dfrac{P}{V}$

So,

$\dfrac{\partial V}{\partial T} \cdot \dfrac{\partial T}{\partial P} \cdot \dfrac{\partial P}{\partial V} = \dfrac{V}{T} \cdot \dfrac{T}{P} \cdot \dfrac{-P}{V} = -1$

81. (a) Cost
= (area bottom)(cost per unit area)
+ (area top)(cost per unit area)
+ (area sides)(cost per unit area)
+ (volume)(cost per unit volume)

$C(R, H)$
$= 0.0005[\pi R^2 + \pi R^2 + 2\pi RH]$
$\qquad\qquad + 0.001(\pi R^2 H)$
$= 0.0005(2\pi)[R^2 + RH] + 0.001\pi R^2 H$
$= 0.001\pi[R^2 + RH + R^2 H]$

(b) $\Delta C \approx \dfrac{\partial C}{\partial R}\Delta R + \dfrac{\partial C}{\partial H}\Delta H$
$= [0.001\pi(2R + H + 2RH)]\Delta R$
$\qquad\qquad + [0.001\pi(R + R^2)]\Delta H$

When $R = 3$, $H = 12$, $\Delta R = 0.3$ and $\Delta H = -0.2$,
$\Delta C \approx [0.001\pi(2 \cdot 3 + 12 + 2 \cdot 3 \cdot 12)](0.3)$
$\qquad\qquad + [0.001\pi(3 + 3^2)](-0.2)$
≈ 0.0773
or an increase of 0.08 cents per scan.

83. To estimate the amount of material used, approximate the change in volume of the cylinder as its radius increases by 4 inches $\left(\dfrac{1}{3}\text{ ft}\right)$ and its height increases by 6 inches $\left(\dfrac{1}{2}\text{ ft}\right)$. Since the volume of a right circular cylinder is $V = \pi r^2 h$,

$\Delta V \approx \dfrac{\partial V}{\partial r}\Delta r + \dfrac{\partial V}{\partial h}\Delta h$

$\approx (2\pi r h)\Delta r + (\pi r^2)\Delta h$

$\approx 2\pi(6)(80)\left(\dfrac{1}{3}\right) + \pi(6)^2\left(\dfrac{1}{2}\right)$

$\approx 1,005.3 + 56.5$

$\approx 1,062$ cubic feet

85. $x^2 + xy + y^3 = 1$

$\dfrac{dy}{dx} = -\dfrac{f_x}{f_y} = -\dfrac{2x + y}{x + 3y^2}$

When $x = -1$ and $y = 1$, the slope is

$= -\dfrac{2(-1) + 1}{-1 + 3(1)^2} = \dfrac{1}{2}$

The equation of the tangent line is

$$y - 1 = \frac{1}{2}(x+1)$$

$$y = \frac{1}{2}x + \frac{3}{2}, \text{ or } x - 2y = -3$$

7.3 Optimizing Functions of Two Variables

1. $f(x, y) = 5 - x^2 - y^2$

$f_x = -2x$ so $f_x = 0$ when $x = 0$

$f_y = -2y$ so $f_y = 0$ when $y = 0$

and only critical point is $(0, 0)$.

$f_{xx} = -2, \ f_{yy} = -2, \ f_{xy} = 0$

$$D = f_{xx}f_{yy} - (f_{xy})^2$$

For the point $(0, 0)$,

$$D = (-2)(-2) - 0^2 > 0$$

$f_{xx} < 0$

So, $(0, 0)$ is a relative maximum.

3. $f(x, y) = xy$

$f_x = y$ and $f_x = 0$ when $y = 0$

$f_y = x$ and $f_y = 0$ when $x = 0$

and only critical point is $(0, 0)$.

$f_{xx} = 0, \ f_{yy} = 0, \ f_{xy} = 1$

For the point $(0, 0)$,

$$D = (0)(0) - (1)^2 < 0$$

So, $(0, 0)$ is a saddle point.

5. $f(x, y) = \dfrac{16}{x} + \dfrac{6}{y} + x^2 - 3y^2$

$$f_x = -\frac{16}{x^2} + 2x$$

So, $f_x = 0$ when $0 = -\dfrac{16}{x^2} + 2x$, or $x = 2$.

$$f_y = -\frac{6}{y^2} - 6y$$

So, $f_y = 0$ when $0 = -\dfrac{6}{y^2} - 6y$, or $y = -1$

and only critical point is $(2, -1)$.

$$f_{xx} = \frac{32}{x^3} + 2, \ f_{yy} = \frac{12}{y^3} - 6, \ f_{xy} = 0$$

For the point $(2, -1)$,

$$D = \left[\frac{32}{(2)^3} + 2\right]\left[\frac{12}{(-1)^3} - 6\right] - 0 < 0$$

So, $(2, -1)$ is a saddle point.

7. $f(x, y) = 2x^3 + y^3 + 3x^2 - 3y - 12x - 4$

$f_x = 6x^2 + 6x - 12 = 6(x+2)(x-1)$

So, $f_x = 0$ when $x = -2, 1$.

$f_y = 3y^2 - 3 = 3(y+1)(y-1)$

So, $f_y = 0$ when $y = -1, 1$ and the critical points are $(-2, -1)$, $(-2, 1)$, $(1, -1)$, and $(1, 1)$.

$f_{xx} = 12x + 6, \ f_{yy} = 6y, \ f_{xy} = 0$

For the point $(-2, -1)$,

$D = [12(-2) + 6][6(-1)] - 0 > 0$

and $f_{xx} < 0$, so $(-2, -1)$ is a relative maximum.

For the point $(-2, 1)$,

$D = [12(-2) + 6][6(1)] - 0 < 0$

So, $(-2, 1)$ is a saddle point.

For the point $(1, -1)$,

$D = [12(1) + 6][6(-1)] - 0 < 0$

So, $(1, -1)$ is a saddle point.

For the point $(1, 1)$,

$D = [12(1) + 6][6(1)] > 0$ and $f_{xx} > 0$, so $(1, 1)$ is a relative minimum.

9. $f(x, y) = x^3 + y^2 - 6xy + 9x + 5y + 2$

$f_x = 3x^2 - 6y + 9$

So, $f_x = 0$ when $0 = 3(x^2 - 2y + 3)$, or $0 = x^2 - 2y + 3$.

$f_y = 2y - 6x + 5$

So, $f_y = 0$ when $0 = 2y - 6x + 5$. Solving this system of equations by adding,

$0 = x^2 - 6x + 8 = (x-2)(x-4)$

So, $x = 2, 4$.

When $x = 2$, $0 = (2)^2 - 2y + 3$, or $y = \dfrac{7}{2}$.

When $x = 4$, $0 = (4)^2 - 2y + 3$, or $y = \dfrac{19}{2}$.

So, the critical points are $\left(2, \dfrac{7}{2}\right)$ and

$\left(4, \dfrac{19}{2}\right)$.

$f_{xx} = 6x$, $f_{yy} = 2$, $f_{xy} = -6$

For the point $\left(2, \dfrac{7}{2}\right)$,

$D = 6(2)(2) - (-6)^2 < 0$.

So, $\left(2, \dfrac{7}{2}\right)$ is a saddle point.

For the point $\left(4, \dfrac{19}{2}\right)$,

$D = 6(4)(2) - (-6)^2 > 0$ and $f_{xx} > 0$, so

$\left(4, \dfrac{19}{2}\right)$ is a relative minimum.

11. $f(x, y) = xy^2 - 6x^2 - 3y^2$

$f_x = y^2 - 12x$

So, $f_x = 0$ when $y^2 - 12x = 0$, or $x = \dfrac{y^2}{12}$

$f_y = 2xy - 6y$

So, $f_y = 0$ when

$$2xy - 6y = 0$$

$$2\left(\frac{y^2}{12}\right)y - 6y = 0$$

$$\frac{1}{6}y^3 - 6y = 0$$

$$y^3 - 36y = 0$$

$$y(y+6)(y-6) = 0$$

or $y = 0$, $y = -6$, $y = 6$.

When $y = 0$, $x = 0$; $y = -6$, $x = 3$; $y = 6$, $x = 3$.

So the critical points are $(0, 0)$, $(3, -6)$ and $(3, 6)$. Now, $f_{xx} = -12$, $f_{yy} = 2x - 6$,

$f_{xy} = 2y$.

For the point $(0, 0)$,

$D = (-12)(-6) - 0^2 > 0$.

Since $f_{xx} < 0$, the point $(0, 0)$ is a relative

maximum. For the point $(3, -6)$,

$D = (-12)(0) - [2(-6)]^2 < 0$.

So, the point $(3, -6)$ is a saddle point.
For the point $(3, 6)$,

$D = (-12)(0) - [2(6)]^2 < 0$

So, the point $(3, 6)$ is a saddle point.

13. $f(x, y) = (x^2 + 2y^2)e^{1-x^2-y^2}$

$f_x = (x^2 + 2y^2)(-2xe^{1-x^2-y^2})$

$\qquad\qquad + (e^{1-x^2-y^2})(2x)$

$\quad = -2xe^{1-x^2-y^2}(x^2 + 2y^2 - 1)$

So, $f_x = 0$ when $x = 0$ or $x^2 + 2y^2 - 1 = 0$

$f_y = (x^2 + 2y^2)(-2ye^{1-x^2-y^2})$

$\qquad\qquad + (e^{1-x^2-y^2})(4y)$

$\quad = -2ye^{1-x^2-y^2}(x^2 + 2y^2 - 2)$

So, $f_y = 0$ when $y = 0$, or $x^2 + 2y^2 - 2 = 0$.
There are no solutions to the system of
equations $x^2 + 2y^2 - 1 = 0$ and

$x^2 + 2y^2 - 2 = 0$. Further, when $x = 0$,

$f_y = 0$ when $0 = -2ye^{1-y^2}(2y^2 - 2)$, or

$y = 0, -1, 1$. When $y = 0$, $f_x = 0$ when

$0 = -2xe^{1-x^2}(x^2 - 1)$ or, $x = 0, -1, 1$.

So, the critical points are $(-1, 0)$, $(0, 0)$,
$(1, 0)$, $(0, -1)$ and $(0, 1)$.
Rewriting f_x as

$f_x = -2e^{1-x^2-y^2}(x^3 + 2xy^2 - x)$

$f_{xx} = -2[e^{1-x^2-y^2}(3x^2 + 2y^2 - 1)$

$\qquad\quad + (x^3 + 2xy^2 - x)(-2xe^{1-x^2-y^2})]$

$f_{yy} = -2[e^{1-x^2-y^2}(x^2 + 6y^2 - 2)$

$\qquad\quad + (x^2y + 2y^3 - 2y)(-2ye^{1-x^2-y^2})]$

$f_{xy} = -2[e^{1-x^2-y^2}(4xy)$

$\qquad\quad + (x^3 + 2xy^2 - x)(-2ye^{1-x^2-y^2})]$

For the point $(-1, 0)$, $D = (-4)(2) - 0 < 0$

So, (−1, 0) is a saddle point.
For the point (0, 0), $D = (2e)(4e) - 0 > 0$ and $f_{xx} > 0$, so (0, 0) is a relative minimum.
For the point (1, 0), $D - (-4)(2) - 0 < 0$
So, (1, 0) is a saddle point.
For the point (0, −1), $D = (-2)(-8) - 0 > 0$ and $f_{xx} < 0$, so (0, −1) is a relative maximum.
For the point (0, 1), $D = (-2)(-8) - 0 > 0$ and $f_{xx} < 0$, so (0, 1) is a relative maximum.

15. $f(x, y) = x^3 - 4xy + y^3$

$f_x = 3x^2 - 4y$

So, $f_x = 0$ when $0 = 3x^2 - 4y$, or $y = \dfrac{3x^2}{4}$.

$f_y = -4x + 3y^2$

So, $f_y = 0$ when $0 = -4x + 3y^2$

$$= -4x + 3\left(\frac{3x^2}{4}\right)^2$$

$$= \frac{27}{16}x^4 - 4x$$

$$= 4x\left(\frac{27}{64}x^3 - 1\right)$$

or $x = 0, \dfrac{4}{3}$.

When $x = 0$, $f_x = 0$ when $y = 0$.

When $x = \dfrac{4}{3}$, $f_x = 0$ when

$0 = 3\left(\dfrac{4}{3}\right)^2 - 4y$, or $y = \dfrac{4}{3}$.

So the critical points are (0, 0) and $\left(\dfrac{4}{3}, \dfrac{4}{3}\right)$.

$f_{xx} = 6x$, $f_{yy} = 6y$, $f_{xy} = -4$

For the point (0, 0),

$D = 6(0)6(0) - (-4)^2 < 0$
So, (0, 0) is a saddle point.
For the point $\left(\dfrac{4}{3}, \dfrac{4}{3}\right)$,

$D = 6\left(\dfrac{4}{3}\right)6\left(\dfrac{4}{3}\right) - (-4)^2 > 0$ and $f_{xx} > 0$,

so $\left(\dfrac{4}{3}, \dfrac{4}{3}\right)$ is a relative minimum.

17. $f(x, y) = 4xy - 2x^4 - y^2 + 4x - 2y$

$f_x = 4y - 8x^3 + 4$

So, $f_x = 0$ when $4y - 8x^3 + 4 = 0$

$f_y = 4x - 2y - 2$

So, $f_y = 0$ when $4x - 2y - 2 = 0$, or

$y = 2x - 1$
Substituting above,

$$4y - 8x^3 + 4 = 0$$
$$4(2x - 1) - 8x^3 + 4 = 0$$
$$-8x^3 + 8x = 0$$
$$-8x(x + 1)(x - 1) = 0$$

or $x = 0$, $x = -1$, $x = 1$
When $x = 0$, $y = -1$; when $x = -1$, $y = -3$; when $x = 1$, $y = 1$.
So, the critical points are (0, −1), (−1, −3) and (1, 1).

Now, $f_{xx} = -24x^2$, $f_{yy} = -2$, $f_{xy} = 4$

For the point (0, −1),

$D = (0)(-2) - [4]^2 < 0$
So, the point (0, −1) is a saddle point.
For the point (−1, −3),

$D = (-24)(-2) - [4]^2 > 0$
Since $f_{xx} < 0$, the point (−1, −3) is a relative maximum.
For the point (1, 1),

$D = (-24)(-2) - [4]^2 > 0$
Since $f_{xx} < 0$, the point (1, 1) is a relative maximum.

19. $f(x, y) = \dfrac{1}{x^2 + y^2 + 3x - 2y + 1}$

$f_x = \dfrac{-(2x+3)}{(x^2 + y^2 + 3x - 2y + 1)^2}$

So, $f_x = 0$ when $x = -\dfrac{3}{2}$.

$f_y = \dfrac{-(2y-2)}{(x^2 + y^2 + 3x - 2y + 1)^2}$

So, $f_y = 0$ when $y = 1$. The only critical point is $\left(-\dfrac{3}{2}, 1\right)$.

$f_{xx} = \dfrac{1}{(x^2 + y^2 + 3x - 2y + 1)^4}((x^2 + y^2 + 3x - 2y + 1)^2(-2) + (2x+3)[(2(x^2 + y^2 + 3x - 2y + 1)(2x+3)])$

$f_{yy} = \dfrac{1}{(x^2 + y^2 + 3x - 2y + 1)^4}((x^2 + y^2 + 3x - 2y + 1)^2(-2) + (2y-2) + [2(x^2 + y^2 + 3x - 2y + 1)(2y-2)])$

$f_{xy} = \dfrac{1}{(x^2 + y^2 + 3x - 2y + 1)^4}(0 + (2x+3)[2(x^2 + y^2 + 3x - 2y + 1)(2y-2)])$

For the point $\left(-\dfrac{3}{2}, 1\right)$, $D = (-4)(-4) - 0 > 0$ and $f_{xx} < 0$, so $\left(-\dfrac{3}{2}, 1\right)$ is a relative maximum.

21. $f(x, y) = x\ln\left(\dfrac{y^2}{x}\right) + 3x - xy^2$

$\qquad = x(\ln y^2 - \ln x) + 3x - xy^2$

$\qquad = x\ln y^2 - x\ln x + 3x - xy^2$

$f_x = \ln y^2 - \left[x\left(\dfrac{1}{x}\right) + \ln x(1)\right] + 3 - y^2$

$\qquad = \ln y^2 - \ln x + 2 - y^2$

So, $f_x = 0$ when $0 = 2\ln y - \ln x + 2 - y^3$

$f_y = 2x\left(\dfrac{1}{y}\right) - 2xy = \dfrac{2x(1-y^2)}{y}$

So, $f_y = 0$ when $x = 0$, $y = -1$, 1. We must reject $x = 0$, since f is undefined when $x = 0$.
When $y = -1$, $f_x = 0$ when

$\qquad 0 = \ln 1 - \ln x + 2 - 1$

$\qquad 0 = 1 - \ln x$

$\qquad \ln x = 1$, or $x = e$.

When $y = 1$, $f_x = 0$ when

$0 = \ln 1 - \ln x + 2 - 1$, or $x = e$.
So, the critical points are $(e, -1)$ and $(e, 1)$.

$f_{xx} = -\dfrac{1}{x}, \quad f_{yy} = -\dfrac{2x}{y^2} - 2x, \quad f_{xy} = \dfrac{2}{y} - 2y$

For the point $(e, -1)$,

$D = \left(\dfrac{-1}{e}\right)(-4e) - 0 > 0$ and $f_{xx} < 0$, so

$(e, -1)$ is a relative maximum.
For the point $(e, 1)$,

$D = \left(-\dfrac{1}{e}\right)(-4e) - 0 > 0$

$(e, 1)$ is a relative maximum.

23. $f(x, y) = xy - x - 3y$; vertices $(0, 0)$, $(5, 0)$, $(5, 5)$
Finding the critical points in R,

$f_x(x, y) = y - 1$ and $f_y(x, y) = x - 3$

$f_x(x, y) = 0$ when $y = 1$

$f_y(x, y) = 0$ when $x = 3$

So, the only critical point in R is $(3, 1)$.
Now, the boundary lines of R have the equations $x = 5$, $y = 0$ and $y = x$. Start with the segment on the line $x = 5$.
The original function becomes a function of y only, or

$u(y) = f(5, y) = 5y - 5 - 3y = 2y - 5$ for $0 \le y \le 5$

$u'(y) = 2 \ne 0$

So, the extremes for this segment can only occur at the endpoints $(5, 0)$ and $(5, 5)$.
Next, consider the segment on the line $y = 0$. The original function becomes a function of x only, or

$v(x) = f(x, 0) = -x$ for $0 \le x \le 5$

$v'(x) = -1 \ne 0$

So again, only the endpoints $(0, 0)$ and $(5, 0)$ can be considered. Lastly, consider the segment on the line $y = x$. Substituting in the original function gives

$w(x) = f(x, x) = x^2 - x - 3x = x^2 - 4x$

for $0 \le x \le 5$

$w'(x) = 2x - 4 = 0$ when $x = 2$ and

$y = x = 2$

So, the point $(2, 2)$ must be considered along with the endpoints $(0, 0)$ and $(5, 5)$.
Collecting these possible extreme points, need to test $(3, 1)$, $(0, 0)$, $(5, 0)$, $(5, 5)$, and $(2, 2)$ in the original function to see which yields the largest and smallest values of f.

$f(3, 1) = (3)(1) - 3 - 3(1) = -3$

$f(0, 0) = 0$

$f(5, 0) = 0 - 5 - 0 = -5$

$f(5, 5) = (5)(5) - 5 - 3(5) = 5$

$f(2, 2) = (2)(2) - 2 - 3(2) = -4$

So, the smallest value is -5 and the largest value is 5.

25. $f(x, y) = 2x^2 + y^2 + xy^2 - 2$; $(5, 5)$, $(-5, 5)$, $(5, -5)$, $(-5, -5)$
Finding the critical points in R,

$f_x(x, y) = 4x + y^2$ and

$f_y(x, y) = 2y + 2xy$

$f_x = 0$ when $4x = -y^2$, or $x = \dfrac{-1}{4} y^2$

$f_y = 0$ when

$2y + 2xy = 0$

$2y + 2\left(-\dfrac{1}{4} y^2\right) y = 0$

$2y - \dfrac{1}{2} y^3 = 0$

$y\left(2 - \dfrac{1}{2} y^2\right) = 0$

$y = 0$ or

$2 - \dfrac{1}{2} y^2 = 0$, $2 = \dfrac{1}{2} y^2$, $y^2 = 4$, $y = \pm 2$

So, the critical points in R are $(0, 0)$, $(-1, -2)$ and $(-1, 2)$.
The boundary lines of R have the equations $x = -5$, $y = 5$, $x = 5$ and $y = -5$.
Using $x = -5$,

$u(y) = f(-5, y) = 50 + y^2 - 5y^2 - 2$

$= 48 - 4y^2$

$u'(y) = -8y = 0$ when $y = 0$

and $x = -5$
So, the extremes can only occur at $(-5, 0)$, $(-5, 5)$ and $(-5, -5)$.
Using $y = 5$,

$v(x) = f(x, 5) = 2x^2 + 25 + 25x - 2$

$v'(x) = 4x + 25 = 0$ when $x = -\dfrac{25}{4}$

and $y = 5$

But, $\left(-\dfrac{25}{4}, 5\right)$ is not in R, so reject it.
Using $x = 5$,

$w(y) = f(5, y) = 50 + y^2 + 5y^2 - 2$

$= 48 + 6y^2$

$w'(y) = 12y = 0$ when $y = 0$

and $x = 5$

So, the points to consider here are (5, 0), (5, −5) and (5, 5).

Using $y = -5$,

$$r(x) = f(x, -5) = 2x^2 + 25 + 25y - 2$$

$$r'(x) = 4x + 25 = 0 \text{ when } x = -\frac{25}{4}$$

and $y = -5$

But, $\left(-\dfrac{25}{4}, -5\right)$ is not in R, so reject it.

Collecting all points to consider,

$$f(0, 0) = 0 + 0 + 0 - 2 = -2$$

$$f(-1, -2)$$

$$= 2(-1)^2 + (-2)^2 + (-1)(-2)^2 - 2 = 0$$

$$f(-1, 2)$$

$$= 2(-1)^2 + (2)^2 + (-1)(2)^2 - 2 = 0$$

$$f(-5, 0)$$

$$= 2(-5)^2 + 0 + 0 - 2 = 48$$

$$f(-5, 5)$$

$$= 2(-5)^2 + (5)^2 + (-5)(5)^2 - 2 = -52$$

$$f(-5, -5)$$

$$= 2(-5)^2 + (-5)^2 + (-5)(-5)^2 - 2 = -52$$

$$f(5, 0)$$

$$= 2(5)^2 + 0 + 0 - 2 = 48$$

$$f(5, -5)$$

$$= 2(5)^2 + (-5)^2 + (5)(-5)^2 - 2 = 198$$

$$f(5, 5)$$

$$= 2(5)^2 + (5)^2 + (5)(5)^2 - 2 = 198$$

So, the smallest value is −52, and the largest value is 198.

27. $f(x, y) = x^4 + 2y^3; \ x^2 + y^2 = 1$

Finding the critical points in R,

$$f_x(x, y) = 4x^3 \text{ and } f_y(x, y) = 6y^2$$

$$f_x(x, y) = 0 \text{ when } x = 0$$

$$f_y(x, y) = 0 \text{ when } y = 0$$

So, the only critical point in R is (0, 0).

The boundary equations are $x = \sqrt{1 - y^2}$

and $x = -\sqrt{1 - y^2}$.

Using $x = \sqrt{1 - y^2}$,

$$u(y) = f\left(\sqrt{1 - y^2}, y\right) = \left(1 - y^2\right)^2 + 2y^3$$

$$= 1 - 2y^2 + y^4 + 2y^3$$

$$u'(y) = 4y^3 + 6y^2 - 4y$$

$$= 2y(2y - 1)(y + 2)$$

$$u'(y) = 0 \text{ when } y = 0, \frac{1}{2}, -2$$

Since $y = -2$ outside of R, reject that value. When $y = 0$, $x = 1$ or −1. When

$$y = \frac{1}{2}, x = \pm\sqrt{1 - \frac{1}{4}} \text{ or, } x = -\frac{\sqrt{3}}{2} \text{ or}$$

$x = \dfrac{\sqrt{3}}{2}$. Using $x = -\sqrt{1 - y^2}$,

$$v(y) = f\left(-\sqrt{1 - y^2}, y\right)$$

$$= \left(1 - y^2\right)^2 + 2y^3$$

so no new points are found. Collecting all points to consider,

$$f(0, 0) = 0$$

$$f(1, 0) = 1$$

$$f(-1, 0) = 1$$

$$f\left(\frac{\sqrt{3}}{2}, \frac{1}{2}\right) = \left(\frac{\sqrt{3}}{2}\right)^4 + 2\left(\frac{1}{2}\right)^3 = \frac{13}{16}$$

$$f\left(-\frac{\sqrt{3}}{2}, \frac{1}{2}\right) = \left(-\frac{\sqrt{3}}{2}\right)^4 + 2\left(\frac{1}{2}\right)^3 = \frac{13}{16}$$

So, the smallest value is 0 and the largest value is 1.

29. Profit = (profit from sales Duncan shirts)
    + (profit from sales James shirts)

$P(x, y) = (x - 2)(40 - 50x + 40y)$
$\qquad + (y - 2)(20 + 60x - 70y)$

$P_x = (x - 2)(-50) + (40 - 50x + 40y)(1)$
$\qquad\qquad + (y - 2)(60) + 0$
$\quad = 20(-5x + 5y + 1)$

So, $P_x = 0$ when $0 = 20(-5x + 5y + 1)$, or
$-5x + 5y + 1 = 0$.

$P_y = (x - 2)(40) + 0 + (y - 2)(-70)$
$\qquad\qquad + (20 + 60x - 70y)(1)$
$\quad = 20(5x - 7y + 4)$

So, $P_y = 0$ when $0 = 20(5x - 7y + 4)$, or
$0 = 5x - 7y + 4$.

Solving this system of equations by adding,

$0 = -2y + 5$, or $y = \dfrac{5}{2} = 2.5$

When $y = 2.5$, $P_x = 0$ when
$0 = -5x + 5(2.5) + 1$, or $x = 2.7$.
So the critical point is $(2.7, 2.5)$
$P_{xx} = -100$, $P_{yy} = -140$, $P_{xy} = 100$

$D = (-100)(-140) - (100)^2 > 0$ and
$P_{xx} < 0$

So, profit is maximized when Duncan shirts sell for $2.70 and James shirts sell for $2.50.

31. Profit = revenue − cost

$P(x, y) = [x(100 - x) + y(100 - y)]$
$\qquad\qquad - [x^2 + xy + y^2]$
$\quad = -2x^2 - 2y^2 + 100x + 100y - xy$

$P_x = -4x + 100 - y$

So, $P_x = 0$ when $0 = -4x + 100 - y$.

$P_y = -4y + 100 - x$

So, $P_y = 0$ when $0 = -4y + 100 - x$.

Solving this system of equations by multiplying the first equation by −4 and adding to second, $0 = 15x - 300$, or
$x = 20$.

When $x = 20$, $P_x = 0$ when
$0 = -4(20) + 100 - y$, or $y = 20$. So, the critical point is $(20, 20)$.
$P_{xx} = -4$; $P_{yy} = -4$; $P_{xy} = -1$

$D = (-4)(-4) - (-1)^2 > 0$ and $P_{xx} < 0$

So, profit is maximized when 20 gallons of each are produced.

33. profit = (profit from domestic market)
    + (profit from foreign market)

$P(x, y) = x\left(60 - \dfrac{x}{5} + \dfrac{y}{20}\right)$
$\qquad + y\left(50 - \dfrac{y}{10} + \dfrac{x}{20}\right) - 10(x + y)$
$\quad = 50x - \dfrac{x^2}{5} + \dfrac{xy}{10} + 40y - \dfrac{y^2}{10}$

$P_x = 50 - \dfrac{2}{5}x + \dfrac{y}{10}$

So, $P_x = 0$ when

$0 = 50 - \dfrac{2}{5}x + \dfrac{y}{10} = 500 - 4x + y$.

$P_y = \dfrac{x}{10} + 40 - \dfrac{y}{5}$

So, $P_y = 0$ when

$0 = \dfrac{x}{10} + 40 - \dfrac{y}{5} = x + 400 - 2y$

Solving this system by multiplying the first equation by two and adding to the second, $0 = 1400 - 7x$, or $x = 200$.

When $x = 200$, $P_y = 0$ when

$0 = 200 + 400 - 2y$, or $y = 300$.

$P_{xx} = -\dfrac{2}{5}$, $P_{yy} = -\dfrac{1}{5}$, $P_{xy} = \dfrac{1}{10}$

$D = \left(-\dfrac{2}{5}\right)\left(-\dfrac{1}{5}\right) - \left(\dfrac{1}{10}\right)^2 > 0$ and $P_{xx} < 0$

So, profit is maximized when 200 machines are supplied to the domestic market and 300 are supplied to the foreign market.

35. Using the hint

$$P(x, y) = \left(\frac{640y}{y+3} + \frac{216x}{x+5}\right)(210-135)$$
$$-1{,}000(x+y)$$

noting that the price and cost per unit are in dollars, while promotion and development are in thousand dollars.

$$P(x, y) = 75\left(\frac{640y}{y+3} + \frac{216x}{x+5}\right) - 1{,}000x$$
$$-1{,}000y$$

$$P_x(x, y)$$

$$= 75\left[\frac{(x+5)(216)-(216x)(1)}{(x+5)^2}\right] - 1{,}000$$

$$= 75\left[\frac{1{,}080}{(x+5)^2}\right] - 1{,}000$$

$$= \frac{81{,}000}{(x+5)^2} - 1{,}000$$

$P_x(x, y) = 0$ when

$$\frac{81{,}000}{(x+5)^2} - 1{,}000 = 0$$

$$\frac{81{,}000}{(x+5)^2} = 1{,}000$$

$$81 = (x+5)^2$$

$$9 = x+5, \text{ or } x = 4$$

$$P_y(x, y)$$

$$= 75\left[\frac{(y+3)(640)-(640y)(1)}{(y+3)^2}\right]$$
$$-1{,}000$$

$$= 75\left[\frac{1{,}920}{(y+3)^2}\right] - 1{,}000$$

$$= \frac{144{,}000}{(y+3)^2} - 1{,}000$$

$P_y(x, y) = 0$ when

$$\frac{144{,}000}{(y+3)^2} - 1{,}000 = 0$$

$$\frac{144{,}000}{(y+3)^2} = 1{,}000$$

$$144 = (y+3)^2$$

$$12 = y+3, \text{ or } y = 9$$

To see if the point (4, 9) corresponds to maximum,

$$P_x(x, y) = 81{,}000(x+5)^{-2} - 1{,}000$$

$$P_y(x, y) = 144{,}000(y+3)^{-2} - 1{,}000$$

$$P_{xx}(x, y) = -162{,}000(x+5)^{-3}$$

$$= \frac{-162{,}000}{(x+5)^3}$$

$$P_{yy}(x, y) = -288{,}000(y+3)^{-3}$$

$$= \frac{-288{,}000}{(y+3)^3}$$

$$P_{xy}(x, y) = P_{yx}(x, y) = 0$$

$$D(4, 9) = \left(\frac{-162{,}000}{9^3}\right)\left(\frac{-288{,}000}{12^3}\right) - 0$$

Since $D(4, 9) > 0$ and $P_{xx}(x, y) < 0$, (4, 9) corresponds to a maximum. So, $4,000 should be spent on development and $9,000 should be spent on promotion.

37. $C(x, y) = x^2 - 280x + y^2 - 380y$
$$+ 60{,}000$$

(a) Since $400 - x - y \geq 0$, $y \leq 400 - x$. Similarly, since $600 - 2x - y \geq 0$, $y \leq 600 - 2x$. Graphing the lines $y = 400 - x$ and $y = 600 - 2x$, the region R enclosed by the intersection of the inequalities is a quadrilateral with vertices (0, 0), (0, 400), (200, 200) and (300, 0).

(b) Finding the critical points in R,

$C_x(x, y) = 2x - 280$ and

$C_y(x, y) = 2y - 380$

$C_x(x, y) = 0$ when $x = 140$

$C_y(x, y) = 0$ when $y = 190$

So, the critical point in R is (140, 190). The boundary equations of R are $y = 400 - x$, $y = 600 - 2x$, $x = 0$ and $y = 0$. Using $y = 400 - x$,

$u(x) = C(x, 400 - x)$

$\quad = x^2 - 280x + (400 - x)^2$

$\quad\quad - 380(400 - x) + 60{,}000$

$\quad = 2x^2 - 700x + 68{,}000$

$u'(x) = 4x - 700$

$u'(x) = 0$ when $x = 175$ and $y = 225$

So, the point (175, 225) must be considered along with the endoints (0, 400) and (200, 200). Using $y = 600 - 2x$,

$v(x) = C(x, 600 - 2x)$

$\quad = x^2 - 280x + (600 - 2x)^2$

$\quad\quad - 380(600 - 2x) + 60{,}000$

$\quad = 5x^2 - 1{,}920x + 192{,}000$

$v'(x) = 10x - 1{,}920$

$v'(x) = 0$ when $x = 192$ and $y = 216$

So, the point (192, 216) must be considered as well as the endpoints (200, 200) and (300, 0). Using $x = 0$,

$w(y) = C(0, y) = y^2 - 380y + 60{,}000$

$w'(y) = 2y - 380$

$w'(y) = 0$ when $y = 190$

So, the point (0, 190) must be considered as well as the endpoints (0, 0) and (0, 400). Using $y = 0$,

$r(x) = C(x, 0) = x^2 - 280x + 60{,}000$

$r'(x) = 2x - 280$

$r'(x) = 0$ when $x = 140$

So, the point (140, 0) must be considered as well as the endpoints (0, 0) and (300, 0).
Collecting all points to consider,

$C(140, 190) = 4{,}300$

$C(175, 225) = 6{,}750$

$C(0, 400) = 68{,}000$

$C(200, 200) = 8{,}000$

$C(192, 216) = 7{,}680$

$C(300, 0) = 66{,}000$

$C(0, 190) = 23{,}900$

$C(0, 0) = 60{,}000$

$C(140, 0) = 40{,}400$

So, the minimum cost is \$4,300 when $x = \$140$ and $y = \$190$.

39. (a) The cost of obtaining each type of shirt is \$2. Since won't sell below cost, $x \geq 2$ and $y \geq 2$. Also, the quantity sold of each must be nonnegative, so $40 - 50x + 40y \geq 0$ and $20 + 60x - 70y \geq 0$. Graphing

$y = \dfrac{5}{4}x - 1$, $y = \dfrac{6}{7}x + \dfrac{2}{7}$, $y = 2$ and

$x = 2$, the region R enclosed is a triangle with vertices (2, 2), $\left(\dfrac{36}{11}, \dfrac{34}{11}\right)$ and $\left(\dfrac{12}{5}, 2\right)$.

(b) $P(x, y) = -50x^2 + 100xy + 20x$
$\quad\quad\quad\quad\quad - 70y^2 + 80y - 120$

$P_x(x, y) = -100x + 100y + 20$

$P_y(x, y) = 100x - 140y + 80$

$P_x(x, y) = 0$ when

$-100x - 100y + 20 = 0$, or $x = y + 0.2$

$P_y(x, y) = 0$ when $100x - 140y + 80 = 0$,

or $x = 1.4y - 0.8$

So,

$$y + 0.2 = 1.4y - 0.8$$
$$1 = 0.4y$$
$$y = 2.5$$

and $x = 2.7$

The only critical point in R is (2.7, 2.5).
Now, the boundary equations of R are

$y = \dfrac{5}{4}x - 1$, $y = \dfrac{6}{7}x + \dfrac{2}{7}$ and $y = 2$.

Using $y = \dfrac{5}{4}x - 1$,

$$u(x) = P\left(x, \frac{5}{4}x - 1\right)$$

$$= -50x^2 + 100x\left(\frac{5}{4}x - 1\right) + 20x$$

$$- 70\left(\frac{5}{4}x - 1\right)^2 + 80\left(\frac{5}{4}x - 1\right) - 120$$

$$= -34.375x^2 + 195x - 270$$

$$u'(x) = -68.75x + 195$$

$u'(x) = 0$ when $-68.75x + 195 = 0$ or,

$x \approx 2.84$ and $y \approx 2.55$

So, the point (2.84, 2.55) must be
considered as well as the endpoints

$\left(\dfrac{12}{5}, 2\right)$ and $\left(\dfrac{36}{11}, \dfrac{34}{11}\right)$. Using

$y = \dfrac{6}{7}x + \dfrac{2}{7}$,

$$v(x) = P\left(x, \frac{6}{7}x + \frac{2}{7}\right)$$

$$= -50x^2 + 100x\left(\frac{6}{7}x + \frac{2}{7}\right) + 20x$$

$$\approx -15.71x^2 + 82.86x - 102.86$$

$$y'(x) = -31.43x + 82.86$$

$$v'(x) = -30.78x + 82.84$$

$v'(x) = 0$ when $-30.78x + 82.84 = 0$

or, $x \approx 2.64$ and $y \approx 2.55$

So, the point (2.69, 2.59) must be
considered as well as the endpoints (2, 2)

and $\left(\dfrac{36}{11}, \dfrac{34}{11}\right)$.

Using $y = 2$,

$$r(x) = P(x, 2)$$

$$= -50x^2 + 200x + 20x - 280 + 160 - 120$$

$$= -50x^2 + 220x - 240$$

$$r'(x) = -100x + 220$$

$r'(x) = 0$ when $-100x + 220 = 0$ or,

$x = 2.2$ and $y = 2$. So, the point (2.2, 2) must
be considered as well as the endpoints

$\left(\dfrac{12}{5}, 2\right)$ and (2, 2). Collecting

all points to consider,

$$P(2.7, 2.5) = 7$$

$$P(2.84, 2.55) = 6.55$$

$$P\left(\frac{12}{5}, 2\right) = 0$$

$$P\left(\frac{36}{11}, \frac{34}{11}\right) = 0$$

$$P(2.64, 2.55) = 6.35$$

$$P(2, 2) = 0$$

$$P(2.2, 2) = 2$$

So, the conclusion is again that Duncan
shirts should sell for \$2.70 and the James

shirts for $2.50 (no change in answer). The smallest profit occurs at each of the three vertices of the triangular region.

41. $f(x, y) = C + xye^{1-x^2-y^2}$

$$f_x = y[x(-2xe^{1-x^2-y^2}) + e^{1-x^2-y^2}(1)]$$
$$= ye^{1-x^2-y^2}(-2x^2 + 1)$$

So, $f_x = 0$ when $y = 0$, or $x = \dfrac{\sqrt{2}}{2}$

(rejecting the negative solution).

$$f_y = x[y(-2ye^{1-x^2-y^2}) + e^{1-x^2-y^2}(1)]$$
$$= xe^{1-x^2-y^2}(-2y^2 + 1)$$

When $y = 0$, $f_y = 0$ when $x = 0$.

When $x = -\dfrac{\sqrt{2}}{2}$, $f_y = 0$ when $y = \dfrac{\sqrt{2}}{2}$.

When $x = \dfrac{\sqrt{2}}{2}$, $f_y = 0$ when $y = \dfrac{\sqrt{2}}{2}$.

Again rejecting the negative solutions, the critical points are $(0, 0)$ and $\left(\dfrac{\sqrt{2}}{2}, \dfrac{\sqrt{2}}{2}\right)$.

Rewriting f_x as

$$f_x = e^{1-x^2-y^2}(-2x^2y + y)$$
$$f_{xx} = (e^{1-x^2-y^2})(-4xy)$$
$$\quad + (-2x^2y + y)(-2xe^{1-x^2-y^2})$$

Similarly,

$$f_{yy} = (e^{1-x^2-y^2})(-4xy)$$
$$\quad + (-2xy^2 + x)(-2ye^{1-x^2-y^2})$$
$$f_{xy} = (e^{1-x^2-y^2})(-2x^2 + 1)$$
$$\quad + (-2x^2y + y)(-2ye^{1-x^2-y^2})$$

For the point $(0, 0)$, $D = (0)(0) - (e)^2 < 0$
So, the point $(0, 0)$ does not correspond to the maximum.

For the point $\left(\dfrac{\sqrt{2}}{2}, \dfrac{\sqrt{2}}{2}\right)$,

$D = (-2)(-2) - 0 > 0$ and $f_{xx} < 0$

So, $\dfrac{\sqrt{2}}{2}$ units of each stimulus maximizes performance.

43. The square of the distance from $S(a, b)$ to each point is:

$$(a+5)^2 + (b-0)^2 = a^2 + 10a + 25 + b^2$$
$$(a-1)^2 + (b-7)^2 = a^2 - 2a + b^2 - 14b + 50$$
$$(a-9)^2 + (b-0)^2 = a^2 - 18a + 81 + b^2$$
$$(a-0)^2 + (b+8)^2 = a^2 + b^2 + 16b + 64$$

So, the sum of the distances is

$$f(a, b) = 4a^2 - 10a + 4b^2 + 2b + 220$$

$f_a = 8a - 10$, so $f_a = 0$ when $a = \dfrac{5}{4}$

$f_b = 8b + 2$, so $f_b = 0$ when $b = -\dfrac{1}{4}$

$f_{aa} = 8$, $f_{bb} = 8$, $f_{ab} = 0$ so,
$D = (8)(8) - 0 > 0$ and $f_{aa} > 0$

The sum is minimized at $\left(\dfrac{5}{4}, -\dfrac{1}{4}\right)$.

45. Since $p + q + r = 1$, $r = 1 - p - q$ and $P(p, q)$

$$= 2pq + 2p(1 - p - q) + 2(1 - p - q)q$$
$$= 2p - 2p^2 - 2pq + 2q - 2q^2$$
$$P_p = 2 - 4p - 2q$$

So, $P_p = 0$ when $0 = 2 - 4p - 2q$, or
$0 = 1 - 2p - q$.
$$P_q = -2p + 2 - 4q$$
So, $P_q = 0$ when $0 = -2p + 2 - 4q$, or
$0 = -p + 1 - 2q$.
Solving this system of equations by multiplying the first equation by negative two and adding to the second,

$0 = -1 + 3p$, or $p = \dfrac{1}{3}$.

When $p = \dfrac{1}{3}$, $P_q = 0$ when

$0 = -\dfrac{1}{3} + 1 - 2q$, or $q = \dfrac{1}{3}$.

$P_{pp} = -4$, $P_{qq} = -4$, $P_{pq} = -2$

$D = (-4)(-4) - (-2)^2 > 0$ and $P_{pp} < 0$

So, P maximized when $P = \dfrac{1}{3}$, $q = \dfrac{1}{3}$,

and $r = \dfrac{1}{3}$. The maximum is

$$P = 2\left(\frac{1}{3}\right)\left(\frac{1}{3}\right) + 2\left(\frac{1}{3}\right)\left(\frac{1}{3}\right) + 2\left(\frac{1}{3}\right)\left(\frac{1}{3}\right)$$

$$= \frac{2}{3}$$

47. The goal is to maximize the livable space subject to a constraint on the surface area. Let s be the length along the floor, at each end, where a 6-foot-tall person cannot stand. Then, the livable space is $L = 6(x - 2s)y$.

From similar triangles, $\dfrac{s}{6} = \dfrac{\frac{x}{2}}{\frac{\sqrt{3}}{2}x}$ or,

$s = \dfrac{6}{\sqrt{3}}$ and

$$L = 6\left(x - \frac{12}{\sqrt{3}}\right)y = 6xy - \frac{72}{\sqrt{3}}y$$

Since the surface area must be 500, the

constraint is $500 = 2xy + 2\left(\dfrac{\sqrt{3}}{4}x^2\right)$ and

$$g(x) = 2xy + \frac{\sqrt{3}}{2}x^2$$

$$L_x = 6y;\ L_y = 6x - \frac{72}{\sqrt{3}}$$

$$g_x = 2y + \sqrt{3}x;\ g_y = 2x$$

So, the three Lagrange equations are

$$6y = \left(2y + \sqrt{3}x\right)\lambda$$

$$6x - \frac{72}{\sqrt{3}} = 2x\lambda$$

$$2xy + \frac{\sqrt{3}}{2}x^2 = 500$$

Solving the second equation for λ and substituting into the first equation gives

$$6y = \left(2y + \sqrt{3}x\right)\left(3 - \frac{36}{\sqrt{3}x}\right)$$

$$6y = 6y + 3\sqrt{3}x - \frac{72y}{\sqrt{3}x} - 36$$

$$\frac{72}{\sqrt{3}x}y = 3\sqrt{3}x - 36$$

$$y = \frac{1}{8}x^2 - \frac{\sqrt{3}}{2}x$$

Substituting into the third equation gives

$$2x\left(\frac{1}{8}x^2 - \frac{\sqrt{3}}{2}x\right) + \frac{\sqrt{3}}{2}x^2 = 500$$

$$\frac{1}{4}x^3 - \sqrt{3}x^2 + \frac{\sqrt{3}}{2}x^2 = 500$$

$$x^3 - 4\sqrt{3}x^2 + 2\sqrt{3}x^2 = 2{,}000$$

$$x^3 - 2\sqrt{3}x^2 - 2{,}000 = 0$$

To use the calculator to solve

$x^3 - 2\sqrt{3}x^2 - 2{,}000 = 0$

Press $\boxed{y=}$.

Input $x \wedge 3 - 2 * \sqrt{\ }(3) * x^2 - 2{,}000$ for

$y_1 =$.

Use window dimensions [0, 20]5 by [-50, 500]150.

Press $\boxed{graph}$.

Use the zero function under the calc menu to find $x = 13.866$.

When $x = 13.866$ feet,

$$y = \frac{1}{8}(13.866)^2 - \frac{\sqrt{3}}{2}(13.866)$$

$$\approx 12.025 \text{ feet.}$$

49. (a) The problem is to minimize the total time $T(x, y)$, where

$$T = \frac{\sqrt{(1.2)^2 + x^2}}{2} + \frac{\sqrt{(2.5)^2 + y^2}}{4}$$

$$+ \frac{4.3 - (x + y)}{6}$$

$$\frac{\partial T}{\partial x} = \frac{1}{2}\left[\frac{1}{2}\frac{2x}{\sqrt{(1.2)^2 + x^2}}\right] - \frac{1}{6}$$

$$\frac{\partial T}{\partial y} = \frac{1}{4}\left[\frac{1}{2}\frac{2y}{\sqrt{(2.5)^2 + y^2}}\right] - \frac{1}{6}$$

$\dfrac{\partial T}{\partial x} = \dfrac{\partial T}{\partial y} = 0$ when

$\dfrac{1}{2} \dfrac{x}{\sqrt{(1.2)^2 + x^2}} = \dfrac{1}{6}$ and

$\dfrac{1}{4} \dfrac{y}{\sqrt{(2.5)^2 + y^2}} = \dfrac{1}{6}$ which leads to

$x = 0.424$ and $y = 2.236$.

In addition to his path, the "boundary" cases must also be considered. That is, a path where Tom moves directly to the river (perpendicular to the river), then Tom swims directly across the river (perpendicular to the river), and Mary runs to the finish. The second boundary path is along the diagonal connection S and F.

Case 1

$x = 0, y = 0$

Time $= \dfrac{1.2}{2} + \dfrac{2.5}{4} + \dfrac{4.3}{6} \approx 1.942$

Case 2

$x = 0.424, y = 2.236$

Time $= \dfrac{1.273}{2} + \dfrac{3.354}{4} + \dfrac{1.64}{6} = 1.748$

Case 3

$x = 1.395, y = 2.905$

Time $= \dfrac{1.84}{2} + \dfrac{3.833}{4} + \dfrac{0}{6} = 1.878$

The minimum time is when $x = 0.424$ miles and $y = 2.236$ miles.

(b) For the second team, the time is

$T = \dfrac{\sqrt{(1.2)^2 + x^2}}{1.7} + \dfrac{\sqrt{(2.5)^2 + y^2}}{3.5}$
$\qquad + \dfrac{4.3 - (x + y)}{6.3}$

$\dfrac{\partial T}{\partial x} = \dfrac{1}{1.7}\left[\dfrac{x}{\sqrt{(1.2)^2 + x^2}} \right] - \dfrac{1}{6.3}$

$\dfrac{\partial T}{\partial y} = \dfrac{1}{3.5}\left[\dfrac{y}{\sqrt{(2.5)^2 + y^2}} \right] - \dfrac{1}{6.3}$

We must find when $\dfrac{\partial T}{\partial x} = \dfrac{\partial T}{\partial y} = 0$

Press $\boxed{y =}$.

Input $\dfrac{\partial T}{\partial x}$ for $y_1 =$.

Use window dimensions $[0, 2]0.5$ by $[-1, 2]0.5$

Press $\boxed{\text{graph}}$.

Use the zero function under the calc menu to find $x \approx 0.3363$.

Repeat process for $\dfrac{\partial T}{\partial y}$ to find

$y \approx 1.6704$.

Repeating the case scenarios as in part (a)

Case	x	y	Time
1	0	0	2.103
2	0.3363	1.6704	1.9562
3	1.395	2.905	2.177

Tom, Dick, and Mary will win by 0.208 hours (12.5 minutes).

(c) Writing Exercise—Answers will vary.

51. Let l, w, h be the dimensions of the box

Cost = (area)(cost per area)

Cost bottom = $(lw)(3)$

Cost top = $(lw)(5)$

Cost 4 sides = $2(lh)(1) + 2(wh)(1)$

$C = 8lw + 2lh + 2wh$

Since volume must be 32, $32 = lwh$, or

$h = \dfrac{32}{lw}$

$C(l, w) = 8lw + 2l\left(\dfrac{32}{lw}\right) + 2w\left(\dfrac{32}{lw}\right)$

$\qquad = 8lw + \dfrac{64}{w} + \dfrac{64}{l}$

$C_l = 8w - \dfrac{64}{l^2}$

So, $C_l = 0$ when $0 = 8w - \dfrac{64}{l^2}$.

$C_w = 8l - \dfrac{64}{w^2}$

So, $C_w = 0$ when $0 = 8l - \dfrac{64}{w^2}$.

Solving each equation for w^2,

$$8w = \frac{64}{l^2}$$

$$w = \frac{8}{l^2}, \quad w^2 = \frac{64}{l^4}$$

$$8l = \frac{64}{w^2}$$

$$w^2 = \frac{8}{l}$$

So, $\dfrac{64}{l^4} = \dfrac{8}{l}, \quad 64l = 8l^4$

$8l(l^3 - 8) = 0$, or $l = 2$.

When $l = 2$, $w = \dfrac{8}{(2)^2} = 2$.

So, $(2, 2)$ is the critical point.

$C_{ll} = \dfrac{128}{l^3}, \quad C_{ww} = \dfrac{128}{w^3}, \quad C_{lw} = 8$

$D = (32)(32) - (8)^2 > 0$ and $C_{ll} > 0$

When $l = 2$ and $w = 2$, $h = \dfrac{32}{(2)(2)}$.

So, cost is minimized when the dimensions of the box are 2 ft × 2 ft × 8 ft.

53. $f(x, y) = x^2 + y^2 - 4xy$, $f_x = 2x - 4y = 0$

when $y = \dfrac{x}{2}$.

$f_y = 2y - 4x = 0$ when $y = 2x$.

So, $(0, 0)$ is a critical point.

$f_{xx} = 2$, $f_{xy} = -4$, and $f_{yy} = 2$, so

$D = 4 - (-4)^2 < 0$ and $(0, 0)$ is a saddle point.

The above is true but not asked for. If $x = 0$, $f(0, y) = y^2$ which is a parabola with a minimum at $(0, 0)$ (in the vertical yz-plane). If $y = 0$, $f(x, 0) = x^2$ which is a parabola with a minimum at $(0, 0)$ (in the vertical xz-plane).

If $y = x$, $f(x, x) = -2x^2$ which is a

parabola with a maximum at $(0, 0)$ (in the vertical plane passing through the z-axis and the line $y = x$ in the xy plane).

55. $f(x, y) = \dfrac{x^2 + xy + 7y^2}{x \ln y}$

To use the graphing utility to determine critical points of the function,

f_x

$= \dfrac{(x \ln y)(2x + y) - (x^2 + xy + 7y^2)(\ln y)}{(x \ln y)^2}$

$= \dfrac{\ln y[2x^2 + xy - x^2 - xy - 7y^2]}{x^2 \ln^2 y}$

$= \dfrac{x^2 - 7y^2}{x^2 \ln y}$

f_y

$= \dfrac{(x \ln y)(x + 14y) - (x^2 + xy + 7y^2)\left(\frac{x}{y}\right)}{(x \ln y)^2}$

$= \dfrac{\frac{(xy \ln y)(x+14y) - x(x^2 + xy + 7y^2)}{y}}{x^2 (\ln y)^2}$

$= \dfrac{x[(y \ln y)(x + 14y) - x^2 - xy - 7y^2]}{x^2 y (\ln y)^2}$

$= \dfrac{(y \ln y)(x + 14y) - x^2 - xy - 7y^2}{xy(\ln y)^2}$

Next, $f_x = 0$ when $\dfrac{x^2 - 7y^2}{x^2 \ln y} = 0$, or

$x^2 - 7y^2 = 0$. $f_y = 0$ when

$(y \ln y)(x + 14y) - x^2 - xy - 7y^2 = 0$.

The critical points are found by solving the system

$$x^2 - 7y^2 = 0$$
$$(y \ln y)(x + 14y) - x^2 - xy - 7y^2 = 0$$

From the first equation, $x = \pm\sqrt{7}\,y$.

Substitute $x = \sqrt{7}\,y$ into the second equation to obtain

$(y \ln y)\left(\sqrt{7}\,y + 14y\right) - 7y^2 - \sqrt{7}y^2 - 7y^2$

$= 0$

$$y^2\left[(\ln y)\left(\sqrt{7}+14\right)-14-\sqrt{7}\right]=0$$

Press $\boxed{y=}$.

Input $x^2\left(\ln(x)*\left(14+\sqrt{7}\right)-14-\sqrt{7}\right)$ for $y_1 =$. (Remember that we are actually solving for y.)

Use window dimensions $[-5, 10]1$ by $[-10, 10]1$.

Press $\boxed{\text{graph}}$.

Using trace and zoom or the zero function under the calc menu to find the zeros are $y \approx 2.7182818$ ($y = e$) and $y = 0$.

If $x = -\sqrt{7}y,$ we also find the zeros to be $y = 0$ and $y = e$. So, the critical points are $\left(\pm\sqrt{7}e, e\right)$. The point $(0, 0)$ cannot be a critical point since $\ln 0$ is not defined.

57. $f(x, y) = 2x^4 + y^4 - 11x^2y + 18x^2$

$f_x = 8x^3 - 22xy + 36x$

$f_y = 4y^3 - 11x^2$

The critical points are found by solving the system $2x(4x^2 - 11y + 18) = 0$

$\qquad\qquad 4y^3 - 11x^2 = 0$

Solving the first equation gives $2x = 0$, or $4x^2 - 11y + 18 = 0$.

If $2x = 0$, $x = 0$ and substituting this into the second equation gives $4y^3 = 0$, or $y = 0$. One critical point is $(0, 0)$.

To solve $4x^2 - 11y + 18 = 0$, solve the second equation to get $x^2 = \dfrac{4}{11}y^2$ and

substitute. Then, $4\left(\dfrac{4}{11}y^3\right) - 11y + 18 = 0$

$\qquad\qquad \dfrac{16}{11}y^3 - 11y + 18 = 0$

Press $\boxed{y=}$.

Input $y = (16/11)x \wedge 3 - 11x + 18$ for $y_1 =$. (Remember, we are actually solving for y.)

Use the window dimensions $[-10, 5]1$ by $[-10, 10]1$.

Press $\boxed{\text{graph}}$.

Use trace and zoom or the zero function under the calc menu to find the zero is $y \approx -3.354$. We find we cannot use this value, however, since $x^2 = \dfrac{4}{11}(-3.354)^3$,

$x^2 \approx -13.72$, which has no solution. The critical point is $(0, 0)$.

7.4 The Method of Least-Squares

1. The sum $S(m, b)$ of the squares of the vertical distances from the three given points is

$S(m, b)$

$= d_1^2 + d_2^2 + d_3^2$

$= (b-1)^2 + (2m+b-3)^2 + (4m+b-2)^2.$

To minimize $S(m, b)$, set the partial derivatives $\dfrac{\partial S}{\partial m} = 0$ and $\dfrac{\partial S}{\partial b} = 0$.

$\dfrac{\partial S}{\partial m} = 2(2m+b-3)(2) + 2(4m+b-2)(4)$

$\qquad = 40m + 12b - 28$

$\qquad = 0$

$\dfrac{\partial S}{\partial b}$

$= 2(b-1) + 2(2m+b-3) + 2(4m+b-2)$

$= 12m + 6b - 12$

$= 0$

Solve the resulting simplified equations $10m + 3b = 7$ and $6m + 3b = 6$ to get

$m = \dfrac{1}{4}$ and $b = \dfrac{3}{2}$. So, the equation of the

least-squares line is $y = \dfrac{1}{4}x + \dfrac{3}{2}$.

3. The sum $S(m, b)$ of the squares of the vertical distances from the four given points is

$S(m, b) = (m+b-2)^2 + (2m+b-4)^2$

$\qquad\qquad + (4m+b-4)^2 + (5m+b-2)^2.$

To minimize $S(m, b)$, set the partial derivatives $\dfrac{\partial S}{\partial m} = 0$ and $\dfrac{\partial S}{\partial b} = 0$.

$$\frac{\partial S}{\partial m} = 2(m+b-2) + 2(2m+b-4)(2)$$
$$+ 2(4m+b-4)(4)$$
$$+ 2(5m+b-2)(5)$$
$$= 92m + 24b - 72$$
$$= 0$$

$$\frac{\partial S}{\partial b} = 2(m+b-2) + 2(2m+b-4)$$
$$+ 2(4m+b-4) + 2(5m+b-2)$$
$$= 24m + 8b - 24$$
$$= 0.$$

Solve the resulting simplified equations $23m + 6b = 18$ and $3m + b = 3$ to get $m = 0$ and $b = 3$. So, the equation of the least-squares line is $y = 3$.

5.

x	y	xy	x^2
1	2	2	1
2	2	4	4
2	3	6	4
5	5	25	25
$\Sigma x = 10$	$\Sigma y = 12$	$\Sigma xy = 37$	$\Sigma x^2 = 34$

Using the formulas with $n = 4$,
$$m = \frac{4(37) - 10(12)}{4(34) - (10)^2} = \frac{7}{9} \text{ and}$$
$$b = \frac{34(12) - 10(37)}{4(34) - (10)^2} = \frac{19}{18}$$
So, the equation of the least-squares line is
$$y = \frac{7}{9}x + \frac{19}{18}.$$

7.

x	y	xy	x^2
−2	5	−10	4
0	4	0	0
2	3	6	4
4	2	8	16
6	1	6	36
$\Sigma x = 10$	$\Sigma y = 15$	$\Sigma xy = 10$	$\Sigma x^2 = 60$

Using the formulas with $n = 5$,
$$m = \frac{5(10) - 10(15)}{5(60) - (10)^2} = -\frac{100}{200} = -\frac{1}{2} \text{ and}$$
$$b = \frac{60(15) - 10(10)}{5(60) - (10)^2} = \frac{800}{200} = 4$$
So, the equation of the least-squares line is
$$y = -\frac{1}{2}x + 4.$$

9.

x	y	xy	x^2
0	1	0	0
1	1.6	1.6	1
2.2	3	6.6	4.84
3.1	3.9	12.09	9.61
4	5	20	16
Σx $= 10.3$	Σy $= 14.5$	Σxy $= 40.29$	Σx^2 $= 31.45$

Using the formulas with $n = 5$,
$$m = \frac{5(40.29) - 14.5(10.3)}{5(31.45) - (10.3)^2}$$
$$= \frac{52.10}{51.16}$$
$$\approx 1.0184$$
$$b = \frac{31.45(14.5) - 10.3(40.29)}{5(31.45) - (10.3)^2}$$
$$= \frac{41.038}{51.16}$$
$$\approx 0.8022$$
So, the equation of the least-squares line is $y = 1.0184x + 0.8022$.

11.

x	y	xy	x^2
−2.1	3.5	−7.35	4.41
−1.3	2.7	−3.51	1.69
1.5	1.3	1.95	2.25
2.7	−1.5	−4.05	7.29

Σx	Σy	Σxy	Σx^2
$= 0.8$	$= 6.0$	$= -12.96$	$= 15.64$

Using the formulas with $n = 4$,

$$m = \frac{4(-12.96) - (0.8)(6.0)}{4(15.64) - (0.8)^2}$$

$$= \frac{-56.64}{61.92}$$

$$\approx -0.915$$

$$b = \frac{(15.64)(6.0) - (0.8)(-12.96)}{4(15.64) - (0.8)^2}$$

$$= \frac{104.208}{61.92}$$

$$\approx 1.683$$

So, the equation of the least-squares line is
$y = -0.915x + 1.683$.

13. Since $y = Ae^{mx}$

$\ln y = \ln A + \ln e^{mx} = \ln A + mx = mx + \ln A$
We can find the least squares line,
$Y = Mx + b$, using $Y = \ln y$. Then, use
$M = m$ and $b = \ln A$.

x	$y = \ln y$	xy	x^2
1	2.75	2.75	1
3	2.83	8.49	9
5	2.91	14.55	25
7	3.00	21	49
10	3.11	31.1	100

$\Sigma x = 26$ $\Sigma y = 14.6$ $\Sigma xy = 77.89$ $\Sigma x^2 = 184$

Using the formulas with $n = 5$,

$$m = \frac{5(77.89) - (26)(14.6)}{5(184) - (26)^2}$$

$$= \frac{9.85}{244}$$

$$\approx 0.04$$

and

$$b = \frac{(184)(14.6) - (26)(77.89)}{5(184) - (26)^2}$$

$$= \frac{661.26}{244}$$

$$\approx 2.710$$

For our exponential model, $y = Ae^{mx}$.

Since $\ln A = b$, $A = e^b = e^{2.71} \approx 15.029$.
So, the exponential function that best fits the
data is $y = 15.029e^{0.04x}$.

15. Since $y = Ae^{mx}$,

$\ln y = \ln A + \ln e^{mx} = mx + \ln A$
We can find the least-squares line,
$Y = Mx + b$, using $Y = \ln y$. Then, use
$M = m$ and $b = \ln A$.

x	$y = \ln y$	xy	x^2
2	2.60	5.20	4
4	2.20	8.80	16
6	1.79	10.74	36
8	1.39	11.12	64
10	0.99	9.9	100

$\Sigma x = 30$ $\Sigma y = 8.97$ $\Sigma xy = 45.76$ $\Sigma x^2 = 220$

Using the formulas with $n = 5$,

$$m = \frac{5(45.76) - (30)(8.97)}{5(220) - (30)^2}$$

$$= \frac{-40.3}{200}$$

$$\approx -0.202$$

and

$$b = \frac{(220)(8.97) - (30)(45.76)}{5(220) - (30)^2}$$

$$= \frac{600.6}{200}$$

$$\approx 3.003$$

For our exponential model, $y = Ae^{mx}$.

Since $\ln A = b$, $A = e^b = e^{3.003} \approx 20.15$.
So, the exponential function that best fits
the data is $y = 20.15e^{-0.202x}$.

17. (a)

(b)

x	$y = \ln y$	xy	x^2
5	44	220	25
10	38	380	100
15	32	480	225
20	25	500	400
25	18	450	625
30	12	360	900
35	6	210	1,225
Σx = 140	Σy = 175	$\Sigma xy = 2,600$	Σx^2 = 3,500

Using the formulas with $n = 7$,

$$m = \frac{7(2,600) - (140)(175)}{7(3,500) - (140)^2}$$

$$= \frac{-6,300}{4,900}$$

$$\approx -1.29$$

and

$$b = \frac{(3,500)(175) - (140)(2,600)}{7(3,500) - (140)^2}$$

$$= \frac{248,500}{4,900}$$

$$\approx 50.71.$$

So, the equation of the least-squares line is $y = -1.29x + 50.71$.

(c) If 4,000 units are produced, $x = 40$ and $y = -1.29(40) + 50.71 = -0.89$. Since this predicted price is negative, all 4,000 units cannot be sold at any price.

19. (a) Since $V(t) = Ae^{rt}$,

$$\ln V = \ln A + \ln e^{rt}$$

$$\ln V = rt + \ln A$$

We can find the least-squares line using $y = \ln V$. Then use $m = r$, $x = t$, and $b = \ln A$.

x	$y = \ln V$	xy	x^2
1	4.04	4.04	1
2	4.09	8.18	4
3	4.13	12.39	9
4	4.17	16.68	16
5	4.13	20.65	25
6	4.17	25.02	36
7	4.25	29.75	49
8	4.32	34.56	64
9	4.37	39.33	81
10	4.44	44.40	100
Σx = 55	Σy = 42.11	$\Sigma xy = 235.0$	$\Sigma x^2 = 385$

Using the formulas with $n = 10$,

$$m = \frac{10(235.0) - (55)(42.11)}{10(385) - (55)^2}$$

$$= \frac{33.95}{825}$$

$$\approx 0.041$$

and

$$b = \frac{(385)(42.11) - (55)(235.0)}{10(385) - (55)^2}$$

$$= \frac{3,287.35}{825}$$

$$\approx 3.985$$

For our exponential model, $V(t) = Ae^{rt}$. Since $\ln A = b$,

$$A = e^b = e^{3.985} \approx 53.785.$$

So, the exponential function that best fits the data is $V(t) = 53.785e^{0.041t}$.

Her account is growing at a rate of approximately 4.1% per year.

(b) When $t = 20$,
$$V(20) \approx 53.785e^{0.041(20)}$$
$$\approx 122.1 \text{ thousand, or } \$122,100$$

(c) To find t when $V(t) \approx 300$ thousand,
$$300 = 53.785e^{0.041t}$$
$$5.5778 = e^{0.041t}$$
$$\ln 5.5778 = 0.041t, \text{ or}$$
$$t \approx \frac{\ln 5.5778}{0.041} \approx 42 \text{ years}$$

(d) Using the two points named by Frank,
$$57 = Ae^{r(1)}$$
$$68 = Ae^{r(10)}$$
Solving the first for A and substituting in the second gives
$$68 = (57e^{-r})e^{10r}$$
$$1.19298 = e^{9r}$$
$$\ln 1.19298 = 9r$$
$$\text{or } r \approx \frac{\ln 1.19298}{9} \approx 0.0196 \text{ and}$$
$A = 57e^{-0.0196} \approx 55.89$. Frank's function fits the first and last data point, but may not be a good fit with the other data points. Frank's function would be less usable to predict other values.

21. (a)

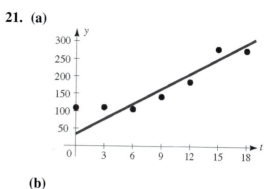

(b)

x	y	xy	x^2
0	109	0	0
3	111	333	9
6	103	618	36
9	142	1,278	81
12	185	2,220	144
15	280	4,200	225
18	276	4,968	324
$\Sigma x = 63$	$\Sigma y = 1,206$	$\Sigma xy = 13,617$	$\Sigma x^2 = 819$

Using the formulas with $n = 7$,
$$m = \frac{7(13,617) - 63(1,206)}{7(819) - (63)^2} \approx 10.96$$
$$b = \frac{819(1,206) - 63(13,617)}{7(819) - (63)^2} \approx 73.61$$
So, the equation of the least squares line is $y = 10.96t + 73.61$, which is a reasonably good fit.

(c) In the year 2015, when $t = 23$, the predicted price is $10.96(23) + 73.61 = 325.69$, or approximately \$3.26 per gallon.

23. (a) Let t denote the number of years after 2004 and y the corresponding GDP. Then,

x	y	xy	x^2
0	15,988	0	0
1	18,494	18,494	1

2	21,631	43,262	4
3	26,581	79,743	9
4	31,405	125,620	16
5	34,051	170,255	25
$\Sigma x = 15$	$\Sigma y = 148{,}150$	$\Sigma xy = 437{,}374$	$\Sigma x^2 = 55$

Using the formulas with $n = 6$,

$$m = \frac{6(437{,}374) - (15)(148{,}150)}{6(55) - (15)^2} \approx 3{,}828.5 \text{ and}$$

$$b = \frac{(55)(148{,}150) - (15)(437{,}374)}{6(55) - (15)^2} \approx 15{,}120.4$$

So, the equation of the least-squares line is $y = 3{,}828.5t + 15{,}120.4$.

(b) In the year 2020, when $t = 16$, the predicted GDP is,

$$y = 3{,}828.5(16) + 15{,}120.4 \approx 76{,}376.4 \text{ or } 76{,}376.4 \text{ billion yuan.}$$

25. (a) Let x be the number of catalogs requested and y the number of applications received (both in units of 1,000). The given points (x, y) are plotted on the accompanying graph.

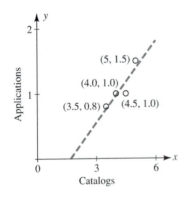

(b)

x	y	xy	x^2
4.5	1.0	4.5	20.25
3.5	0.8	2.8	12.25
4.0	1.0	4.0	16.00
5.0	1.5	7.5	25.00
$\Sigma x = 17.0$	$\Sigma y = 4.3$	$\Sigma xy = 18.8$	$\Sigma x^2 = 73.50$

Using the formulas with $n = 4$,

$$m = \frac{4(18.8) - 17(4.3)}{4(73.5) - (17)^2} \approx 0.42 \text{ and}$$

$$b = \frac{73.5(4.3) - 17(18.8)}{4(73.5) - (17)^2} \approx -0.71$$

So, the equation of the least-squares line is $y = 0.42x - 0.71$.

(c) If 4,800 catalogs are requested by December 1, $x = 4.8$ and $y = 0.42(4.8) - 0.71 = 1.306$, which means that approximately 1,306 completed applications will be received by March 1.

27. (a) Let x denote the number of decades after 1950 and y the corresponding population (in millions). Then,

x	0	1	2	3	4	5
y	150.7	179.3	203.2	226.5	248.7	291.4

Since $y = Ae^{mx}$,

$\ln y = \ln A + \ln e^{mx}$

$\ln y = mx + \ln A$

We can find the least-squares line, $Y = Mx + b$, using $Y = \ln y$. Then, use $M = m$ and $b = \ln A$.

x	$y = \ln y$	xy	x^2
0	5.02	0	0
1	5.19	5.19	1
2	5.31	10.62	4
3	5.42	16.26	9
4	5.52	22.08	16
5	5.67	28.35	25

$\Sigma x = 15 \quad \Sigma y = 32.13 \quad \Sigma xy = 82.5 \quad \Sigma x^2 = 55$

Using the formulas with $n = 6$, $m = \dfrac{6(82.5) - (15)(32.13)}{6(55) - (15)^2} = \dfrac{13.05}{105} \approx 0.124$ and

$b = \dfrac{(55)(32.13) - (15)(82.5)}{6(55) - (15)^2} = \dfrac{529.65}{105} \approx 5.044$. For our exponential model, $P = Ae^{mx}$.

Since $\ln A = b$, $A = e^b = e^{5.044} \approx 155.089$.

So, the exponential function that best fits the data is $P = 155.089e^{0.124x}$. So, the population is growing approximately 12.4% per decade.

(b) In the year 2005, $x = 5.5$ and $P = 155.089e^{0.124(5.5)} \approx 306.74$ million.

In the year 2010, $x = 6$ and $P = 155.089e^{0.124(6)} \approx 326.36$ million.

29. (a) Let x denote the number of hours after the polls open and y the corresponding percentage of registered voters that have already cast their ballots. Then

x	2	4	6	8	10
y	12	19	24	30	37

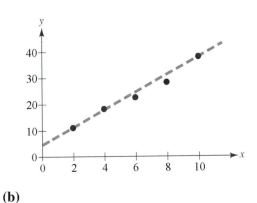

(b)

x	y	xy	x^2
2	12	24	4
4	19	76	16
6	24	144	36
8	30	240	64
10	37	370	100

$\Sigma x = 30 \quad \Sigma y = 122 \quad \Sigma xy = 854 \quad \Sigma x^2 = 220$

Using the formulas with $n = 5$,

$$m = \frac{5(854) - (30)(122)}{5(220) - (30)^2} = \frac{610}{200} = 3.05$$

and $b = \dfrac{(220)(122) - (30)(854)}{5(220) - (30)^2}$

$$= \frac{1,220}{200}$$

$$= 6.10$$

So, the equation of the least-squares line is $y = 3.05x + 6.10$.

(c) When the polls close at 8:00 P.M., $x = 12$ and $y = 3.05(12) + 6.1 = 42.7$, which means that approximately 42.7% of the registered voters can be expected to vote.

31. (a)

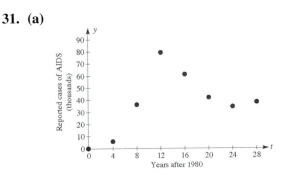

(b)

x	y	xy	x^2
0	99	0	0
4	6,360	25,440	16
8	36,064	288,512	64
12	79,477	953,724	144
16	61,109	977,744	256
20	42,156	843,120	400
24	37,726	905,424	576
28	37,991	1,063,748	784
$\Sigma x =$ 112	Σy = 300,982	Σxy = 5,057,712	Σx^2 = 2,240

Using the formulas with $n = 8$,

$$m = \frac{8(5,057,712) - 112(300,982)}{8(2,240) - (112)^2} \approx 1,255.9$$

$$b = \frac{2,240(300,982) - 112(5,057,712)}{8(2,240) - (112)^2} \approx 20,040.2$$

So, the equation of the least squares line is $y = 1,255.9t + 20,040.2$.

(c) In the year 2012, when $t = 32$, the predicted number of cases is $y = 1,255.9(32) + 20,040.2 = 60,229$.

(d) No. The slope of the least squares line is positive, but since 1992, the number of cases is decreasing. (Writing Exercise—Answers will vary.)

33. (a)

ln W	4.054	4.693	5.297	5.704	5.873	6.040	6.284	6.611
ln C	1.668	2.617	3.645	4.358	4.649	4.905	5.276	5.766

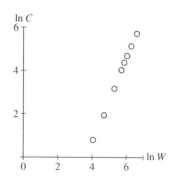

(b)

x	y	xy	x^2
4.054	1.668	6.762	16.435
4.693	2.617	12.282	22.024
5.297	3.645	19.308	28.058
5.704	4.358	24.858	32.536
5.873	4.649	27.304	34.492
6.040	4.905	29.626	36.482
6.284	5.276	33.154	39.489
6.611	5.766	38.119	43.705
Σx $= 44.556$	Σy $= 32.884$	Σxy $= 191.413$	Σx^2 $= 253.221$

Using the formulas with $n = 8$, $m = \dfrac{8(191.413) - (44.556)(32.884)}{8(253.221) - (44.556)^2} \approx \dfrac{66.124}{40.531} \approx 1.631$ and

$b = \dfrac{(253.221)(32.884) - (44.556)(191.413)}{8(253.221) - (44.556)^2} = \dfrac{-201.68}{40.531} \approx -4.976.$

So, the equation of the least-squares line is $y = 1.631x - 4.976$.

(b) $\ln C = 1.631 \ln W - 4.976$

$e^{\ln C} = e^{1.631 \ln W - 4.976}$

$C = e^{\ln W^{1.631}} e^{-4.976}$

$C = e^{-4.976} W^{1.631}$

$C(W) = 0.00690 W^{1.631}$

7.5 Constrained Optimization: The Method of Lagrange Multipliers

1. $f(x, y) = xy$
$g(x, y) = x + y$
$f_x = y;\ f_y = x;\ g_x = 1;\ g_y = 1$
The three Lagrange equations are:
$y = \lambda;\ x = \lambda;\ x + y = 1$
Fom the first two equations, $x = y$ which, when substituted into the third equation
gives $2x = 1$, or $x = \dfrac{1}{2}$.
Since $x = y$, the corresponding value for y is
$y = \dfrac{1}{2}$. So, the constrained maximum is
$f\left(\dfrac{1}{2}, \dfrac{1}{2}\right) = \dfrac{1}{4}$.

3. $f(x, y) = x^2 + y^2$
$g(x, y) = xy$
$f_x = 2x;\ f_y = 2y;\ g_x = y;\ g_y = x$
The three Lagrange equations are:
$2x = \lambda y;\ 2y = \lambda x;\ xy = 1$
Multiply the first equation by y and the
second by x to get $2xy = \lambda y^2$ and
$2xy = \lambda x^2$. Set the two expressions for $2xy$
equal to each other to get $\lambda y^2 = \lambda x^2$,
$y^2 = x^2$, or $x = \pm y$. (Note that another
solution of the equation $\lambda y^2 = \lambda x^2$ is
$\lambda = 0$, which implies that $x = 0$ and $y = 0$,
which is not consistent with the third
equation.)
If $y = x$, the third equation becomes
$x^2 = 1$, which implies that $x = \pm 1$ and
$y = \pm 1$.
If $y = -x$, the third equation becomes
$-x^2 = 1$, which has no solutions. So, the
two points at which the constrained
extrema can occur are $(1, 1)$ and $(-1, -1)$.
Since $f(1, 1) = 2$ and $f(-1, -1) = 2$, the
constrained minimum is 2.

5. $f(x, y) = x^2 - y^2$
$g(x, y) = x^2 + y^2$
$f_x = 2x;\ f_y = -2y;\ g_x = 2x;\ g_y = 2y$
The three Lagrange equations are:
$2x = 2\lambda x;\ -2y = 2\lambda y;\ x^2 + y^2 = 4$
From the first equation, either $\lambda = 1$ or
$x = 0$. If $x = 0$, the third equation becomes
$y^2 = 4$ or $y = \pm 2$. From the second equation,
either $\lambda = -1$ or $y = 0$. If $y = 0$, the third
equation becomes $x^2 = 4$ or
$x = \pm 2$.
If neither $x = 0$ nor $y = 0$, the first equation
implies $\lambda = 1$ while the second equation
implies $\lambda = -1$, which is impossible.
So, the only points at which the constrained
extrema can occur are $(0, -2)$, $(0, 2)$, $(-2, 0)$,
and $(2, 0)$.
Now, $f(0, -2) = -4$, $f(0, 2) = -4$,
$f(-2, 0) = 4$, and $f(2, 0) = 4$. So, the
constrained minimum is -4.

7. $f(x, y) = x^2 - y^2 - 2y$
$g(x, y) = x^2 + y^2$
$f_x = 2x;\ f_y = -2y - 2;\ g_x = 2x;\ g_y = 2y$
The three Lagrange equations are:
$2x = 2\lambda x;\ -2y - 2 = 2\lambda y;\ x^2 + y^2 = 1$
From the first equation, either $\lambda = 1$ or
$x = 0$. If $\lambda = 1$, the second equation
becomes $2y - 2 = 2y$, $4y = -2$, or $y = -\dfrac{1}{2}$.
From the third equation, $x^2 + \left(-\dfrac{1}{2}\right)^2 = 1$,
or $x = \pm \dfrac{\sqrt{3}}{2}$.
If $x = 0$, the third equation becomes
$0^2 + y^2 = 1$ or $y = \pm 1$. So, the only points
at which the constrained extrema can
occur are $\left(-\dfrac{\sqrt{3}}{2}, -\dfrac{1}{2}\right)$, $\left(\dfrac{\sqrt{3}}{2}, -\dfrac{1}{2}\right)$,
$(0, -1)$, and $(0, 1)$. Now,

$f\left(\dfrac{\sqrt{3}}{2}, -\dfrac{1}{2}\right) = f\left(-\dfrac{\sqrt{3}}{2}, -\dfrac{1}{2}\right) = \dfrac{3}{2}$,

$f(0, -1) = 1$, and $f(0, 1) = -3$. So, the

constrained maximum is $\dfrac{3}{2}$ and the

constrained minimum is -3.

9. $f(x, y) = 2x^2 + 4y^2 - 3xy - 2x - 23y + 3$
$g(x, y) = x + y - 15 = 0$
$f_x = 4x - 3y - 2$
$f_y = 8y - 3x - 23$

$g_x = g_y = 1$
The three Lagrange equations are:
$4x - 3y - 2 = \lambda$
$-3x + 8y - 23 = \lambda$
$x + y = 15$
The first two lead to $7x - 11y = -21$.
Substitute $y = 15 - x$ to obtain $18x = 144$
or $x = 8$ and $y = 7$.
The constrained minimum is $f(8, 7) = -18$.

11. $f(x, y) = e^{xy}$

$g(x, y) = x^2 + y^2 - 4 = 0$

$f_x = ye^{xy}$, and $f_y = xe^{xy}$

$g_x = 2x$ and $g_y = 2y$
The three Lagrange equations are:
$ye^{xy} = 2\lambda x$
$xe^{xy} = 2\lambda y$
$x^2 + y^2 - 4 = 0$

Dividing the first two leads to $\dfrac{y}{x} = \dfrac{x}{y}$, or

$x^2 = y^2$. Substitute in $x^2 + y^2 = 4$ to
obtain $x = \pm\sqrt{2}$ and $y = \pm\sqrt{2}$.
Now, $f\left(\sqrt{2}, -\sqrt{2}\right) = f\left(-\sqrt{2}, \sqrt{2}\right) = e^{-2}$

and $f\left(\sqrt{2}, \sqrt{2}\right) = f\left(-\sqrt{2}, -\sqrt{2}\right) = e^{2}$.

So, the constrained maximum is e^2 and

the constrained minimum is e^{-2}.

13. $f(x, y, z) = xyz$
$g(x, y, z) = x + 2y + 3z - 24 = 0$
$f_x = yz,\ f_y = xz,\ \text{and}\ f_z = xy$

$g_x = 1,\ g_y = 2,\ \text{and}\ g_z = 3$
The three Lagrange equations are:
$yz = \lambda;\ xz = 2\lambda;\ xy = 3\lambda$

Dividing the first two leads to $y = \dfrac{x}{2}$,

dividing the first by the third leads to

$z = \dfrac{x}{3}$. Substitute in $x + 2y + 3z = 24$ to

obtain $x = 8$, $y = 4$, and $z = \dfrac{8}{3}$.

The maximum is $f\left(8, 4, \dfrac{8}{3}\right) = \dfrac{256}{3}$.

15. $f(x, y, z) = x + 2y + 3z$
$g(x, y, z) = x^2 + y^2 + z^2 - 16 = 0$
$f_x = 1,\ f_y = 2,\ \text{and}\ f_z = 3$

$g_x = 2x,\ g_y = 2y,\ \text{and}\ g_z = 2z$
The three Lagrange equations are:
$1 = 2\lambda x;\ 2 = 2\lambda y;\ 3 = 2\lambda z$
Dividing the first two leads to $y = 2x$,
dividing the first by the third leads to
$z = 3x$. Substitute in $x^2 + y^2 + z^2 = 16$ to

obtain $x = \pm\dfrac{4}{\sqrt{14}}$, $y = \pm\dfrac{8}{\sqrt{14}}$, and

$z = \pm\dfrac{12}{\sqrt{14}}$.

Now, $f\left(\dfrac{4}{\sqrt{14}}, \dfrac{8}{\sqrt{14}}, \dfrac{12}{\sqrt{14}}\right) = \dfrac{56}{\sqrt{14}} = 4\sqrt{14}$

and

$f\left(\dfrac{-4}{\sqrt{14}}, \dfrac{-8}{\sqrt{14}}, \dfrac{-12}{\sqrt{14}}\right) = -\dfrac{56}{\sqrt{14}} = -4\sqrt{14}$.

So, the constrained maximum is $4\sqrt{14}$
and the constrained minimum is $-4\sqrt{14}$.

17. $P(x, y) = -0.3x^2 - 0.5xy - 0.4y^2 + 85x$
$\qquad\qquad\qquad\qquad\qquad + 125y - 2,500$

Since the constraint is $x + y = 300$
(hundred units), $g(x, y) = x + y$.

$f_x = -0.6x - 0.5y + 85$
$f_y = -0.5x - 0.8y + 125$
$g_x = 1; g_y = 1$
The three Lagrange equations are:
$-0.6x - 0.5y + 85 = \lambda$
$-0.5x - 0.8y + 125 = \lambda$
$x + y = 300$
Equating λ leads to
$-0.6x - 0.5y + 85 = -0.5x - 0.8y + 125$
$$y = \frac{1}{3}x + \frac{400}{3}$$
Substituting in the constraint equation
leads to $x + \frac{1}{3}x + \frac{400}{3} = 300$
$$x = 125 \text{ hundred}$$
$y = 175$ hundred
So, 12,500 Deluxe sets and
17,500 Standard sets should be produced.

19. $S(x, y) = 20x^{3/2}y$

(a) Since the constraint is $x + y = 60$
(thousand dollars), $g(x, y) = x + y$.
$f_x = 30x^{1/2}y; f_y = 20x^{3/2}; g_x = 1;$
$g_y = 1$
The three Lagrange equations are:
$30x^{1/2}y = \lambda, 20x^{3/2} = \lambda, x + y = 60$
Equating λ leads to
$30x^{1/2}y = 20x^{3/2}$
$$y = \frac{2}{3}x$$
Substituting in the constraint equation
leads to $x + \frac{2}{3}x = 60$
$$x = 36, y = 24$$
So, $36,000 should be spent on
development and $24,000 should be
spent on promotion.
$$S(36, 24) = 20(36)^{3/2}(24) = 103,680$$
So, the maximum sales level is 103,680
copies.

(b) Using $20x^{3/2} = \lambda$ with $x = 36$ gives
$\lambda = 4,320$ so, with an extra $1,000

approximately 4,320 more copies will
be sold.

21. $Q(x, y) = 60x^{1/3}y^{2/3}$

(a) Since the constraint is $x + y = 120$
(thousand dollars), $g(x, y) = x + y$.
$f_x = 20x^{-2/3}y^{2/3}; f_y = 40x^{1/3}y^{-1/3};$
$g_x = 1; g_y = 1$
The three Lagrange equations are
$20x^{-2/3}y^{2/3} = \lambda$
$40x^{1/3}y^{-1/3} = \lambda$
$x + y = 120$
Equating λ leads to
$20x^{-2/3}y^{2/3} = 40x^{1/3}y^{-1/3}$
$$y = 2x$$
Substituting in the constraint equation
leads to $x + 2x = 120$
$$x = 40, y = 80$$
So, $40,000 should be spent on labor
and $80,000 should be spent on
equipment.

(b) Using $40x^{1/3}y^{-1/3} = \lambda$ with $x = 40$
and $y = 80$ gives
$$\frac{40(40)^{1/3}}{(80)^{1/3}} = \lambda \approx 31.75.$$
So, with an additional $1,000
approximately 31.75 more units will be
produced.

23. From problem #36, $P_x = P_y = \lambda$. Using
$$P_y = \frac{6,400}{(y+2)^2} - 1,000 \text{ and } y = 5,$$
$$\lambda = P_y = \frac{64,000}{49} - 1,000 = 306.122 \text{ (for}$$
each $1,000)
Since the change in this
promotion/development is $100, the
corresponding change in profit is $30.61
(Remember that the Lagrange multiplier is
the change in maximum profit for a
1 (thousand) dollar change in the constraint.

25. (a) The goal is to maximize utility,
$U(x, y) = 100x^{0.25}y^{0.75}$ subject to the constraint $2x + 5y = 280$, so
$g(x, y) = 2x + 5y$.
$U_x = 25x^{-0.75}y^{0.75}$
$U_y = 75x^{0.25}y^{-0.25}$

$g_x = 2; g_y = 5$
The three Lagrange equations are
$25x^{-0.75}y^{0.75} = 2\lambda$
$75x^{0.25}y^{-0.25} = 5\lambda$
$2x + 5y = 280$

Solving the first two equations for λ and equating yields
$15x^{0.25}y^{-0.25} = 12.5x^{-0.75}y^{0.75}$, or
$y = 1.2x$.
Substituting in the third equation yields
$x = 35$ and $y = 42$.

(b) $\lambda = (15)(35^{0.25})(42^{-0.25}) \approx 14.33$
which approximates the change in maximum utility due to an additional $1.00 in available funds.

27. $\lambda \approx \Delta u$ if $\Delta k = \$1$. Since $U(x, y) = x^\alpha y^\beta$,
$\alpha x^{\alpha-1}y^\beta = \lambda a$ and $k = ax + by$, it follows

that
$$\lambda = \frac{\alpha x^{\alpha-1}y^\beta}{a}$$
$$= \frac{\alpha y^\beta}{ax^{1-\alpha}}$$
$$= \left(\frac{\alpha}{a}\right)\left(\frac{k\beta}{b}\right)^\beta \left(\frac{\alpha}{k\alpha}\right)^{1-\alpha}$$
$$= \left(\frac{\alpha}{a}\right)\left(\frac{k\beta}{b}\right)^\beta \left(\frac{\alpha}{k\alpha}\right)^\beta$$
$$= \left(\frac{\alpha}{a}\right)\left(\frac{k\beta a}{bk\alpha}\right)^\beta$$
$$= \frac{\alpha\beta^\beta a^{\beta-1}}{\alpha^\beta b^\beta}$$
$$= \frac{\alpha^{\beta-1}\beta^\beta}{\alpha^{\beta-1}b^\beta}$$
$$= \left(\frac{\alpha}{a}\right)^\alpha \left(\frac{\beta}{b}\right)^\beta$$

29. Let $Q(x, y)$ be the production level curve subject to $px + qy = k$. The three Lagrange equations then are $Q_x = \lambda p$, $Q_y = \lambda q$, and $px + qy = k$. From the first two equations
$$\frac{Q_x}{p} = \frac{Q_y}{q}.$$

31. Need to find extrema of
$Q(K, L) = 55[0.6K^{-1/4} + 0.4L^{-1/4}]^{-4}$
subject to $g(K, L) = 2K + 5L - 150 = 0$.
$Q_K = -220[0.6K^{-1/4} + 0.4L^{-1/4}]^{-5}$
$$(-0.15K^{-5/4})$$
$Q_L = -220[0.6K^{-1/4} + 0.4L^{-1/4}]^{-5}$
$$(-0.1L^{-5/4})$$

$g_K = 2; g_L = 5$
The three Lagrange equations are
$-220(0.6K^{-1/4} + 0.4L^{-1/4})^{-5}(-0.15K^{-5/4})$
$= 2\lambda$
$-220(0.6K^{-1/4} + 0.4L^{-1/4})^{-5}(-0.1L^{-5/4})$
$= 5\lambda$
$2K + 5L - 150 = 0$
Solving the first two equations for λ gives

$$\frac{33K^{-5/4}(0.6K^{1/4}+0.4L^{-1/4})^{-5}}{2}=\lambda$$

$$\frac{22L^{-5/4}(0.6K^{1/4}+0.4L^{-1/4})^{-5}}{5}=\lambda$$

Setting these equal,

$$\frac{33K^{-5/4}(0.6K^{1/4}+0.4L^{-1/4})^{-5}}{2}$$

$$=\frac{22L^{-5/4}(0.6K^{1/4}+0.4L^{-1/4})^{-5}}{5}$$

$$\frac{33K^{-5/4}}{2}=\frac{22L^{-5/4}}{5}$$

$$165L^{5/4}=44K^{5/4}$$

$$L=\left(\frac{44K^{5/4}}{165}\right)^{4/5}=\left(\frac{44}{165}\right)^{4/5}K$$

Use the third equation,

$$2K+5\left(\frac{44}{165}\right)^{4/5}K-150=0$$

$$K\approx 40.14$$

$$L\approx\left(\frac{44}{165}\right)^{4/5}K\approx 13.89$$

$$Q(40.14, 13.89)$$

$$\approx 55[0.6(40.14)^{-1/4}+0.4(13.89)^{-1/4}]^{-4}$$

$$\approx 1395.4$$

33. $Q(K, L) = A[\alpha K^{-\beta}+(1-\alpha)L^{-\beta}]^{-1/\beta}$

Since the constraint is $c_1K+c_2L=B$,

$g(K, L) = c_1K+c_2L$.

Q_K

$$=-\frac{A}{\beta}[\alpha K^{-\beta}+(1-\alpha)L^{-\beta}]^{-1/\beta-1}$$

$$\qquad\qquad(-\alpha\beta K^{\beta-1})$$

$$=A\alpha K^{\beta-1}[\alpha K^{-\beta}+(1-\alpha)L^{-\beta}]^{-1/\beta-1}$$

$$Q_L=-\frac{A}{\beta}[\alpha K^{-\beta}+(1-\alpha)L^{-\beta}]^{-1/\beta-1}$$

$$\qquad\qquad(-\beta(1-\alpha)L^{-\beta-1})$$

$$=A(1-\alpha)L^{-\beta-1}[\alpha K^{-\beta}+(1-\alpha)L^{-\beta}]^{-1/\beta-1}$$

$g_K=c_1;\ g_L=c_2$

The three Lagrange equations are:

$$A\alpha K^{-\beta-1}[\alpha K^{-\beta}+(1-\alpha)L^{-\beta}]^{-1/\beta-1}$$
$$=c_1\lambda$$

$$A(1-\alpha)L^{-\beta-1}[\alpha K^{-\beta}+(1-\alpha)L^{-\beta}]^{-1/\beta-1}$$
$$=c_2\lambda$$

$$c_1K+c_2L=B$$

Solving both for λ and equating gives

$$\frac{A\alpha K^{-\beta-1}[\alpha K^{-\beta}+(1-\alpha)L^{-\beta}]^{-1/\beta-1}}{c_1}$$

$$=\frac{A(1-\alpha)L^{-\beta-1}[\alpha K^{-\beta}+(1-\alpha)L^{-\beta}]^{-1/\beta-1}}{c^2}$$

$$\frac{\alpha K^{-\beta-1}}{c_1}=\frac{(1-\alpha)L^{-\beta-1}}{c_2}$$

$$\left(\frac{K}{L}\right)^{-\beta-1}=\frac{c_1}{c_2}\left(\frac{1-\alpha}{\alpha}\right)$$

35. $\dfrac{x^2}{4}+\dfrac{y^2}{9}=1$

(a) Need to maximize

$$f(x, y)=(x-1)^2+(y-1)^2$$

subject to

$$g(x, y)=\frac{x^2}{4}+\frac{y^2}{9}.\text{ So,}$$

$$f_x(x, y)=2(x-1)$$

$$f_y(x, y)=2(y-1)$$

$$g_x(x, y)=\frac{x}{2}$$

$$g_y(x, y)=\frac{2}{9}y$$

and the three Lagrange equations are

$$2(x-1)=\frac{x}{2}\lambda$$

$$2(y-1)=\frac{2}{9}y\lambda$$

$$g(x, y)=\frac{x^2}{4}+\frac{y^2}{9}=1$$

Equating λ leads to

$$\frac{4(x-1)}{x} = \frac{9(y-1)}{y}$$

$$4xy - 4y = 9xy - 9x$$

$$9x = 5xy + 4y$$

$$y = \frac{9x}{5x+4}$$

Substituting into the constraint equation leads to

$$\frac{x^2}{4} + \frac{\left(\dfrac{9x}{5x+4}\right)^2}{9} = 1$$

$$\frac{x^2}{4} + \frac{81x^2}{9(5x+4)^2} = 1$$

$$\frac{x^2}{4} + \frac{9x^2}{(5x+4)^2} = 1$$

Multiplying both sides by $4(5x+4)^2$,

$$x^2(5x+4)^2 + 36x^2 = 4(5x+4)^2$$

$$x^2(25x^2 + 40x + 16) + 36x^2$$
$$= 4(25x^2 + 40x + 16)$$

$$25x^4 + 40x^3 + 52x^2$$
$$= 100x^2 + 160x + 64$$

$$25x^4 + 40x^3 - 48x^2 - 160x - 64$$
$$= 0$$

Solving using a calculator or computer software, $x \approx -0.49$ and

$$y = \frac{9(-0.49)}{5(-0.49)+4} \approx -2.91$$

$x \approx 1.82$ and

$$y = \frac{9(1.82)}{5(1.82)+4} \approx 1.25$$

To find the maximum,
$f(-0.49, -2.91) \approx 17.5082$
$f(1.82, 1.25) \approx 0.7349$
Since f measures the square of the

distance, the radius is
$$\sqrt{17.5082} \approx 4.18 \text{ miles.}$$

(b) Writing exercise - Answers will vary.

37. Let S denote the surface area of the bacterium. Then, $S(R, Y) = 2\pi R^2 + 2\pi RH$. The goal is to maximize this function subject to the constraint $\pi R^2 H = C$ (volume is fixed, C is a constant), so

$g(R, Y) = \pi R^2 H$.
$S_R = 4\pi R + 2\pi H$; $S_H = 2\pi R$;

$g_R = 2\pi RH$; $g_H = \pi R^2$
The three Lagrange equations are:
$4\pi R + 2\pi H = 2\pi\lambda RH$

$$2\pi R = \pi\lambda R^2$$
$$\pi R^2 H = C$$

The second equation leads to $\lambda = \dfrac{2}{R}$, which leads to $2R = H$, using the first equation.

39. $(R-20)^2 + 25(F-5)^2 = 234$

The goal is to maximize $F + R$, subject to the above constraint. Let T be the function for the total number of foxes plus rabbits. Then,

$$T(F, R) = F + R$$

$$g(F, R) = (R-20)^2 + 25(F-5)^2$$
and
$$T_F = 1; T_5 = 1$$
$$g_F = 50(F-5); g_R = 2(R-20)$$
The three Lagrange equations are
$$1 = 50(F-5)\lambda$$
$$1 = 2(R-20)\lambda$$
$$(R-20)^2 + 25(F-5)^2 = 234$$
Equating λ leads to

$$\frac{1}{50(F-5)} = \frac{1}{2(R-20)}$$

$$2(R-20) = 50(F-5)$$

$$(R-20) = 25(F-5)$$

Substituting in the constraint equation leads to

$$\left[25(F-5)\right]^2 + 25(F-5)^2 = 234$$

$$625(F-5)^2 + 25(F-5)^2 = 234$$

$$650(F-5)^2 = 234$$

$$(F-5)^2 = 0.36$$

$$F-5 = 0.6$$

$$F = 5.6$$

and $R - 20 = 25(5.6-5)$

$$R - 20 = 15$$

$$R = 35$$

So, the largest total number is 560 foxes + 3,500 rabbits = 4,060.

41. Let f denote the amount of fencing needed to enclose the pasture, x the side parallel to the river and y the sides perpendicular to the river. Then, $f(x, y) = x + 2y$.
The goal is to minimize this function subject to the constraint that the area $xy = 3,200$, so $g(x, y) = xy$.
The partial derivatives are $f_x = 1$, $f_y = 2$, $g_x = y$, and $g_y = x$.
The three Lagrange equations are
$$1 = \lambda y; \ 2 = \lambda x; \ xy = 3,200$$

From the first equation, $\lambda = \dfrac{1}{y}$. From the second equation $\lambda = \dfrac{2}{x}$. Setting the two expressions for λ equal to each other gives $\dfrac{1}{y} = \dfrac{2}{x}$ or $x = 2y$, and substituting this into the third equation yields $2y^2 = 3,200$, $y^2 = 1,600$, or $y = \pm 40$.
Only the positive value is meaningful in the context of this problem. So, $y = 40$, and (since $x = 2y$), $x = 80$. That is, to minimize the amount of fencing, the dimensions of the field should be
40 meters by 80 meters.

43. Let f denote the volume of the parcel. Then,
$$f(x, y) = x^2 y.$$
The girth $4x$ plus the length y can be at most 108 inches. The goal is to maximize this function $f(x, y)$ subject to the constraint $4x + y = 108$, so
$g(x, y) = 4x + y$.
The partial derivatives are $f_x = 2xy$, $f_y = x^2$, $g_x = 4$, and $g_y = 1$.
The three Lagrange equations are
$$2xy = 4\lambda; \ x^2 = \lambda; \ 4x + y = 108$$
From the first equation, $\lambda = \dfrac{xy}{2}$, which, combined with the second equation, gives
$\dfrac{xy}{2} = x^2$ or $y = 2x$. (Another solution is $x = 0$, which is impossible in the context of this problem.)
Substituting $y = 2x$ into the third equation gives $6x = 108$ or $x = 18$, and since $y = 2x$, the corresponding value of y is $y = 36$.
So, the largest volume is
$$f(18, 36) = (18)^2(36) = 11,664 \text{ cubic inches.}$$

45. Let M denote the amount of metal used to construct the can. Then,
$$M(R, H) = 2\pi R^2 + 2\pi RH.$$
The goal is to maximize this function $M(R, H)$ subject to the constraint that (volume) $\pi R^2 H = 6.89\pi$, so
$g(R, H) = \pi R^2 H$. The partial derivatives are $M_R = 4\pi R + 2\pi H$; $M_H = 2\pi R$,
$g_R = 2\pi RH$; $g_H = \pi R^2$
The three Lagrange equations are:
$$4\pi R + 2\pi H = 2\pi\lambda RH$$
$$2\pi R = \pi\lambda R^2$$
$$\pi R^2 H = 6.89\pi$$

The second equation leads to $\lambda = \dfrac{2}{R}$,

which leads to $2R = H$, using the first equation. Using the third equation yields

$H = \sqrt[3]{27.56} \approx 3.02$, and $R = \dfrac{H}{2} \approx 1.51$.

So, the amount of metal is minimized when the can's radius is 1.51 inches and its height is 3.02 inches.

47. The goal is to maximize $s(d_o, d_i) = d_o + d_i$ subject to the constraint $\dfrac{1}{d_o} + \dfrac{1}{d_i} = \dfrac{1}{L}$, so

$g(d_o, d_i) = \dfrac{1}{d_o} + \dfrac{1}{d_i}$.

$s_{d_o} = 1; s_{d_i} = 1; \; g_{d_o} = -\dfrac{1}{d_o^2}; \; g_{d_i} = -\dfrac{1}{d_i^2}$

The three Lagrange equations are:

$1 = \lambda \cdot \dfrac{-1}{d_o^2}; \quad 1 = \lambda \cdot \dfrac{-1}{d_i^2}; \quad \dfrac{1}{d_o} + \dfrac{1}{d_i} = L$

This leads to $\lambda = -(d_o^2); \; \lambda = -(d_i)^2$, or

$d_o = d_i$

Substituting into the third equation, $d_o = d_i = 2L$ and the maximum value of s is $4L$.

49. Let k be the cost per square cm of the bottom and sides. Then the cost of the top is $2k$ per square cm and the cost of the interior partitions is $\dfrac{2k}{3}$ per square cm.

The goal is to minimize the cost of the box,

$C(x, y) = k(x^2 + 4xy) + 2kx^2 + \dfrac{2k}{3}(2xy)$

subject to the constraint $x^2 y = 800$, so

$g(x, y) = x^2 y$.

$C_x = 6kx + \dfrac{16}{3}ky$

$C_y = \dfrac{16}{3}kx, \; g_x = 2xy, \; g_y = x^2$

The three Lagrange equations are:

$6kx + \dfrac{16}{3}ky = 2\lambda xy; \; \dfrac{16}{3}kx = \lambda x^2;$

$x^2 y = 800$

Solving the first two equations for λ and equating yields $x = \dfrac{8}{9}y$. Substituting into the third equation yields $\dfrac{9}{8}x^3 = 800$, or

$x = 8.93$ and $y = 10.04$.

51. $E(x, y, z) = \dfrac{k^2}{8m}\left(\dfrac{1}{x^2} + \dfrac{1}{y^2} + \dfrac{1}{z^2}\right)$

$g(x, y) = xyz$

$E_x = \dfrac{k^2}{8m}\left(-\dfrac{2}{x^3}\right)$

$E_y = \dfrac{k^2}{8m}\left(-\dfrac{2}{y^3}\right)$

$E_z = \dfrac{k^2}{8m}\left(-\dfrac{2}{z^3}\right)$

$g_x = yz, \; g_y = xz,$ and $g_z = xy$

The three Lagrange equations are:

$\dfrac{k^2}{8m}\left(-\dfrac{2}{x^3}\right) = \lambda yz$

$\dfrac{k^2}{8m}\left(-\dfrac{2}{y^3}\right) = \lambda xz$

$\dfrac{k^2}{8m}\left(-\dfrac{2}{z^3}\right) = \lambda xy$

Dividing the first two leads to $y^2 = x^2$, or $y = x$; dividing the first by the third leads to $z^2 = x^2$, or $z = x$. Substitute in $xyz = V_0$ to obtain $x = y = z = V_0^{1/3}$. The minimum E is $\dfrac{3k^2}{8m}V_0^{-2/3}$.

53. Let x denote the length of the shed, y the width, and z the height. the goal is to maximize the volume, $V = xyz$ subject to the constraint
$15xy + 12(2yz + xz) + 20xz = 8,000$

so $g(x, y, z) = 15xy + 12(2yz + xz) + 20xz$.

$f_x = yz$, $f_y = xz$, and $f_z = xy$

$g_x = 15y + 32z$

$g_y = 15x + 24z$

$g_z = 24y + 32x$

The three Lagrange equations are:

$yz = \lambda(15y + 32z)$

$xz = \lambda(15x + 24z)$

$xy = \lambda(24y + 32x)$

Dividing the first two leads to $y = \dfrac{4x}{3}$ and

dividing the first by the third leads to

$z = \dfrac{5x}{8}$. Substitute in

$15xy + 24yz + 32xz = 8{,}000$ to obtain

$x = \dfrac{20\sqrt{3}}{3} \approx 11.55$ ft

$y = \dfrac{80\sqrt{3}}{9} \approx 15.40$ ft

$z = \dfrac{25\sqrt{3}}{6} \approx 7.22$ ft

55. Need to find extrema of $f(x, y) = x - y$

subject to $g(x, y) = x^5 + x - 2 - y = 0$.

$f_x = 1$, $f_y = -1$, $g_x = 5x^4 + 1$, $g_y = -1$

The three Lagrange equations are:

$1 = \lambda(5x^4 + 1)$

$-1 = \lambda(-1)$

$x^5 + x - 2 - y = 0$

From the second equation, $\lambda = 1$. Then, from
the first equation, $1 = 5x^4 + 1$

$5x^4 = 0$

$x = 0$

Finally, from the third equation,

$-2 - y = 0$

$y = -2$

Therefore, a possible extremum occurs at the
point $(0, -2)$. However, $f(1, 0) = 1$ and
$f(-1, -4) = 3$, which shows $f(0, -2) = 2$ is
not a local maximum or minimum point.
Press $\boxed{y=}$.

Input $x \wedge 5 + x - 2$ for $y_1 =$ and input

$x - L_1$ for $y_2 =$.

From the home screen, input $\{2, 1, 0, -1\}$
$\boxed{\text{sto}\rightarrow}$ $\boxed{\text{2nd}}$ L_1.

Use window dimensions $[-4, 4]1$ by
$[-4, 4]1$.

Press $\boxed{\text{Graph}}$.

From the graphs that the point $(0, -2)$ is an
inflection point.

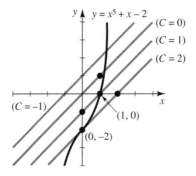

57. $F(x, y) = xe^{xy^2} + \dfrac{y}{x} + x\ln(x + y)$

(a) $0 = (x)\left[e^{xy^2}\left(x \cdot 2y\dfrac{dy}{dx} + y^2 \cdot 1\right)\right] + (e^{xy^2})(1) + \dfrac{x\frac{dy}{dx} - y \cdot 1}{x^2} + (x)\left[\dfrac{1}{x+y}\left(1 + \dfrac{dy}{dx}\right)\right] + \ln(x+y) \cdot 1$

$0 = 2x^2 y e^{xy^2}\dfrac{dy}{dx} + xy^2 e^{xy^2} + e^{xy^2} + \dfrac{1}{x}\dfrac{dy}{dx} - \dfrac{y}{x^2} + \dfrac{x}{x+y} + \dfrac{x}{x+y}\dfrac{dy}{dx} + \ln(x+y)$

$\dfrac{y}{x^2} - \dfrac{x}{x+y} - xy^2 e^{xy^2} - e^{xy^2} - \ln(x+y) = \left(2x^2 y e^{xy^2} + \dfrac{1}{x} + \dfrac{x}{x+y}\right)\dfrac{dy}{dx}$

$\dfrac{dy}{dx} = \dfrac{-xy^2 e^{xy^2} - e^{xy^2} + \frac{y}{x^2} - \frac{x}{x+y} - \ln(x+y)}{2x^2 y e^{xy^2} + \frac{1}{x} + \frac{x}{x+y}} = -\dfrac{xy^2 e^{xy^2} + e^{xy^2} - \frac{y}{x^2} + \frac{x}{x+y} + \ln(x+y)}{2x^2 y e^{xy^2} + \frac{1}{x} + \frac{x}{x+y}}$

(b) $F_x = xy^2 e^{xy^2} + e^{xy^2} - \dfrac{y}{x^2} + \dfrac{x}{x+y} + \ln(x+y)$

$F_y = 2x^2 y e^{xy^2} + \dfrac{1}{x} + \dfrac{x}{x+y}$

$\dfrac{dy}{dx} = -\dfrac{F_x}{F_y} = -\dfrac{xy^2 e^{xy^2} + e^{xy^2} - \frac{y}{x^2} + \frac{x}{x+y} + \ln(x+y)}{2x^2 y e^{xy^2} + \frac{1}{x} + \frac{x}{x+y}}$

59. Minimize $f(x, y) = \ln(x + 2y)$ subject to $xy + y = 5$.

$f(x, y) = \ln(x + 2y)$; $g(x, y) = xy + y - 5 = 0$

$f_x = \dfrac{1}{x + 2y}$; $f_y = \dfrac{2}{x + 2y}$

$g_x = y$; $g_y = x + 1$

The three Lagrange equations are:

$\dfrac{1}{x + 2y} = \lambda y$

$\dfrac{2}{x + 2y} = \lambda(x + 1)$

$xy + y = 5$

From the first equation, $\lambda = \dfrac{1}{y(x + 2y)}$.

From the second equation,

$\lambda = \dfrac{2}{(x+1)(x + 2y)}$.

Equating these two gives

$(x + 1)(x + 2y) = 2y(x + 2y)$ or $y = \dfrac{(x+1)}{2}$.

Substituting $y = \dfrac{1}{2}(x + 1)$ into the third equation,

$\dfrac{1}{2}x(x+1) + \dfrac{1}{2}(x+1) = 5$

$x(x+1) + (x+1) = 10$

$(x+1)[x+1] = 10$

$(x+1)^2 = 10$

This gives $x \approx -4.1623$, $x \approx 2.1623$. This leads to the points $(-4.1623, -1.5811)$ and $(2.1623, 1.5811)$. We cannot use $(-4.1623, -1.5811)$ since this point leads $f(x, y)$ to be undefined.
Find
$f(2.1623, 1.5811)$
$= \ln[2.1623 + 2(1.5811)]$
$\approx 1.6724.$

61. $f(x, y) = xe^{x^2 - y}$ and

$g(x, y) = x^2 + 2y^2 - 1 = 0$

$f_x = e^{x^2 - y} + (x)(e^{x^2 - y})(2x)$

$$f_y = xe^{x^2-y}(-1) = -xe^{x^2-y}$$

$$f_x = (2x^2+1)(e^{x^2-y})$$

$$g_x = 2x \qquad g_y = 4y$$

The three Lagrange equations are:

$$(2x^2+1)(e^{x^2-y}) = \lambda(2x)$$

$$-xe^{x^2-y} = \lambda(4y)$$

$$x^2 + 2y^2 = 1$$

From the first equation,

$$\lambda = \frac{(2x^2+1)(e^{x^2-y})}{2x}.$$

From the second equation, $\lambda = \dfrac{-xe^{x^2-y}}{4y}$.

Equating these and simplifying,

$$8x^2y + 4y = -2x^2$$

$$2x^2(1+4y) = -4y$$

$$x^2 = -\frac{2y}{4y+1}$$

Substituting this into the third equation,

$$-\frac{2y}{4y+1} + 2y^2 = 1$$

$$-2y + 2y^2(4y+1) = 4y+1$$

$$8y^2 + 2y^2 - 6y - 1 = 0$$

To solve, press y= and input

$8x\wedge 3 + 2x^2 - 6x - 1$ for $y_1 =$ (remember we are actually solving for y).

Use window dimensions $[-4, 4]1$ by $[-4, 4]1$.

Press Graph.

Use the zero function under the calc menu to find that $y \approx -0.9184$, $y \approx -0.1636$, and $y \approx 0.832$. We reject $y \approx -0.9184$ and $y \approx 0.832$ since these would result in x being undefined. If $y = -0.1636$, then $x = \pm 0.9729$.

The two points for consideration are $(0.9729, -0.1636)$ and $(-0.9729, -0.1636)$.

Press y=.

Input $xe\wedge(x^2 - L_1)$ for $y_1 =$.

From the home screen, input $\{-0.1636\}$

sto→ 2nd L_1.

Press Graph.

Use the value function under the calc menu to find $f(0.9729, -0.1636) \approx 2.952$ and $f(-0.9729, -0.1636) \approx -2.952$

The maximum point is $(0.9729, -0.1636)$.

7.6 Double Integrals

1.
$$\int_0^1 \int_1^2 x^2 y \, dx\, dy = \int_0^1 \left[\int_1^2 x^2 y \, dx \right] dy$$

$$= \int_0^1 \left[\frac{x^3}{3} y \bigg|_1^2 \right] dy$$

$$= \int_0^1 \left[\frac{8}{3} y - \frac{1}{3} y \right] dy$$

$$= \frac{7}{6} y^2 \bigg|_0^1$$

$$= \frac{7}{6}$$

3.
$$\int_0^{\ln 2} \int_{-1}^0 2xe^y \, dx\, dy = \int_0^{\ln 2} \left[\int_{-1}^0 2xe^y \, dx \right] dy$$

$$= \int_0^{\ln 2} \left[x^2 e^y \bigg|_{-1}^0 \right] dy$$

$$= \int_0^{\ln 2} [-e^y] \, dy$$

$$= -e^y \bigg|_0^{\ln 2}$$

$$= -1$$

5.
$$\int_1^3 \int_0^1 \frac{2xy}{x^2+1} \, dx\, dy = \int_1^3 \left[\int_0^1 \frac{2xy}{x^2+1} \, dx \right] dy$$

$$= \int_1^3 \left[y \ln(x^2+1) \bigg|_0^1 \right] dy$$

$$= \int_1^3 y \ln 2 \, dy$$

$$= \ln 2 \left(\frac{1}{2} \right) y^2 \bigg|_1^3$$

$$= 4 \ln 2$$

7. $\int_0^4 \int_{-1}^1 x^2 y \, dx \, dy = \int_0^4 \left[\int_{-1}^1 x^2 y \, dy \right] dx$

$$= \int_0^4 \left[\frac{y^2}{2} x^2 \Big|_{-1}^1 \right] dx$$

$$= 0$$

9. $\int_2^3 \int_1^2 \frac{x+y}{xy} \, dy \, dx = \int_2^3 \left[\int_1^2 \left(\frac{1}{y} + \frac{1}{x} \right) dy \right] dx$

$$= \int_2^3 \left[\ln(y) + \frac{y}{x} \right]_1^2 dx$$

$$= (x \ln 2 + \ln x) \Big|_2^3$$

$$= \ln 2 + \ln \frac{3}{2}$$

$$= \ln 3$$

11. $\int_0^4 \int_0^{\sqrt{x}} x^2 y \, dy \, dx = \int_0^4 \left[\int_0^{\sqrt{x}} x^2 y \, dy \right] dx$

$$= \int_0^4 \left[\frac{x^2 y^2}{2} \Big|_0^{\sqrt{x}} \right] dx$$

$$= \int_0^4 \frac{x^3}{2} \, dx$$

$$= \frac{x^4}{8} \Big|_0^4$$

$$= 32$$

13. $\int_0^1 \int_{y-1}^{1-y} (2x+y) \, dx \, dy$

$$= \int_0^1 \left[\int_{y-1}^{1-y} (2x+y) \, dx \right] dy$$

$$= \int_0^1 \left[(x^2 + xy) \Big|_{y-1}^{1-y} \right] dy$$

$$= \int_0^1 [[(1-y)^2 + (1-y)y]$$

$$\qquad\qquad - [(y-1)^2 + (y-1)y]] dy$$

$$= \int_0^1 2y - 2y^2 \, dy$$

$$= y^2 - \frac{2y^3}{3} \Big|_0^1$$

$$= \frac{1}{3}$$

15. $\int_0^1 \int_0^4 \sqrt{xy} \, dy \, dx = \int_0^1 \left[\int_0^4 x^{1/2} y^{1/2} \, dy \right] dx$

$$= \int_0^1 \left[\frac{2x^{1/2} y^{3/2}}{3} \Big|_0^4 \right] dx$$

$$= \int_0^1 \frac{16 x^{1/2}}{3} \, dx$$

$$= \frac{32 x^{3/2}}{9} \Big|_0^1$$

$$= \frac{32}{9}$$

17. $\int_1^e \int_0^{\ln x} xy \, dy \, dx = \int_1^e \left[\int_0^{\ln x} xy \, dy \right] dx$

$$= \int_1^e \left[\frac{xy^2}{2} \Big|_0^{\ln x} \right] dx$$

$$= \int_1^e \frac{x(\ln x)^2}{2} \, dx$$

Using integration by parts with $u = (\ln x)^2$

and $dV = \frac{x}{2} dx$

$$= \frac{x^2}{4}(\ln x)^2 \Big|_1^e - \int_1^e \frac{x}{2} \ln x \, dx$$

$$= \frac{e^2}{4} - \int_1^e \frac{x}{2} \ln x \, dx$$

Using integration by parts again, with

$u = \ln x$ and $dV = \frac{x}{2} dx$

$$= \frac{e^2}{4} - \left[\frac{x^2}{4} \ln x \Big|_1^e - \int_1^e \frac{x}{4} dx \right]$$

$$= \frac{e^2}{4} - \left[\left(\frac{x^2}{4} \ln x - \frac{x^2}{8} \right) \Big|_1^e \right]$$

$$= \frac{e^2}{4} - \left[\left(\frac{e^2}{4} - \frac{e^2}{8} \right) - \left(0 - \frac{1}{8} \right) \right]$$

$$= \frac{e^2 - 1}{8}$$

19. Solving $x^2 = 3x$ yields $x = 0$ and $x = 3$. Similarly, after solving each equation for x,

$\sqrt{y} = \frac{y}{3}$ when $y = 0$ and $y = 9$. So, R can be

described in terms of vertical cross sections
by $0 \le x \le 3$ and $x^2 \le y \le 3x$ and in terms of
horizontal cross sections by

$0 \le y \le 9$ and $\frac{y}{3} \le x \le \sqrt{y}$.

21. The given points form a rectangle. So, R can be described in terms of vertical cross sections by $-1 \le x \le 2$ and $1 \le y \le 2$ and in terms of horizontal cross sections by $1 \le y \le 2$ and $-1 \le x \le 2$.

23. Solving $\ln x = 0$ yields $x = 1$, with the second boundary given as $x = e$. Similarly, solving $y = \ln x$ for x yields $x = e^y$, with the second boundary given as $y = 0$. So, R can be described in terms of vertical cross sections by $1 \le x \le e$ and $0 \le y \le \ln x$ and in terms of horizontal cross sections by $0 \le y \le 1$ and $e^y \le x \le e$.

25.
$$\iint_R 3xy^2 \, dA = \int_{-1}^0 \int_{-1}^2 3xy^2 \, dx \, dy$$

$$= \int_{-1}^0 \left[\int_{-1}^2 3xy^2 \, dx \right] dy$$

$$= \int_{-1}^0 \left[\frac{3x^2 y^2}{2} \Big|_{-1}^2 \right] dy$$

$$= \int_{-1}^0 \frac{9y^2}{2} \, dy$$

$$= \frac{3y^3}{2} \Big|_{-1}^0$$

$$= \frac{3}{2}$$

Note: problem can be equivalently worked

as $\int_{-1}^2 \int_{-1}^0 3xy^2 \, dy \, dx$.

27. Since the line joining the points $(0, 0)$ and $(1, 1)$ is $y = x$,

$$\iint_R xe^y \, dA = \int_0^1 \int_0^x xe^y \, dy \, dx$$

$$= \int_0^1 \left[\int_0^x xe^y \, dy \right] dx$$

$$= \int_0^1 \left[xe^y \Big|_0^x \right] dx$$

$$= \int_0^1 (xe^x - x) \, dx$$

$$= \int_0^1 xe^x \, dx - \int_0^1 x \, dx$$

$$= \int_0^1 xe^x \, dx - \frac{x^2}{2} \Big|_0^1$$

$$= \int_0^1 xe^x \, dx - \frac{1}{2}$$

Using integration by parts with $u = x$ and

$dV = e^x \, dx$

$$= xe^x \Big|_0^1 - \int_0^1 e^x \, dx - \frac{1}{2}$$

$$= (xe^x - e^x) \Big|_0^1 - \frac{1}{2}$$

$$= \frac{1}{2}$$

Note: problem can be equivalently worked as $\int_0^1 \int_0^y x e^y \, dx \, dy$.

29. Solving $x^2 = 2x$ yields $x = 0$ and $x = 2$, so

$$\iint_R (2y - x)dA$$

$$= \int_0^2 \int_{x^2}^{2x} (2y - x) dy\, dx$$

$$= \int_0^2 \left[\int_{x^2}^{2x} (2y - x) dy \right] dx$$

$$= \int_0^2 \left[(y^2 - xy) \Big|_{x^2}^{2x} \right] dx$$

$$= \int_0^2 [[(2x)^2 - x(2x)] - [(x^2)^2 - x(x^2)]] dx$$

$$= \int_0^2 (2x^2 - x^4 + x^3) dx$$

$$= \left(\frac{2x^3}{3} - \frac{x^5}{5} + \frac{x^4}{4} \right) \Big|_0^2$$

$$= \frac{44}{15}$$

Note: problem can be equivalently worked as $\int_0^4 \int_{\sqrt{y}}^{y/2} (2y - x) dx\, dy$.

31. The line joining the points $(-1, 0)$ and $(0, 1)$ is $y = x + 1$, or $x = y - 1$. Similarly, the line joining the points $(0, 1)$ and $(1, 0)$ is $y = 1 - x$, or $x = 1 - y$. So,

$$\iint_R (2x + 1)dA = \int_0^1 \int_{y-1}^{1-y} (2x + 1) dx\, dy$$

$$= \int_0^1 \left[\int_{y-1}^{1-y} (2x + 1) dx \right] dy$$

$$= \int_0^1 \left[(x^2 + x) \Big|_{y-1}^{1-y} \right] dy$$

$$= \int_0^1 [[(1 - y)^2 + (1 - y)]$$

$$\quad - [(y - 1)^2 + (y - 1)]] dy$$

$$= \int_0^1 (2 - 2y) dy$$

$$= (2y - y^2) \Big|_0^1$$

$$= 1$$

33. After solving each equation for x, $2y = -y$ when $y = 0$, with the other boundary given as $y = 2$. So,

$$\iint_R \frac{1}{y^2 + 1} dA = \int_0^2 \int_{-y}^{2y} \frac{1}{y^2 + 1} dx\, dy$$

$$= \int_0^2 \left[\int_{-y}^{2y} \frac{1}{y^2 + 1} dx \right] dy$$

$$= \int_0^2 \left[\frac{x}{y^2 + 1} \Big|_{-y}^{2y} \right] dy$$

$$= \int_0^2 \left[\frac{2y}{y^2 + 1} - \frac{-y}{y^2 + 1} \right] dy$$

$$= \int_0^2 \frac{3y}{y^2 + 1} dy$$

$$= 3 \int_0^2 \frac{y}{y^2 + 1} dy$$

Using substitution with $u = y^2 + 1$,

$$= 3 \int_1^5 \frac{1}{u} \cdot \frac{1}{2} du$$

$$= \frac{3}{2} \int_1^5 \frac{1}{u} du$$

$$= \frac{3}{2} \left(\ln |u| \Big|_1^5 \right)$$

$$= \frac{3}{2} (\ln 5 - \ln 1)$$

$$= \frac{3 \ln 5}{2}$$

35. After solving each equation for x, $y^{1/3} = y$ when $y = 0$ and $y = 1$. So,

$$\iint\limits_R 12x^2 e^{y^2}\, dA$$

$$= \int_0^1 \int_y^{y^{1/3}} 12x^2 e^{y^2}\, dx\, dy$$

$$= \int_0^1 \left[\int_y^{y^{1/3}} 12x^2 e^{y^2}\, dx \right] dy$$

$$= \int_0^1 \left[4x^3 e^{y^2} \Big|_y^{y^{1/3}} \right] dy$$

$$= \int_0^1 (4y e^{y^2} - 4y^3 e^{y^2})\, dy$$

$$= 4\int_0^1 y e^{y^2}\, dy - 4\int_0^1 y^2 (y e^{y^2})\, dy$$

Using substitution for the first integral with $u = y^2$, and using integration by parts for the second integral with $u = y^2$ and

$dV = y e^{y^2}\, dy$ (where solving for V requires substitution as well)

$$= 4\int_0^1 \frac{e^u}{2}\, du - 4\left[\frac{y^2}{2} e^{y^2} \Big|_0^1 - \int_0^1 y e^{y^2}\, dy \right]$$

$$= 4\left(\frac{1}{2} e^{y^2} \Big|_0^1 \right) - 4\left[\left(\frac{y^2}{2} e^{y^2} \Big|_0^1 \right) - \frac{1}{2} e^{y^2} \Big|_0^1 \right]$$

$$= 2(e^{y^2} - y^2 e^{y^2} + e^{y^2}) \Big|_0^1$$

$$= 2(2e^{y^2} - y^2 e^{y^2}) \Big|_0^1$$

$$= 2(e - 2)$$

37. The region for $\int_0^2 \int_0^{4-x^2} f(x, y)\, dy\, dx$ is

bounded above by $y = 4 - x^2$ and below by $y = 0$. It is bounded on the left by $x = 0$ and on the right by $x = 2$. So, the region is:

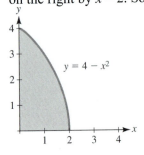

Reversing the integration yields

$$\int_0^4 \int_0^{\sqrt{4-y}} f(x, y)\, dx\, dy.$$

39. The region for $\int_0^1 \int_{x^3}^{\sqrt{x}} f(x, y)\, dy\, dx$ is

bounded above by $y = \sqrt{x}$ and below by $y = x^3$. It is bounded on the left by $x = 0$ and on the right by $x = 1$. So, the region is:

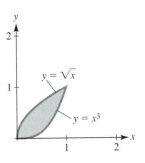

Reversing the integration yields

$$\int_0^1 \int_{y^2}^{y^{1/3}} f(x, y)\, dx\, dy.$$

41. The region for $\int_1^{e^2} \int_{\ln x}^2 f(x, y)\, dy\, dx$ is

bounded above by $y = 2$ and below by $y = \ln x$. It is bounded on the left by $x = 1$ and on the right by $x = e^2$. So, the region is:

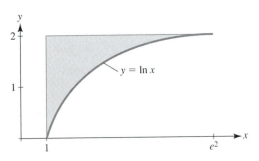

Reversing the integration yields

$$\int_0^2 \int_1^{e^y} f(x, y)\, dx\, dy.$$

43. The region for $\int_{-1}^1 \int_{x^2+1}^2 f(x, y)\, dy\, dx$ is

bounded above by $y = 2$ and below by

$y = x^2 + 1$. It is bounded on the left by $x = -1$ and on the right by $x = 1$. So, the region is:

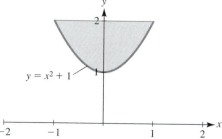

Reversing the integration yields

$$\int_1^2 \int_{-\sqrt{y-1}}^{\sqrt{y-1}} f(x, y)\, dx\, dy.$$

45. The line joining the points $(-4, 0)$ and $(2, 6)$ is $y = x + 4$, with the bottom boundary being $y = 0$. So, the area of R is

$$\int_{-4}^2 \int_0^{x+4} (1)\, dy\, dx = \int_{-4}^2 \left[\int_0^{x+4} 1\, dy \right] dx$$

$$= \int_{-4}^2 \left[y \Big|_0^{x+4} \right] dx$$

$$= \int_{-4}^2 (x+4)\, dx$$

$$= \left(\frac{x^2}{2} + 4x \right) \Big|_{-4}^2$$

$$= 18$$

47. Solving $\frac{1}{2} x^2 = 2x$ yields $x = 0$ and $x = 4$.

So, the area of R is

$$\int_0^4 \int_{x^2/2}^{2x} (1)\, dy\, dx = \int_0^4 \left[\int_{x^2/2}^{2x} 1\, dy \right] dx$$

$$= \int_0^4 \left[y \Big|_{x^2/2}^{2x} \right] dx$$

$$= \int_0^4 \left(2x - \frac{x^2}{2} \right) dx$$

$$= \left(x^2 - \frac{x^3}{6} \right) \Big|_0^4$$

$$= \frac{16}{3}$$

49. Solving $x^2 - 4x + 3 = 0$ yields $x = 1$ and $x = 3$. So, the area of R is

$$\int_1^3 \int_0^{x^2-4x+3} (1)\, dy\, dx = \int_1^3 \left[\int_{x^2-4x+3}^0 1\, dy \right] dx$$

$$= \int_1^3 \left[y \Big|_{x^2-4x+3}^0 \right] dx$$

$$= \int_1^3 (-x^2 + 4x - 3)\, dx$$

$$= \left(-\frac{x^3}{3} + 2x^2 - 3x \right) \Big|_1^3$$

$$= \frac{4}{3}$$

51. Solving $\ln x = 0$ yields $x = 1$, with the other boundary given as $x = e$. So, the area of R is

$$\int_1^e \int_0^{\ln x} (1)\, dy\, dx = \int_1^e \left[\int_0^{\ln x} 1\, dy \right] dx$$

$$= \int_1^e \left[y \Big|_0^{\ln x} \right] dx$$

$$= \int_1^e \ln x\, dx$$

Using integration by parts with $u = \ln x$ and $dV = dx$

$$= x \ln x \Big|_1^e - \int_1^e 1\, dx$$

$$= (x \ln x - x) \Big|_1^e$$

$$= 1$$

53. After solving each equation for x,

$$\sqrt{4 - y} = \frac{y}{3} \text{ when } y = -12 \text{ and } y = 3.$$

However, the region is also bounded by $y = 0$, making the limits $y = 0$ and $y = 3$. So, the area of R is

$$\int_0^3\int_{y/3}^{\sqrt{4-y}}(1)dx\,dy = \int_0^3\left[\int_{y/3}^{\sqrt{4-y}}1\,dy\right]dx$$

$$= \int_0^3\left[x\Big|_{y/3}^{\sqrt{4-y}}\right]dy$$

$$= \int_0^3\left(\sqrt{4-y}-\frac{y}{3}\right)dy$$

$$= \int_0^3\sqrt{4-y}\,dy - \int_0^3\frac{y}{3}\,dy$$

$$= \int_0^3\sqrt{4-y}\,dy - \frac{y^2}{6}\Big|_0^3$$

$$= \int_0^3\sqrt{4-y}\,dy - \frac{3}{2}$$

Using substitution with $u = 4-y$,

$$= \int_4^1 u^{1/2}-du-\frac{3}{2}$$

$$= -\int_4^1 u^{1/2}du-\frac{3}{2}$$

$$= \int_1^4 u^{1/2}du-\frac{3}{2}$$

$$= \frac{2}{3}u^{3/2}\Big|_1^4 - \frac{3}{2}$$

$$= \frac{19}{6}$$

55. $V = \int_0^1\int_0^2(6-2x-2y)dy\,dx$

$$= \int_0^1\left[\int_0^2(6-2x-2y)dy\right]dx$$

$$= \int_0^1\left[(6y-2xy-y^2)\Big|_0^2\right]dx$$

$$= \int_0^1(8-4x)dx$$

$$= (8x-2x^2)\Big|_0^1$$

$$= 6$$

Note: problem can be equivalently worked

as $\int_0^2\int_0^1(6-2x-2y)dx\,dy$.

57. $V = \int_1^2\int_1^3\frac{1}{xy}dy\,dx$

$$= \int_1^2\left[\int_1^3\frac{1}{x}\cdot\frac{1}{y}dy\right]dx$$

$$= \int_1^2\left[\frac{1}{x}\ln|y|\Big|_1^3\right]dx$$

$$= \int_1^2\frac{1}{x}\ln 3\,dx$$

$$= \left[\ln 3\ln|x|\right]_1^2$$

$$= (\ln 3)(\ln 2)$$

Note: problem can be equivalently worked

as $\int_1^3\int_1^2\frac{1}{xy}dx\,dy$.

59. $V = \int_0^1\int_0^2 xe^{-y}dy\,dx$

$$= \int_0^1\left[\int_0^2 xe^{-y}dy\right]dx$$

$$= \int_0^1\left[-xe^{-y}\Big|_0^2\right]dx$$

$$= \int_0^1(-xe^{-2}+x)dx$$

$$= \int_0^1(1-e^{-2})x\,dx$$

$$= (1-e^{-2})\frac{x^2}{2}\Big|_0^1$$

$$= \frac{e^2-1}{2e^2}$$

$$= \frac{1}{2}\left(1-\frac{1}{e^2}\right)$$

61. After solving both equations for x, $y = 2 - y$ when $y = 1$, with the other boundary given as $y = 0$. So,

$$V = \int_0^1 \int_y^{2-y} (2x + y)\,dx\,dy$$

$$= \int_0^1 \left[\int_y^{2-y} (2x + y)\,dx \right] dy$$

$$= \int_0^1 \left[(x^2 + xy) \Big|_y^{2-y} \right] dy$$

$$= \int_0^1 [[(2-y)^2 + (2-y)y]$$
$$\qquad\qquad - [(y)^2 + (y)y]]\,dy$$

$$= \int_0^1 (4 - 2y - 2y^2)\,dy$$

$$= \left(4y - y^2 - \frac{2y^3}{3} \right) \Big|_0^1$$

$$= \frac{7}{3}$$

63. Solving $8 - x^2 = x^2$ yields $x = -2$ and $x = 2$. So,

$$V = \int_{-2}^2 \int_{x^2}^{8-x^2} (x+1)\,dy\,dx$$

$$= \int_{-2}^2 \left[\int_{x^2}^{8-x^2} (x+1)\,dy \right] dx$$

$$= \int_{-2}^2 \left[(x+1)y \Big|_{x^2}^{8-x^2} \right] dx$$

$$= \int_{-2}^2 [(x+1)(8-x^2) - (x+1)(x^2)]\,dx$$

$$= \int_{-2}^2 (8 + 8x - 2x^2 - 2x^3)\,dx$$

$$= \left(8x + 4x^2 - \frac{2x^3}{3} - \frac{x^4}{2} \right) \Big|_{-2}^2$$

$$= \frac{64}{3}$$

65. The area of the rectangular region is 15.

$$f_{av} = \frac{1}{15} \int_{-2}^3 \int_{-1}^2 xy(x - 2y)\,dy\,dx$$

$$= \frac{1}{15} \int_{-2}^3 \left(\frac{x^2 y^2}{2} - \frac{2xy^3}{3} \right) \Big|_{-1}^2 dx$$

$$= \frac{1}{15} \int_{-2}^3 \left(2x^2 - \frac{16x}{3} - \frac{x^2}{2} - \frac{2x}{3} \right) dx$$

$$= \frac{1}{15} \int_{-2}^3 (1.5x^2 - 6x)\,dx$$

$$= \frac{1}{15} \left(\frac{x^3}{2} - 3x^2 \right) \Big|_{-2}^3$$

$$= \frac{1}{6} \approx 0.1667$$

67. The area of the rectangular region is 2.

$$f_{av} = \frac{1}{2} \int_0^2 \int_0^1 xy e^{x^2 y}\,dx\,dy$$

Using substitution with $u = x^2 y$,

$$= \frac{1}{2} \int_0^2 \left[\int_0^y e^u \cdot \frac{1}{2}\,du \right] dy$$

$$= \frac{1}{4} \int_0^2 \left(e^u \Big|_0^y \right) dy$$

$$= \frac{1}{4} \int_0^2 (e^y - 1)\,dy$$

$$= \frac{1}{4} (e^y - y) \Big|_0^2$$

$$= \frac{1}{4} (e^2 - 3) \approx 1.0973$$

69. The area of the rectangular region is $\frac{3}{2}$. The line joining the points $(0, 0)$ and $(3, 1)$ is $y = \frac{x}{3}$.

$$f_{av} = \frac{1}{\frac{3}{2}} \int_0^3 \int_{x/3}^1 6xy \, dy \, dx$$

$$= \frac{2}{3} \int_0^3 \left[3xy^2 \Big|_{x/3}^1 \right] dx$$

$$= \frac{2}{3} \int_0^3 \left(3x - \frac{x^3}{3} \right) dx$$

$$= \frac{2}{3} \left(\frac{3x^2}{2} - \frac{x^4}{12} \right) \Big|_0^3$$

$$= \frac{9}{2}$$

Note: problem can be equivalently worked as $\frac{2}{3} \int_0^1 \int_0^{3y} 6xy \, dx \, dy$.

71. The area of the given region is

$$\int_{-2}^2 4 - x^2 \, dx = \left(4x - \frac{x^3}{3} \right) \Big|_{-2}^2 = \frac{32}{3}$$

$$f_{av} = \frac{1}{\frac{32}{3}} \int_{-2}^2 \int_0^{4-x^2} x \, dy \, dx$$

$$= \frac{3}{32} \int_{-2}^2 (xy) \Big|_0^{4-x^2} dx$$

$$= \frac{3}{32} \int_{-2}^2 (4x - x^3) dx$$

$$= \frac{3}{32} \left(2x^2 - \frac{x^4}{4} \right) \Big|_{-2}^2$$

$$= 0$$

73. $\int_1^3 \int_2^5 \frac{\ln xy}{y} \, dy \, dx$

$$= \int_1^3 \int_2^5 \ln xy \cdot \frac{1}{y} \, dy \, dx$$

Using substitution with $u = \ln xy$,

$$= \int_1^3 \left[\int_{\ln 2x}^{\ln 5x} u \, du \right] dx$$

$$= \int_1^3 \left(\frac{u^2}{2} \right) \Big|_{\ln 2x}^{\ln 5x} dx$$

$$= \frac{1}{2} \int_1^3 (\ln^2 5x - \ln^2 2x) dx$$

$$= \frac{1}{2} \int_1^3 (\ln 5x + \ln 2x)(\ln 5x - \ln 2x) dx$$

$$= \frac{1}{2} \int_1^3 (\ln 10x^2) \left(\ln \frac{5}{2} \right) dx$$

$$= \frac{\ln 2.5}{2} \int_1^3 \ln 10x^2 \, dx$$

Using integration by parts with

$$u = \ln 10x^2 \quad \text{and} \quad dV = dx$$
$$\qquad\qquad\qquad\qquad V = x$$
$$du = \frac{2}{x} dx$$

$$= \frac{\ln 2.5}{2} \left[x \ln 10x^2 \Big|_1^3 - \int_1^3 x \cdot \frac{2}{x} dx \right]$$

$$= \frac{\ln 2.5}{2} [x \ln 10x^2 - 2x]_1^3$$

$$= \frac{\ln 2.5}{2} [(3 \ln 90 - 6) - (\ln 10 - 2)]$$

$$\approx 3.297$$

75. $\int_0^1 \int_0^1 x^3 e^{x^2 y} \, dy \, dx = \int_0^1 \int_0^1 x e^{x^2 y} x^2 \, dy \, dx$

Using substitution with $u = x^2 y$,

$$= \int_0^1 \int_0^{x^2} x e^u \, du \, dx$$

$$= \int_0^1 \left[x \left(e^u \Big|_0^{x^2} \right) \right] dx$$

$$= \int_0^1 (x e^{x^2} - x e^0) dx$$

$$= \int_0^1 x e^{x^2} \, dx - \int_0^1 x \, dx$$

Using substitution with $u = x^2$,

$$= \frac{1}{2}\int_0^1 e^u \, du - \int_0^1 x \, dx$$

$$= \frac{1}{2}\left(e^u \Big|_0^1\right) - \frac{x^2}{2}\Big|_0^1$$

$$= \frac{1}{2}(e^1 - e^0) - \left(\frac{1}{2} - 0\right)$$

$$= \frac{1}{2}e - 1$$

$$= \frac{e-2}{2}$$

77. $Q_{av} = \dfrac{1}{35}\int_0^7 \int_0^5 (2x^3 + 3x^2 y + y^3)\,dx\,dy$

$$= \frac{1}{35}\int_0^7 (0.5x^4 + x^3 y + xy^3)\Big|_0^5 \, dy$$

$$= \frac{1}{35}\int_0^7 x(0.5x^3 + x^2 y + y^3)\Big|_0^5 \, dy$$

$$= \frac{1}{7}\int_0^7 \left(62.5 + 25y + y^3\right) dy$$

$$= \frac{1}{7} y \left(62.5 + \frac{25}{2} y + \frac{y^3}{4}\right)\Big|_0^7$$

$$= 62.5 + \frac{175}{2} + \frac{343}{4} = 235.75$$

79. $P(xy) = \displaystyle\int_{70}^{89} \int_{100}^{125} [(x-30)(70+5x-4y) + (y-40)(80-6x+7y)]\,dx\,dy$

$$= \int_{70}^{89} \int_{100}^{125} [5x^2 + 7y^2 + 160x - 10xy - 80y - 5{,}300]\,dx\,dy$$

$$= \int_{70}^{89} [1.6667x^3 + 7xy^2 + 80x^2 - 5x^2 y - 80xy - 5{,}300x]\Big|_{100}^{125} \, dy$$

$$= [1{,}906{,}041.67y + 58.33y^3 - 15{,}062.5y^2]\Big|_{70}^{89}$$

$$= 1.1826(10^7)$$

The area is $(125 - 100)(89 - 70) = 475$.

The average profit is $\dfrac{1.1826(10^7)}{475} = 24{,}896.8$ or \$2,489,800.

81. Value $= \displaystyle\int_{-1}^1 \int_{-1}^1 (300 + x + y)e^{-0.01x}\,dx\,dy$

$$= \int_{-1}^1 \int_{-1}^1 [(300 + y)e^{-0.01x} + xe^{-0.01x}]\,dx\,dy$$

$$= \int_{-1}^1 \left[\frac{(300+y)}{-0.01}e^{-0.01x} - 100xe^{-0.01x} - 10{,}000e^{-0.01x}\right]\Big|_{x=-1}^{x=1} \, dy$$

$$= \int_{-1}^1 [39{,}900e^{0.01} - 40{,}100e^{-0.01} + (100e^{0.01} - 100e^{-0.01})y]\,dy$$

$$= 79{,}800e^{0.01} - 80{,}200e^{-0.01}$$

$$= 1{,}200.007$$

or roughly 1.2 million dollars.

83. $p(x, y) = x \ln y$

Need to find

$$\int_1^5 \int_0^3 x \ln y \, dx \, dy$$

$$= \int_1^5 \left(\frac{x^2}{2} \ln y \right) \Big|_0^3 \, dy$$

$$= \int_1^5 \left(\frac{9}{2} \ln y - 0 \right) dy$$

Using integration by parts,

$$u = \ln y \quad \text{and} \quad dV = \frac{9}{2} dy$$

$$du = \frac{1}{y} dy \qquad V = \frac{9}{2} y$$

So,

$$\int_1^5 \frac{9}{2} \ln y \, dy = \frac{9}{2} y \ln y \Big|_1^5 - \int_1^5 \frac{9}{2} dy$$

$$= \frac{9}{2} y \ln y \Big|_1^5 - \frac{9}{2} y \Big|_1^5 = \frac{9}{2} (y \ln y - y) \Big|_1^5$$

$$= \frac{9}{2} (5 \ln 5 - 5) - (0 - 1) = \frac{9}{2} (5 \ln 5 - 4)$$

$$\approx 18.212$$

So, approximately 18,212 people are positively influenced.

85. $E(x, y) = \frac{90}{5,280} (2x + y^2)$ miles

$$E_{av} = \frac{0.01705}{12} \int_0^3 \int_0^4 (2x + y^2) \, dx \, dy$$

$$= 0.00142 \int_0^3 (16 + 4y^2) \, dy$$

$$= 0.00142 (16y + 1.333 y^3) \Big|_0^3$$

$$= 630 \text{ ft.}$$

87. Solving $4 - x^2 = 0$ yields $x = -2$ and $x = 2$. So,

$$V = \int_{-2}^{2} \int_{0}^{4-x^2} (20 - x^2 - y^2) \, dy \, dx$$

$$= \int_{-2}^{2} \left[\left(20y - x^2 y - \frac{y^3}{3} \right)_{0}^{4-x^2} \right] dx$$

$$= \int_{-2}^{2} \left[20(4 - x^2) - x^2(4 - x^2) - \frac{(4 - x^2)^3}{3} \right] dx$$

$$= \int_{-2}^{2} \left[80 - 20x^2 - 4x^2 + x^4 - \frac{64 - 48x^2 + 12x^4 - x^6}{3} \right] dx$$

$$= \int_{-2}^{2} \left[\frac{176}{3} - 8x^2 - 3x^4 + \frac{x^6}{3} \right] dx$$

$$= \left[\frac{176}{3} x - \frac{8}{3} x^3 - \frac{3}{5} x^5 + \frac{1}{21} x^7 \right]_{-2}^{2}$$

$$= \left[\left(\frac{352}{3} - \frac{64}{3} - \frac{96}{5} + \frac{128}{21} \right) - \left(-\frac{352}{3} + \frac{64}{3} + \frac{96}{5} - \frac{128}{21} \right) \right]$$

$$= \frac{17,408}{105} \approx 165.79 \text{ m}^3$$

89. $E = \int_{-2}^{2} \int_{-2}^{2} \left(1 - \frac{1}{9}(x^2 + y^2) \right) dy \, dx$

$$= \int_{-2}^{2} \left[y - \frac{1}{9} \left(x^2 y + \frac{y^3}{3} \right) \right] \Bigg|_{y=-2}^{y=2} dx$$

$$= \left(4x - \frac{4x^3}{27} - \frac{16x}{27} \right) \Bigg|_{x=-2}^{x=2}$$

$$= 2 \left(8 - \frac{64}{27} \right)$$

$$= \frac{304}{27}$$

91. $f(x, y) = 2,500e^{-0.01x - 0.02y}$

First, the triangular regions should be divided into two sections, using the y-axis as the dividing line. Then, the left region is bounded above by $y = x + 3$ and below by $y = -2$. The values of x are from $x = -5$ to $x = 0$. Similarly, the right region is bounded above by $y = -x + 3$ and below by $y = -2$. The values of x are from $x = 0$ to $x = 5$. So, the population is

$$2,500\int_{-5}^{0}\int_{-2}^{x+3} e^{-0.01x-0.02y}\,dy\,dx + 2,500\int_{0}^{5}\int_{-2}^{-x+3} e^{-0.01x-0.02y}\,dy\,dx$$

$$= 2,500\int_{-5}^{0}\left[\frac{1}{-0.02}e^{-0.01x-0.02y}\Big|_{-2}^{x+3}\right]dx + 2,500\int_{0}^{5}\left[\frac{1}{-0.02}e^{-0.01x-0.02y}\Big|_{-2}^{-x+3}\right]dx$$

$$= -125,000\int_{-5}^{0}[e^{-0.01x-0.02(x+3)} - e^{-0.01x-0.02(-2)}]dx$$

$$\qquad\qquad -125,000\int_{0}^{5}[e^{-0.01x-0.02(-x+3)} - e^{-0.01x-0.02(-2)}]dx$$

$$= -125,000\left[\int_{-5}^{0}e^{-0.03x-0.06}\,dx - \int_{-5}^{0}e^{-0.01x+0.04}\,dx + \int_{0}^{5}e^{0.01x-0.06}\,dx - \int_{0}^{5}e^{-0.01x+0.04}\,dx\right]$$

$$= -125,000\left[\frac{1}{-0.03}\left(e^{-0.03x-0.06}\Big|_{-5}^{0}\right) + \frac{1}{0.01}\left(e^{-0.01x+0.04}\Big|_{-5}^{0}\right) + \frac{1}{0.01}\left(e^{0.01x-0.06}\Big|_{0}^{5}\right)\right.$$

$$\left. + \frac{1}{0.01}\left(e^{-0.01x+0.04}\Big|_{0}^{5}\right)\right]$$

$$= -125,000\left[-\frac{1}{0.03}(e^{-0.06} - e^{0.09}) + \frac{1}{0.01}(e^{0.04} - e^{0.09}) + \frac{1}{0.01}(e^{-0.01} - e^{-0.06})\right.$$

$$\left. + \frac{1}{0.01}(e^{-0.01} - e^{0.04})\right]$$

$$\approx -125,000[5.080325 - 5.336351 + 4.828530 - 5.076094]$$

$$\approx 62,949 \text{ people}$$

93. $p(x, y) = xy$

Need to find

$$\iint_{R} p(x, y)\,dy\,dx$$

To determine the limits of integration, find the *x*-value where the graphs intersect and solve the graph equations for *y*. Then,

$$= \int_0^4 \int_x^{\sqrt{32-x^2}} xy\,dy\,dx$$

$$= \int_0^4 \left(\frac{x}{2} y^2 \Big|_x^{\sqrt{32-x^2}} \right) dx$$

$$= \int_0^4 \left[\frac{x}{2}(32 - x^2) - \frac{x^3}{2} \right] dx$$

$$= \int_0^4 (16x - x^3)\,dx$$

$$= 8x^2 - \frac{x^4}{4} \Big|_0^4$$

$$= \left[8(4)^2 - \frac{(4)^4}{4} \right] - 0 = 64$$

So, approximately 64,000 people are susceptible.

95. (a) $S_{av} = \dfrac{0.0072}{(142)(76.8)} \displaystyle\int_{3.2}^{80} \int_{38}^{180} W^{0.425} H^{0.725}\,dH\,dW$

$$= \frac{0.0072}{(142)(76.8)} \int_{3.2}^{80} W^{0.425} \left(\int_{38}^{180} H^{0.725}\,dH \right) dW$$

$$= \frac{0.0072}{(142)(76.8)} \int_{3.2}^{80} W^{0.425} \left(\frac{H^{1.725}}{1.725} \Big|_{38}^{180} \right) dW$$

$$\approx \frac{0.0072}{(142)(76.8)} \int_{3.2}^{80} W^{0.425} (4195.71)\,dW$$

$$\approx 0.00277 \int_{3.2}^{80} W^{0.425}\,dW$$

$$= 0.00277 \left(\frac{W^{1.425}}{1.425} \Big|_{3.2}^{80} \right)$$

$$\approx 0.00277(357.802)$$

$$\approx 0.991 \text{ sq meters}$$

(b) No; it could only be interpreted as the person's average surface area from birth until his/her adult weight and height was first reached.

97. $\displaystyle\int_0^2\int_0^3 x^2 e^{-xy}\,dy\,dx = \int_0^2 x^2\left[-\frac{1}{x}e^{-xy}\Big|_0^3\right]dx$

$\displaystyle = \int_0^2 -x[e^{-3x}-1]dx$

$\displaystyle = \int_0^2 x\,dx - \int_0^2 xe^{-3x}dx$

Using integration by parts for the second integral, with $u = x$ and $dV = e^{-3x}dx$ gives

$\displaystyle \frac{x^2}{2}\Big|_0^2 - \left[-\frac{1}{3}xe^{-3x}\Big|_0^2 + \frac{1}{3}\int_0^2 e^{-3x}dx\right]$

$\displaystyle = 2 - \left[-\frac{1}{3}xe^{-3x} - \frac{1}{9}e^{-3x}\Big|_0^2\right]$

$\displaystyle = 2 - \left[\left(-\frac{2}{3}e^{-6} - \frac{1}{9}e^{-6}\right) - \left(0 - \frac{1}{9}\right)\right]$

$\displaystyle = 2 - \left(-\frac{7}{9}e^{-6} + \frac{1}{9}\right)$

$\displaystyle = \frac{7}{9}e^{-6} + \frac{17}{9}$

Checkup for Chapter 7

1. (a) $f(x, y) = x^2 + 2xy^2 - 3y^4$

The domain is the set of all real pairs (x, y).

$f_x = 3x^2 + 2y^2;\ f_y = 4xy - 12y^3;$

$f_{xx} = 6x;\ f_{yx} = 4y$

(b) $\displaystyle f(x, y) = \frac{2x+y}{x-y}$

The domain is the set of all real pairs (x, y) such that $x - y \neq 0$, or $y \neq x$.

$\displaystyle f_x = \frac{(x-y)(2)-(2x+y)(1)}{(x-y)^2}$

$\displaystyle = -\frac{3y}{(x-y)^2}$

$\displaystyle f_y = \frac{(x-y)(1)-(2x+y)(-1)}{(x-y)^2}$

$\displaystyle = \frac{3x}{(x-y)^2}$

$\displaystyle f_{xx} = (-3y)\left[-2(x-y)^{-3}(1)\right] = \frac{6y}{(x-y)^3}$

$\displaystyle f_{yx} = \frac{(x-y)^2(3)-(3x)2(x-y)(1)}{(x-y)^4}$

$\displaystyle = -\frac{3(x+y)}{(x-y)^3}$

(c) $f(x, y) = e^{2x-y} + \ln(y^2 - 2x)$

The domain of e^{2x-y} is the set of all real pairs (x, y), but the domain of $\ln(y^2 - 2x)$ is the set of all real pairs such that $y^2 - 2x > 0$, or $y^2 > 2x$.

$\displaystyle f_x = 2e^{2x-y} - \frac{2}{y^2-2x}$

$\displaystyle f_y = -e^{2x-y} + \frac{2y}{y^2-2x}$

$\displaystyle f_{xx} = 4e^{2x-y} + 2(y^2-2x)^{-2}(-2)$

$\displaystyle = 4e^{2x-y} - \frac{4}{(y^2-2x)^2}$

$\displaystyle f_{yx} = -2e^{2x-y} - 2y(y^2-2x)^{-2}(-2)$

$\displaystyle = -2e^{2x-y} + \frac{4y}{(y^2-2x)^2}$

2. (a) $f(x, y) = x^2 + y^2$

Level curves are of the form $x^2 + y^2 = C$, which are circles having the origin as their center and radius $\sqrt{C}$, and also the single point $(0, 0)$, when $C = 0$.

(b) $f(x, y) = x + y^2$

Level curves are of the form $x + y^2 = C$, which are parabolas having a horizontal axis, opening to the left, and a vertex on the x-axis.

3. (a) $f(x, y) = 4x^3 + y^3 - 6x^2 - 6y^2 + 5$

$f_x = 12x^2 - 12x = 12x(x-1)$

$f_x = 0$ when $x = 0, 1$

$f_y = 3y^2 - 12y = 3y(y - 4)$

$f_y = 0$ when $y = 0, 4$

So, the critical points are $(0, 0)$, $(0, 4)$, $(1, 0)$, and $(1, 4)$.

$f_{xx} = 24x - 12$; $f_{yy} = 6y - 12$; $f_{xy} = 0$

For the point $(0, 0)$,

$D = [24(0) - 12][6(0) - 12] - 0 > 0$

and $f_{xx} = 0$

So, $(0, 0)$ is a relative maximum. For the point $(0, 4)$,

$D = [24(0) - 12][6(4) - 12] - 0 < 0$

So, $(0, 4)$ is a saddle point. For the point $(1, 0)$,

$D = [24(1) - 12][6(0) - 12] - 0 < 0$

So, $(1, 0)$ is a saddle point. For the point $(1, 4)$,

$D = [24(1) - 12][6(4) - 12] - 0 > 0$

and $f_{xx} > 0$

So, $(1, 4)$ is a relative minimum.

(b) $f(x, y) = x^2 - 4xy + 3y^2 + 2x - 4y$

$f_x = 2x - 4y + 2$

$f_x = 0$ when $2x - 4y = -2$

$f_y = -4x + 6y - 4$

$f_y = 0$ when $-4x + 6y = 4$

Solving this system of equations, by multiplying the first by two and adding to the second, gives $y = 0$ and $x = -1$. So, the only critical point is $(-1, 0)$.

$f_{xx} = 2$; $f_{yy} = 6$; $f_{xy} = -4$

$D = (2)(6) - (-4)^2 < 0$

So, $(-1, 0)$ is a saddle point.

(c) $f(x, y) = xy - \dfrac{1}{y} - \dfrac{1}{x}$

$f_x = y + \dfrac{1}{x^2}$

$f_x = 0$ when $y = -\dfrac{1}{x^2}$, or $y^2 = \dfrac{1}{x^4}$

$f_y = y + \dfrac{1}{y^2}$

$f_y = 0$ when $0 = x + \dfrac{1}{y^2}$

$0 = x + x^4$

$0 = x(x^3 + 1)$

or, $x = -1$ (rejecting $x = 0$ since f undefined for $x = 0$) and $y = -1$. So, the only critical point is $(-1, -1)$.

$f_{xx} = -\dfrac{2}{x^3}$; $f_{yy} = -\dfrac{2}{y^3}$; $f_{xy} = 1$

$D = \left[-\dfrac{2}{(-1)^3}\right]\left[-\dfrac{2}{(-1)^3}\right] - (1)^2 > 0$

and $f_{xx} > 0$

So, $(-1, -1)$ is a relative minimum.

4. (a) $f(x, y) = x^2 + y^2$

$g(x, y) = x + 2y$

$f_x = 2x$; $f_y = 2y$; $g_x = 1$; $g_y = 2$

The three Lagrange equations are

$2x = \lambda$; $2y = 2\lambda$; $x + 2y = 4$

Equating λ from the first two equations gives $2x = y$.

Substituting in the third equation gives $x = \dfrac{4}{5}$. Then, $y = \dfrac{8}{5}$ and the minimum value of the function is

$f\left(\dfrac{4}{5}, \dfrac{8}{5}\right) = \dfrac{16}{5}$.

(b) $f(x, y) = xy^2$

$g(x, y) = 2x^2 + y^2$

$f_x = y^2$; $f_y = 2xy$; $g_x = 4x$; $g_y = 2y$

The three Lagrange equations are

$y^2 = 4\lambda x$; $2xy = 2\lambda y$

$2x^2 + y^2 = 6$

Solving the first two equations for λ and equating gives $y^2 = 4x^2$.

Substituting into the third equation gives $x = -1, 1$. When $x = -1, y = -2$

or 2. When $x = 1$, $y = -2$, or 2. So, the critical points are $(-1, -2)$, $(-1, 2)$, $(1, -2)$, and $(1, 2)$.

$f(-1, -2) = f(-1, 2) = -4$ and
$f(1, -2) = f(1, 2) = 4$

So, the maximum value of f is 4, and the minimum value of f is -4.

5. (a) $\displaystyle\int_{-1}^{3}\int_{0}^{2} x^3 y\, dx\, dy = \int_{-1}^{3}\left(\frac{x^4 y}{4}\Big|_{0}^{2}\right) dy$

$\displaystyle = \int_{-1}^{3} 4y\, dy$

$\displaystyle = \left(2y^2\right)\Big|_{-1}^{3}$

$= 16$

(b) $\displaystyle\int_{0}^{2}\int_{-1}^{1} x^2 e^{xy}\, dx\, dy = \int_{-1}^{1}\int_{0}^{2} x^2 e^{xy}\, dy\, dx$

$\displaystyle = \int_{-1}^{1}\int_{0}^{2} xe^{xy} x\, dy\, dx$

Using substitution with $u = xy$,

$\displaystyle = \int_{-1}^{1} x\int_{0}^{2x} e^u\, du\, dx$

$\displaystyle = \int_{-1}^{1} x\left(e^u\Big|_{0}^{2x}\right) dx$

$\displaystyle = \int_{-1}^{1} x(e^{2x} - e^0)\, dx$

$\displaystyle = \int_{-1}^{1} (xe^{2x} - x)\, dx$

$\displaystyle = \int_{-1}^{1} xe^{2x}\, dx - \int_{-1}^{1} x\, dx$

Using integration by parts with

$u = x$ and $dV = e^{2x} dx$

$du = dx$ $\qquad V = \frac{1}{2} e^{2x}$

$\displaystyle = \frac{x}{2} e^{2x}\Big|_{-1}^{1} - \int_{-1}^{1}\frac{1}{2} e^{2x}\, dx - \int_{-1}^{1} x\, dx$

$\displaystyle = \left(\frac{x}{2} e^{2x} - \frac{1}{4} e^{2x} - \frac{x^2}{2}\right)\Big|_{-1}^{1}$

$\displaystyle = \left(\frac{1}{2} e^2 - \frac{1}{4} e^2 - \frac{1}{2}\right)$
$\displaystyle \qquad - \left(-\frac{1}{2} e^{-2} - \frac{1}{4} e^{-2} - \frac{1}{2}\right)$

$\displaystyle = \frac{1}{4} e^2 + \frac{3}{4} e^{-2}$

$\displaystyle = \frac{1}{4}(e^2 + 3e^{-2})$

$\displaystyle = \frac{1}{4}\left(e^2 + \frac{3}{e^2}\right)$

$\displaystyle = \frac{1}{4}\left(\frac{e^4 + 3}{e^2}\right)$

$\displaystyle = \frac{e^4 + 3}{4e^2}$

(c) $\displaystyle\int_{1}^{2}\int_{1}^{y}\frac{y}{x}\, dx\, dy = \int_{1}^{2} y\left(\int_{1}^{y}\frac{1}{x}\, dx\right) dy$

$\displaystyle = \int_{1}^{2} y\left(\ln|x|\Big|_{1}^{y}\right) dy$

$\displaystyle = \int_{1}^{2} y\ln y\, dy$

Using integration by parts with
$u = \ln y$ and $dV = y\, dy$

$\displaystyle = \frac{y^2}{2}\ln y\Big|_{1}^{2} - \int_{1}^{2}\frac{y}{2}\, dy$

$\displaystyle = \left(\frac{y^2}{2}\ln y - \frac{y^2}{4}\right)\Big|_{1}^{2}$

$\displaystyle = (2\ln 2 - 1) - \left(0 - \frac{1}{4}\right)$

$\displaystyle = 2\ln 2 - \frac{3}{4}$

(d) $\int_0^2 \int_0^{2-x} xe^{-y}\,dy\,dx$

$= \int_0^2 (-xe^{-y})\Big|_0^{2-x}\,dx$

$= \int_0^2 -xe^{x-2} + x\,dx$

Using integration by parts with

$u = -x$ and $dV = e^{x-2}\,dx$

$= -xe^{x-2}\Big|_0^2 - \int_0^2 -e^{x-2}\,dx + \int_0^2 x\,dx$

$= -xe^{x-2}\Big|_0^2 + \int_0^2 e^{x-2}\,dx + \int_0^2 x\,dx$

$= \left(-xe^{x-2} + e^{x-2} + \frac{x^2}{2}\right)\Big|_0^2$

$= (-2e^0 + e^0 + 2) - (0 + e^{-2} + 0)$

$= 1 - \frac{1}{e^2}$

$= \frac{e^2 - 1}{e^2}$

6. $Q(K, L) = 120K^{3/4}L^{1/4}$

$Q_K = \frac{90L^{1/4}}{K^{1/4}};\ Q_L = \frac{30K^{3/4}}{L^{3/4}}$

When $K = 1{,}296$ thousand dollars and $L = 20{,}736$ worker-hours, $Q_K = 180$ and $Q_L = 3.75$.

7. $U(x, y) = \ln\left(x^2\sqrt{y}\right)$; $g(x, y) = 20x + 50y$

$U_x = \frac{1}{x^2\sqrt{y}} \cdot 2x\sqrt{y} = \frac{2}{x}$

$U_y = \frac{1}{x^2\sqrt{y}} \cdot \frac{1}{2}x^2 y^{-1/2} = \frac{1}{2y}$

$g_x = 20;\ g_y = 50$

The three Lagrange equations are

$\frac{2}{x} = 20\lambda;\ \frac{1}{2y} = 50\lambda;\ 20x + 50y = 500$

Solving the first two equations for λ and equating gives $x = 10y$. Substituting in the third equation gives $y = 2$, from which follows that $x = 20$. So, Everett should buy 20 DVDs and 2 video games.

8. $E = 0.05(xy - 2x^2 - y^2 + 95x + 20y)$

$E_x = 0.05(y - 4x + 95)$

$E_x = 0$ when $4x - y = 95$

$E_y = 0.05(x - 2y + 20)$

$E_y = 0$ when $-x + 2y = 20$

Solving the system of equations by multiplying the first by two and adding to the second gives $x = 30$ units of A, so $y = 25$ units of B. Since the combined dosage is less than 60 units, there will not be a risk of side effects. Further, this is an equivalent dosage of $E(30, 25) = 83.75$ units, it will be effective.

9. The area of the rectangular region is 2.

$T_{AV} = \frac{1}{2}\int_0^1 \int_0^2 10ye^{-xy}\,dx\,dy$

Using substitution with $u = -xy$ and $-du = y\,dy$,

$= \frac{1}{2}\int_0^1 \left[-10e^u\Big|_0^{-2y}\right]dx$

$= -5\int_0^1 (e^{-2y} - 1)\,dy$

$= -5\left(-\frac{1}{2}e^{-2y} - y\right)\Big|_0^1$

$= -5\left[\left(-\frac{1}{2}e^{-2} - 1\right) - \left(-\frac{1}{2}e^0 - 0\right)\right]$

$= -5\left(-\frac{1}{2}e^{-2} - 1 + \frac{1}{2}\right)$

$= -5\left(-\frac{1}{2}e^{-2} - \frac{1}{2}\right)$

$= \frac{5}{2}(e^{-2} + 1)°C$

10. Let x denote the year of operation and y the corresponding profit, in millions of dollars.

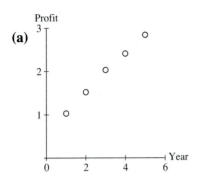

(a)

(b)

x	y	xy	x^2
1	1.03	1.03	1
2	1.52	3.04	4
3	2.03	6.09	9
4	2.41	9.64	16
5	2.84	14.20	25

$$\Sigma x = 15 \quad \Sigma y = 9.83 \quad \Sigma xy = 34.00 \quad \Sigma x^2 = 55$$

Using the formulas with $n = 5$,

$$m = \frac{5(34) - (15)(9.83)}{5(55) - (15)^2} = \frac{22.55}{50} \approx 0.451$$

$$b = \frac{(55)(9.83) - (15)(34)}{5(55) - (15)^2} = \frac{30.65}{30} \approx 0.613$$

So, the equation of the least squares line is $y = 0.451x + 0.613$.

(c) When $x = 6$, $y = 0.451(6) + 0.613 = 3.319$ so the prediction is \$3,319,000.

Review Exercises

1. $f(x, y) = 2x^3 y + 3xy^2 + \dfrac{y}{x}$

$$f_x = 6x^2 y + 3y^2 - \frac{y}{x^2}$$

$$f_y = 2x^3 + 6xy + \frac{1}{x}$$

3. $f(x, y) = \sqrt{x}(x - y^2) = x^{3/2} - x^{1/2} y^2$

$$f_x = \frac{3}{2} x^{1/2} - \frac{1}{2} x^{-1/2} y^2$$

$$= \frac{3}{2}\sqrt{x} - \frac{y^2}{2\sqrt{x}}$$

$$= \frac{3x - y^2}{2\sqrt{x}}$$

$$f_y = -2x^{1/2} y = -2y\sqrt{x}$$

5. $f(x, y) = \sqrt{\dfrac{x}{y}} + \sqrt{\dfrac{y}{x}}$

$$= x^{1/2}y^{-1/2} + x^{-1/2}y^{1/2}$$

$$f_x = \frac{1}{2}x^{-1/2}y^{-1/2} - \frac{1}{2}x^{-3/2}y^{1/2}$$

$$= \frac{1}{2\sqrt{xy}} - \frac{\sqrt{y}}{2x^{3/2}}$$

$$f_y = -\frac{1}{2}x^{1/2}y^{-3/2} + \frac{1}{2}x^{-1/2}y^{-1/2}$$

$$= -\frac{\sqrt{x}}{2y^{3/2}} + \frac{1}{2\sqrt{xy}}$$

7. $f(x, y) = \dfrac{x^3 - xy}{x + y}$

$$f_x = \frac{(x+y)(3x^2 - y) - (x^3 - xy)(1)}{(x+y)^2}$$

$$= \frac{2x^3 + 3x^2 y - y^2}{(x+y)^2}$$

$$f_y = \frac{(x+y)(-x) - (x^3 - xy)(1)}{(x+y)^2}$$

$$= \frac{-x^3 - x^2}{(x+y)^2}$$

$$= \frac{-x^2(x+1)}{(x+y)^2}$$

9. $f(x, y) = \dfrac{x^2 - y^2}{2x + y}$

$$f_x = \frac{(2x+y)(2x) - (x^2 - y^2)(2)}{(2x+y)^2}$$

$$= \frac{2x^2 + 2xy + 2y^2}{(2x+y)^2}$$

$$= \frac{2(x^2 + xy + y^2)}{(2x+y)^2}$$

$$f_y = \frac{(2x+y)(-2y) - (x^2 - y^2)(1)}{(2x+y)^2}$$

$$= \frac{-x^2 - 4xy - y^2}{(2x+y)^2}$$

11. $f(x, y) = e^{x^2 + y^2}$

$$f_x = 2xe^{x^2 + y^2}$$

$$f_{xx} = (2x)(2xe^{x^2 + y^2}) + (e^{x^2 + y^2})(2)$$

$$= 2e^{x^2 + y^2}(2x^2 + 1)$$

$$f_y = 2ye^{x^2 + y^2}$$

$$f_{yy} = (2y)(2ye^{x^2 + y^2}) + (e^{x^2 + y^2})(2)$$

$$= 2e^{x^2 + y^2}(2y^2 + 1)$$

$$f_{xy} = f_{yx} = 4xye^{x^2 + y^2}$$

13. $f(x, y) = x \ln y$

$$f_x = \ln y \qquad\qquad f_y = \frac{x}{y}$$
$$f_{xx} = 0$$
$$\qquad\qquad\qquad\qquad f_{yy} = -\frac{x}{y^2}$$

$$f_{xy} = f_{yx} = \frac{1}{y}$$

15. (a) When $f = 2$, the level curve $x^2 - y = 2$ is a parabola, with vertical axis, opening up, and having the vertex $(0, -2)$.

When $f = -2$, the level curve $x^2 - y = -2$ is a parabola, with vertical axis, opening up, and having the vertex $(0, 2)$.

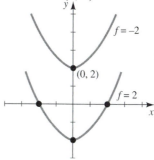

(b) When $f = 0$, the level curve is $6x + 2y = 0$, or $y = -3x$, which is a line through the origin with slope -3.
When $f = 1$, the level curve is

$$6x + 2y = 1, \text{ or } y = -3x + \frac{1}{2}, \text{ which is}$$

the same line translated up $\dfrac{1}{2}$ a unit.

When $f = 2$, the level curve is
$6x + 2y = 2$, or $y = -3x + 1$, which is the
same line translated up one unit.

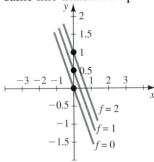

17. $f(x, y) = (x + y)(2x + y - 6)$

$f_x = (x + y)(2) + (2x + y - 6)(1)$
$\quad = 4x + 3y - 6$

$f_y = (x + y)(1) + (2x + y - 6)(1)$
$\quad = 3x + 2y - 6$

To find the critical points, set $f_x = 0$ and
$f_y = 0$ and solve the system of equations.

$4x + 3y - 6 = 0$
$3x + 2y - 6 = 0$

Multiply the first equation by 2 and the
second equation by -3. Then, add the two
resulting equations

$8x + 6y - 12 = 0$
$\underline{-9x - 6y + 18 = 0}$
$\qquad -x + 6 = 0$
$\qquad\quad x = 6$

When $x = 6$, $4(6) + 3y - 6 = 0$
$\qquad\qquad\qquad\qquad y = -6$

So, the only critical point is $(6, -6)$.
$f_{xx} = 4;\ f_{yy} = 2;\ f_{xy} = 3$

$D = (4)(2) - (3)^2 < 0$

So, the point $(6, -6)$ is a saddle point.

19. $f(x, y) = x^3 + y^3 + 3x^2 - 3y^2$

$f_x = 3x^2 + 6x;\ f_y = 3y^2 - 6y$

$f_x = 0$ when $3x(x + 2) = 0$, $x = 0$ and
$x = -2$

$f_y = 0$ when $3y(y - 2) = 0$, $y = 0$ and
$y = 2$

So, the critical points are $(0, 0)$, $(0, 2)$, $(-2,$
$0)$ and $(-2, 2)$.

$f_{xx} = 6x + 6;\ f_{yy} = 6y - 6;\ f_{xy} = 0$

For the point $(0, 0)$, $D = (6)(-6) - 0 < 0$
So, it is a saddle point.
For the point $(0, 2)$, $D = (6)(6) - 0 > 0$
Since $f_{xx} > 0$, it is a relative minimum.
For the point $(-2, 0)$, $D = (-6)(-6) - 0 > 0$
Since $f_{xx} < 0$, it a relative maximum.
For the point $(-2, 2)$, $D = (-6)(6) - 0 > 0$
So it is a saddle point.

21. $f(x, y) = x^2 + y^3 + 6xy - 7x - 6y$

$f_x = 2x + 6y - 7$

$f_y = 3y^2 + 6x - 6$

To find the critical points, set $f_x = 0$ and
$f_y = 0$. So, $2x + 6y - 7 = 0$ and

$3y^2 + 6x - 6 = 0$, or $2x + 6y - 7 = 0$ and

$2x + y^2 - 2 = 0$.

Subtracting the two equations gives

$y^2 - 6y + 5 = 0$, $(y - 1)(y - 5) = 0$, or

$y = 1$ and $y = 5$.

When $y = 1$, the first equation gives $x = \dfrac{1}{2}$

and when $y = 5$, the first equation gives

$x = -\dfrac{23}{2}$.

So, the critical points of f are $\left(\dfrac{1}{2}, 1\right)$,

$\left(-\dfrac{23}{2}, 5\right)$.

Since $f_{xx} = 2$, $f_{yy} = 6y$, and $f_{xy} = 6$,

$D = f_{xx} f_{yy} - (f_{xy})^2$
$\quad = (2)(6y) - 36$
$\quad = 12(y - 3)$

For the point $\left(\dfrac{1}{2}, 1\right)$,

$D = 12(-2) = -24 < 0$ and f has a saddle

point at $\left(\dfrac{1}{2}, 1\right)$.

For the point $\left(-\dfrac{23}{2}, 5\right)$,

$D = 12(2) = 24 > 0$ and $f_{xx} > 0$

So, f has a relative minimum at $\left(-\dfrac{23}{2}, 5\right)$.

23. $f(x, y) = xe^{2x^2+5xy+2y^2}$

$f_x = (x)[e^{2x^2+5xy+2y^2}(4x+5y)$
$\qquad\qquad + (e^{2x^2+5xy+2y^2})(1)$

$\quad = e^{2x^2+5xy+2y^2}[x(4x+5y)+1]$

$\quad = e^{2x^2+5xy+2y^2}(4x^2+5xy+1)$

$f_x = 0$ when $4x^2+5xy+1 = 0$

$f_y = x[e^{2x^2+5xy+2y^2}(5x+4y)]$

$f_y = 0$ when $x(5x+4y)e^{2x^2+5xy+2y^2} = 0$

So, $f_y = 0$ when $x = 0$ and when

$5x + 4y = 0$, or $x = -\dfrac{4}{5}y$

When $x = 0$, substituting into $f_x = 0$ yields no solution.

When $x = -\dfrac{4}{5}y$, $f_x = 0$ when

$0 = 4\left(-\dfrac{4}{5}y\right)^2 + 5\left(-\dfrac{4}{5}y\right)(y) + 1$ or,

$y = \pm\dfrac{5}{6}$

When $y = \dfrac{5}{6}$, $x = -\dfrac{4}{5}\left(\dfrac{5}{6}\right) = -\dfrac{2}{3}$

When $y = -\dfrac{5}{6}$, $x = -\dfrac{4}{5}\left(-\dfrac{5}{6}\right) = \dfrac{2}{3}$

So, the critical points are $\left(-\dfrac{2}{3}, \dfrac{5}{6}\right)$ and

$\left(\dfrac{2}{3}, -\dfrac{5}{6}\right)$.

$f_{xx} = (e^{2x^2+5xy+2y^2})(8x+5y)$
$\qquad + (4x^2+5xy+1)[e^{2x^2+5xy+2y^2}(4x+5y)]$

$f_{yy} = x[(e^{2x^2+5xy+2y^2})(4)$
$\qquad\qquad + (5x+4y)e^{2x^2+5xy+2y^2}(5x+4y)]$

$f_{xy} = (e^{2x^2+5xy+2y^2})(5x)$
$\qquad + (4x^2+5xy+1)e^{2x^2+5xy+2y^2}(5x+4y)$

For the point $\left(-\dfrac{2}{3}, \dfrac{5}{6}\right)$,

$D \approx (-0.7076)(-2.0218) - (-1.6174)^2 < 0$

So, $\left(-\dfrac{2}{3}, \dfrac{5}{6}\right)$ is a saddle point.

For the point $\left(\dfrac{2}{3}, -\dfrac{5}{6}\right)$,

$D \approx (0.7076)(2.0218) - (1.6174)^2 < 0$

So, $\left(\dfrac{2}{3}, -\dfrac{5}{6}\right)$ is also a saddle point.

25. $f(x, y) = x^2 + 2x + y^2 - 4y + 12$;
$(-4, 0), (1, 0), (0, 4)$
Finding the critical points in R,

$f_x(x, y) = 2x + 2$ and $f_y(x, y) = 2y - 4$

$f_x(x, y) = 0$ when $x = -1$

$f_y(x, y) = 0$ when $y = 2$

So, the only critical point in R is $(-1, 2)$.
The boundary equations are $y = 0$,
$y = x + 4$ and $y = -4x + 4$. Using $y = 0$,

$u(x) = f(x, 0) = x^2 + 2x + 12$

$u'(x) = 2x + 2$

$u'(x) = 0$ when $x = -1$

and $y = 0$

So, the point $(-1, 0)$ must be considered along with the endpoints $(-4, 0)$ and $(1, 0)$. Using $y = x + 4$,

$$v(x) = f(x, x+4)$$
$$= x^2 + 2x + (x+4)^2 - 4(x+4)$$
$$+ 12$$
$$= 2x^2 + 6x + 12$$
$$v'(x) = 4x + 6$$

$$v'(x) = 0 \text{ when } x = -\frac{3}{2}$$

$$\text{and } y = -\frac{3}{2} + 4 = \frac{5}{2}$$

So, the point $\left(-\frac{3}{2}, \frac{5}{2}\right)$ must be considered along with the endpoints $(-4, 0)$ and $(0, 4)$. Using $y = -4x + 4$,

$$w(x) = f(x, -4x+4)$$
$$= x^2 + 2x + (-4x+4)^2$$
$$- 4(-4x+4) + 12$$
$$= 17x^2 - 14x + 12$$
$$w'(x) = 34x - 14$$

$$w'(x) = 0 \text{ when } x = \frac{7}{17}$$

$$\text{and } y = -4\left(\frac{7}{17}\right) + 4 = \frac{40}{17}$$

So, the point $\left(\frac{7}{17}, \frac{40}{17}\right)$ must be |

considered along with the endpoints $(0, 4)$ and $(1, 0)$. Collecting all points to consider,

$$f(-1, 2)$$
$$= (-1)^2 + 2(-1) + (2)^2 - 4(2) + 12$$
$$= 7$$
$$f(-1, 0)$$
$$= (-1)^2 + 2(-1) + 12 = 11$$
$$f(-4, 0)$$
$$= (-4)^2 + 2(-4) + 12 = 20$$
$$f(1, 0)$$
$$= (1)^2 + 2(1) + 12 = 15$$
$$f\left(-\frac{3}{2}, \frac{5}{2}\right)$$
$$= \left(-\frac{3}{2}\right)^2 + 2\left(-\frac{3}{2}\right) + \left(\frac{5}{2}\right)^2 - 4\left(\frac{5}{2}\right)$$
$$+ 12 = 7.5$$
$$f(0, 4) = (4)^2 - 4(4) + 12 = 12$$
$$f\left(\frac{7}{17}, \frac{40}{17}\right) = \left(\frac{7}{17}\right)^2 + 2\left(\frac{7}{17}\right)$$
$$+ \left(\frac{40}{17}\right)^2 - 4\left(\frac{40}{17}\right) + 12 \approx 9.1$$

So, the smallest value is 7 and the largest value is 20.

27. $f(x, y) = x^3 - 4xy + 4x + y^2$; $(1, 2)$, $(4, 2)$, $(1, 5)$, $(4, 5)$
Finding the critical points in R,
$$f_x(x, y) = 3x^2 - 4y + 4 \text{ and}$$
$$f_y(x, y) = -4x + 2y$$
$$f_x(x, y) = 0 \text{ when}$$
$$0 = 3x^2 - 4y + 4$$
$$4y = 3x^2 + 4$$
$$y = \frac{3x^2 + 4}{4} = \frac{3}{4}x^2 + 1$$

$f_y(x, y) = 0$ when

$$0 = -4x + 2y$$

Substituting,

$$0 = -4x + 2\left(\frac{3}{4}x^2 + 1\right)$$

$$0 = -4x + \frac{3}{2}x^2 + 2$$

$$0 = 3x^2 - 8x + 4$$

$$0 = (3x - 2)(x - 2)$$

or, $x = \frac{2}{3}$ and $x = 2$. Reject $x = \frac{2}{3}$, since

it is not in R. Then, the only critical point in R is (2, 4). The boundary equations are $y = 2$, $y = 5$, $x = 1$, and $x = 4$.
Using $y = 2$,

$$u(x) = f(x, 2)$$

$$= x^3 - 4x(2) + 4x + (2)^2$$

$$= x^3 - 4x + 4$$

$$u'(x) = 3x^2 - 4$$

$$u'(x) = 0 \text{ when } x = \sqrt{\frac{4}{3}}, \text{ or } \frac{2\sqrt{3}}{3}$$

(rejecting $x = -\sqrt{\frac{4}{3}}$ since not in R)

So, the point $\left(\dfrac{2\sqrt{3}}{3}, 2\right)$ must be

considered along with the endpoints (1, 2) and (4, 2).
Using $y = 5$,

$$v(x) = f(x, 5) = x^3 - 4x(5) + 4x + (5)^2$$

$$= x^3 - 16x + 25$$

$$v'(x) = 3x^2 - 16$$

$$v'(x) = 0 \text{ when } x = \frac{4\sqrt{3}}{3}$$

So, the point $\left(\dfrac{4\sqrt{3}}{3}, 5\right)$ must be

considered along with the endpoints (1, 5) and (4, 5). Using $x = 1$,

$$w(y) = f(1, y) = y^2 - 4y + 5$$

$$w'(y) = 2y - 4$$

$$w'(y) = 0 \text{ when } y = 2$$

So, the endpoint (1, 2) is considered along with the other endpoint (1, 5).
Using $x = 4$,

$$r(y) = f(4, y) = (4)^3 - 4(4)y$$

$$+ 4(4) + y^2$$

$$= y^2 - 16y + 80$$

$$r'(y) = 2y - 16$$

$$r'(y) = 0 \text{ when } y = 8$$

(reject since not in R)
Collecting all points to consider,

$$f(2, 4) = (2)^3 - 4(2)(4) + 4(2)$$

$$+ (4)^2 = 0$$

$$f\left(\frac{2\sqrt{3}}{3}, 2\right) = \left(\frac{2\sqrt{3}}{3}\right)^3 - 4\left(\frac{2\sqrt{3}}{3}\right)(2)$$

$$+ 4\left(\frac{2\sqrt{3}}{3}\right) + (2)^2 \approx 0.92$$

$f(1, 2) = (1)^3 - 4(1)(2) + 4(1)$

$+ (2)^2 = 1$

$f(4, 2) = (4)^3 - 4(4)(2) + 4(4)$

$+ (2)^2 = 52$

$f\left(\dfrac{4\sqrt{3}}{3}, 5\right) = \left(\dfrac{4\sqrt{3}}{3}\right)^3 - 4\left(\dfrac{4\sqrt{3}}{3}\right)(5)$

$+ 4\left(\dfrac{4\sqrt{3}}{3}\right) + (5)^2 \approx 0.37$

$f(1, 5) = (1)^3 - 4(1)(5) + 4(1)$

$+ (5)^2 = 10$

$f(4, 5) = (4)^3 - 4(4)(5) + 4(4)$

$+ (5)^2 = 25$

So, the smallest value is 0 and the largest value is 52.

29. $f(x, y) = e^{x^2 + 4x + y^2}$; $x^2 + 4x + y^2 = 0$

Finding the critical points in R,

$f_x(x, y) = (2x + 4)e^{x^2 + 4x + y^2}$

$f_y(x, y) = 2ye^{x^2 + 4x + y^2}$

$f_x(x, y) = 0$ when $2x + 4 = 0$, or $x = -2$

($e^{x^2 + 4x + y^2}$ is never zero)

$f_y(x, y) = 0$ when $2y = 0$, or $y = 0$

So, the only critical point in R is $(-2, 0)$.
The boundary equations are

$y = -\sqrt{-x^2 - 4x}$ and $y = \sqrt{-x^2 - 4x}$.

Using $y = -\sqrt{-x^2 - 4x}$,

$u(x) = f\left(x, -\sqrt{-x^2 - 4x}\right)$

$= e^{x^2 + 4x - x^2 - 4x} = e^0 = 1$

$u'(x) = 0$ for all x. So, all points on the

boundary $y = -\sqrt{-x^2 - 4x}$ must be

considered.

Similarly, using $y = \sqrt{-x^2 - 4x}$,

$v(x) = f\left(x, \sqrt{-x^2 - 4x}\right) = e^0$

$v'(x) = 0$ for all x. So, all points on the

boundary $y = \sqrt{-x^2 + 4x}$ must be
considered. Collecting the points to be
considered,

$f(-2, 0) = e^{-4} \approx 0.018$

$f\left(x, -\sqrt{-x^2 - 4x}\right) = e^0 = 1$

$f\left(x, \sqrt{-x^2 + 4x}\right) = e^0 = 1$

So, the smallest value is 0.018 and the
largest value is 1.

31. $f(x, y) = x^2 + 2y^2 + 2x + 3$; $x^2 + y^2 = 4$

Since the constraint is $x^2 + y^2 = 4$,

$g(x, y) = x^2 + y^2$.

$f_x = 2x + 2 \qquad\qquad f_y = 4y$
$g_x = 2x \qquad\qquad\quad g_y = 2y$

The three Lagrange equations are

$2x + 2 = 2x\lambda$

$4y = 2y\lambda$

$x^2 + y^2 = 4$

From the second equation, $\lambda = 2$, or $y = 0$.
Substituting $\lambda = 2$ in the first equation
gives $2x + 2 = 4x$, or $x = 1$. Using the third
equation $y = -\sqrt{3}$ or $y = \sqrt{3}$.
Substituting $y = 0$ into the third equation
gives $x = -2$ or $x = 2$. So, the critical
points are $(2, 0)$, $(-2, 0)$, $\left(1, -\sqrt{3}\right)$ and

$\left(1, \sqrt{3}\right)$.

Testing all points in the original function
yields:

$f(2, 0) = 11 \qquad\qquad f(-2, 0) = 3$

$f\left(1, -\sqrt{3}\right) = 12 \qquad f\left(1, \sqrt{3}\right) = 12$

So, the maximum value is 12, and it

occurs at the points $\left(1, -\sqrt{3}\right)$ and

$\left(1, \sqrt{3}\right)$. The minimum value is 3 and it occurs at $(-2, 0)$.

33. $f(x, y) = x + 2y, \; 4x^2 + y^2 = 68$

Since the constraint is $4x^2 + y^2 = 68$,

$g(x, y) = 4x^2 + y^2$.

$f_x = 1; \; f_y = 2; \; g_x = 8x; \; g_y = 2y$

The three Lagrange equations are

$1 = 8x\lambda, \; 2 = 2y\lambda, \; 4x^2 + y^2 = 68$

Solving the first two equations for λ and equating gives $y = 8x$. Substituting in the third equation gives $68x^2 = 68$, or $x = -1$ and $x = 1$. It follows that $y = -8$ and $y = 8$. So, the critical points are $(-1, -8)$ and $(1, 8)$. Testing these points in the original function gives $f(-1, -8) = -17$, $f(1, 8) = 17$.
So, the maximum value is 17 and it occurs at $(1, 0)$. The minimum value is -17 and it occurs at $(-1, -8)$.

35. $Q = 40K^{1/3}L^{1/2}$

The marginal product of capital is

$\dfrac{\partial Q}{\partial K} = \dfrac{40}{3}K^{-2/3}L^{1/2} = \dfrac{40L^{1/2}}{3K^{2/3}}$ which is

approximately the change ΔQ in output due to one (thousand dollar) unit increase in capital. When $K = 125$ (thousand) and L

$= 900, \; \Delta Q \approx \dfrac{\partial Q}{\partial K} = \dfrac{40(900)^{1/2}}{3(125)^{2/3}} = 16$ units

37. $Q(x, y) = 60x^{1/3}y^{2/3}$

For any value of x, the slope of the level curve $Q = k$ is an approximation of the change in unskilled labor y that should be made to offset a one-unit increase in skilled labor x so that the level of output will remain constant. So,

ΔQ = change in unskilled labor

$\approx \dfrac{dQ}{dx}$

$= -\dfrac{Q_x}{Q_y}$

$= -\dfrac{20x^{-2/3}y^{2/3}}{40x^{1/3}y^{-1/3}}$

$= -\dfrac{y}{2x}$

When $x = 10$ and $y = 40$,

$\Delta Q \approx \dfrac{dQ}{dx} = -\dfrac{40}{2(10)} = -2$

That is, the level of unskilled labor should be decreased by approximately 2 workers.

39. The goal is to maximize the area of a rectangle $A(l, w) = lw$ subject to the constraint $2l + 2w = k$, where k is some positive constant. So, $g(l, w) = 2l + 2w$.

$A_l = w; \; A_w = l; \; g_l = 2; \; g_w = 2$

The three Lagrange equations are

$w = 2\lambda; \; l = 2\lambda; \; 2l + 2w = k$

Solving the first two for λ and equating

gives $\dfrac{w}{2} = \dfrac{l}{2}$, or $w = l$. So, the rectangle

having the greatest area is a square.

41. From problem 18, the profit function is

$P(x, y) = \dfrac{50y}{y+2} + \dfrac{20x}{x+5} - x - y$

The constraint is $x + y = 11$ thousand dollars, so $g(x, y) = x + y$.

$P_x = \dfrac{100}{(x+5)^2} - 1; \quad P_y = \dfrac{100}{(y+2)^2} - 1;$

$g_x = 1; \; g_y = 1$

The three Lagrange equations are

$\dfrac{100}{(x+5)^2} - 1 = \lambda$

$\dfrac{100}{(y+2)^2} - 1 = \lambda$

$x + y = 11$

From the first two equations,

$(x+5)^2 = (y+2)^2$ or $y = x + 3$ (rejecting

the negative solution). Substituting into the third equation gives $x = 4$, and the corresponding value of y is 7.

So, to maximize profit, \$4,000 should be spent on development and \$7,000 should be spent on promotion.

43. $f(x, y) = \dfrac{12}{x} + \dfrac{18}{y} + xy$

Suppose y is fixed (say at $y = 1$), then f is very large when x is quite small.

f is also large when x is large, with smaller values of f occurring between these extremes. The same reasoning applies to y when x is fixed.

$$f_x = -\frac{12}{x^2} + y; \quad f_y = -\frac{18}{y^2} + x$$

To find the critical points, set $f_x = 0$ and $f_y = 0$. Then $y = \dfrac{12}{x^2}$ and $x = \dfrac{18}{y^2}$.

Substituting leads to

$$y = \frac{12}{x^2} = \frac{12}{\left(\frac{18}{y^2}\right)^2} = \frac{12y^4}{18^2} \text{ or } y = 0 \text{ (which}$$

is not in the domain of the function) and $12y^3 = 18^2$, $y^3 = 27$, $y = 3$. The

corresponding value for $x = \dfrac{18}{3^2} = 2$. So,

the critical point of f is (2, 3).

$$f_{xx} = \frac{24}{x^3}; \quad f_{yy} = \frac{36}{y^3}; \quad f_{xy} = 1$$

For the point (2, 3), $D = \dfrac{(24)(36)}{(2^3)(3^3)} - 1 > 0$

and $f_{xx}(2, 3) > 0$, so the minimum is $f(2, 3) = 18$.

45. $\displaystyle\int_0^1\int_0^2 e^{-x-y}\,dy\,dx = \int_0^1\int_0^2 e^{-x}e^{-y}\,dy\,dx$

$$= \int_0^1 (-e^{-x}e^{-y})\Big|_0^2\,dx$$

$$= \int_0^1 (-e^{-x}e^{-2} + e^{-x})\,dx$$

$$= (1 - e^{-2})\int_0^1 e^{-x}\,dx$$

$$= (1 - e^{-2})(-e^{-x})\Big|_0^1$$

$$= (1 - e^{-2})(-e^{-1} + 1)$$

$$= 0.5466$$

47. $\displaystyle\int_0^1\int_{-1}^1 xe^{2y}\,dy\,dx = \int_0^1 \left(\frac{1}{2}\right)xe^{2y}\Big|_{-1}^1\,dx$

$$= \int_0^1 \left(\frac{xe^2}{2} - \frac{xe^{-2}}{2}\right)dx$$

$$= \frac{e^2 - e^{-2}}{2}\int_0^1 x\,dx$$

$$= \frac{e^2 - e^{-2}}{4} \approx 1.8134$$

49. $I = \displaystyle\int_1^e\int_1^e (\ln x + \ln y)\,dy\,dx$

$$= \int_1^e [y(\ln x) + (y\ln y - y)]\Big|_1^e\,dx$$

$$= \int_1^e [(e - 1)\ln x + 1]\,dx$$

$$= [(e - 1)(x\ln x - x) + x]\Big|_1^e$$

$$= 2(e - 1)$$

$$= 3.4366$$

51. $\displaystyle\int_1^2\int_0^x e^{y/x}\,dy\,dx = \int_1^2\int_0^x e^{\frac{1}{x}y}\,dy\,dx$

$\displaystyle = \int_1^2 xe^{y/x}\Big|_0^x dx$

$\displaystyle = \int_1^2 xe - x\,dx$

$\displaystyle = \int_1^2 (e-1)x\,dx$

$\displaystyle = (e-1)\frac{x^2}{2}\Big|_1^2$

$\displaystyle = (e-1)\left(2-\frac{1}{2}\right)$

$\displaystyle = \frac{3}{2}(e-1)$

53. $\displaystyle\iint_R (x+2y)\,dA = \int_0^1\int_{-2}^2 (x+2y)\,dy\,dx$

$\displaystyle = \int_0^1 (xy + y^2)\Big|_{-2}^2 dx$

$\displaystyle = \int_0^1 4x\,dx$

$\displaystyle = 2x^2\Big|_0^1$

$\displaystyle = 2$

55. $\displaystyle V = \int_1^2\int_2^3 xe^{-y}\,dy\,dx$

$\displaystyle = \int_1^2 (-xe^{-y})\Big|_2^3 dx$

$\displaystyle = \int_1^2 (x)(e^{-2} - e^{-3})\,dx$

$\displaystyle = (e^{-2} - e^{-3})\frac{3}{2}$

$\displaystyle = 0.1283$

57. The sum of the three numbers is
$x + y + z = 20$, so $z = 20 - x - y$. Their product is
$P = xyz$

$\quad = xy(20 - x - y)$

$\quad = 20xy - x^2 y - xy^2$

$P_x = 20y - 2xy - y^2$

$P_x = 0$ when $y(20 - 2x - y) = 0$

$P_y = 20x - x^2 - 2xy$

$P_y = 0$ when $x(20 - x - 2y) = 0$

Since the numbers must be positive, reject the solution $x = 0$ or $y = 0$. Solving the system of equations by multiplying the first by -2 and adding to the second gives

$x = \dfrac{20}{3}$. When $x = \dfrac{20}{3}$, $20 - \dfrac{20}{3} - 2y = 0$, or

$y = \dfrac{20}{3}$. Then, $z = 20 - \dfrac{20}{3} - \dfrac{20}{3} = \dfrac{20}{3}$. So, the product is maximized when

$x = y = z = \dfrac{20}{3}$.

59. Using the hint in the problem, let D denote the square of the distance from the origin to the surface. Then, $D = x^2 + y^2 + z^2$.

Since $y^2 - z^2 = 10$, $y^2 = 10 + z^2$ and

$D = x^2 + 10 + 2z^2$

$D_x = 2x$, so $D_x = 0$ when $x = 0$

$D_z = 4z$, so $D_z = 0$ when $z = 0$

When $z = 0$, $y^2 = 10$ or $y = \pm\sqrt{10}$.

So, the critical points are $\left(0, -\sqrt{10}, 0\right)$

and $\left(0, \sqrt{10}, 0\right)$.

$D_{xx} = 2$, $D_{zz} = 4$, $D_{xz} = 0$

For the point $\left(0, -\sqrt{10}, 0\right)$,

$D = (2)(4) - 0 > 0$ and $D_{xx} > 0$

So, $\left(0, -\sqrt{10}, 0\right)$ is a relative minimum.

For the point $\left(0, \sqrt{10}, 0\right)$, $D > 0$ and

$D_{xx} > 0$

So, it is also a relative minimum. The square of the distance, using either point, is $D = 0 + 10 + 0 = 10$.

So, the minimum distance $= \sqrt{10}$.

61. (a) Let x denote the monthly advertising expenditure and y the corresponding sales (both measured in units of $\$1,000$). Then

x	3	4	7	9	10

y	78	86	138	145	156

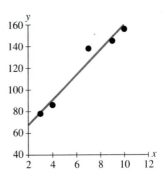

(b)

x	y	xy	x^2
3	78	234	9
4	86	344	16
7	138	966	49
9	145	1,305	81
10	156	1,560	100
$\Sigma x = 33$	$\Sigma y = 603$	$\Sigma xy = 4,409$	$\Sigma x^2 = 255$

Using the formulas with $n = 5$,

$$m = \frac{5(4,409) - 33(603)}{5(255) - (33)^2} = 11.54$$

$$b = \frac{255(603) - 33(4,409)}{5(255) - (33)^2} = 44.45$$

So, the equation of the least-squares line is $y = 11.54x + 44.45$.

(c) $y = 11.54(5) + 44.45 = 102.15$ thousand, or $102,150.

63. $Q(x, y) = 200 + 10x^2 - 20y$

$x(t) = 18 + 0.02t$

$y(t) = 21 + 0.4\sqrt{t}$

$$\frac{dQ}{dt} = \frac{\partial Q}{\partial x}\frac{dx}{dt} + \frac{\partial Q}{\partial y} \cdot \frac{dy}{dt}$$

$$= (20x)(0.02) + (-20)\left(\frac{0.2}{\sqrt{t}}\right)$$

When $t = 9$, $x(9) = 18.18$ and

$$\frac{dQ}{dt} = 20(18.18)(0.02) + (-20)\left(\frac{0.2}{\sqrt{9}}\right)$$

$$\approx 5.94$$

So, demand is decreasing at a rate of approximately 6 quarts per month.

65. $P(x, y) = \frac{1}{4}x^{1/3}y^{1/2}$

$x = 129 - \sqrt{8t}$

$y = 15.60 + 0.2t$

$Q = \frac{4184}{p}$

$\frac{dQ}{dt} = \frac{dQ}{dp} \cdot \frac{dp}{dt}$ where

$\frac{dp}{dt} = \frac{\partial p}{\partial x} \cdot \frac{dx}{dt} + \frac{\partial p}{\partial y} \cdot \frac{dy}{dt}$

$\frac{dQ}{dt} = -\frac{4184}{p^2}\left[\left(\frac{1}{12}x^{-2/3}y^{1/2}\right)\left(-\frac{\sqrt{8}}{2\sqrt{t}}\right)\right.$

$\left. + \left(\frac{1}{8}x^{1/3}y^{-1/2}\right)(0.2)\right]$

When $t = 2$, $x = 125$, $y = 16$ and
$p(125, 16) = 5$ so

$\frac{dQ}{dt} = -\frac{4184}{(5)^2}\left[\left(\frac{1}{12}\cdot\frac{1}{25}\cdot 4\right)\left(-\frac{\sqrt{8}}{2\sqrt{2}}\right)\right.$

$\left. + \left(\frac{1}{8}\cdot 5\cdot\frac{1}{4}\right)(0.2)\right]$

≈ -3.00

or demand is decreasing at a rate of 3 pies per week.

67. $Q(E, T) = 125E^{2/3}T^{1/2}$

$\frac{dQ}{dt} = \frac{\partial Q}{\partial E} \cdot \frac{dE}{dt} + \frac{\partial Q}{\partial T} \cdot \frac{dT}{dt}$

$= \left(\frac{250}{3}E^{-1/3}T^{1/2}\right)\left(\frac{1}{11}\right)$

$\quad + \left(\frac{125}{2}E^{2/3}T^{-1/2}\right)(-0.21)$

$= \left[\frac{250}{3}(151)^{-1/3}(10)^{1/2}\right]\left(\frac{1}{11}\right)$

$\quad + \left[\frac{125}{2}(151)^{2/3}(10)^{-1/2}\right](-0.21)$

≈ -113.19

or decreasing at a rate of 113 units per day.

69. $N(r, s) = 40e^{-r/2}e^{-s/3}$

Pollution

$= \int_2^3\int_1^2 40e^{-r/2}e^{-s/3}\,ds\,dr$

$= \int_2^3\left(40e^{-r/2}\cdot -3e^{-s/3}\Big|_1^2\right)dr$

$= -120\int_2^3[e^{-r/2}(e^{-2/3} - e^{-1/3})]dr$

$= -120(e^{-2/3} - e^{-1/3})\int_2^3 e^{-r/2}\,dr$

$= -120(e^{-2/3} - e^{1/3})\left[-2e^{-r/2}\Big|_2^3\right]$

$= 240(e^{2/3} - e^{-1/3})(e^{-3/2} - e^{-1})$

≈ 7.056 units

71. With $Q = x^a y^b$, $Q_x = ax^{a-1}y^b$ and
$Q_y = bx^a y^{b-1}$.

$xQ_x + yQ_y = x(ax^{a-1}y^b) + y(bx^a y^{b-1})$

$= (a + b)x^a y^b$

$= (a + b)Q$

If $b = 1 - a$, then

$xQ_x + yQ_y = (a + b)Q = Q.$

NOTES

NOTES

NOTES

NOTES

NOTES

NOTES

NOTES

NOTES

NOTES

NOTES

NOTES